Materials for Engineering

A FUNDAMENTAL DESIGN APPROACH

B DERBY

D A HILLS

C RUIZ

Longman
Scientific &
Technical

Copublished in the United States with
JOHN WILEY AND SONS, INC
New York

MATERIALS FOR ENGINEERING
A Fundamental Design Approach

Longman Scientific & Technical,
Longman Group UK Limited,
Longman House, Burnt Mill, Harlow,
Essex CM20 2JE, England
and Associated Companies throughout the world.

Copublished in the United States with
John Wiley & Sons, Inc., 605 Third Avenue, New York, NY 10158

First published 1992

British Library Cataloguing in Publication Data
Derby, B.
  Materials for Engineering: A Fundamental Design Approach
  I. Title   II. Hills, D.A.   III. Ruiz, C.
  620.1

  ISBN 0-582-03185-0

Library of Congress Cataloging-in-Publication Data
Derby, B. (Brian), 1957–
  Materials for engineering: a fundamental design approach / B.
  Derby, D.A. Hills, C. Ruiz.
    p.   cm.
  Includes bibliographical references and index.
  ISBN 0-470-21791-X
    1. Strength of materials. I. Hills, D. A. (David Anthony), 1955–
  . II. Ruiz, C. (Carlos), 1933–   . III. Title.
  TA405.D44   1992
  620.1'12--dc20   91-15762
                                                          CIP

Set in 10/12 pt Garamond.

Produced by Longman Group (FE) Limited
Printed in Hong Kong

LIST OF CHAPTERS

CONTENTS

PREFACE

Materials for Engineering: A Fundamental Design Approach is written for use by second and third year mechanical, civil or materials engineering students. A knowledge of basic strength of materials, e.g. simple tension, compression, bending of beams of symmetrical cross section and torsion of cylindrical shafts, is assumed. Also assumed is some knowledge of materials science at an elementary level. The book builds on both disciplines and provides a link between them. It covers the design of engineering components using metals, ceramics, polymers and composites distinguishing between fundamental modes of failure, i.e. elastic and plastic stable or unstable deformation, brittle fracture, fatigue and creep. It is illustrated with numerous exercises, most of them with solutions.

WHY THIS BOOK?

With literally dozens of other books on engineering materials you may be asking yourself why you should choose this one. The fact is, that most other authors limit themselves to describing the mechanical behaviour of observed laboratory tests and offer some explanation for it. Here, the mechanical properties are treated from the viewpoint of the designer, and differences in mechanical behaviour are then related to the appropriate design procedures. Thus the book combines materials science with mechanical engineering emphasising the interactive features of design with modern materials. Other books have been written by materials scientists or by engineers whilst this one is the result of collaboration between both.

The treatment of engineering materials given here is comprehensive and up-to-date. It includes conventional metals, the new high-strength ceramics and composites.

UNIQUENESS OF THE APPROACH ADOPTED

Design of complex components entails a detailed understanding of the possible failure mechanisms which can occur, and these in turn demand some knowledge of micromechanics. The successful design engineer is one who can exploit a material's qualities whilst avoiding its limitations and in this regard new materials are not only less forgiving than conventional metals but they can be tailored to suit the product. Our approach throughout the book has been to integrate traditional solid mechanics — elasticity, plasticity — and materials science. This is the approach followed in our integrated materials engineering course in Oxford.

At all points in this text we have tried to avoid the pitfalls of making the mathematics unnecessarily complicated, of ignoring the way in which materials fail at an atomic level and of becoming too engrossed in specific examples. We have tried to emphasise the physical principles needed to understand the function of materials and components because a mastery of them will enable the reader to apply the ideas to a much wider range of problems.

GUIDE FOR CLASSROOM USE

Failure which may be considered to be geometric limitation rather than a material breakdown such as elastic or plastic instability, or simply deflexion limited designs are considered only very briefly. They are much more profitably included in a solid mechanics text. What we have done is to include a chapter on elastic design philosophy (Chapter 2) and how it may be used successfully for the efficient use of material over a range of applications. This chapter, together with the introductory chapter (Chapter 1) which describes the various modes of failure, should prove of general interest. Both chapters provide a general introduction to the principles of structural design.

The limiting factor in some designs may well be the onset of yield. Yield criteria are treated within a separate chapter (Chapter 3), together with other design philosophies such as shakedown which may permit a more efficient use of the material in cyclically loaded structures or the limit state philosophy. These concepts are more familiar to structural engineers than to mechanical engineers or materials scientists but their application to structures other than traditional frames, plates and shells offers many possibilities. Brittle fracture cannot, of course, be predicted by bulk stress analysis alone and is considered in the science known as fracture mechanics. This too has a chapter to itself (Chapter 4), which includes micromechanics, applied mechanics and experimentally determined treatments of the subject. Often, this is misunderstood by undergraduates. We continue by treating failure under cyclic loads and under maintained load, i.e. fatigue and creep in metals and in non-metals (Chapter 5). We conclude with a treatment of two new kinds of materials, high strength ceramics (Chapter 7) and composites (Chapter 8).

No concession has been made to what is fashionable now, instead, we have chosen to emphasise permanent fundamental principles whilst showing how these are applied to modern materials engineering.

ACKNOWLEDGEMENTS

We are grateful to the following for permission to reproduce copyright material:

ASM International for Figs 1.10–1.12; the author, Dr L. Boniface for Fig. 8.28; Cambridge University Press and the author, Prof. D. Hull for Figs 8.3 & 8.29 (Hull, 1981); Centro Nacional de Investigaciones Metalúrgicas for Fig. 1.1 (Calvo, 1964); Cookson Group plc for Fig. 7.1; Elsevier Applied Science Publishers Ltd. for Fig. 6.13 (Harding, 1986); the Controller of Her Majesty's Stationery Office for Tables 5.1 & 5.2 (Pook, 1978); ICI Ltd. for Fig. 6.17; the Council of the Institution of Mechanical Engineers for Fig. 1.5 (Parsons, 1947); Marlow Ropes for Fig. 8.2.

CHAPTER ONE

General Introduction

1.1 A BRIEF HISTORICAL NOTE

The very fact that the names of materials have been given to the ages of mankind is an indication of their paramount importance. From stone through bronze and iron we have now reached the age of strong ceramics, non-ferrous and ferrous alloys, polymers and composites when it has become possible to engineer materials so as to match their properties with the shape and function of the required product. In comparison with the slow evolution that has characterized the earlier ages, as shown in Table 1.1, the current advanced materials age has seen a truly vertiginous progress in the production of new materials and in their utilization.

Until the late 19th century, ceramics in the form of bricks, stone and mortar were the main structural materials. With a few noteworthy exceptions, the use of metals was limited to weapons, tools, fasteners, ornaments and other small objects. Cast iron was too brittle to be used in other than compression members, while the high cost of wrought iron and the shortage of skilled craftsmen capable of producing it with consistently good mechanical properties prevented its widespread use. A medieval ironworks, for example, would be housed in a stone and mortar building (Fig. 1.1). The structural components of furnaces and machines would be bricks, clay and wood and only the reinforcing bands and fasteners would be wrought iron. Other naturally available materials — leather, hemp — would be used for the manufacture of the bellows. The brittleness and low tensile strength of the ceramics coupled with the poor dimensional stability of wood required the main load-bearing components to be under compression, with large cross-sectional areas to maintain sufficiently low stress levels. The typical massive structures that met these requirements is illustrated in Fig. 1.2.

1

Figure 1.1   A medieval forge (Ferreria de Mirandaola, Spain)[2]

Stresses induced by self-weight were often more important than those induced by external loading. Once erected, the structure would stand for ever, provided that the foundations remained undisturbed and that corrosion or erosion by the environment did not significantly alter the original shape. The only mode of failure that concerned the designers was therefore static instability.

The Coalbrookdale bridge, erected in 1779, was one of the first major structures made entirely with iron (Fig. 1.3) but it was only with the advent of the Bessemer and open hearth processes in 1860 that steel established itself as a widely used constructional material. The relatively high strength of steel permitted an increased stress level in tension as well as in compression. Stress analysis, which had progressed very little since Leonardo da Vinci, took a giant step forward with the work of Lagrange, Coulomb and Navier at the same time as the behaviour of materials was studied in a systematic and scientific manner by Hooke, Young and others.[4] The slenderness of the new structures, in particular, brought to light earlier work on elastic instability by Euler, to explain the failure of bridges through buckling of the upper compressed chord (Fig. 1.4). Such unexpected failures proved that design could no longer be based entirely on experience, and created the need for analytical tools.

Materials have been developed to meet the demands imposed by new or improved processes for the manufacture of chemical products, more efficient power plant, structural requirements, etc. A typical example

Figure 1.2   A typical stone structure consisting of stone arches (Aqueduct of Segovia, Spain)

Figure 1.3   Darby's bridge at Coalbrookdale[i]

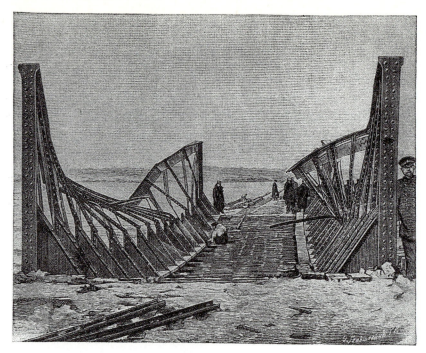

Figure 1.4   Failure of an open bridge in the 1890s[4]

is offered by the evolution of steam-generating plant, from the early
tank boiler (Fig. 1.5) that produced saturated steam at temperatures
below 200 °C with a notorious record of explosions, to the modern
supercritical boilers. In this case, the material must not weaken at the
high temperatures and in the otherwise aggressive environment to
which it will be exposed in service. In other cases, low temperatures,
repeated loading, wear, corrosion or erosion may be the principal
causes of failure. In the past, the goal has been to combine ductility
with tensile strength in metals with approximately isotropic properties.
The design methods followed when dealing with these metals are
clearly not applicable to the fine ceramics used, for example, to make
turbine blades operating at 1400 °C (Fig. 1.6), since their high tensile
strength at elevated temperature is achieved at the cost of a severe
reduction in ductility. Equally, the anisotropy introduced in fibre-
reinforced composites in which fibres may be aligned in the direction
of the transmitted loads presents yet another challenge to the designer
while it offers substantial benefits.

The developments reflected in Table 1.1 can be explained by the
continued quest for an ideal material that combines high tensile strength
with ductility, toughness, resistance to aggressive environments and
ease of shaping into finished products. In the absence of such material,
engineers have been forced to compromise, using whatever happened
to be most suitable for each particular application from the materials

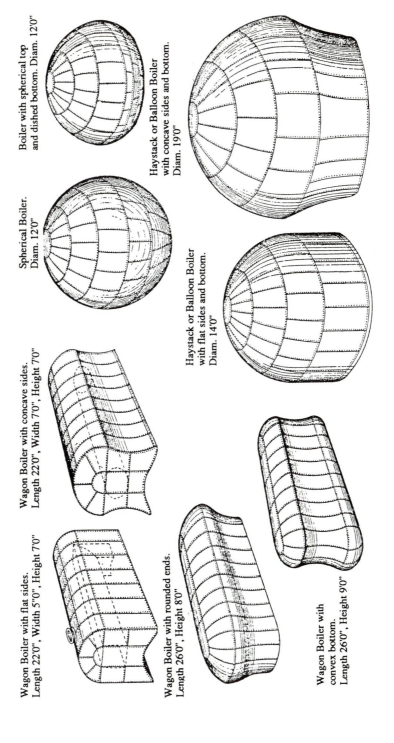

Boiler with spherical top and dished bottom. Diam. 12'0"

Haystack or Balloon Boiler with concave sides and bottom. Diam. 19'0"

Spherical Boiler. Diam. 12'0"

Haystack or Balloon Boiler with flat sides and bottom. Diam. 14'0"

Wagon Boiler with concave sides. Length 22'0", Width 7'0", Height 7'0"

Wagon Boiler with flat sides. Length 22'0", Width 5'0", Height 7'0"

Wagon Boiler with rounded ends. Length 26'0", Height 8'0"

Wagon Boiler with convex bottom. Length 26'0", Height 9'0"

Figure 1.5 Various early designs of tank boilers[5]

Figure 1.6   Silicon nitride turbine blade for 1400 °C (courtesy of Rolls Royce plc)

that were available to them. Materials selection is still at the heart of the design process but it is rapidly becoming more common to tailor the material to the finished product. The evolution in the design process has mirrored that in materials. Until the Renaissance, perhaps even later, a designer was usually a craftsman, who applied his natural flair and experience to the design of products little different from those with which he was familiar. Today's designs, while still containing a large and important element of practical experience, embody the application of the fundamental principles of the mechanics of solids. They combine materials science with stress analysis to match shape, function and material.

1.2   ELEMENTARY MODES OF FAILURE

Before discussing in any detail the relationship between materials and design, it is instructive to review the main features of those modes of stress-dependent failure which may be regarded as elementary. These are categorized in Table 1.2. They may act together. For example, a beam designed to withstand static loading may also be subjected to a cyclic load so that fatigue may be added as a secondary cause of failure to excessive stable plastic deformation. Secondary causes can accelerate the process of failure and must not be neglected when assessing the strength of a structural component.

Table 1.1
Historical
development of
materials[1]

Period	Material
Palaeolithic	Flint and obsidian. Ochre, emery. Wood, bones
Neolithic (to 3500 BC)	Granite, diorite, limestone, sandstone (materials less brittle than in previous period)
Egyptian Predynastic (3500−3000 BC)	Alabaster, marble, rocksalt. Native metals (gold, silver, meteoric iron, copper)
Metal Age I (3000−2000 BC)	Copper oxides and carbonates used as copper ores. Copper alloyed with lead, antimony and tin. Silver from galena. First experimental smelting
Metal Age II (2000−1200 BC)	Copper from copper sulphides. Tin produced from ore
Early Iron Age (1200−500 BC)	Hardened (carburized) iron, quenching and tempering
Late Iron Age (500−50 BC)	Brass from copper and calamine
Middle Ages	Cast iron (pig iron). Blister steel
1740	Crucible process for steelmaking
1843	Nickel plating (Boettger process)
1860	Bessemer and open hearth processes. *General use of steel as a constructional material*
1879	First application of ductile nickel
1888	First application of aluminium
1905	First application of Monel (corrosion resistance)
1912	First application of stainless steels
1920−1945	Discovery and application of synthetic rubber, PVC, polyethylene. Large-scale use of alloy steel
1945−	Industrial use of reinforced plastics and new materials (titanium, beryllium, zirconium . . .). High-strength and toughness structural steels. Fabrication processes, e.g. powder metallurgy, resulting in entirely new materials

In metals, yielding, brittle fracture, fatigue and creep may be regarded as the elementary modes of failure. Other ductile materials, in particular some polymers, may also fail due to yielding but most non-metals tend to fail in a brittle manner. The process of failure, of course, depends on the nature of the material and will be discussed in detail in the appropriate chapters. Here only the main characteristics will be described from a purely phenomenological standpoint. Fatigue and creep are also found in non-metals. Composites exhibit the elementary

Table 1.2 Stress−dependent modes of failure[1]

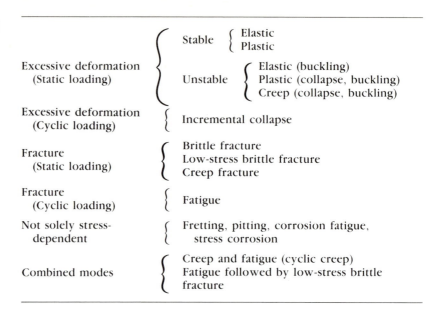

Excessive deformation (Static loading)
- Stable
  - Elastic
  - Plastic
- Unstable
  - Elastic (buckling)
  - Plastic (collapse, buckling)
  - Creep (collapse, buckling)

Excessive deformation (Cyclic loading)
- Incremental collapse

Fracture (Static loading)
- Brittle fracture
- Low-stress brittle fracture
- Creep fracture

Fracture (Cyclic loading)
- Fatigue

Not solely stress-dependent
- Fretting, pitting, corrosion fatigue, stress corrosion

Combined modes
- Creep and fatigue (cyclic creep)
- Fatigue followed by low-stress brittle fracture

modes of failure of their constituents plus other modes peculiar to their structure.

1.2.1 YIELDING

Most of the metals and many non-metals commonly used can withstand large deformations without breaking. As a result, the machines or structures made of such materials may cease to function or to satisfy the design conditions long before there is an actual rupture, failure being ascribed to the setting up of excessive deformations under load. Of particular concern is the situation that arises when the elastic limit is reached and plastic deformation sets in. The typical force−extension curve for a rod in uniaxial tension (Fig. 1.7) may be interpreted by defining the engineering stress,

$$\sigma_e = \frac{F}{A_0} \tag{1.1}$$

and the engineering strain,

$$\epsilon_e = \frac{l - l_0}{l_0} \tag{1.2}$$

where the subindex 0 denotes the initial dimensions. The *true stress* is defined as

$$\sigma = \frac{F}{A} \text{ or } F = \sigma A \tag{1.3}$$

8

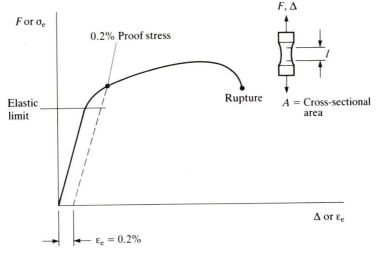

Figure 1.7   Typical force–extension curve

where A is the current cross-sectional area at the instant when the force F is sustained.

If the force is increased by dF, the gauge length increases by dl and the corresponding strain increment is

$$d\epsilon = \frac{dl}{l} \tag{1.4}$$

The *natural strain* is defined as

$$\epsilon = \int_{l_0}^{l} d\epsilon = \ln \frac{l}{l_0} \tag{1.5}$$

For small deformations, A is approximately equal to A_0 and

$$\sigma \simeq \frac{F}{A_0} \tag{1.6}$$

while Eqn (1.5) may be expanded in series form:

$$\epsilon = \ln \left(1 + \frac{l - l_0}{l} \right) \simeq \frac{l - l_0}{l_0} \tag{1.7}$$

so that for small strains the true and engineering strains are coincident.

When plastic deformation occurs, it is observed that the change in volume becomes negligible as dislocation glide does not involve a change in volume. Hence,

$$Al = \text{constant} \tag{1.8}$$

$$A\,dl + l\,dA = 0, \quad \frac{dl}{l} = -\frac{dA}{A} = d\epsilon$$

The maximum load will be achieved when, from Eqn (1.3),

$$dF = \sigma dA + A d\sigma = 0 \tag{1.9}$$

where the first term represents the effect of Poisson contraction and the second that of work-hardening, giving the condition

$$\frac{d\sigma}{\sigma} = -\frac{dA}{A} = d\epsilon \tag{1.10}$$

Hence instability occurs when $d\sigma/d\epsilon = \sigma$.

Given a true stress–natural strain curve as shown in Fig. 1.8(a), this condition is satisfied at point C.

Normally, stress–strain curves are presented in terms of the engineering stress (F/A_0) and the engineering strain, $(l-l_0)/l_0$. It is a simple matter to change the variables. Since the total change in volume, including the elastic dilatation, is small,

$$Al \simeq A_0 l_0$$

and from the definition of the various terms,

$$\sigma = \left(\frac{F}{A_0}\right)\frac{A_0}{A} = \left(\frac{F}{A_0}\right)\left(1 + \frac{l-l_0}{l_0}\right)$$

i.e. $\dfrac{\sigma}{\sigma_e} = 1 + \epsilon_e$ \hfill (1.11)

while ϵ is given by Eqn (1.7).

An alternative interpretation to the construction of Fig. 1.8(a) is based on Eqn (1.11) which, combined with Eqn (1.10) gives

$$\frac{d\sigma}{d\epsilon_e} = \frac{\sigma}{1 + \epsilon_e}$$

at instability. This is represented in Fig. 1.8(b). It may be shown that the value of the stress at point C corresponds to the engineering stress at instability, i.e. the tensile strength.

A compressive load would, in principle, have the same effect, except that the cross-section would increase instead of decreasing and the strains would be negative instead of positive. A cylindrical specimen tested in compression between the two platens of a press could therefore be used to obtain a true stress–strain curve in perference to the conventional tensile test. Instabilities are avoided and, indeed, such tests are often used for metalworking processes. There are however practical difficulties in ensuring the correct alignment of the specimen to avoid buckling and the friction between specimen and platens results in barrelling, i.e. a non-uniform complex state of triaxial stress and strain. Of course, the same non-uniformity of strain sets in the tensile specimen when necking occurs, after the maximum load has been reached. The formation of highly strained regions will be discussed in relation to the problem of creep.

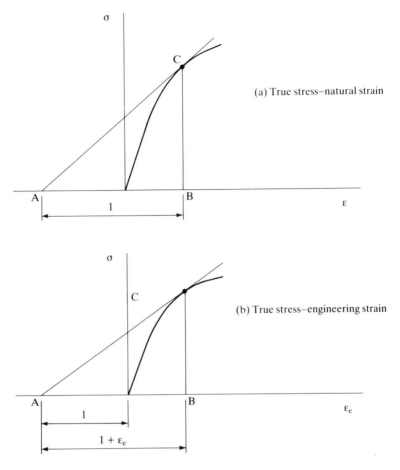

Figure 1.8   Graphical determination of the conditions at instability in tension

It is possible to express mathematically the stress–strain behaviour of a material by means of constitutive equations but normally, this is too complicated for most practical problems. A simplification is possible by observing that the material compliance is several orders of magnitude larger in the plastic regime than under elastic conditions. Three approximations are commonly used:

(a)   *Rigid–ideally plastic material* (FIG. 1.9a)   The elastic deformation is regarded as negligible and only the deformation in the plastic regime is taken into consideration. It is also assumed that once σ_Y has been reached, the material flows without an increase in stress.

(b)   *Elastic–ideally plastic material* (FIG. 1.9b)   This is a refinement of the first approximation in which the elastic deformation is not neglected. It is a safe assumption for

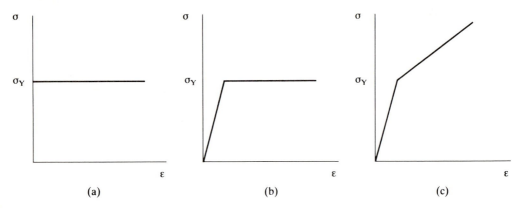

Figure 1.9   Simplified models of material behaviour: (a) rigid—ideally plastic; (b) elastic—ideally plastic; (c) elastic—linear strain-hardening

structural calculations since most metals work-harden and have therefore extra strength. In metalworking processes a more accurate model is necessary.

(c)   *Rigid or elastic—linear strain-hardening material* (FIG. 1.9c)   After yielding, the stress required to increase the deformation increased linearly with the strain.

As the deformation increases, microvoids are formed, as shown in Fig. 1.10.

At any given instant, the fraction of microvoids, f, may be taken to defne a 'damage factor', increasing the true stress in accordance with the expression

$$\sigma^* = \frac{\sigma}{1-f} \qquad (1.12)$$

Fracture finally takes place when the microvoids coalesce and results in the dimpled appearance of the fracture surfaces. These dimples are equiaxed, in case of pure tension, or elongated when in the presence of shear, as is normally found near free surfaces where fracture follows from the slip along the planes of maximum shear stress. The ligaments between large clusters of microvoids may also tear, forming sharp ridges.

Metals are normally built up from crystals of atoms in the face-centred cubic, hexagonal close-packed or body-centred cubic structures. Of these, the face-centred cubic are the most ductile and always fail by the void coalescence mechanism after yield. Such metals include gold, copper, and alloys like α-brass and aluminium alloys. Metals with the other crystal structures do not possess the easy

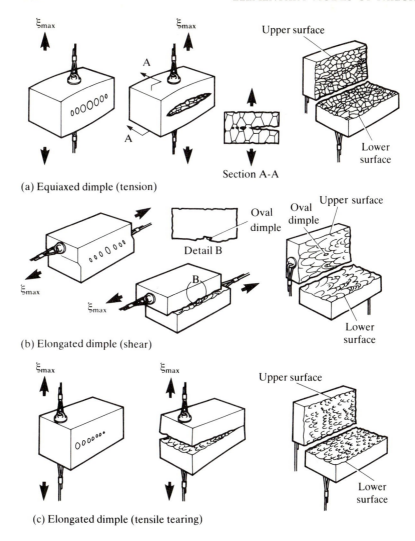

(a) Equiaxed dimple (tension)

(b) Elongated dimple (shear)

(c) Elongated dimple (tensile tearing)

Figure 1.10   Appearance of fracture surfaces resulting from microvoid coalescence[6]

dislocation motion of face-centred alloys and under low temperature or high strain rate they may fail by cleavage, e.g. separation occurs along well-defined crystal planes. The cleavage is not related to the total plastic deformation of the specimen. The random orientation of the crystals within the metal means that, as fracture progresses from one crystal to the next, the fracture plane tilts or twists creating distinctive patterns of steps on the fracture surfaces or 'river patterns' where the steps intersect dislocation concentrations (Fig. 1.11). This behaviour is very important in mild steels and zinc and magnesium alloys.

Yielding is also found in some polymers, for example polycarbonate.

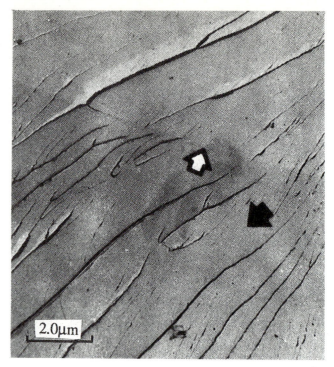

Figure 1.11   Cleavage fracture: river marks on the fracture surface of Alnico alloy. The white arrows point to a cleavage step, the black arrow shows the direction of crack propagation[6]

The fracture surfaces, however, do not exhibit the features found in metals. Normally, in brittle plastics, fracture starts from a region showing a close, irregular network of lines (crazing) and is strongly strain-rate-dependent. Time, regarded as irrelevant in the preceding description of the plastic deformation of metals, cannot be ignored in the case of polymers, in which the process of failure, as distinct from the mode of failure, is elasto-viscous rather than elasto-plastic and is therefore closer to creep than to time-dependent yielding.

1.2.2   BRITTLE FRACTURE

From an engineering or macroscopic standpoint, brittle fracture is said to occur in a tensile specimen if no permanent deformation is observed in the two specimen halves, which may be fitted back together so that the appearance is that of the original specimen. The fracture mechanism may be cleavage, i.e. separation of planes within a grain, since cracks are associated with limited crystallographic slip and hence small strains, but the most characteristic mechanism is intergranular fracture. The

14

Figure 1.12   Intergranular fracture in steel[6]

fracture surfaces reveal the outline of the grains (Fig. 1.12) that have separated due to the presence of weak or brittle phases between them or to environmental effects such as stress corrosion or hydrogen embrittlement.

Brittle fracture is the main mode of failure of glass and ceramics. Fracture starts from a small flaw — typically a few microns — and develops into a fracture mirror before entering into a mist (Fig. 1.13) characterised by the presence of fine radial ridges. This is followed by a hackle, with longer, better defined radial lines. Finally, a major crack or cracks sweep through the rest of the surface.

It will be shown in Chapter 4 that failure depends on the size of the starter flaw, on the applied stress and on the *fracture toughness* of the material, in accordance with the general expression

$$Y \sigma \sqrt{\pi a} = K_c \tag{1.13}$$

where Y is a dimensionless parameter depending on geometry, σ is the nominal applied stress, a the characteristic size of the flaw and K_c the fracture toughness. For a ceramic, K_c may be of the order of 5 MPa $\sqrt{m}$ when the stress is normal to the plane of the flaw. In a steel, it may be as high as 150 MPa$\sqrt{m}$ so that, for the same applied stress, the flaw needed to start a brittle fracture would be of the order of millimetres rather than of microns. The fracture surfaces then show the features illustrated in Fig. 1.14(b). Starting from a thumbnail crack

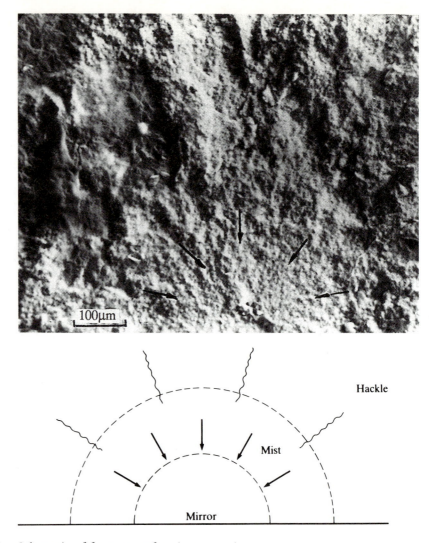

Figure 1.13   Schematic of fracture surface in a ceramic

ahead of the flaw, a crack has propagated from left to right leaving a herringbone pattern and, at the edges, shear lips that indicate some measure of ductility. If the material is sufficiently tough, the crack may stop, tunnelling its way into the plate as shown in Fig. 1.14(c) while the shear lips eventually extend over its whole thickness. The final fracture in such cases is a combination of the brittle and ductile modes and is found in normally ductile materials when a triaxial state of stress inhibits yielding. This phenomenon, known as plastic constraint, is responsible for the initiation of cracks from deep notches, welded joints, etc., and will be considered in detail in Chapters 3 and 4.

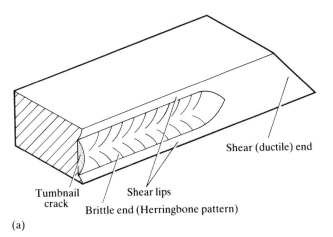

Shear (ductile) end

Tumbnail
crack

Shear lips

Brittle end (Herringbone pattern)

(a)

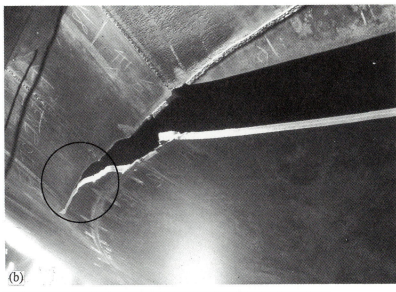

(b)

(c)

Figure 1.14 Quasi-brittle fracture in steel: (a) schematic of fracture; (b) brittle region; (c) ductile end

1.2.3   FATIGUE

The term 'fatigue' was first applied by Poncelet in 1839 to the failure under cyclic loading of machines and structures, and in 1843 Rankine described the process of fatigue crack initiation and subsequent growth from the fracture surfaces observed in locomotive axles.[5] The role of stress concentrations at notches, transitions in section, shoulder, etc., had already been identified at that time. A fatigue fracture in a shaft rotating under constant load, and hence subjected to alternating bending, is shown in Fig. 1.15. After a period of initiation, the crack propagates in jerks, causing a series of marks called beachmarks, whose curvature changes as they extend over the cross-section of the shaft. Depending on the relative magnitude of the peak stress (the stress concentration factor) and the average stress, the general aspect of the fracture surface changes as indicated in the diagram. Multiple initiation is commonly associated with high average stresses. Final separation may follow from the sudden propagation of the crack in a brittle manner or from ductile fracture, depending on the stress level, the crack size and the mechanical properties of the material.

The ridges formed as the fatigue crack grows are also clearly visible in the scanning electron micrograph of Fig. 1.16. They are preceded by an initiation stage by slip along planes of maximum shear. Plastic

Figure 1.15   Typical fatigue fracture: rotating shaft under constant load (alternating bending)[1]

Figure 1.16   Fatigue striations revealed by scanning electron micrography

deformation is seen to play a vital role in the process of fatigue and it may therefore be expected that only those materials that yield may also fail due to cyclic loading. Brittle materials, such as ceramics, suffer from slow crack growth under maintained loading, assisted by environmental effects. In them, it is the total time under load rather than the load or stress amplitude that affects life, although the subject is still under active discussion. In polymers, crazing plays the same role as plastic deformation in the growth of cracks under stress reversals.

1.2.4   CREEP

The progressive deformation under constant tension of lead and glass has been known for centuries. The same phenomenon occurs in the materials used for the construction of power plants, chemical vessels, etc., where the nature of the thermodynamic process requires exposure at elevated temperatures. A typical creep curve, Fig. 1.17, exhibits three stages: a primary stage in which elastic and plastic deformation sets are followed by time-dependent deformation at a decreasing rate of strain; a secondary stage at constant strain rate; and finally a tertiary stage of accelerated deformation leading to failure. Similar curves apply to polymers even at low temperatures and, at sufficiently elevated temperatures, to ceramics.

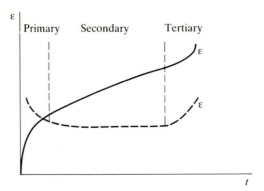

Figure 1.17   Typical creep curve

1.3   FAILURE OF COMPOSITES

Composites, in which a matrix is reinforced with particles or fibres, are engineered to suit specific applications and do not follow any general modes of failure. Perhaps the only modes that may be regarded as specific to these materials are fibre pull-out and delamination. The first mode is illustrated in Fig. 1.18. Depending on the adhesion between fibre and matrix and on the relative elongation-to-fracture of each, the fibres may break while the matrix itself retains its integrity. The stress concentration induced at the tips of the broken fibres starts a crack around the fibre that propagates in shear along the fibre/matrix interface. The load is eventually transferred to the matrix that breaks, exposing the fibres. Alternatively, the matrix may break first, in which case the fibres will stretch under the applied load until they fracture. The features of the fracture surface are essentially similar, even though the length of fibre that is pulled out of the matrix differs.

In composites consisting of a stack of layers, delamination occurs due to the presence of flaws in the adhesive layer or simply to the presence of an excessively high shear stress at the interfaces between adjacent layers. It can be detected only by non-destructive testing or by indirect means, since it will be reflected in a drop in the flexural stiffness.

1.4   THE 'DESIGN-FOR-FITNESS' APPROACH

Once the likely mode or modes of failure under the specified service conditions have been identified, the designer's function consists in deciding on the shape and dimensions of the structural elements so

Figure 1.18 Failure of fibre-reinforced composite by crack propagation and fibre pull-out: scanning electron micrograph of the fracture of glass–epoxy composite. Magnification ×20

as to eliminate or minimize the risk of failure. This is known as the 'design-for-fitness' approach. At the core of the design process is the determination of the service loads and of the stresses, strains and deformations that they produce. In some cases, for example in the massive stone structures of Figs 1.1 and 1.2, only a simple analysis will be sufficient to check the stability of the whole, provided that the foundations can be relied upon to take the load at the footings without yielding. In general, however, the magnitude of the stresses will be taken as a figure of merit to assess structural integrity. In most cases the material remains elastic and there is a linear relationship between loads, stresses, strains and structural deformation. An exception to this general rule is elastic buckling, when the maximum acceptable value of the compressive stresses depends on the structural shape and form of loading as well as on the material properties.

Consider first the elementary case of a bar of cross-sectional area A under an external tension F, giving a stress

$$\sigma = \frac{F}{A} \tag{1.14}$$

and assume that the material is elastic–ideally plastic with a yield stress

σ_Y. It is obvious that failure would occur if σ were allowed to reach σ_Y or if F were allowed to exceed the product $A\sigma_Y$. The *safety factor* is defined as

$$\text{SF} = \frac{\sigma_Y}{\sigma_{\text{design}}} > 1 \tag{1.15}$$

and the *load factor*, LF, is defined, in general, as the ratio between the load capable of causing the structure to fail and the load in service, assumed to be applied in a similar manner. In this case,

$$\text{LF} = \frac{A\sigma_Y}{F_{\text{design}}} > 1 \tag{1.16}$$

It is evident that in this trivial example they are equal. The stress σ is entirely induced by the force F. It is called a *primary stress* and will be denoted as σ_I. It is important to note that only primary stresses exist when the structure is statically determinate. The presence of redundancies introduces self-equilibrating stresses that arise from the need to satisfy the structural constraints. This key point is best understood by considering thermal stressing.

Assume now that the same bar is fixed to a rigid frame, as in Fig. 1.19. When the temperature of the frame is increased by T, it stretches by αTl, and the strain in the bar is αT corresponding to an elastic stress $E\alpha T$. This stress *only* arises because of the constraint. It would not exist if the bar were free. Of course, such a stress is only physically possible if it is less than the yield point. Once the bar has yielded, it will continue to stretch and will only fail when the strain reaches the critical value at fracture, ϵ_F. In a ductile material, this is so large that it does not constitute a realistic limitation. Normally, the assembly will have deformed so much by then as to cease to serve whatever function it was originally designed to fulfil. The calculated elastic stress,

$$\sigma_{II} = E\alpha T \tag{1.17}$$

is a *secondary stress* that arises from the structural constraints, unlike

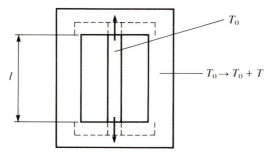

Figure 1.19 Bar fixed to a rigid frame. The thermal expansion of the frame forces the bar to stretch

the primary stress that was due to the need to balance an externally applied load. It would be unnecessarily restrictive to maintain σ_{II} below the yield point. Values of σ_{II} above σ_Y should be interpreted in terms of strains, rather than stresses. In this particular example, the load factor is

$$\text{LF} = \frac{\epsilon_F}{\alpha \Delta T} = \frac{E\epsilon_F}{\sigma_{\text{II}}} > 1 \tag{1.18}$$

The safety factor has no physical meaning whatsoever.

A second example will serve to illustrate this matter more fully. Consider the three-bar system of Fig. 1.20. The bars are rigidly fixed at both ends to rigid cross-members to which is applied the load F.

It can easily be shown that, if all bars are loaded elastically, the stresses are

$$\sigma_i = \frac{nF}{A(1+nq)} < \sigma_Y \quad \sigma_0 = \frac{F}{A(1+nq)} \tag{1.19}$$

with the notation of Fig. 1.20. For the elastic−ideally plastic material, when the inner bar yields the stresses become

$$\sigma_i = \sigma_Y, \quad \sigma_0 = \frac{F}{A} - \sigma_Y q < \sigma_Y, \quad F < \sigma_Y(1+q)A \tag{1.20}$$

Finally, the system fails when the limit load

$$F_{\max} = \sigma_Y(1+q)A \tag{1.21}$$

is reached. The *limit load* is a useful concept in that it provides an indication of how much load, distributed in a pre-specified form, can be carried by the structure just before it becomes unsafe, and of the pattern of the deformation that precedes failure. It could be argued that a knowledge of the limit load is preferable to the determination of the elastic stresses and, indeed, this is recognized by the gradual

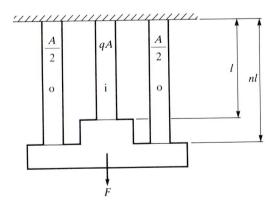

Figure 1.20   Three-bar system

transition from the traditional design, based on the limitation of the stresses calculated for an assumed elastic behaviour of the material, and the now-common approach based on the limit load. The limit load can be calculated directly from the equilibrium conditions, ignoring the constraints imposed by the initial elastic deformations. The *primary stresses* will be

$$\sigma_{Ii} = \sigma_{I0} = \frac{F}{A(1+q)} \tag{1.22}$$

They must be maintained below σ_Y. The load factor or the safety factor defined in terms of these primary stresses is

$$\frac{1}{SF} = \frac{1}{LF} = \frac{F}{F_{max}} = \frac{\sigma_I}{\sigma_Y} \tag{1.23}$$

The total stress will be the difference between the total calculated elastic stress and the primary stress. From Eqns (1.19) and (1.22),

$$\sigma_{IIi} = \frac{F(n-1)}{A(1+q)(1+nq)} \qquad \sigma_{II0} = -\frac{Fq(n-1)}{A(1+q)(1+nq)} \tag{1.24}$$

At first sight, no limit needs be placed on these secondary stresses provided that the material is sufficiently ductile. Problems may arise, however, if the load is removed and then applied several times, as will be discussed in greater detail in Chapter 3. At this stage it is sufficient to follow the process of unloading from A to B and C to D (Fig. 1.21). When the load is removed, the stresses are found to be

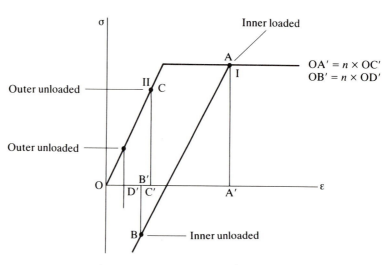

Figure 1.21 State of stress in the three-bar system

$$(\sigma_i)_{\text{unloaded}} = \sigma_Y - \frac{nF}{A(1 + nq)}$$

$$(\sigma_0)_{\text{unloaded}} = \frac{nqF}{A(1 + nq)} - \sigma_Y q$$

$$(1.25)$$

While yielding just once may be acceptable, repeated yielding will eventually cause the system to fail (plastic or high-strain fatigue) and cannot be tolerated. This means that a limit must be set to the stresses that have just been calculated:

$$(\sigma_i)_{\text{unloaded}} > -\sigma_Y, \quad F < 2\sigma_Y \frac{A(1 + nq)}{n}$$

$$(\sigma_0)_{\text{unloaded}} < \sigma_Y, \quad F < \frac{1 + q}{q} \sigma_Y \frac{A(1 - nq)}{n}$$

$$(1.26)$$

or, in terms of the primary and secondary stresses in the most highly stressed bar,

$$\sigma_i = \sigma_{iI} + \sigma_{iII} < m\sigma_Y \qquad (1.27)$$

where $m = 2$ or $(1 + q)/q$, whichever is smaller. In other words, the sum of the primary and secondary stresses, calculated under assumed elastic behaviour, may be allowed to exceed the yield point. After the first load/unload cycle, the system will *shake-down* and behave elastically, having acquired a certain set deformation.

Limiting these stresses will prevent failure through yielding under constant or cyclic loading. It clearly may not avoid buckling, nor has any attempt been made to safeguard against creep failure at elevated temperature. The design implications of these modes will be discussed in the relevant chapters. The elastic analysis will also be used when designing against the brittle propagation of cracks, following the general principles of fracture mechanics. It will also be shown to be effective in the design against fatigue of components containing small notches, shoulders, sharp radii, etc. The common feature of these raisers is that their effect is limited to a small area, of similar dimensions to those of the stress raiser itself. The peak stress is often referred to as *tertiary* and added to the primary and secondary stresses. Limits are then based on the fatigue life.

EXAMPLES

1 Determine the ultimate tensile strength of a material whose true stress—natural strain curve is given by

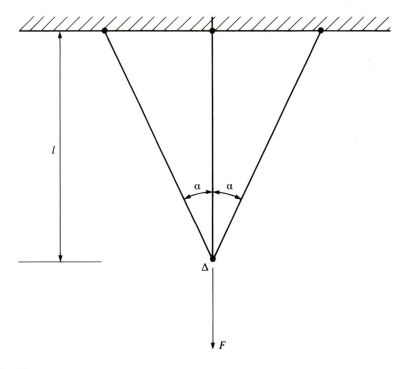

Figure 1.22   Three-bar truss

$$\sigma = Y + C\epsilon^m$$

where $Y = 350$ MN m^{-2}, $C = 1400$ MN m^{-2} and $m = 0.8$.

2   Plot the extension of the truss in Fig. 1.22 against the load F. Take the cross-sectional area of the bars to be unity and assume elastic−ideally plastic behaviour.

3   Plot the curvature of a beam of rectangular cross-section, 250 mm deep by 100 mm wide with $E = 200$ GN m^{-2} and $\sigma_Y = 200$ MN m^{-2} against bending moment, assuming the material to be elastic−ideally plastic.

4   List the materials used for the construction of a modern motor car, giving approximate values for their mechanical properties, reasons for choosing them, and possible alternatives; identify possible modes of failure.

REFERENCES

1   RUIZ C, KOENIGSBERGER F 1970 *Design for Strength and Production* Macmillan, London
2   CALVO F A 1964 *La España de los Metales* CENIM, Madrid
3   SINGER C, HOLMYARD E J, HALL A R, WILLIAMS T I 1975 *A History of Technology* Vol IV, The Industrial Revolution, Clarendon Press, Oxford
4   TIMOSHENKO S P 1953 *History of Strength of Materials* McGraw-Hill, New York
5   PARSONS R H 1947 *History of the Institution of Mechanical Engineers* IMechE, London
6   FELLOWS J A (ed) 1974 *Metals Handbook* Vol 9, Fractography and Atlas of Fractographs, Amer Soc Metals Ohio

CHAPTER TWO

Elastic Deformation

2.1 INTRODUCTION

Excessive elastic deformation is the main failure criterion when designing structures such as machine-tool frames where stiffness is required to ensure the accuracy of the machined product, engines and machines in which small deformations may cause seizure of kinematic pairs, airframes, etc. In such cases, stresses must be maintained below the yield point to ensure that the deformation is recoverable upon unloading, and pretensioning is sometimes used to extend the service range and reduce the weight. The engineer looks for structural stiffness rather than for material strength, and the manner in which the material is placed — structural shape — is as important as the material itself, as will be shown in this chapter by means of a few examples.

2.2 STABLE AND UNSTABLE ELASTIC DEFORMATION

The deflection of a beam under normal loading and the extension of a bar under tension are simple examples of stable elastic deformation. If the material obeys Hooke's law, the deformation is proportional to the load. We also know, from elementary strength of materials, that problems may arise when slender struts are loaded in compression, as the designers of the first steel bridges found to their cost (see Fig. 1.4). Even when the material is linear elastic and the stress is lower than the yield point, it may be possible for the structural deformation to increase rapidly as a critical value of the load is approached. The familiar Euler treatment and the more general methods of Rayleigh and

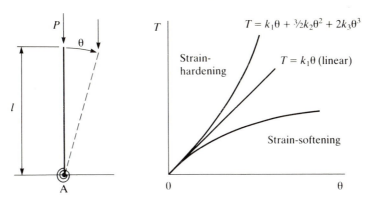

Figure 2.1   Simple rigid strut, hinged at A and fixed to a torsional spring

Timoshenko[1,2] can be extended to non-linear material behaviour of structural deformation. This will be done by considering the system illustrated in Fig. 2.1 consisting of a rigid strut, hinged at A and held vertical by a spiral spring. Under a vertical load P, the strut rotates by a small angle θ. The work of the external load is

$$W = Pl\,(1 - \cos\,\theta) \simeq Pl\,\frac{\theta^2}{2} \tag{2.1}$$

Taking the torque−rotation characteristics of the spring to be described by

$$T = k_1\,\theta + \tfrac{3}{2}\,k_2\,\theta^2 + 2\,k_3\,\theta^3 \tag{2.2}$$

where $k_1 > 0$ is the linear component of the stiffness and k_2 and k_3 are non-linear components that may be positive or negative depending on whether the material hardens or softens with strain, the potential energy stored is

$$U = \int_0^\theta T\,\mathrm{d}\theta = \tfrac{1}{2}\,k_1\,\theta^2 + \tfrac{1}{2}\,k_2\,\theta^3 + \tfrac{1}{2}\,k_3\,\theta^4 \tag{2.3}$$

The equilibrium condition is obtained by balancing the energy,

$$W = U \tag{2.4}$$

which can be written in the following form:

$$\frac{P}{k_1/l} = 1 + \frac{k_2}{k_1}\theta + \frac{k_3}{k_1}\theta^2 \tag{2.4a}$$

For equilibrium to be stable,

$$\frac{\mathrm{d}W}{\mathrm{d}\theta} \le \frac{\mathrm{d}U}{\mathrm{d}\theta} \tag{2.5}$$

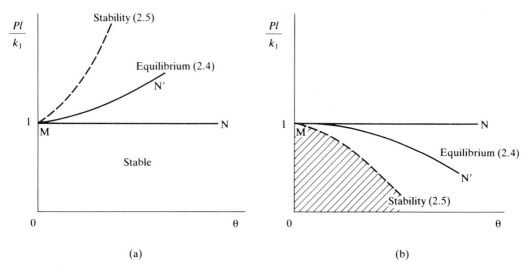

Figure 2.2 Equilibrium and stability in the post-buckling behaviour of the simple strut: (a) strain-hardening spring; (b) strain-softening spring

or

$$\frac{P}{k_1/l} \leq 1 + \frac{3}{2}\frac{k_2}{k_1}\theta + 2\frac{k_3}{k_1}\theta^2 \qquad (2.5a)$$

Equations (2.4) and (2.5) are represented in Fig. 2.2(a) for a strain-hardening spring and in Fig. 2.2(b) for a strain-softening spring. In either case, the strut rotates when the dimensionless load (Pl/k_1) exceeds unity, following the lines MN for a linear spring, MN′ for non-linear springs. The difference between the two plots is that while the equilibrium is stable in Fig. 2.2(a), it becomes unstable as soon as the branching point M is reached in Fig. 2.2(b). The material properties are thus seen to contribute not only to the load-carrying capacity of the structure, i.e. its buckling strength, but to the post-buckling behaviour of the structure. It is obvious that the value chosen for the safety or load factor will depend on whether the structure remains stable, as in Fig. 2.2(a), or not.

In the example that has been considered the structure is under a dead load. There are other situations in which it is not the load but the end displacement that is controlled. For example, in the system of Fig. 2.3 the central strut is held between two rigid beams to which are fixed two bars. When the bars are cooled by ΔT while the temperature of the strut is maintained constant, self-equilibrating forces $-P, P/2$ will appear to ensure that the deformations of strut and bars are compatible. There is no external work, only internal potential energy and, if we assume that the material is linear elastic, we can write that, for the bars,

$$\Delta = \frac{Pl}{AE} - \alpha l \Delta T \tag{2.6}$$

The bending moment in the strut is

$$Pu(x) = -EI \frac{d^2u}{dx^2} \tag{2.7}$$

as is known from elementary bending theory where $u(x)$ is the transverse deflection. Integrating Eqn (2.7) we find that the deflection is given by

$$u = A \sin \sqrt{\left(\frac{P}{EI}\right)} x + B \cos \sqrt{\left(\frac{P}{EI}\right)} x \tag{2.8}$$

where A and B are constants whose values depend on the boundary conditions. These are

$$u = 0 \quad \text{when} \quad x = \pm \tfrac{l}{2}$$

$$\frac{du}{dx} = 0 \quad \text{and} \quad u = u_0 \quad \text{when} \quad x = 0$$

These require

$$A = 0, \quad B = u_0 \quad \text{and} \quad \sqrt{\frac{P}{EI}} = \frac{\pi}{l} \tag{2.9}$$

The vertical shortening of the strut is

$$-\Delta = 2 \int_0^{l/2} \frac{1}{2} \left(\frac{du}{dx}\right)^2 dx = \frac{\pi^2 u_0^2}{4l} \tag{2.10}$$

Eliminating Δ between Eqns (2.6), (2.9) and (2.10) we find that the maximum deflection of the strut is

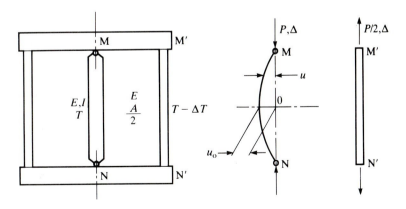

Figure 2.3   Thermal stressing of a strut

$$u_0 = 2 \left(\frac{\alpha l^2 \Delta T}{\pi^2} - \frac{I}{A} \right)^{1/2} \qquad (2.11)$$

This is an interesting expression. Contrary to what might be expected, the only material property that appears in it is the coefficient of thermal expansion. The Young's modulus is quite irrelevant. Also, if

$$\Delta T < \frac{\pi^2 I}{A \alpha l^2}$$

the term within the square root is negative which means that u_0 is zero for a temperature drop less than $(\pi^2 I/A\alpha l^2)$. Note that we have assumed that the strut is infinitely rigid in compression. A correction to the analysis will be needed when the compression of the strut becomes as significant as the shortening due to bending.

These two examples show that the material properties and the structural shape and loading conditions are intimately related and cannot be regarded as entirely independent factors — a truth that, while universally acknowledged, is sometimes forgotten in engineering practice.

2.3 DESIGN OF SHAFTS AND STRUCTURAL MEMBERS IN TORSION

An important group of structural components — shafts, fan blades — are subjected to torsional loads and, while it is not the intention to treat the problem of elastic torsion in detail — see, for example, Ref. 3 — the implications of material behaviour in the design of structural members in torsion will be considered in this section.

2.3.1 SHAFTS OF CIRCULAR CROSS-SECTION UNDER PURE TORQUE

The shear stress at a radius r is given by the familiar expression,

$$\tau = \frac{Tr}{J} \qquad (2.12)$$

where T is applied torque and $J = \frac{1}{2}\pi (r_0^4 - r_i^4)$ is the second polar moment of area (see Fig. 2.4). The angle of twist per unit length is

$$\theta = \frac{T}{GJ} \qquad (2.13)$$

assuming that the material is linear elastic and isotropic. G is the modulus of elasticity in shear, equal to $E/2(1 + \nu)$.

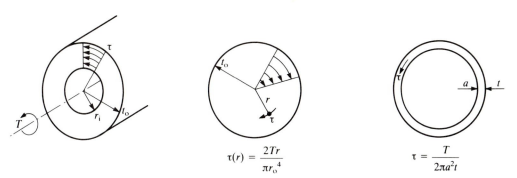

$$\tau(r) = \frac{2Tr}{\pi r_o{}^4}$$

$$\tau = \frac{T}{2\pi a^2 t}$$

Figure 2.4 Circular shaft under torsion and stress distribution in a solid shaft and in a thin-walled tubular shaft

It is well known that a hollow shaft is preferable to a solid shaft because the material is nearer to the periphery, where it is most effective. For a thin-walled tubular shaft, of mean radius a and thickness t,

$$J \simeq 2\pi a^3 t \quad \text{where} \quad t \ll a \tag{2.14}$$

In the solid shaft, the shear stress increases linearly from 0 at the axis to a maximum at the circumference, in accordance with Eqn (2.12). In the tubular shaft, the stress may be taken to be constant. When the two types of shaft are under equal torques, the maximum stress will be the same if

$$2\pi a^2 t = \pi \frac{r_0^3}{2}, \quad a^2 t = \frac{r_0^3}{4} \tag{2.15}$$

The stiffness ratio will then be

$$k = \frac{\text{Stiffness of hollow shaft}}{\text{Stiffness of solid shaft}} = \frac{a}{r_0}$$

Equation (2.15) may be expressed as follows:

$$\left(\frac{a}{r_0}\right) = k = \left(\frac{1}{4}\frac{a}{t}\right)^{1/3} \tag{2.16}$$

Thus, it follows that the stiffness ratio increases as the wall thickness-to-radius ratio decreases. The ideal shaft would be a paper-thin tube of very large radius, but a limitation is imposed on how thin the wall can be that prevents the designer from going to such an extreme. Considering an element of the shaft (Fig. 2.5), the state of torsion is equivalent to one of tension along AC and compression along BD and, while the tensile stresses will not cause any harm provided that they

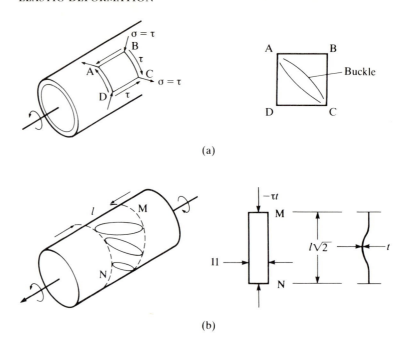

(a)

(b)

Figure 2.5 Buckling of thin-walled tube under torsion

remain below the yield point, the compression may cause buckling of the thin wall. A detailed treatment of this problem is found in Ref. 4 but, for our purpose, a simpler approach is sufficient. If the buckles are confined within a band of width equal to l, as in Fig. 2.5(b), strips of unit width such as MN may be considered as columns buckling under the critical load $(-\tau t)$. We know that the critical (Euler) load is proportional to (EI/l^2). In this case, the second moment of area of the section, I, is $(t^3/12)$. Therefore,

$$(\tau t)_{\text{crit.}} \simeq \frac{Et^3}{l^2} \qquad \tau_{\text{crit.}} \simeq \frac{Et^2}{l^2} \qquad (2.17)$$

or combined with Eqns (2.12) and (2.14),

$$t_{\text{crit.}} \simeq \left(\frac{Tl^2}{a^2E}\right)^{1/3} \qquad (2.18)$$

The minimum wall thickness for a given torque is seen to be proportional to the ratio $(l/a)^{2/3}$ where l is, so far, undefined and to the material compliance $(1/E)$. An increase in Young's modulus will permit a reduction in the thickness and thus in the stiffness ratio k. The characteristic length of the buckle can be fixed by, for example, fixing internal diaphragms or external flanges on a regular pitch, an effective way of increasing the stiffness of the shaft without an undue

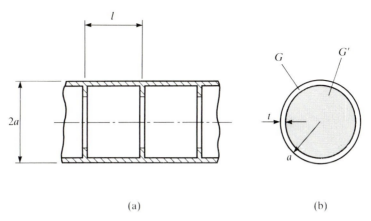

(a)   (b)

Figure 2.6   Stiffening of thin-walled tubular shafts: (a) diaphragms; (b) porous core

increase in weight, as shown in Fig. 2.6(a). In a long shaft of uniform wall thickness the characteristic length of the buckle will depend on the radius, increasing with it, and to a lesser extent on the thickness. The only way open to improve the design would then be to use a material with a higher Young's modulus.

It should be noted that the exact expression given in Ref. 4 for Eqn (2.17) is

$$\tau_{\text{crit.}} = A \frac{E}{1 - \nu^2} \frac{t^2}{l^2},$$

where A depends on the end conditions and, to a lesser extent, on the ratio $(l/\sqrt{at})$.

In tubes made of fibre-reinforced composites, the advantage of running the fibres along the directions of the principal stresses is obvious. The material is then orthotropic, with a maximum value of the Young's modulus along the direction of maximum compression while at right angles the tensile stress is resisted by the second set of fibres. This subject will be treated in Chapter 8.

An alternative form of stiffening consists in filling the tubular shaft with a light, porous core (Fig. 2.6(b)). Since the main purpose of the core is to maintain the circularity of the skin, which is where the structural strength resides, it may be entirely ignored in calculating J from Eqn (2.14). The skin itself may be wrapped round the core under tension, as in Fig. 2.7(a). The state of stress when the torque is applied consists of the superposition of the pretension σ_0 and the shear τ (Fig. 2.7(b)). From the Mohr circle, the pretension is equivalent to stresses in the axial and hoop directions x, y respectively, given by

$$\sigma_x = \tfrac{1}{2}\sigma_0 \left(1 - \cos 2\alpha\right) \quad \sigma_y = \tfrac{1}{2}\sigma_0 \left(1 + \cos 2\alpha\right) \quad \tau_{xy} = \tfrac{1}{2}\sigma_0 \sin 2\alpha$$

(see Fig. 2.7(c)). When added to the shear stresses τ, there will be a

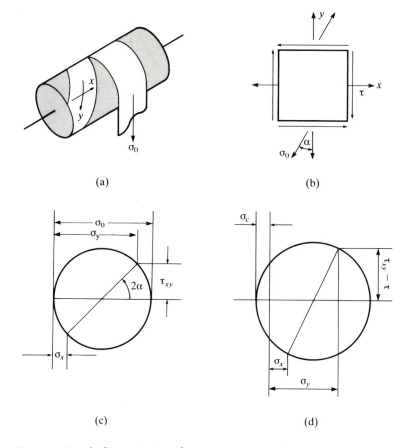

(a)

(b)

(c)

(d)

Figure 2.7 A composite shaft consisting of a porous core and a pre-tensioned skin

compressive stress (Fig. 2.7(d)) given by

$$\sigma_c = \frac{\sigma_x + \sigma_y}{2} - \sqrt{\left(\frac{\sigma_x - \sigma_y}{2}\right)^2 + (\tau_{xy} - \tau)^2}$$

The risk of skin buckling will be eliminated when $\sigma_c = 0$, a condition that may be written as follows:

$$\sigma_0 \sin 2\alpha = \tau$$

Of course, the measures taken to combat buckling will be effective only if the torque is always applied in one direction.

If the core material, instead of being perfectly compliant in shear, has an average shear modulus G', it can easily be shown that the combined stiffness of the shaft is

$$(GJ)_{\text{combined}} = 2\pi a^3 tG \left(1 + \frac{G'a}{4Gt}\right) \quad \text{where } t \ll a$$

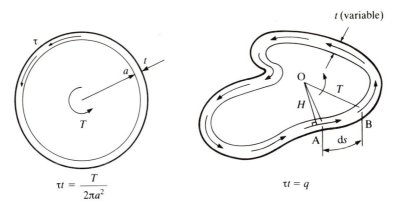

$$\tau t = \frac{T}{2\pi a^2}$$ $$\tau t = q$$

Figure 2.8   Comparison between a circular shaft and a non-circular thin-walled section under torsion

2.3.2   TORSION OF PRISMATIC MEMBERS OF THIN-WALLED SECTION

While the circular cross-section is the most efficient in resisting torsion, non-circular sections are often needed for reasons other than structural strength — for example, in turbine blades, wings, etc. A thin membrane will then be preferable to a solid section for the same reasons that apply to the circular shaft. In both cases, the torque is resisted by shear stresses shown in Fig. 2.8 and it is convenient to define the 'shear flow' q as

$$\text{Shear flow } (= q) \ = \ \tau t \qquad\qquad\qquad (2.19)$$

Continuity demands that the shear flow be constant at all points of the membrane. With respect to an arbitrary point, O, the torque exerted by the shear flow acting on an element AB is

$$dT \ = \ q \times H \times ds \ = \ 2q \times (\text{area OAB})$$

Integrating over the whole area,

$$T \ = \ 2qA \qquad\qquad\qquad (2.20)$$

where A is the total area within the perimeter of the membrane.

The strain energy per unit volume, assuming linear elastic behaviour, is

$$U_1 \ = \ \frac{1}{2}\tau\,\frac{\tau}{G} \ = \ \frac{q^2}{2t^2G}$$

and, integrating over the whole membrane section, the strain energy for a member of unit length is

$$U \ = \ \int U_1\,d(\text{vol.}) \ = \ \int \frac{q^2}{2t^2G}\,tds \ = \ \frac{q^2}{2G}\int \frac{ds}{t}$$

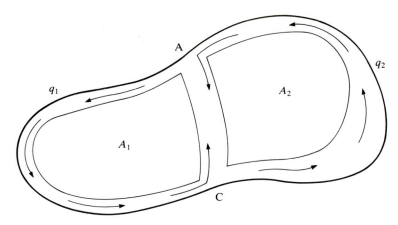

Figure 2.9   Torsion of a thin-walled section with an internal partition

where the integral is extended over the perimeter. The external work is

$$W = \tfrac{1}{2} T\theta$$

Writing that the external work is equal to the strain energy in the section and eliminating q by means of Eqn (2.20), we finally obtain the following expression:

$$\theta = \frac{T}{4A^2G} \int \frac{\mathrm{d}s}{t} \qquad (2.21)$$

As before, it is interesting to reduce the thickness to a minimum, compatible with the need to avoid buckling. Internal diaphragms, external flanges or directional reinforcement could all be used. Alternatively, longitudinal partitions, splitting the section into two or more cells, may be preferred. The principles underlying the analysis of these cellular members are best explained by considering a single partition (Fig. 2.9). The shear flow along the walls that define cells 1 and 2 are q_1 and q_2 when the torque is applied in the anti-clockwise direction. Along the partition wall, the net shear flow is $q_1 - q_2$. Cell 1 takes a torque $T_1 = 2q_1A_2$, likewise the torque taken by cell 2 is $T_2 = 2q_2A_2$ and, for equilibrium,

$$T = T_1 + T_2 = 2q_1A_1 + 2q_2A_2 \qquad (2.22)$$

To find angles of twist of the cells, we follow the same approach as before, observing that in cell 1, for example, the shear flow is q over the length ABC and $(q_1 - q_2)$ over the partition, CA,

$$\theta_1 = \frac{1}{2A_1G} \left[\int_{\mathrm{ABC}} q_1 \frac{\mathrm{d}s}{t} + \int_{\mathrm{CA}} (q_1 - q_2) \frac{\mathrm{d}s}{t} \right] \qquad (2.23)$$

A similar expression applies to cell 2. Writing that, for compatibility of deformation,

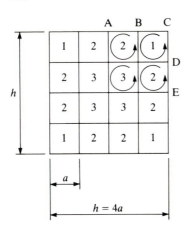
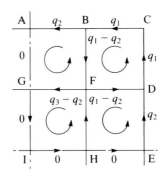

Figure 2.10   Cellular square tube (4 × 4 cells) under torsion

$$\theta = \theta_1 = \theta_2 \tag{2.24}$$

from Eqns (2.22), (2.23) and (2.24) it is possible to obtain q_1, q_2 and θ for a given value of the torque. The same reasoning can be applied to a multi-cell section divided into 4×4 identical cells of area a^2 and equal wall thickness (Fig. 2.10). Due to symmetry, we only need to consider one-quarter of the whole section. The constitutive cells are identified by the numbers 1, 2 and 3. For equilibrium,

$$T = 4T_1 + 8T_2 + 4T_3 \quad \text{with } T_1 = 2q_1a^2 \; T_2 = 2q_2a^2 \; T_3 = 2q_3a^2 \tag{2.25}$$

$$\theta_1 = \frac{1}{2a^2Gt} [2q_1a + 2(q_1 - q_2)a]$$

$$\theta_2 = \frac{1}{2a^2Gt} [q_2a + (q_2 - q_1)a + (q_2 - q_3)a] \quad \theta_1 = \theta_2 = \theta_3 = \theta \tag{2.26}$$

$$\theta_3 = \frac{1}{2a^2Gt} [2(q_3 - q_2)a]$$

From Eqns (2.25) and (2.26), we find

$$q_1 = 0.345 \frac{T}{b^2} \quad q_2 = 0.483 \frac{T}{b^2} \quad q_3 = 0.690 \frac{T}{b^2}$$

$$\theta = \frac{T}{G} \times \frac{1}{1.21b^3t}$$

These expressions are to be compared with the values for the hollow square section, from Eqns (2.20) and (2.21):

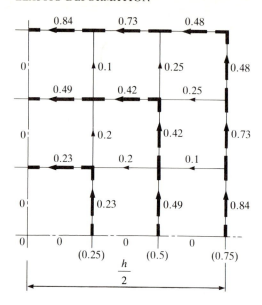

Figure 2.11 Shear flow in a cellular square tube (6 × 6) under torsion, constant wall thickness t

$$q = 0.5 \frac{T}{h^2} \qquad \theta = \frac{T}{G} \frac{1}{h^3 t}$$

Partitioning has increased the stiffness by 21%. The shear flow along the cell walls is lower than in the hollow square, as might be expected.

It is left as an exercise to show that the stiffness of a square tube with 6 × 6 cells is 44% higher than for the hollow tube of equal overall dimensions. The distribution of shear flow is presented in Fig. 2.11 in terms of the ratio between the value of the shear flow at a given position and the shear flow in the hollow tube under equal torque. It should be noted that the shear flow in the three concentric tubes highlighted in bold lines increases with distance from the centre line, decreasing as the corners are approached. The walls linking these tubes carry very little shear and the walls along the two axes of symmetry can be removed without causing any adverse effects on the structural strength or stiffness. These observations suggest a simple approximate analysis. Assume that the section consists of the three concentric tubes, linked in such a way as to force them to twist together. Then, from Eqns (2.20) and (2.21),

$$T = 2q_1 \times \frac{h^2}{9} + 2q_2 \times \frac{4h^2}{9} + 2q_3 \times h^2$$

$$\theta_1 = \frac{6q_1}{Ght} \qquad \theta_2 = \frac{3q_2}{Ght} \qquad \theta_3 = \frac{2q_3}{Ght} \; (=\theta)$$

obtaining for the shear flows the value in parentheses in Fig. 2.11 (referred to the hollow tube) and a stiffness 33% superior to that of the hollow tube. This simple approach is therefore seen to yield results that are sufficiently accurate for most purposes and certainly entirely adequate for design.

2.4   DESIGN OF BEAMS

2.4.1   BEAMS OF RECTANGULAR CROSS-SECTION UNDER PURE BENDING

In a beam of rectangular cross-section under pure bending, the stress is given by the familiar expression, when the material is uniform and isotropic,

$$\sigma = \frac{Mz}{I_y} \tag{2.27}$$

where z is the distance from the neutral axis and I_y the second moment of area with respect to the y axis, equal to $(bb_3/12)$ (Fig. 2.12).

The bending stiffness is

$$EI_y = \frac{Ebb^3}{12} \tag{2.28}$$

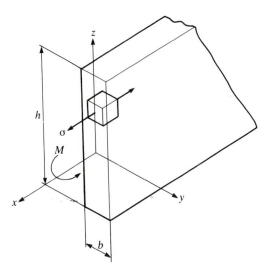

Figure 2.12   Beam of rectangular cross-section under pure bending

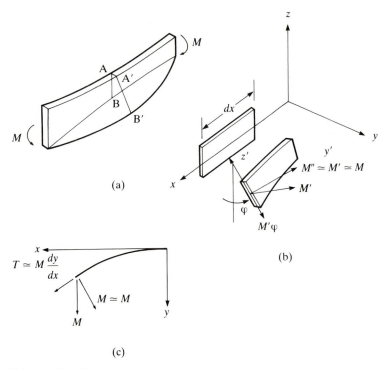

Figure 2.13 Sideways buckling of a beam

For a given cross-sectional area (bh) or mass per unit length, the stiffness increases as $h \to \infty$ and $b \to 0$. A reduction in the maximum stress accompanies this increase in stiffness. As in the case of torsion, the problem here is that buckling considerations make it unwise to use excessively thin sections. The best design for a given mass and bending moment is the one that exhibits the highest stiffness while it has adequate margins of safety with regards to the relevant modes of failure. In this case, these are plastic deformation of the outermost fibres of the beam, when the maximum stress reaches the yield point, and sideways buckling, as shown in Fig. 2.13, in which sections such as AB rotate and deflect to A′B′. An element of length dx deforms as shown in Fig. 2.13(b) and the beam centre line as in Fig. 2.13(c). The vertical deflection is assumed to be very much smaller and is neglected. The bending moment M is equivalent to a moment M' on the xy plane along the normal to the deflection curve and a torque T tangential to that curve. Since the beam is twisted as well as bent, M' produces bending about the two axes of symmetry of the cross-section, which is then under the combined effect of the torque T, a bending moment B about the z' axis equal to $-M\phi$ and a bending moment M'' about the y' axis approximately equal to M. Taking into account the static properties of the section,

$$B = -M\phi = EI_z \frac{d^2y}{dx^2} = E \frac{b^3h}{12} \frac{d^2y}{dx^2}$$

$$T = M \frac{dy}{dx} = GC \frac{d\phi}{dx} = \frac{E}{2(1+\nu)} \frac{bh^3}{3} \frac{d\phi}{dx}$$

where C is equivalent second polar moment of area of the section (see, for example, Refs 1,3). These equations reduce to

$$\frac{d^2\phi}{dx^2} + \left[\frac{72(1+\nu)M^2}{E^2 b^4 h^4} \right] \phi = 0$$

whose solution, with the boundary conditions $\phi = 0$ at $x = 0$ and at $x = 1$ and M constant, is

$$\phi = C \sin \frac{\pi x}{l}$$

with

$$M = \frac{\pi E b^2 h^2}{6l\sqrt{2(1+\nu)}}$$

being the critical value of the bending moment. With a safety factor on buckling f_b, the admissible bending moment in service becomes

$$M = \frac{1}{f_b} \frac{\pi E b^2 h^2}{6l\sqrt{2(1+\nu)}}$$

The bending moment is also limited by the need to maintain the whole section within the elastic limit, σ_e, i.e.

$$M = \frac{1}{f_e} \frac{\sigma_e b h^2}{6}$$

In a well-balanced design, the two limits will be equal. It follows that

$$\frac{b}{l} = \left(\frac{f_b}{f_e} \right) \frac{\sigma_e}{\pi E} \sqrt{2(1+\nu)} \tag{2.29}$$

Combining Eqns (2.28) and (2.29), it is found that for a given cross-sectional area and beam span, the bending stiffness of the design is proportional to (E^3/σ_e^2). Equation (2.29) is interesting in that it shows that the width-to-span ratio depends on the material properties. As a result, it follows that materials cannot be selected solely by considering the material itself since it affects the structural shape in a manner that must be recognized by the designer. When choosing between two candidate materials denoted by I and II, the preceding equations may be presented in the following forms

$$\frac{b_I}{b_{II}} = \frac{\sigma_{eI}}{\sigma_{eII}} \frac{E_{II}}{E_I}; \quad \frac{b_I}{b_{II}} = \frac{\sigma_{eI}}{\sigma_{eII}} \left(\frac{E_{II}}{E_I} \right)^{1/2}$$

$$\frac{\text{(Stiffness-to-weight ratio) I}}{\text{(Stiffness-to-weight ratio) II}} = \frac{\rho_{II}}{\rho_I} \left(\frac{\sigma_{eI}}{\sigma_{eII}} \right)^2 = \frac{\rho_{II}}{\rho_I} \left(\frac{E_I}{E_{II}} \right)^{3/2}$$

that define the width-to-height ratio of the optimum rectangular cross-section and the stiffness-to-weight of the beam. These equations have, unfortunately, little more than an academic interest — besides illustrating the nature of the designer's task, since the rectangular section itself is unlikely to make the best possible use of the material. Indeed, as in the case of a solid versus a hollow shaft, the most efficient beam is the one in which most of the material will be placed as far away from the centre line (neutral axis) as possible.

2.4.2 EFFECT OF SHEAR

The parabolic distribution of shear stress found when the beam is loaded by a normal force or non-constant bending moment instead of under constant bending moment (Fig. 2.14) must also be taken into account. As in the case of torsion, the state of pure shear at the neutral axis is equivalent to biaxial tension/compression and it is this compressive component that can cause buckling of the beam section. To analyse this problem is outside the scope of this book — the reader is referred to Ref. 4, for example — but, intuitively, it might be expected that buckling will occur as the slenderness of the section increases. Treating the web as an Euler strut, the critical compression component of the shear stress will be proportional to (Et^2/h^2). This imposes another design limitation when deciding on the dimensions of the beam.

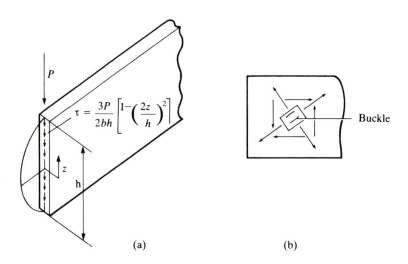

(a)   (b)

Figure 2.14   Effect of shear force: (a) shear stress distribution; (b) local shear buckling

2.4.3   I Sections and composite beams

In view of the preceding remarks, the advantages of I sections, in which a thin web links two larger flanges, are clear. Tension and compression are resisted mainly by the flanges, the function of the web being to maintain them at the correct position and to resist the shear. Having split the beam into components with separate functions, the way is open for improving the design still further. The flange under compression must not be so slender as to buckle and, because its mode of failure may well be unrelated to the actual strength of the material as characterized by the compressive yield strength, it may consist of a large section of low-strength material. Slenderness does not play any part in the choice of shape for the flange under tension, whose section is solely governed by the need to maintain the stress below the yield point. Concerning the web, its thickness depends on the admissible shear stress and on the need to avoid shear buckling and compression buckling under the external normal load. The web may be continuous or discontinuous. These considerations are reflected in the development of many possible designs, some of which are illustrated in Fig. 2.15. The first development is the standard rolled I section. By cutting it into two halves along the stepped line ABCD and joining the two sections, the total depth of the beam is increased without increasing the weight (Fig. 2.15(c)).

Two remarks must be made concerning the detailed design of such beams. One is that the top flange is subject to localized normal crushing stress where the load is applied. The second is that in the castellated beams the welds are in regions of high shear stress and their integrity is therefore vital to that of the beam as a whole.

The ideal I section will have a very thin web and wide, thin flanges. To avoid buckling, reinforcing plates may be welded, thus limiting the unsupported length of the three constitutive elements (Fig. 2.25(d)). Alternatively, the compression flange may be more compact than the one in tension (Fig. 2.25(e)). In all these examples, the web may consist of a lightweight material with a low Young's modulus provided its strength in shear and crushing is adequate, the material for the flange in tension must be characterized by a high strength and Young's modulus, and for the flange under compression a high Young's modulus is the main requirement.

The most efficient use of the material in a structure happens when all of it is either under uniform tension or under uniform compression, unlike in the web, where the region around the neutral axis may be virtually stress-free. This ideal situation can be achieved in an open-lattice girder, Fig. 2.15(f). The two chords and the diagonal bracing need not be of the same material. Indeed, tension members may be fibres, ropes or wires, whilst those in compression can be low-strength, low-density materials in the form of short, compact struts.

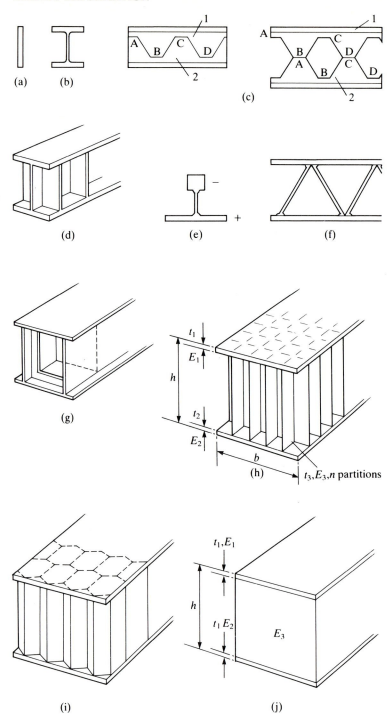

Figure 2.15 Evolutive design of a beam section

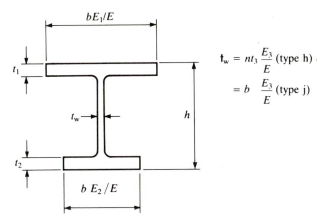

$$t_w = nt_3 \frac{E_3}{E} \ (\text{type h})$$

$$= b \ \frac{E_3}{E} \ (\text{type j})$$

Figure 2.16   I section equivalent to composite beam

The stiffened design of Fig. 2.15(d) develops into the box beam of Fig. 2.15(g), with internal diaphragms to prevent shear buckling of the webs, and compression buckling of the top flange.

From two webs in the box beam, we can go to any number of thin webs, linked by diaphragms arranged in a small pitch and forming a large number of square cells, Fig. 2.15(i). Indeed, there is no reason why the evolution should stop there, since a beam consisting of two flanges fixed to a core made of a porous material, as in Fig. 2.15(j), may be regarded as an extension of type (h) or (i) when horizontal partitions are added, or of (b). Such a structure of stiff outer skins containing a porous core has been adapted by evolution to provide the light, stiff bones that support our bodies.

Any of the designs in Fig. 2.15, with the exception of the open lattice, may be treated by defining an equivalent section of equal depth. For example, for types (h) and (j), the equivalent section is defined in Fig. 2.16 by reference to a material of Young's modulus E.

2.5   RESIDUAL STRESSES: PRETENSIONING

2.5.1   BARS IN TENSION OR COMPRESSION

Pretensioning can increase the elastic range of a given structure and material. As an example, consider the three-bar assembly of Fig. 2.17 in which a residual stress system is created by stretching the two outer bars until their length is equal to that of the central bar before clamping all three to the rigid cross-head. The residual stresses must be such that

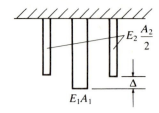

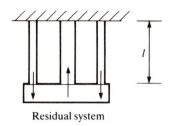

Residual system

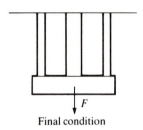

Final condition

Figure 2.17 Residual stresses introduced in a three-bar system through lack-of-fit

$$\sigma_{1R}A_1 + \sigma_{2R}A_2 = 0 \text{ for equilibrium}$$

$$\frac{\sigma_{1R}}{E_1} = \frac{\sigma_{2R}}{E_2} - \frac{\Delta}{l} \text{ for compatibility of displacements}$$

The stresses induced by a force F applied subsequently must satisfy the conditions

$$\sigma_1 A_1 + \sigma_2 A_2 = F$$

$$\frac{\sigma_1}{E_1} = \frac{\sigma_2}{E_2}$$

and the total stresses are obtained by adding σ_{1R} to σ_1, σ_{2R} to σ_2, obtaining finally

$$\sigma_{1T} = \frac{FE_1}{A_1E_1 + A_2E_2}\left(1 - \frac{A_2E_2\Delta}{Fl}\right) < (\sigma_{\text{yield}})_1$$

$$\sigma_{2T} = \frac{FE_2}{A_1E_1 + A_2E_2}\left(1 + \frac{A_1E_1\Delta}{Fl}\right) < (\sigma_{\text{yield}})_2 \qquad (2.30)$$

Crushing and buckling under compression must, of course, be avoided. This consideration is particularly important at the pretensioning stage since the compression is relieved by the external force.

The stiffness of the assembly is

$$k_T = \frac{A_1E_1 + A_2E_2}{l} \tag{2.31}$$

against the stiffness of the constituents,

$$k_1 = \frac{A_1E_1}{l} \quad k_2 = \frac{A_2E_2}{l} \tag{2.32}$$

i.e. it is the same as the stiffness of the same set of components without prestressing.

Consider first the case when the central column cannot take any tension, because it is made of either a number of loose blocks or of a material such as concrete, rock or a low-strength brittle ceramic with a very low tensile strength. It is assumed that the compressive stress must be limited to a value that depends on the crushing strength of the material, the yield point or the critical buckling load of the struts:

$$|\sigma_1| < |\sigma_c|$$

Taking $E_2 = nE_1$, $A_1 = mA_2$,

$$k_T = \frac{A_2E_1}{l}(m+n) = k_2\frac{m+n}{n} \tag{2.33}$$

and

$$(\sigma_{yield})_2 = \sigma_F$$

Substituting in Eqns (2.30) and (2.31), we find

$$\Delta = \sigma_c\frac{l(m+n)}{nE_1} \tag{2.34}$$

$$\sigma_{1T} = \frac{F}{A_2(m+n)} - \sigma_c \leq 0 \quad F \leq \sigma_cA_2(m+n) \tag{2.35}$$

$$\sigma_{2T} = \frac{Fn}{A_2(m+n)} + m\sigma_c \leq \sigma_F \quad F \leq (\sigma_F - m\sigma_c)A_2\frac{m+n}{n} \tag{2.36}$$

For the ideal combination of materials and cross-sectional areas, both limits will be the same, i.e.

$$\sigma_F = (m+n)\sigma_c \tag{2.37}$$

in which case,

$$F \leq \sigma_F A_2 \tag{2.38}$$

These relationships are plotted in Fig. 2.18(a) for the total assembly and in Fig. 2.18(b) for its components. In Fig. 2.18(a), the stiffness is constant between O and A, until the force is such as to cause the breakdown of the central column (Eqn 2.35). From then on, only the

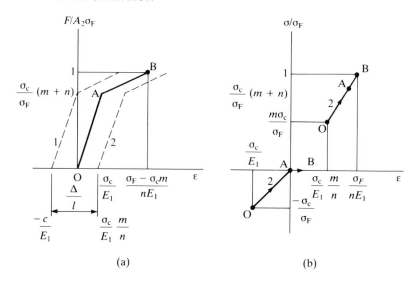

Figure 2.18 Extension of the system of Fig. 2.17 under load: (a) combined system, (b) components

two outer bars continue to take any load, deforming elastically until the force reaches the value given by Eqn (2.38). The strains in the three components follow the broken lines of Fig. 2.18(a) and the corresponding stress–strain curves are shown in Fig. 2.18(b).

As an example, take the case of a concrete column and steel bars for which

$$\sigma_c = \frac{1}{20} \sigma_F \quad n = 15$$

For the optimum design, Eqn (2.37) gives $m = 5$. The combined stiffness, from Eqn (2.33), is 25% higher than for the steel bars alone. We may also choose to reduce the precompression in the column, increasing its area to increase the combined stiffness.

Taking for example

$$\sigma_c = \frac{1}{40} \sigma_F$$

we obtain $m = 25$, whereupon the stiffness is increased by a factor of 5/3.

As the ratio between the values of the Young's modulus for the two materials decreases, the benefits of prestressing become less apparent. It does not make any sense to pretension a stack of steel blocks with steel wires or a stack of carbon blocks with carbon fibres, since taking

$$\sigma_c = \sigma_F \quad E_1 = E_2$$

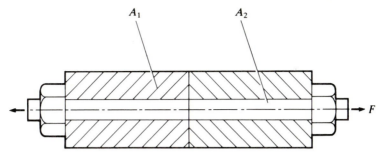

Figure 2.19   Bolted joint

in Eqns (2.33) to (2.38) gives $m = 0$. There are, however, situations when the pretensioning is not carried out for structural reasons but in order to maintain contact between the faces of two components, for example two flanges, a cylinder head and the block in an internal combustion engine. In those cases, the service limitation is given by Eqn (2.35), corresponding to separation between the contacting surfaces. The stress in the components in tension, usually bolts, studs or tendons, is $\sigma_c(m + n)$, with the proviso that this must be less than the yield or failure stress σ_F. As an example, take a bolted joint between two collars (Fig. 2.19), for which

$$m = 40 \quad n = 1$$

The assembly behaves as a continuous solid, with a stiffness equal to 41 times that of the bolt until separation occurs. With a precompression given by Eqn (2.37), the instant of separation occurs as the bolt begins to fail,

$$\sigma_c = \frac{\sigma_F}{41} \quad F = \sigma_F \ A_2 = 41 \quad \sigma_c \ A_2$$

Pretensioning also helps when designing with two materials of different yield strength, or when, as is the case with ceramics, the scatter of properties is such that it is wise to apply a higher safety factor to one of the two materials. It is a fairly trivial exercise to extend the preceding treatment to this problem.

2.5.2   PRETENSIONING OF BEAMS

The principle of pretensioning has found its widest and most important application in the prestressing of reinforced concrete beams. A concrete beam of rectangular cross-section would crack under a very small bending moment since the material is brittle and has a negligible tensile strength. Precompressing the section by, for example, a tendon AB held

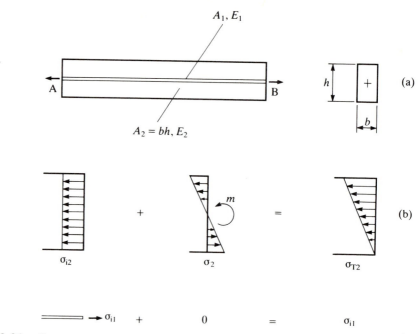

Figure 2.20 Beam prestressed with a central tendon (a) and state of stress (b)

in tension between the ends (Fig. 2.20(a)) means that the beam can take a certain bending moment before the stress in the bottom outermost fibre becomes tensile and failure is imminent. In this design, which could be applied equally well to any other combination of materials with either a monolithic beam or one made up of separate blocks, the stress in the tendon remains unchanged when the bending moment is applied since it is positioned exactly on the neutral axis of the beam. The precompression in the beam is

$$\sigma_{i2} = -\sigma_{i1} \frac{A_1}{A_2} \quad \text{where } \sigma_{i1} < \sigma_{F1}$$

and the additional stresses due to bending in the beam are

$$-\frac{6M}{A_2 b} < \sigma_2 < \frac{6M}{A_2 b}$$

Failure occurs when

$$\sigma_{i1} \frac{A_1}{A_2} = \frac{6M}{A_2 b} \quad \text{(tensile failure)}$$

or

$$\sigma_{i1} \frac{A_1}{A_2} = \frac{6M}{A_2 b} = \sigma_{c2} \quad \text{(crushing)}$$

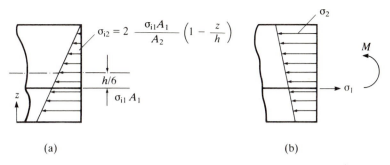

Figure 2.21   Pretensioning of a beam by a tendon offset with respect to the centre line: (a) pretension; (b) final stresses

In the optimum design both modes give the same load while $\sigma_{i1} = \sigma_{F1}$. Therefore,

$$\frac{A_1}{A_2} = \frac{1}{2} \frac{\sigma_{c2}}{\sigma_{F1}} \qquad M_{max} = \frac{\sigma_{F1} A_1 h}{6} \qquad (2.39)$$

Note that in this case the Young's modulus does not appear in the equations. On the other hand, if the tendon is away from the neutral axis it will stretch during bending by an amount which will depend on the ratio between the moduli of tendon and bulk material. When it is at $h/6$ from the centre line, Fig. 2.21, the initial state of stress is as shown in (a). It is left as an exercise to show that the final stress system (b) is, for the beam,

$$\sigma_2 \cong \frac{-2\sigma_{i1} A_1}{A_2} + \frac{6M}{A_2 h} \quad \text{at the bottom}$$

$$\sigma_2 \cong \frac{-6M}{A_2 h} \quad \text{at the top}$$

and for the tendon,

$$\sigma_1 \cong \sigma_{i1} + \frac{2M}{A_2 h}$$

2.5.3   PRETENSIONING ACHIEVED BY OVERLOLADING

A simple and effective way of introducing a residual stress system which will help to extend the range of elastic behaviour in service is to apply an overload, of the same nature as the service load that takes the structure into the plastic regime prior to service. In a beam (Fig. 2.22), as the bending moment increases, the outermost fibres yield until the whole section deforms plastically. For an elastic–ideally plastic material and a rectangular cross-section this happens when $M =$

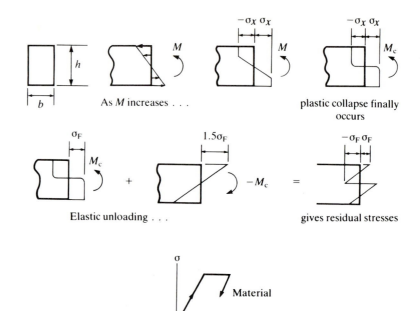

Figure 2.22   Pretensioning of a beam by overloading

$\sigma_y bh^2/4$. This is referred to as the collapse moment, M_c. Since, in a beam free from any residual stresses, the maximum value of the bending moment that can be allowed before plastic deformation sets in is $\sigma_y bh^2/6$, pretensioning in this manner has extended the service range by 50%. Reference may be made to Chapter 3, where plastic deformation is studied.

The same process is applied to rods of circular cross-section under torsion — torsion bars and helical springs in vehicle suspensions, for example. The stress distribution is τ_y everywhere on application of a torque $T_c = 2\pi r^3 \tau_y/3$. Applying $(-T_c)$ elastically gives a final state of prestress similar to that of Fig. 2.22 and extends the service range by 33%. Other applications to thick-walled pressure vessels, guns, rotating discs, etc., exploit the same principle. Of course, the materials used must have adequate ductility and be capable of retaining the residual stresses throughout the service life.

EXAMPLES

2.1   A cantilever beam of length L under a point load P at the end and of circular cross-section is to be designed for a given stiffness

$$\frac{P}{\Delta} = \frac{3}{4} \pi E \frac{r^4}{L^3}$$

where r is the section radius. Show that the ratio $(E^{1/2}/\rho)$ may be taken as a figure of merit and, from the materials listed below, select the best based only on stiffness and based on cost and stiffness.

	E (GN m^{-2})	ρ (kg m^{-3})	Cost (£/ton)
Steel	200	7800	300
Wood	9	400	300
Concrete	20	2400	200
Aluminium	70	2700	1200
CFRP	100	1500	120000

2.2 The strut of Fig. 2.1 is inclined by a small angle θ_0 to the vertical before the load is applied. The spring is made of a strain-softening material ($k_1 > 0$, $k_2 < 0$) with $k_3 = 0$. Show that the maximum load is reached when the strut has tilted by an additional angle

$$\theta = \left[2 \left(1 - \frac{k_1}{2k_2} \right)^{1/2} - 1 \right] \theta_0$$

2.3 Find the shear flow and stiffness in the triangular section of Fig. 2.23(a). What is the effect of introducing the partitions shown in Fig. 2.23(b)?

2.4 Extend the approximate treatment described for the torsion of a square tube with 6×6 cells to one with $n \times n$ cells where n is any number larger than 6. Apply it to $n = 6$, 8 and 10.

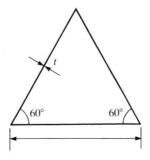
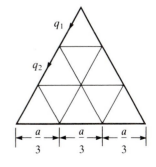

Figure 2.23

2.5 Find the ratio (depth/length) for the beam of Fig. 2.13 if sideways buckling forces the beam to deform in such a way that the top fibre remains straight.

2.6 A composite beam of overall depth h consists of a porous core sandwiched between two thin shells of thickness t, density ρ and Young's modulus E. The Young's modulus of the core material, E_c, depends on its density ρ_c according to the expression

$$E_c = A\rho_c \quad \text{for} \quad 500 < \rho_c < 800 \text{ kg m}^{-3}$$

where A is a constant, equal to 1.5×10^7 Nm kg^{-1}. Taking $E = 100$ GN m^{-2} and $\rho = 1500$ kg m^{-3}, suggest values of t/h and ρ_c based on the maximum stiffness for a given mass per unit width and a given depth. If the cost of the sheet material is £50 kg^{-1} and that of the core is £0.50 kg^{-1}, would you change the design?

2.7 Figure 2.24 shows alternative designs for a bracket supporting a horizontal shaft (a) and a vertical shaft (b). Which of the two alternatives is preferable? Why?

2.8 A shaft, 50 mm diameter, is twisted until full plasticity is reached. Assuming that the yield point is 210 MN m^{-2} and that no strain-hardening takes place, determine the magnitude of the required torque and the final stress distribution obtained upon unloading.

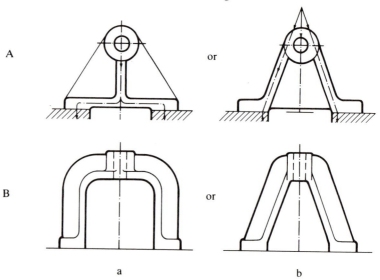

A

or

B

or

a

b

Figure 2.24

REFERENCES

1   DEN HARTOG J P 1952 *Advanced Strength of Materials* McGraw-Hill, New York

2   TIMOSHENKO S P 1955 *Strength of Materials* Vols 1 and 2, Van Nostrand, Princeton

3   TODD J D 1981 *Structural Theory and Analysis* Macmillan, London

4   TIMOSHENKO S P, GERE J M 1961 *Theory of Elastic Stability* McGraw-Hill, New York

Table 2.1 Typical mechanical and physical properties of materials

Ceramic	E (GNm^{-2})	H (GNm^{-2})	ρ (Mgm^{-3})	σ_b (MNm^{-2})	α $(\text{K}^{-1} \times 10^6)$
Diamond	1000	50	—	—	—
Tungsten carbide, WC	450–650	6	14–17	—	—
Borides of Ti, Zr, Hf	500	5–12	—	—	—
Silicon carbide, SiC	210–480	4–10	2.6–3.2	98–172	3.8
Alumina, Al_2O_3	220–400	3–6	3.4–4	157–315	6.5
Beryllia, BeO	300–386	3–5	2.8–3	175–266	7.4
Silicon nitride, Si_3N_4	97–245	3–8	2–3.2	200–700	3.2
Magnesia, MgO	250	3	3.5	100	14.8
Zirconia, ZrO_2	156–200	3–4	5.5	245	
Mullite, $3Al_2O_3.2SiO_2$	150	4	3.1	175	4.6
Cordierite, $2MgO.2Al_2O_3.5SiO_2$	60–120	3–4	2–2.3	56–105	2.4
Rock	20–60	—	2.2–3	—	—
Glass	40–60	2–7	2.5–2.7	50–80	10
Concrete	30–50	—	2.4–2.5	—	7

(a)

H = hardness; σ_b = bending strength

Fibre	E (GN m^{-2})	ρ (Mg m^{-3})	σ_T (MN m^{-2})
E-Glass	72.5	2.58	3450
A-Glass	69	2.50	3040
S-Glass	86	2.48	4590
C (PAN precursor)	480–230	1.74–1.96	1500–4500
Aramid (Kevlar 49)	138	1.44	2800
Al_2O_3	76–380	3–4	1300–2070
SiO_2	70	2.2	200–400
Boron	400	2.5	3600
Nicalon (SiC90)	180	2.55	3000
Silar (SiC)	690	3.2	6900

(b)

σ_T = tensile strength

Table 2.1
(continued)

Polymer	E (GN m^{-2})	ρ (Mg m^{-3})	σ_Y (MN m^{-2})	T_g (°C)
Thermoplastics				
Styrenic				
ABS	2.5	1.2	50	80
Polystyrene (PS)	2.7	1.1	50	100
Polycarbonate (PC)	2.5	1.2–1.3	68	150
Polyethylene				
High-density	0.56	0.96	30	
Low-density	0.18	0.91	11	−20
Poly(ethylene terephthalate) (PET)	8	0.94	135	67
Poly(methyl methacrylate) (PMMA)	2.8	1.2	70	100
Poly(phenylene oxide) (PPD)	2.4	0.9	62	180
Poly(phenylene sulphide) (PPS)	3.3	1.2	75	150
Polypropylene (PP)	1.3	0.9	35	0
Poly(vinyl chloride) (PVC)	2.0	1.2	25	80
Polyamide	2.8	1.1–1.2	70	60
Polytetrafluorethylene (PTFE)	0.4	2.3	25	120
Thermosets				
Epoxides	3	1.1–1.4	60	(150–250)
Phenolics	18	1.5	80	(200–300)
Polyesters (glass-filled)	14	1.1–1.5	70	(200)
Polyimides (glass-filled)	21	1.3	190	(350)
Silicones (glass-filled)	8	1.25	40	(300)
Ureas	7	1.3	60	(80)
Urethanes	7	1.2–1.4	70	(100)

(c)

ABS = Acrylonitrile–butadiene–styrene; σ_Y = yield strength; T_g = transition temperature (Maximum temperature)

Table 2.1
(continued)

Polymeric composites	E_1 (GN m^{-2})	ρ (Mg m^{-3})	σ_T (MN m^{-2})	T_f (°C)
ABS/20 Glass	6.2	1.18	90	105
ABS/40 Glass	10.5	1.38	125	110
PS/40 Glass	14	1.38	100	100
PC/40 Glass	11	1.52	140	150
PET/55 Glass	20	1.8	196	230
PPS/Glass 20	9	1.5	100	260
PPS/Glass 40	14	1.67	140	260
PPS/C 40	33	1.46	180	260
PP/Glass 40	7.5	1.23	110	150
Polyetheretherketone PEK/C 40	30	1.46	270	285
Epoxy/Glass 60	15	1.86−1.96	245	120
Epoxy/C 60	15	1.48−1.54	220	120
Polyester/Glass 60	20	1.84−1.9	56	245

(d)

T_f = deflection temperature

Table 2.1
(continued)

Metallic materials	E (GN m^{-2})	ρ (Mg m^{-3})	σ_Y (MN m^{-2})	σ_T (MN m^{-2})	α (K$^{-1} \times 10^6$)
Ferrous					
Iron	196	7.9	50	200	—
Cast iron					
Grey	83	6.9	—	140	10.8
Malleable	176	6.9–7.4	225–315	350–450	10.6–13.5
Nodular	154–172	7.1	420–525	630	12
White	—	7.6–7.7	—	150–350	9–9.5
Low-C steel	210	7.85	250–300	450–600	14–15
High-C steel (Q&T)	210	7.85	600–1300	800–1800	14–15
Pressure vessel (low-C)	210	7.85	<1200	<1500	14–15
Low alloy	210	7.85	<1200	<1500	14–15
Stainless					
Ferritic	200	7.75	280	525	10
Austenitic	193	8	270	560	17.2
Matensitic	200	7.75	240–600	450–770	10
Maraging steel	193	7.83	1200–1300	1300–1500	11.5
Non-ferrous					
Aluminium	70	2.7	30–150	80–150	24
Al alloys	70	2.7–2.8	35–500	90–580	24
Mg alloys	45–50	1.35–1.8	70–140	250–450	25
Copper	124	8.9	70–250	200–300	17
Cu alloys	120–150	8.9	200–400	300–500	$\simeq 17$
Nickel	214	8.9	150	450	13.3
Ni alloys	130–230	7.8–9.2	210–1200	550–1500	$\simeq 13$
Titanium	116	4.5	450	550	9
Ti alloys	80–130	4.3–5.1	800–1200	1000–1300	$\simeq 9$

(e)

CHAPTER THREE

Yielding

3.1   INTRODUCTION

The concept of plasticity is intuitively clear in some metals, where beyond a particular stress large strains ensue, and removal of the applied load leaves residual deformation. Up to this stress (the yield stress), imposed deformation is accommodated by an increase in the interatomic distance (i.a.d.), a process which is largely conservative, and which is macroscopically manifested as an elastic response. If the applied loads are low the extension (or strain) may be proportional to the applied load, and the material is then Hookean.

Consider a simple cubic material subjected to a pure shear stress τ, as shown in Fig. 3.1. The resistive shear imposed by the lattice must be of period a, and although its precise form is speculative, Frenkel assumed it was sinusoidal, i.e.

$$\tau(x) = \tau_0 \sin \left\{ \frac{2\pi x}{a} \right\} \tag{3.1}$$

Differentiating gives

$$\frac{d\tau}{dx} = \tau_0 \frac{2\pi}{a} \cos \left\{ \frac{2\pi x}{a} \right\} \tag{3.2}$$

and the shear strain induced is given by

$$\gamma = \frac{x}{a} \tag{3.3}$$

Combining Eqns (3.2) with (3.3) noting that $\tau/\gamma = G$ (the modulus of rigidity or shear modulus) for small x gives

$$G = \frac{d\tau}{d\gamma} = \tau_0 \, 2\pi \tag{3.4}$$

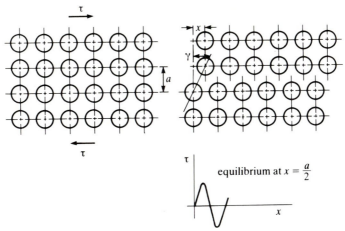

Figure 3.1   Theoretical strength of a perfect crystal: (a) undisturbed lattice subject to shear stress; (b) resultant deformation

Hence the theoretical shear strength ought to be of the order of $G/2\pi$. Although common metals have other crystallographies a similar calculation may be performed and the implied strength in shear is still of the order of one-tenth of the modulus of rigidity. The Young's modulus and yield strength (in tension) of some common metals are given in Chapter 2, Table 2.1, where it may be seen that the predicted yield strength is of the order of 100 times the value realized. It was this discrepancy which led, during the 1930s, to speculation of the existence of dislocations which was subsequently confirmed by careful electron microscopy.

The classic edge dislocation in a simple cubic lattice is shown in Fig. 3.2. The way in which it reduces the crystal strength may be viewed in two ways: first, at the atomic level, it may be seen that in the neighbourhood of the core of the dislocation the bonds are severely stretched, simply because of the lattice misfit, and that in order for the dislocation to move through the lattice it is only necessary to rupture one row of bonds at a time. Thus, the applied shear stress may be thought of as being, in some ways, concentrated in the neighbourhood of the dislocation, rather than averaged out over all the bonds across the slip plane, as implied for a perfect crystal. Secondly, the influence of the dislocation may be viewed macroscopically, by considering a thin strip of material (whose thickness is equal to one atom plane for the dislocation of Fig. 3.2) inserted in the material. In metals the displacement discontinuity found by making a circuit around the dislocation core is equivalent to one atom spacing along the direction of closest atom approach (the close-packed direction), and we refer to it as the Burger's vector, $\boldsymbol{b}_x$, Fig. 3.3. This gives rise to the following state of stress:

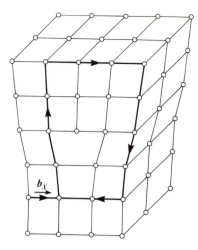

Figure 3.2 Classical view of edge dislocation in a simple cubic lattice. Burger's vector, $\boldsymbol{b}_x$, is shown

$$\sigma_{xx} = - \frac{\mu \boldsymbol{b}_x}{2\pi(1-\nu)} \left(\frac{y}{r^2} + \frac{2x^2y}{r^4} \right) \tag{3.5}$$

$$\sigma_{yy} = - \frac{\mu \boldsymbol{b}_x}{2\pi(1-\nu)} \left(\frac{y}{r^2} - \frac{2x^2y}{r^4} \right) \tag{3.6}$$

$$\tau_{xy} = - \frac{\mu \boldsymbol{b}_x}{2\pi(1-\nu)} \left(\frac{x}{r^2} - \frac{2x^3}{r^4} \right) \tag{3.7}$$

$$r^2 = x^2 + y^2$$

where μ is the modulus of rigidity and ν is Poisson's ratio. These stress components assume that the material is an elastic continuum, and may be expected beyond a distance of about 20 atom diameters from the dislocation core. They highlight the extremely severe residual stress state associated with the displacement discontinuity. These stresses are the ones manifested in a micromechanics explanation of strength reduction as stretched bonds, and indicate that the theoretical shear strength of the material may well be attained without *any* external force. The purpose of the external load is then to *propel* the dislocation along the slip planes, a process which is not conservative: the external stress must overcome this reluctance of the dislocation to move, which is called the Peierls–Nabarro force. This quantity is very much less than the theoretical shear strength, and depends not only on the material but the precise type of dislocation present. Also, the practically realized yield stress would only be the same as the Peierls–Nabarro force providing the dislocation could move freely without hindrance. In practice thousands of dislocations must glide for appreciable strains

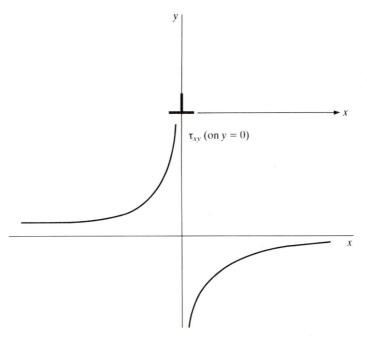

Figure 3.3   Applied mechanics view of a dislocation. Make a cut from $y = \infty$ to $y = 0$ along $x = 0$ and insert a thin strip of material of width b_x. Shear stress induced on $y = 0$

to develop. Dislocations glide along distinct planes defined by the material's crystal structure; in ductile materials there are many of these planes and during plastic flow glide planes may intersect frequently. Sometimes dislocations may flow freely 'through' one another whilst other types may either annihilate or otherwise become fixed ('sessile'). In the latter case further dislocations being propelled along the same slip planes will become held up; they will be unable to travel further until the applied stress is increased, and this feature is interpreted macroscopically as work-hardening. It is a property of most metals that the greater the strains induced, the more dislocation tangles will ensue and hence the more work-hardening will occur.

Hindrance to dislocation motion is also caused by grain boundaries, and hence components having large grains are softer (i.e. yield at a lower stress) than those having small grains. In the latter case dislocations may glide only a short distance before approaching a boundary. A well-known experimental correlation of this phenomenon is the Hall−Petch relation:

$$\sigma_Y = \sigma_f + \frac{k}{\sqrt{d}} \tag{3.8}$$

where d is the characteristic grain diameter, k is a constant and σ_f is

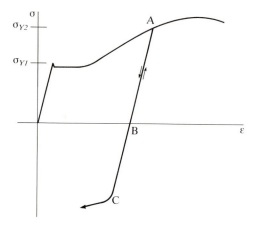

Figure 3.4   Uniaxial stress—strain curve. Work-hardening and the Bauschinger effect are shown

the friction stress. This last quantity is the stress needed to propel a dislocation through a large crystal, and is composed of the Peierls—Nabarro force together with other resistive influences such as precipitate particles and subgrain boundaries.

The elastic design strategy relies on maintaining the state of stress everywhere within a component in the elastic state. If stress (rather than displacement) is the limiting factor, high strength is desirable. Assuming that the basic choice of material has already been made (perhaps on the grounds of cost or manufacturing qualities), the only way in which the strength may be changed significantly by mechanical means is by exploiting the effects of plasticity associated with forming (e.g. pressing, folding). Figure 3.4 shows a typical uniaxial stress—strain curve. If a component is loaded until point A is reached and then permitted to relax, it will unload elastically (as the original i.a.d. is recovered) to point B. Upon reloading it will respond elastically, again until point A is attained, so that the yield stress has been elevated from σ_{Y1} to σ_{Y2}. If the loading is multiaxial it is much more difficult to anticipate the subsequent flow stress, and indeed if the direction of loading is *reversed*, i.e. towards C (Fig. 3.4), it is possible that the strength may be less than expected. This is known as the Bauschinger effect, and occurs because all active dislocations within the component are now being propelled in the opposite direction to that occurring during the first phase of loading. Therefore they are moving *away* from the grain boundaries and locked regions so that there are far less work-hardening effects.

3.2   CHEMISTRY OF STRENGTHENING

It is valuable to understand the ways in which pure metals may be rendered stronger by influencing their crystallography, i.e. by producing alloys, and the means by which the strengthening occurs. Invariably, the underlying idea is a distortion of the lattice, preferably over significant volumes, so that a dislocation is hindered in its glide, and must either turn on to a different (perhaps less preferred) glide direction, or climb over the obstacle. These processes require a larger propelling force, which is manifested macroscopically as a higher yield stress.

One of the simplest theoretical means of producing the distortion is by dissolving small atoms within the interstitial spaces of the parent lattice, to produce an interstitial solid solution. However, the amount of distortion which the lattice can accommodate is limited, and hence the solute atoms must in practice be less than half the radius of the metal, restricting the choice to carbon, boron, nitrogen and oxygen. In iron, carbon can dissolve interstitially to a significant extent (about 2%) at elevated temperatures, but at room temperatures a second phase precipitates out. This alloy of iron and carbon is the basis of steel.

Even the small amount of carbon in solution in steel at room temperature affects its mechanical properties. The lattice distortion caused by the interstitial solution is minimized close to dislocations; hence carbon atoms move slowly through the iron lattice by diffusion to form a greater concentration in the vicinity of the dislocations. This local 'atmosphere' pins the dislocation more effectively than a general dispersion of solute, and a greater yield strength occurs. However, once yield occurs, the dislocations move away from the concentrations of solute and subsequent deformation is easier. This is clearly visible on the stress—strain curve of mild steel, shows a peak strength at the onset of plastic flow, a phenomenon known as yield drop. This behaviour leads to a localization of deformation because the first region of the material to yield can flow at a reduced stress. It is a serious problem in sheet metalworking which is alleviated by deforming the entire sheet by rolling just prior to final forming.

Substitutional solid solution, in which atoms of the metal are *replaced* by those of the solute, is a second mechanism. Since the distortion cannot be excessive the difference in radii cannot exceed about 20%. Some pairs of metals, such as the copper—nickel, molybdenum—tungsten and cobalt—nickel systems, show complete solubility of one within the other, so that all compositions are possible, whilst others show only a limited solubility. For example, copper may dissolve up to about 38% zinc, giving rise to the brasses.

Substitutional solid solutions do not show the yield drop experienced by interstitial solution alloys because diffusion of substitutional solutes is much slower than that of interstitials. Thus no local atmospheres form around dislocations increasing their pinning force.

When multiple phases result, i.e. when there is incomplete solubility of one metal within the other, a microstructure results which can have one of many forms, depending on the metal-pair and on cooling conditions. Phase boundaries form effective barriers to dislocation motion for two reasons: firstly because the i.a.d. is perturbed near the boundary, and secondly because the glide plane and Burger's vector will change as a dislocation passes from one phase to the next.

If the second phase appears as very fine particles it is often found to have a considerable strengthening effect. This is because small dispersions act as obstacles to dislocation motion and, to allow the passage of a dislocation, they must either break or be by-passed, leaving a loop of dislocation around each particle. It has been shown that the restraining force is directly proportional to the Burger's vector and inversely proportional to the mean separation of the particles, l. If the particles are in a volume fraction of f and their mean radius is r, $l = r/f^{1/3}$. Thus, the smaller r, the greater the retarding force and hence the yield stress of the material. This principle is exploited in aluminium alloys. These contain alloying elements such as copper, magnesium, silicon and zinc in quantities of up to a few per cent. At high temperatures (in excess of 500 °C) these elements are in solution but at room temperatures less than 1% can be held in solution. By rapidly cooling or quenching to room temperature supersaturated solutions are formed. By carefully reheating to temperatures in the region 100−200 °C for a period of hours, the supersaturated solution can be induced to precipitate as a very finely dispersed phase. These small precipitates result in yield stress in excess of 300 MPa. Such high-strength alloys are known as precipitation-hardened or age-hardened alloys. The reheating precipitation cycle is known as the ageing process and must be carefully controlled to achieve peak hardness in a material. If the process goes on too long, overageing occurs, the particles become too large and their strengthening effect is diminished.

Of special interest in engineering is the martensitic transformation which occurs in mid- to high-carbon steels. This occurs because at elevated temperatures iron exists with a face-centred cubic configuration which can dissolve up to 2.1% carbon. At room temperatures it becomes body-centred cubic, and can dissolve a negligible amount of carbon. Equilibrium cooling results in the precipitation of cementite (an intermediate compound containing 6.69% carbon), but quenching suppresses diffusion of the carbon, which becomes trapped in certain sites giving rise to a cell similar to body-centred cubic, but where one lattice parameter is much greater than the other two. This heavily strained lattice is called martensite, and it is both extremely hard and brittle. Some of the toughness may be regained by heating to well below the lower critical temperature (727 °C) and cooling again, a process known as tempering. This is analogous to ageing in aluminium alloys. During tempering the super-saturated carbon solution precipitates out as a fine dispersion of small, hard carbide particles. By using different alloying additions very high-strength 'quench and tempered' alloys steels can be designed.

It is essential to remember that elastic response is by bond distortion, a conservative process, that plastic flow occurs by glide which is not recovered, and that the micromechanics of hardening rely on means of inhibiting dislocation motion. Real materials may have a complex stress—strain curve which shows continuously yielding, or a non-linear elastic, response: there may not be a sharp transition from one mode to another. Non-metals may be very prone to creep (particularly polymers and wood composites) or they may show different properties in different directions (particularly heavily rolled metals, and certainly in woven-matrix composites). These and other phenomena make the mathematical modelling of the material's response difficult, and idealizations are needed. It is certainly feasible to incorporate some features such as anisotropic elasticity, but this is complex. Regarding isotropic materials the following are the most common models:

(a) *Rigid—ideally plastic material (Fig. 3.5(a))*   Elastic deformation is neglected and only plastic flow is considered. No work-hardening is taken into account. The model is useful for representing processes where deformation is great (e.g. metalworking, punching, drawing, extrusion) where neglect of the elastic response is not important. Unfortunately, the absence of work-hardening means that the predicted working loads may be exceeded. By the same token it means that a safe bound is obtained when the model is used to find collapse loads for a structure.

(b) *Elastic—ideally plastic material (Fig. 3.5(b))*   This refinement does not neglect elastic strains, and is needed when modelling processes which are essentially elastic but where some overload into the plastic region may occur, e.g. shrink fits, shakedown, bolted joints, heavy-duty rolling element bearings.

(c) *Linear strain-hardening material (Fig. 3.5(c))*   This model is quite realistic for some materials, although the work-hardening portion is inevitably an idealization. It is not easy to implement with the slip-line field or upper-bound methods often used for materials in category (a), but may be used in conjunction with classical plastic flow rules.

YIELDING

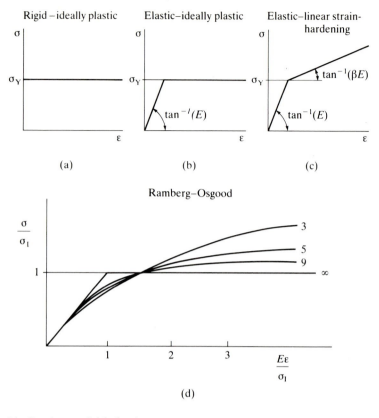

Figure 3.5   Idealized material behaviour

(d)   *Ramberg–Osgood material (Fig. 3.5(d))*   This constitutive
law was proposed in the 1940s. The stress–strain relation is
modelled by a curve of the form

$$\epsilon \simeq \frac{\sigma}{E} \left[1 + \frac{3}{7} \left(\frac{\sigma}{\sigma_{\mathrm{I}}} \right)^{n-1} \right] \qquad (3.9)$$

where $E, n, \sigma_{\mathrm{I}}$ are parameters. Here, E has its usual meaning,
n is chosen to give a best fit and σ_{I} may be interpreted as the
stress at which the secant modulus E_{s} is equal to $0.7\,E$.
Although it looks unwieldy at first sight, it will be seen that
by suitable choice of parameters any response from ideally
elastic–plastic to continuously yielding behaviour may be
modelled (Fig. 3.5(d)). Also, for $\sigma \gg \sigma_{\mathrm{I}}$ the equation is
asymptotic to a pure power law, $\epsilon = k\sigma^{n}$. Some exact
solutions (no stress concentration near holes and crack-tip
behaviour) have been found.

3.3   YIELD CRITERIA

Even under uniaxial tensile loading, it is apparent that the strain field
is triaxial, consisting of stretching along the loading direction and
contraction in the other two directions. In general, triaxial conditions
will apply to both stress and strain. It is then necessary to know how
to apply the stress–strain curve obtained by means of, say, a
conventional tensile test, to the triaxial conditions encountered. The
first question to be answered is, given a material with a uniaxial yield
stress σ_Y, will it yield under a state of stress defined by the values of
the three principal stresses $(\sigma_1, \sigma_2, \sigma_3)$? Yield criteria have been
proposed to answer this question.

3.3.1   TRESCA'S YIELD CRITERION

In a uniaxial tensile test, bands of intense deformation are observed
to appear at about $45°$ to the direction of the tensile stress. The state
of stress is then $(\sigma_0, 0, 0)$, as shown in Fig. 3.6 and, from the well-
known Mohr circle construction (see Chapter 8), the $45°$ direction
corresponds to planes of maximum shear (slip planes). It follows that
shear, rather than tensile stress, governs yield, a conclusion that
supports what has already been said in previous sections. The Tresca
yield criterion recognizes that conclusion, and states that yield occurs
when the maximum shear stress reaches a critical value. In Fig. 3.7,
the shear stress on planes ADFH, ABGH and BDFG is obtained from
the corresponding Mohr circles and the yield criterion is expressed by

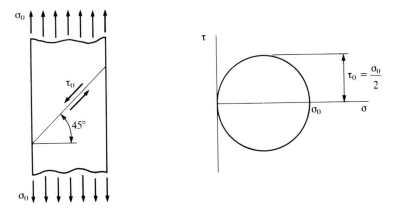

Figure 3.6   Bar under uniform tension has maximum shear stress at $45°$

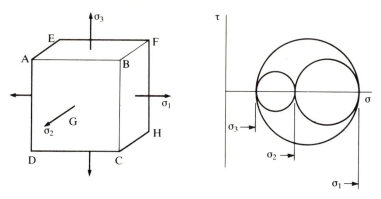

Figure 3.7 Maximum shear stress under triaxial stress and corresponding Mohr's circle

$$\text{Maximum of } \left| \frac{\sigma_1 - \sigma_2}{2} \right|, \left| \frac{\sigma_1 - \sigma_3}{2} \right|, \left| \frac{\sigma_2 - \sigma_3}{2} \right| = \tau_Y = \frac{\sigma_Y}{2} \quad (3.10)$$

Where the subindex Y refers to the yield condition.

The Tresca yield criterion distinguishes between stress systems in which at least two of the three principal stresses are different in magnitude and hydrostatic systems, in which $\sigma_1 = \sigma_2 = \sigma_3 (= \sigma)$. Yielding is theoretically impossible in a hydrostatic system, since all three shear components are zero. No matter how large the value of σ, the shear stress will never reach the critical value for yielding. Although the first impression is one of surprise, on reflection this is consistent with the assumption inherent in the development of the criterion, namely that yielding occurs as a result of the shear along certain preferred slip directions. It will be noted that such directions are independent of the crystallographic slip planes only because the material is treated as physically homogeneous and isotropic, yet another assumption is the formulation of the model that can only be accepted if yielding occurs over a region containing a sufficiently large number of randomly oriented crystals.

A simple graphical representation of the Tresca yield criterion is possible when the state of stress is biaxial (or plane stress obtains), i.e. $\sigma_3 = 0$. This is the hexagon shown in Fig. 3.8(a). Points inside the hexagon correspond to elastic behaviour, points on the hexagon to the condition of plastic flow that results from having reached the yield point. What happens at the onset of yielding? Consideration of the deformation experienced by a bar under uniaxial tension provides a first indication. As the load increases, the material follows Hooke's law,

$$\epsilon_1 = \frac{\sigma_1}{E} \quad \epsilon_2 = \epsilon_3 = -\frac{\nu \sigma_1}{E} \quad (3.11)$$

until it yields when $\sigma_1 = \sigma_Y$. Up to this point, the relative change of volume is given by

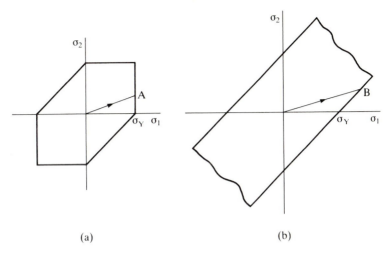

(a)   (b)

Figure 3.8   Tresca yield criterion in two dimensions, (a) plane stress conditions,
(b) plane strain conditions

$$\frac{\Delta V}{V} = \epsilon_1 + \epsilon_2 + \epsilon_3 = \frac{(1-2\nu)}{E}\,\sigma_1 = \frac{\sigma_1}{K} \qquad (3.12)$$

but, once plastic flow starts, it is observed that the deformation
proceeds without any further change of volume, and it consists solely
of change of shape (distortion). This implies a sudden change in the
bulk modulus K, equivalent to a change in the Poisson's ratio ν from
its elastic value (0.3–0.4 for most materials) to $\frac{1}{2}$. We must not, of
course, forget that an elastic dilatation will precede the purely
distortional plastic deformation. *To avoid any misunderstandings*, it
may be preferred to write Eqn (3.12) in terms of strain increments:

$$\frac{\mathrm{d}V}{V} = \mathrm{d}\epsilon_1 + \mathrm{d}\epsilon_2 + \mathrm{d}\epsilon_3 = 0 \text{ under plastic flow} \qquad (3.13)$$

In the uniaxial loading case, $\mathrm{d}\epsilon_2 + \mathrm{d}\epsilon_3 = -(\mathrm{d}\epsilon_1)/2$, i.e. the bar
stretches in the axial direction twice as fast as it contracts in the other
two directions. Equations (3.11) can also be rewritten for the plastic
strain increments in the form

$$\frac{\mathrm{d}\epsilon_1}{2\sigma_1} = \frac{\mathrm{d}\epsilon_2}{-\sigma_1} = \frac{\mathrm{d}\epsilon_3}{-\sigma_1} = \frac{\mathrm{d}\epsilon_1 + \mathrm{d}\epsilon_2 + \mathrm{d}\epsilon_3}{0} = \frac{0}{0} \qquad (3.14)$$

which may be extended to the general triaxial case in the form

$$\frac{\mathrm{d}\epsilon_1}{2\sigma_1 - \sigma_2 - \sigma_3} = \frac{\mathrm{d}\epsilon_2}{2\sigma_2 - \sigma_1 - \sigma_3} = \frac{\mathrm{d}\epsilon_3}{2\sigma_3 - \sigma_1 - \sigma_2} = \frac{0}{0} \qquad (3.15)$$

An alternative interpretation of these equations is offered by splitting stresses and strains into a hydrostatic (or dilatational) component (σ, ϵ) and a deviatoric (or distortional) component ($s_i, d\Delta_i$):

$$\sigma_1 = \frac{\sigma_1 + \sigma_2 + \sigma_3}{3} + \frac{2\sigma_1 - \sigma_2 - \sigma_3}{3} = \sigma + s_1$$

$$\sigma_2 = \frac{\sigma_1 + \sigma_2 + \sigma_3}{3} + \frac{2\sigma_2 - \sigma_1 - \sigma_3}{3} = \sigma + s_2 \qquad (3.16)$$

$$\sigma_3 = \frac{\sigma_1 + \sigma_2 + \sigma_3}{3} + \frac{2\sigma_3 - \sigma_1 - \sigma_2}{3} = \sigma + s_3$$

and

$$d\epsilon_1 = \frac{d\epsilon_1 + d\epsilon_2 + d\epsilon_3}{3} + \frac{2d\epsilon_1 - d\epsilon_2 - d\epsilon_3}{3} = 0 + d\Delta_1$$

$$d\epsilon_2 = \frac{d\epsilon_1 + d\epsilon_2 + d\epsilon_3}{3} + \frac{2d\epsilon_2 - d\epsilon_1 - d\epsilon_3}{3} = 0 + d\Delta_2 \qquad (3.17)$$

$$d\epsilon_3 = \frac{d\epsilon_1 + d\epsilon_2 + d\epsilon_3}{3} + \frac{2d\epsilon_3 - d\epsilon_1 - d\epsilon_2}{3} = 0 + d\Delta_3$$

Equations (3.15) express a condition of proportionality between the deviatoric stress components and the corresponding distortional strain increments, in agreement with the observed behaviour in the case of pure tension. In a thin-walled tube under pure torsion, $\sigma_1 = -\sigma_2$, $\sigma_3 = 0$, $d\epsilon_1 = -d\epsilon_2$ and $d\epsilon_3 = 0$, i.e. deformation occurs without any reduction in thickness. This is also in agreement with the observations.

The distinction between the plane stress condition represented in Fig. 3.8(a) and the plane strain condition $d\epsilon_3 = 0$ is of great interest. Referring to Eqns (3.15), plane strain implies that

$$\sigma_3 = \frac{\sigma_1 + \sigma_2}{2} \qquad (3.18)$$

and substituting this equation into the general form of Tresca's yield criterion, the yield condition is reduced to the condition

$$\left| \frac{\sigma_1 - \sigma_2}{2} \right| = \tau_Y \qquad (3.19)$$

since the other two shear stress components are equal to $\pm(\sigma_1 - \sigma_2)/4$. The new condition is shown in Fig. 3.8(b). Consider a situation in which as the load is increased the biaxiality ratio (σ_1/σ_2) remains fixed and equal, for example, to 1/2. Under plane stress conditions, yielding occurs when $\sigma_1 = \sigma_Y$ (Fig. 3.8(a), point A) while under plane strain it will occur when $\sigma_1 = 2\sigma_Y$ (Fig. 3.8(b), point B). The difference between the two increases as the biaxiality ratio tends towards unity when $\sigma_1 = \infty$ for yielding under plane strain. Note that Fig. 3.8(b) describes the condition for continued plastic flow, not the point of first yield.

3.3.2 VON MISES' YIELD CRITERION

Tresca's criterion is based on the deformation observed during yielding. An alternative criterion, known as von Mises', is based on the strain energy involved in plastic flow. Consider again the cubic element under stresses (σ_1, σ_2, σ_3), yielding by ($d\epsilon_1$, $d\epsilon_2$, $d\epsilon_3$). The work done is

$$dW = \sigma_1 \, d\epsilon_1 + \sigma_2 \, d\epsilon_2 + \sigma_3 d\epsilon_3 \tag{3.20}$$

It must be emphasized that, given that the volume remains constant under conditions of plastic flow, this energy is solely associated with distortion.

In the uniaxial loading case,

$$dW = \sigma d\epsilon = \text{critical value} \tag{3.21}$$

Combining Eqns (3.15), (3.20) and (3.21), a condition of equivalence in terms only of stresses is obtained,

$$\sigma_1^2 + \sigma_2^2 + \sigma_3^2 - \sigma_1\sigma_2 - \sigma_1\sigma_3 - \sigma_2\sigma_3 = \overset{*}{\sigma}{}^2 \tag{3.22}$$

At yielding, $\overset{*}{\sigma} = \sigma_Y$ and the yield criterion may be written in the form,

$$(\sigma_1^2 + \sigma_2^2 + \sigma_3^2 - \sigma_1\sigma_2 - \sigma_1\sigma_3 - \sigma_2\sigma_3)^{1/2} = \sigma_Y \tag{3.23}$$

This is the *von Mises' yield criterion*.

When $\sigma_3 = 0$ (plane stress condition), Eqn (3.23) is represented as

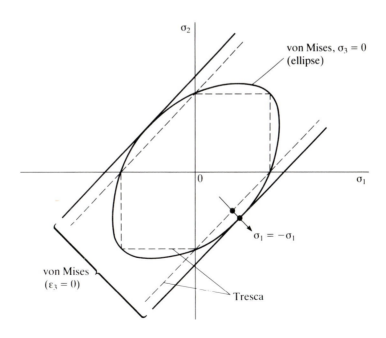

Figure 3.9   von Mises' yield curve and comparison with Tresca's

an ellipse, Fig. 3.9, while under plane strain (Eqn 3.18) it degenerates into two parallel lines.

The Tresca, or maximum shear stress, criterion and the Von Mises' criterion give yield conditions that are not entirely dissimilar, as shown in Fig. 3.9. The maximum difference is found for $\sigma_1 = -\sigma_2$, when, according to Tresca, yield occurs at

$$\sigma_1 = -\sigma_2 = 0.5\sigma_Y$$

and according to von Mises, it occurs at

$$\sigma_1 = -\sigma_2 = 0.577\,\sigma_Y$$

The equation of the ellipse offers a third interpretation of the fundamental relationship between stresses and plastic strain increments. At any given point, the slope of the tangent to the ellipse is obtained by differentiating Eqn (3.23) with $\sigma_3 = 0$.

$$2\sigma_1 d\sigma_1 + 2\sigma_1 d\sigma_2 - \sigma_1 d\sigma_2 - \sigma_2 d\sigma_1 = 0$$

Therefore $\quad \dfrac{d\sigma_2}{d\sigma_1} = -\dfrac{2\sigma_1 - \sigma_2}{2\sigma_2 - \sigma_1}$

while, from Eqns (3.15),

$$\frac{d\epsilon_2}{d\epsilon_1} = \frac{2\sigma_1 - \sigma_2}{2\sigma_1 - \sigma_1} = -\frac{d\sigma_1}{d\sigma_2}$$

this means that the vector of components $(d\epsilon_1, d\epsilon_2)$ is normal to the yield locus. What has been said for the biaxial stresses is also true for the general case, when the ellipse becomes an ellipsoid.

This normality condition between the strain increment vector and the yield locus is often referred to as the *flow rule* and means that the deformation under plastic flow is such as to maximize the work or energy dissipation.

The above results in the celebrated Levy–Mises equations. For further explanation of this topic, the reader is referred to the books by Hill[1] and by Hoffman and Sachs[2].

3.3.3   UNIAXIAL STRESS AND STRAIN EQUIVALENT TO THE GENERAL
TRIAXIAL SITUATION

When discussing the Tresca criterion we already indicated the necessity of establishing an equivalence between the uniaxial loading system characteristic of the conventional tensile test and the general triaxial state of stress and strain. To simplify the treatment of this problem, we define the principal stress direction such as to ensure that

$$\sigma_1 > \sigma_2 > \sigma_3 \tag{3.24}$$

in which case the flow rule will tell us that

$$d\epsilon_1 > d\epsilon_2 > d\epsilon_3 \tag{3.25}$$

Tresca's criterion will be written as

$$(\sigma_1 - \sigma_3) = \overset{*}{\sigma} \; (= \sigma_Y) \tag{3.26}$$

where $\overset{*}{\sigma}$ may be regarded as a uniaxial stress equivalent to $(\sigma_1, \sigma_2, \sigma_3)$. Also,

$$(d\epsilon_1 - d\epsilon_3) = [d\overset{*}{\epsilon} - (-\tfrac{1}{2} d\overset{*}{\epsilon})] = \tfrac{3}{2} d\overset{*}{\epsilon} \tag{3.27}$$

where $d\overset{*}{\epsilon}$ is the equivalent uniaxial strain increment, accompanied of course by a strain increment $-d\overset{*}{\epsilon}/2$ in the other two normal directions.

Von Mises' criterion already provides an expression for the equivalent stress,

$$(\sigma_1^2 + \sigma_2^2 + \sigma_3^2 - \sigma_1\sigma_3 - \sigma_2\sigma_3 - \sigma_1\sigma_3)^{1/2} = \overset{*}{\sigma} \tag{3.28}$$

and the corresponding expression from the strain increments is

$$(d\epsilon_1^2 + d\epsilon_2^2 + d\epsilon_3^2 - d\epsilon_1 d\epsilon_3 - d\epsilon_2 d\epsilon_3 - d\epsilon_1 d\epsilon_2)^{1/2} = \tfrac{3}{2} d\overset{*}{\epsilon} \tag{3.29}$$

The right-hand side of this equation is obtained by taking

$$d\epsilon_1 = d\overset{*}{\epsilon}, \; d\epsilon_2 = d\epsilon_3 = d\overset{*}{\epsilon}/2$$

3.4   REPRESENTATION OF THREE-DIMENSIONAL STRESS STATES

The two yield criteria introduced in the previous section may be interpreted in a geometric way. Qualitatively, the *magnitude* of the state of stress at a point is given by the three principal stresses, σ_1, σ_2, σ_3 (Fig. 3.7). Note that six pieces of information are required to specify the complete state, but the other three may be viewed as specifying the *orientation* of the principal axes, and this information is not needed in dealing with isotropic material. It would seem possible, therefore, to think of a stress space, whose axes represented the principal stress values (Fig. 3.10). An additional axis is added whose direction cosines are $1/\sqrt{3}$ with respect to the base axes. Any point on this line therefore experiences a state of stress $\sigma_1 = \sigma_2 = \sigma_3 \, (= 3\bar{\sigma})$, and hence this is the hydrostatic axis. A general state of stress (represented by $\overline{OP}$) may then be thought of as having two components $\overline{OQ} + \overline{QP}$. The latter may be seen to be the hydrostatic component, and the former is given by

$$\overline{OQ} = \overline{OP} - \overline{QP}$$
$$= (\sigma_1 \boldsymbol{i} + \sigma_2 \boldsymbol{j} + \sigma_3 \boldsymbol{k}) - \bar{\sigma} (\boldsymbol{i} + \boldsymbol{j} + \boldsymbol{k})$$

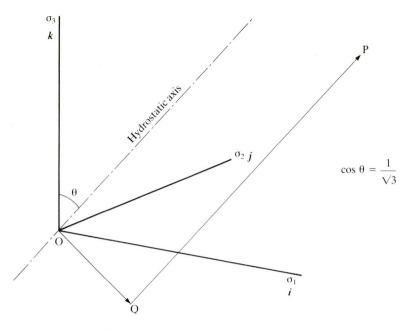

Figure 3.10 Representation of a state of stress in principal stress space

$$= (\sigma_1 - \bar{\sigma})\, \boldsymbol{i} + (\sigma_2 - \bar{\sigma})\, \boldsymbol{j} + (\sigma_3 - \bar{\sigma})\, \boldsymbol{k}$$

$$= s_1\, \boldsymbol{i} + s_2\, \boldsymbol{j} + s_3\, \boldsymbol{k} \tag{3.30}$$

which from Eqn (3.16) are seen to be the deviatoric stress components. The physical interpretation of the diagram is that the hydrostatic component controls dilatation whilst the deviatoric components are associated with shear.

The stress space diagram is useful in visualizing the yield criteria already developed, but from another viewpoint. Experimentally, it has been found that yielding of a polycrystalline metal is completely independent of the hydrostatic stress component. Thus, any surface delineating the yield condition must be independent of the hydrostatic component, i.e. it must extend infinitely parallel with the hydrostatic axis. If the material is isotropic an interchange of the three principal stress values must predict the same tendency to yield, i.e. there must be three-way symmetry. Further, if the material is to have the same yield stress in tension and compression there must be six-fold symmetry. Lastly, it can be shown from energy considerations that the yield surface must be convex. These four observations are sufficient to tie down the permissible surface to a cylinder (in the mathematical sense) centred on the hydrostatic axis and bounded by the hexagons shown in Fig. 3.11, which is a view along the cylinder looking down the hydrostatic axis.

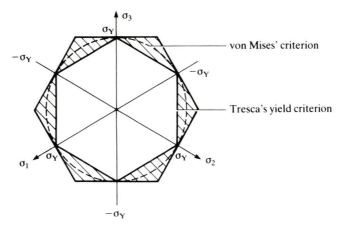

Figure 3.11   Shaded area indicates acceptable cross-section for yield criterion, looking down hydrostatic axis

It may be readily verified that the inner hexagon represents Tresca's yield criterion, already treated in detail. The outer hexagon is less useful, other than as a physical bound: it corresponds to the notion that yielding will occur when the biggest principal deviatoric stress (in magnitude) reaches a critical value, i.e.

$$\text{Maximum } |s_1|, |s_2|, |s_3| \le \sigma_Y \tag{3.31}$$

It is little used. Of far more interest is the von Mises' criterion. Consider a circle, centred on the origin, and passing through the yield points, Fig. 3.11. It has already been shown (Eqn 3.30) that the radius vector OQ has the vector equation $s_1 \boldsymbol{i} + s_2 \boldsymbol{j} + s_3 \boldsymbol{k}$. Thus, a yield locus of constant radius implies that

$$s_1^2 + s_2^2 + s_3^2 = a_0 \, \sigma_Y^2 \tag{3.32}$$

where substitution of uniaxial stress conditions gives $a_0 = 2/3$. The familiar form of Eqn (3.23) may be shown to be equivalent in the following way. From Eqn (3.23) we have

$$(\sigma_1 - \sigma_2)^2 + (\sigma_2 - \sigma_3)^2 + (\sigma_3 - \sigma_1)^2 = 2\sigma_Y^2 \tag{3.33}$$

$$(s_1 - s_2)^2 + (s_2 - s_3)^2 + (s_3 - s_1)^2 = 2\sigma_Y^2$$

$$2 \, (s_1^2 + s_2^2 + s_3^2 - s_1 s_2 - s_2 s_3 - s_3 s_1) = 2\sigma_Y^2$$

$$3 \, (s_1^2 + s_2^2 + s_3^2) - \left\{ s_1^2 + s_2^2 + s_3^2 + 2s_1 s_2 + 2s_2 s_3 + 2s_3 s_1 \right\} = 2\sigma_Y^2$$

This is the same as Eqn (3.32) if the term in braces is zero. This may be shown by expanding

$$(s_1 + s_2 + s_3)^2 \quad (= 0)$$

$$= s_1^2 + s_2^2 + s_3^2 + 2s_1 s_2 + 2s_2 s_3 + 2s_3 s_1$$

which is recognized as being the same.

We shall return to the concept of stress space and the graphical representation of yield criteria when considering work-hardening and shakedown. First, however, we should consider the usefulness of stress invariants.

One of the difficulties with the above equation is that if a body is suffering a three-dimensional state of stress all six components may be non-zero. A first step in deciding whether yielding has occurred would therefore be to find the principal values. This is lengthy. However, we can show that if we arrange the elements of the stress tensor in the form of a matrix

$$\begin{bmatrix} \sigma_{xx} & \tau_{yx} & \tau_{zx} \\ \tau_{xy} & \sigma_{yy} & \tau_{zy} \\ \tau_{xz} & \tau_{yz} & \sigma_{yy} \end{bmatrix} \equiv \sigma_{ij} \tag{3.34}$$

the three principal stresses σ_n are given by the eigenvalues, i.e.

$$[\sigma_{ij}] - \sigma_n [I] = 0 \tag{3.35}$$

Now the principal stresses do not depend on the orientation of the original axis set (in two dimensions Mohr's circle is unique even though the rotation from an arbitrary set of axes may vary), and hence it follows that the coefficients of the cubic (3.35) are themselves independent of the coordinate set chosen. They are therefore *invariant*. This may be made clear by examining the two-dimensional case, when Eqn (3.35) becomes

$$\begin{bmatrix} \sigma_{xx} - \sigma_n & \tau_{yx} \\ \tau_{xy} & \sigma_{yy} - \sigma_n \end{bmatrix} = 0 \tag{3.36}$$

Expanding,

$$\sigma_n^2 - \sigma_n (\sigma_{xx} + \sigma_{yy}) - \tau_{xy}^2 + \sigma_{xx}\sigma_{yy} = 0 \tag{3.37}$$

i.e.

$$\sigma_n = \tfrac{1}{2} (\sigma_{xx} + \sigma_{yy}) \pm \sqrt{\tau_{xy}^2 + \tfrac{1}{4} (\sigma_{xx} - \sigma_{yy})^2} \tag{3.38}$$

It will be recognized that in Eqn (3.38) the first term corresponds to the location of the centre of Mohr's circle (the average stress) whilst the second is the radius of the circle (the maximum shear stress). These are invariants, as are, from Eqn (3.37),

$$\begin{aligned} I_1 &= \sigma_{xx} + \sigma_{yy} \\ I_2 &= \sigma_{xx}\sigma_{yy} - \tau_{xy}^2 \end{aligned} \tag{3.39}$$

The corresponding quantities in the three-dimensional case are

$$I_1 = \sigma_{xx} + \sigma_{yy} + \sigma_{zz}$$

$$I_2 = \sigma_{xx}\sigma_{yy} + \sigma_{yy}\sigma_{zz} + \sigma_{zz}\sigma_{xx} - \tau_{xy}^2 - \tau_{yz}^2 - \tau_{zx}^2$$

$$I_3 = \begin{vmatrix} \sigma_{xx} & \tau_{xy} & \tau_{zx} \\ \tau_{xy} & \sigma_{yy} & \tau_{yz} \\ \tau_{zx} & \tau_{yz} & \sigma_{zz} \end{vmatrix} \tag{3.40}$$

Although the physical significance of the first invariant is that it is proportional to the hydrostatic component of stress the other two have no physical meaning. However, the same concept may be used to find invariants of the deviatoric tensor, where (3.34) is replaced by

$$\begin{bmatrix} s_{xx} & s_{yx} & s_{zx} \\ s_{xy} & s_{yy} & s_{zy} \\ s_{xz} & s_{yz} & s_{zz} \end{bmatrix} \equiv s_{ij} \tag{3.41}$$

where $s_{xy} \equiv \tau_{xy}$ etc.
and $s_{xx} \equiv \sigma_{xx} - \bar{\sigma}$ etc.

The invariants of Eqn (3.41) are then

$$J_1 = s_{xx} + s_{yy} + s_{zz} \equiv 0$$

$$\begin{aligned} J_2 &= s_{xx}s_{yy} + s_{yy}s_{zz} + s_{zz}s_{xx} - s_{xy}^2 - s_{yz}^2 - s_{zx}^2 \\ &= \tfrac{1}{6}\{(\sigma_{xx} - \sigma_{yy})^2 + (\sigma_{yy} - \sigma_{zz})^2 + (\sigma_{zz} - \sigma_{xx})^2\} \\ &\quad + \tau_{xy}^2 + \tau_{yz}^2 + \tau_{zx}^2 \end{aligned} \tag{3.42}$$

$$J_3 = \begin{vmatrix} s_{xx} & s_{xy} & s_{zx} \\ s_{xy} & s_{yy} & s_{yz} \\ s_{zx} & s_{yz} & s_{zz} \end{vmatrix} \tag{3.43}$$

The usefulness of these expressions in the analysis of yielding is this: yielding depends *only* on the deviatoric stress tensor. All of the information about the *magnitude* of the deviatoric stress is contained within the invariants, only two of which are non-zero. Therefore, all yield criteria ought to be expressible in terms of J_2 and J_3, whose values may be found without recourse to finding the principal stresses. For example, by comparison of Eqn (3.42) with Eqn (3.33) it is easy to see that von Mises' criterion may simply be expressed as

$$J_2 \leq \sigma_Y^2/3 \tag{3.44}$$

It is possible to express Tresca's criterion in terms of invariants but this is very complicated, and it is for this reason (rather than von Mises' yield surface being a better approximation to the experimentally determined one) that von Mises' criterion is more often used in theoretical work, when the principal stresses are not known in advance.

3.5 POST-YIELD BEHAVIOUR

For a component which remains entirely elastic, determining the internal state of stress is classical, and it is then easy (using the results of the previous section) to ascertain if any portion has exceeded the yield condition, i.e. if it has become plastic. It is important to recognize that if any significant zone of plasticity exists there will be a redistribution of stress within the entire object, and this is very difficult to solve by hand except for the few problems in which the strain distribution is kinematically determined, e.g. a beam or thick-walled tube. Previously the flow rules were discussed for the Tresca and von Mises yield criteria. It is useful to visualize what happens to an element of material as it is loaded by considering the trajectory of the stress state in stress space, as in the previous section. In general we are not concerned with the hydrostatic stress, but need to understand the deviatoric component, i.e. OQ, Fig. 3.10. The so-called π plane (or deviatoric plane) passes through the origin and has the hydrostatic axis as its normal.

From Fig. 3.10 we may see that the projection of the principal stress σ_3 (say) on to the π plane will give rise to a deviatoric component $\sigma_3 \sin \theta = \sqrt{(2/3)}\, \sigma_3$, and hence we obtain the axis set shown in Fig. 3.12. We know that von Mises' criterion will appear as a circle on the π plane and substitution of uniaxial stress conditions shows that its radius must be $\sqrt{(2/3)}\, \sigma_Y$ or $\sqrt{(2)}k$ where k is the yield stress in pure shear. Similarly, Tresca's criterion leads to a hexagon of side $\sqrt{(8/3)}k$. To plot a particular trajectory it is helpful to introduce Cartesian coordinates ζ, η as shown, given by

$$\zeta = \sqrt{\left(\frac{2}{3}\right)}\, (\sigma_1 - \sigma_2) \cos 30° = \frac{1}{\sqrt{2}} (\sigma_1 - \sigma_2) \qquad (3.45)$$

$$\eta = \sqrt{\left(\frac{2}{3}\right)} \{\sigma_3 - (\sigma_1 + \sigma_2) \sin 30°\} = \sqrt{\left(\frac{2}{3}\right)} \left\{\sigma_3 - \frac{1}{2}(\sigma_1 + \sigma_2)\right\}$$

Some typical loading trajectories are shown in Fig. 3.12: curve A is wholly within the yield surface and hence only elastic. Curve B shows some plastic behaviour, during which the appropriate flow rules (Eqn (3.15), for example) would need to be used. Curve C depicts a special type of loading regime known as proportional loading, where

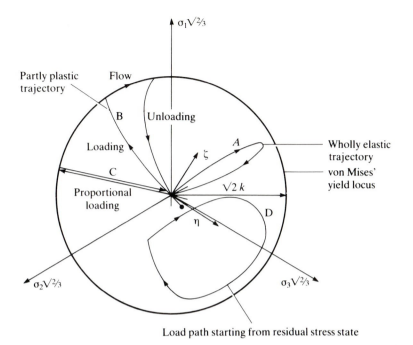

Figure 3.12   Stress trajectories projected on to deviatoric (π) plane

$$\frac{\mathrm{d}\sigma_1}{\sigma_1} = \frac{\mathrm{d}\sigma_2}{\sigma_2} = \frac{\mathrm{d}\sigma_3}{\sigma_3} \tag{3.46}$$

This commonly occurs when the applied loading increases in magnitude but otherwise remains unchanged, so that the principal stresses increase in proportion. Note that upon unloading we would not necessarily expect curves B and C to return to a state of zero stress, since there are often residual stresses remaining. Curve D shows a loading path which, from the unloaded state (the origin), would intersect the yield surface, but which does not do so when starting from the residual stress state shown. The starting point may be displaced as a result of treatment (e.g. shot peening, electrodeposition) or because an earlier loading cycle produced plastic flow.

When a component is loaded beyond the yield condition, work-hardening will occur, so that upon unloading and reloading a greater load may be sustained while the material remains elastic. Also, except for the simplest of geometries where elastic relaxation to a stress-free state occurs, there will, upon removal of the load, be an internal residual state of stress. Reloading from this state to the same load as was previously experienced may produce a trajectory entirely within the yield envelope: shakedown is said to have occurred. Both phenomena may be interpreted using the π plane.

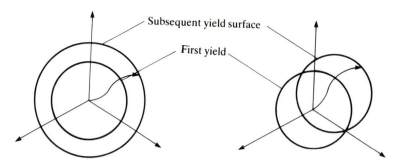

Subsequent yield surface

First yield

Figure 3.13 Theories of strain-hardening: (a) isotropic; (b) kinematic

Work-hardening is a complex process and the best way to model it for any given material is still the subject of current research. Two types of model have been proposed: the first is called isotropic hardening, as it is assumed that the yield strength is increased uniformly in all directions. Thus, when the stress trajectory touches the yield surface it is pushed out (in accordance with whatever work-hardening law is used), as shown in Fig. 3.13(a). The second is called kinematic hardening, and it assumes that the yield surface remains of the same shape and size but is simply pushed along by the tip of the trajectory: Fig. 3.13(b). This theory has the benefit of showing some degree of a generalized Bauschinger effect in so far as reloading with the signs of the principal stresses reversed will induce yielding at a lower load. In practice, however, it is probable that most real metals show an element of both types of hardening.

Suppose some material has the uniaxial stress–strain curve given by Fig. 3.5(c) and exhibits isotropic hardening. Below the yield stress Hooke's law applies, and above it we note that, from Fig. 3.14,

$$E = \frac{\sigma_0 - \sigma_Y}{\Delta\epsilon} \tag{3.47}$$

$$\beta E = \frac{\sigma - \sigma_Y}{\Delta\epsilon} \tag{3.48}$$

i.e.

$$\frac{E\beta}{1-\beta} = \frac{\sigma - \sigma_Y}{(\sigma_0 - \sigma)/E} = \frac{\Delta\sigma}{\Delta\epsilon_p} \tag{3.49}$$

and since the post-yield response is linear we may write

$$\frac{d\overset{*}{\sigma}}{d\overset{*}{\epsilon}_p} = \frac{\beta E}{1-\beta} \tag{3.50}$$

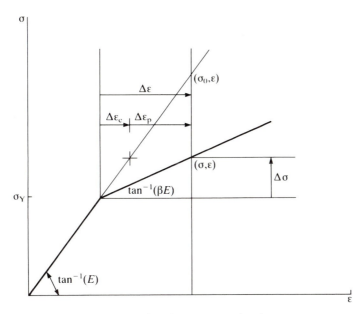

Figure 3.14   Enlargement of yield surface by isotropic hardening

where the asterisk indicates that we are now working in general effective quantities and the subscript p denotes the plastic component of strain. The flow rule has already been derived and may be written in general terms as

$$d\epsilon_{p_i} = \frac{3}{2} d\overset{*}{\epsilon}_p \frac{s_i}{\overset{*}{\sigma}} \quad i = 1,2,3 \tag{3.51}$$

Physically, Eqn (3.50) may be interpreted as ensuring that during any strain increment the resulting stress is such as to remain on the yield surface (which is expanding if the load is increasing), whilst Eqn (3.51) tells us that the plastic strain increment during an increase in load takes place in a direction dictated by the ratio of the deviatoric stresses. Note that the total strain (if needed) may be found by adding a term given by Hooke's law.

Now suppose that we have material with the *same* linearly hardening stress–strain curve (Fig. 3.5c) but that it exhibits purely kinematic hardening. Yielding will occur, when, by von Mises' criterion,

$$\sigma_Y = \overset{*}{\sigma} = \sqrt{\tfrac{3}{2}(s_1^2 + s_2^2 + s_3^2)} \tag{3.52}$$

If now $\overset{*}{\sigma}$ exceeds σ_Y the yield surface translates while remaining the same size and shape so that subsequently the flow condition is given by

$$\sigma_Y = \sqrt{\tfrac{3}{2}[(s_1 - \alpha_1)^2 + (s_2 - \alpha_2)^2 + (s_3 - \alpha_3)^2]} \tag{3.53}$$

where the direction of motion is governed by the strain rate vector, i.e.

85

$$d\acute{\epsilon}^*_{p_i} = \frac{3(1-\beta)}{2\beta E}d\alpha_i \quad i = 1,2,3 \tag{3.54}$$

The above mechanisms physically represent the increasing difficulty of causing plastic flow due to work-hardening, i.e. the influence of dislocation pile-ups and other obstacles. In stress space they are associated with the growth or translation of the yield locus projected on to the π plane. A second type of phenomenon which may mean that when components are loaded cyclically they attain a steady-state elastic response, even though the initial cycle causes some plasticity, is shakedown. This is the self-generation of residual stresses, which, together with the applied stresses, may lie within the yield condition.

A very simple example of elastic shakedown is afforded by the case of a rectangular beam of breadth b and depth h, loaded by a bending moment M, which varies cyclically with time (Fig. 3.15(a)), where the onset of plasticity, the elastic limit, is given by

$$M \leq \frac{bh^2}{6\sigma_Y} \tag{3.55}$$

If the moment is increased beyond this point the beam eventually becomes fully plastic, and the limit state is given by

$$M = \frac{bh^2}{4\sigma_Y} \tag{3.56}$$

which is 50% higher. If the moment is then reduced so that elastic relaxation occurs, the residual stress is as shown in Fig. 3.15(a)(iii). It will be recognized that from this state an *elastic* stress distribution may be imposed, in the same sense as the original moment, but with σ_{max} limited to 3/2 σ_Y. This corresponds to a moment (which for this simple problem coincides with the limit state) given by Eqn (3.56). In general the shakedown limit will be less than the limit state condition. The path of the stress vector for a point on the top of the beam is shown in Fig. 3.15(b).

It should be stressed that in most problems it is not quite so easy to determine the residual stress state, and hence the shakedown limit. One possibility for complex shapes is actually to take the component through several stress cycles, as we did above, and to deduce the residual stress state directly. A better method, however, is to appeal to the two shakedown theorems:

(a) *Statical (lower bound) theorem (Melan's theorem)* This states that if *any* self-equilibrating set of residual stresses can be found which, together with the extended elastic stresses due to the applied load, always lie within the yield condition, elastic shakedown will occur. Note that the residual stress field does not have to be the real one, although if that is found the shakedown limit determined will be the true one.

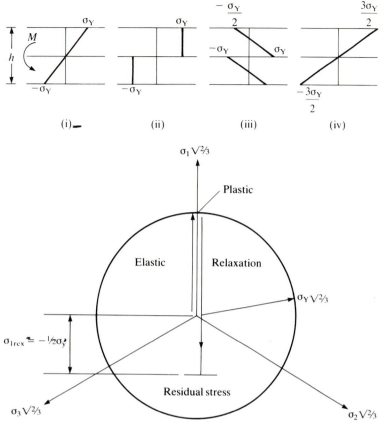

Figure 3.15 (a) Beam subjected to various bending moments M: (i) elastic limit; (ii) limit state (fully plastic); (iii) residual stress state; (iv) maximum elastic stress distribution which may subsequently be added to (iii). (b) Response plotted in the π plane

(b) *Kinematical (upper bound) theorem (Koiter's theorem)* This states that if any kinematically acceptable collapse mechanism can be found in which the rate of doing work by the elastic stresses exceeds the plastic dissipation, incremental collapse will take place. Note that the ratio of the work done by the elastic stresses to the plastic deformation has a maximum in the true mechanism of collapse, so any other mechanism must provide an upper bound to the true elastic limit.

At this point it is probably valuable to record the different types of response which can occur in a cyclically loaded component. These are:

(a) *Elastic response* All elements of material remain within the yield criterion. The elastic limit is not exceeded.

(b) *Elastic shakedown*   The first few cycles of loading are plastic but in the steady state all material deforms elastically throughout the loading cycle.

(c) *Plastic shakedown, or cyclic plasticity*   Some material remains plastic but there is no accumulation of strain. Collapse will not, therefore, occur although conditions may be ideal for crack initiation.

(d) *Ratcheting*   Some material remains plastic in the steady state and plastic strains are accumulated giving rise to incremental collapse.

The shakedown theorems provide information about the highest load at which response (b) may be expected. If the loading exceeds the shakedown limit detailed analysis of the plastic flow is needed to determine whether cyclic plasticity or ratcheting occurs, and there are no easy short cuts.

For simple configurations such as a tube subjected to internal pressure and cyclic heating, it is possible to determine what mode of response (from those classified above) occurs in closed form. For the example quoted, it is possible to plot these responses as a function of thermal and internal pressure on what has become known as a Bree diagram.[3-5]

The question arises: What happens when we build a component with real material which exhibits some work-hardening, but might also be expected to shake down? Two principles which can be intuitively seen to occur are:

(a) that a material displaying continuous isotropic hardening will always eventually shake down to an elastic state; and

(b) that a material exhibiting continuous kinematic hardening can never end in the ratcheting regime, but will, instead, show cyclic plasticity.

Another useful result which is simple to apply is Ponter's shakedown theorem,[6,7] which applies to kinematically hardening material having the pure shear stress–strain curve shown in Fig. 3.16(a). The initial yield is at a stress k_e, when work-hardening commences, but above a stress k_p there is perfect plasticity. In stress space this process may be viewed as the free motion of a yield cylinder of radius k_e within a limiting outer cylinder of radius k_p (Fig. 3.16(b)). Ponter's theorem states that if *any* residual stress system (which does not even have to obey equilibrium) can be found, which together with the applied stress lies within the yield condition, shakedown will occur.

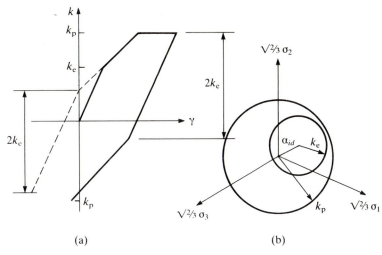

Figure 3.16 (a) Model stress–strain curve used by Ponter. Note that steady-state range is $2k_e$. (b) Motion of yield locus on π plane

(a) It provides an *upper bound* if the material is elastic–perfectly plastic ($k_p = k_e = k$); and

(b) for a material having the stress–strain curve of Fig. 3.16(a) Ponter's theorem must be satisfied with k set to k_e and Melan's theorem obeyed if $k = k_p$.

The practical usefulness of the above theorems is that they enable a component subject to cyclic loading (or a mixture of static and cyclic loading) to be designed so that in the steady state it is wholly elastic, but which may, for a few cycles, enter the plastic regime. These few cycles are not likely to cause any permanent damage although there will naturally be some sort of permanent deformation. Note that the shakedown theorems cannot provide any information about the magnitude of the permanent set, nor can they say how many cycles of loading may be needed to achieve shakedown. The latter can vary from a single cycle of loading to thousands of cycles, if a steady state is approached only asymptotically.

3.6 APPLICATIONS TO PLANE STRESS PROBLEMS

3.6.1 UNIAXIAL STRESS: BEAM IN BENDING

Consider a beam of rectangular cross-section made of a material whose constitutive equation is

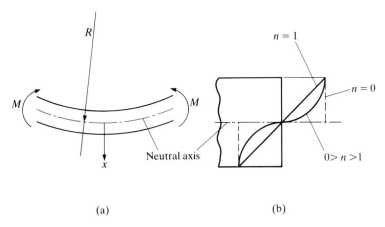

Figure 3.17 Beam in bending: (a) deformation; (b) stress distribution

$$\sigma = k\epsilon^n \tag{3.57}$$

This is similar to the Ramberg−Osgood equation for large strains.

In bending, plane sections remain plane and the strain at a distance x from the centre line (neutral axis) (see Fig. 3.17) is

$$\epsilon = \frac{x}{R} \tag{3.58}$$

where R is the radius of curvature of the deformed beam. For equilibrium,

$$M = 2 \int_0^{b/2} \sigma\, x\, b\, \mathrm{d}x = \frac{bkh^{n+2}}{2^{n+1}(n+2)R^n}$$

and the stress distribution is given by

$$\sigma = \frac{2^{n+1}(n+2)}{bh^2} \left(\frac{x}{h}\right)^n M \tag{3.59}$$

When $n = 1$ the material is linear−elastic and the familiar equation,

$$\sigma = \frac{M}{I} x$$

where I is the second moment of area, is obtained.

The strain-hardening coefficient n varies between 0.1 and 0.5, and the stress distribution follows the curve drawn in Fig. 3.17(b). The rigid−ideally plastic approximation to the constitutive equation may be regarded as a limiting condition, when $n = 0$ and $\sigma \leq \sigma_Y$, in which case the strain can take any value. The stress distribution is then stepped as shown in Fig. 3.17 and the beam can then take any curvature, no matter how large. Figure 3.18 illustrates the possible constitutive equations for the material for various values of n.

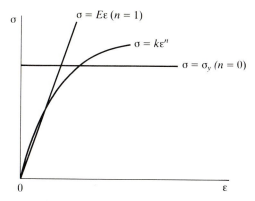

Figure 3.18   Material constitutive equations used in the beam analysis

3.6.2   COMBINED AXIAL AND BENDING LOADS ON A BEAM

The rigid–ideally plastic approximation is particularly useful when only an estimate of the maximum load-carrying capacity of the beam is needed but not the deflected shape. The stress distribution will now be as shown in Fig. 3.19 and, for equilibrium, we can write

$$\sigma_Y b \,(h - 2x) = F \tag{3.60}$$

and taking moments about O,

$$2\sigma_Y \, bx \left(\frac{h-x}{2}\right) = M \tag{3.61}$$

Defining $F_0 = \sigma_Y bh$ and $M_0 = \sigma_Y \dfrac{bh^2}{4}$ as being the force and moment required to produce overall yielding of the beam in tension or in bending respectively, and eliminating x between the two equations (3.60) and (3.61), we find

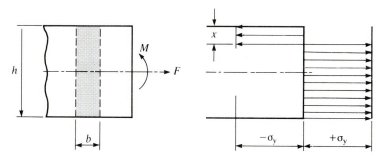

Figure 3.19   Stress distribution at overall yielding in a beam of rectangular cross-section under axial force and bending moment

$$\frac{M}{M_0} + \left(\frac{F}{F_0}\right)^2 = 1 \tag{3.62}$$

as the limit state.

3.6.3 BURSTING OF A THIN-WALLED CYLINDRICAL PRESSURE VESSEL

In a thin-walled cylindrical pressure vessel, the hoop stress and longitudinal stress are given by the well-known expressions,

$$\sigma_\theta = \frac{pr}{t} \quad \sigma_x = \frac{pr}{2t} \tag{3.63}$$

where the subscripts θ and x refer respectively to the hoop and longitudinal directions, p is the internal pressure, r the mean radius and t the thickness. The uniaxial stress equivalent to this biaxial state is, following Tresca's criterion,

$$\sigma = \sigma_\theta = \frac{pr}{t}$$

and following von Mises' criterion (Eqn 3.23),

$$\sigma = \frac{\sqrt{3}}{2} \sigma_\theta = \frac{\sqrt{3}}{2} \frac{pr}{t}$$

The maximum pressure corresponds to the condition

$$dp = d\left(\sigma_\theta \frac{t}{r}\right) = 0$$

Expanding,

$$\sigma_\theta \frac{dt}{r} - \sigma_\theta \frac{t}{r^2} dr + \frac{t}{r} d\sigma_\theta = 0 \tag{3.64}$$

which may be written in the following form:

$$\frac{t}{r}\left[d\sigma_\theta + \sigma_\theta\left(\frac{dt}{t} - \frac{dr}{r}\right)\right] = 0$$

or in terms of the equivalent stress,

$$d\sigma + \sigma\left(\frac{dt}{t} - \frac{dr}{r}\right) = 0$$

The strain increments are (dr/r) in the hoop direction, (dl/l) in the longitudinal direction and (dt/t) in the radial direction. We have already anticipated that

$$\frac{dr}{r} = -\frac{dt}{t} \quad \text{and} \quad \frac{dl}{l} = 0$$

a condition that follows from the flow rule. This implies that when the cylinder yields, plastic flow occurs without any change in the length, the radial expansion being accompanied by the same relative thinning of the wall. The uniaxial strain equivalent to this state is, from the Tresca criterion (Eqn 3.27):

$$d\bar{\epsilon}^* = \frac{4}{3} \frac{dr}{r}$$

and from von Mises' criterion (Eqn 3.29):

$$d\bar{\epsilon}^* = \frac{2\sqrt{3}}{3} \frac{dr}{r}$$

Expressing the strain increments in terms of the uniaxial equivalent strain, we obtain the following equation:

$$\frac{d\sigma}{d\bar{\epsilon}^*} = m \sigma, \quad m = \frac{3}{2} \text{ (Tresca)} \quad \text{or} \quad m = \sqrt{3} \text{ (von Mises)} \quad (3.65)$$

This may be compared with the equation which was obtained for the purely uniaxial cases, in which m is equal to unity. The difference between the two yield criteria is only 15% approximately. If the constitutive equation is (3.57), at maximum load,

$$\frac{d\sigma}{d\bar{\epsilon}^*} = m\sigma = k \, n \, \epsilon^{(n-1)}$$

Therefore

$$\epsilon = \frac{n}{m} \quad \text{and} \quad \sigma = k \left(\frac{n}{m}\right)^n$$

Assuming the material obeys von Mises' yield criterion,

$$p = \frac{2}{\sqrt{3}} \frac{t}{r} \sigma = \frac{2}{\sqrt{3}} k \left(\frac{n}{m}\right)^n \frac{t_0}{r_0} \left(\frac{t}{t_0} \frac{r_0}{r}\right)$$

where the subscript 0 denotes the initial dimensions. Given the definition of the strain and neglecting volume changes in the elastic regime,

$$\int_{r_0}^{r} \frac{dr}{r} = -\int_{t_0}^{t} \frac{dt}{t} = \frac{\sqrt{3}}{2} \epsilon$$

$$\frac{t}{t_0} = \frac{r_0}{r} = \exp\left(-\frac{\sqrt{3}}{2}\epsilon\right) = \exp\left(-\frac{n}{2}\right)$$

Therefore,

$$p = \frac{2}{\sqrt{3}} k \left(\frac{n}{m}\right)^n \frac{t_0}{r_0} \exp(-n)$$

is the bursting pressure.

3.6.4 BURSTING OF A SPHERICAL PRESSURE VESSEL

The preceding treatment can be applied to a spherical vessel, in which the two main components of the stress, in the hoop and in the meridional directions, are equal. It is easy to show that for a von Mises material the maximum pressure is

$$p = 2k \left(\frac{2n}{3} \right)^n \frac{t_0}{r_0} \exp (-n)$$

3.6.5 THICK-WALLED CYLINDER UNDER INTERNAL PRESSURE

Consider now a thick-walled cylinder sealed with a free-moving piston. For simplicity, we assume that the material can adequately be modelled as rigid−ideally plastic and that it obeys Tresca's criterion. For equilibrium of an element (Fig. 3.20),

$$(\sigma_r + \mathrm{d}\sigma_r)(r + \mathrm{d}r)\mathrm{d}\theta = \sigma_r r \mathrm{d}\theta + 2\sigma_\theta \mathrm{d}r \frac{\mathrm{d}\theta}{2}$$

Expanding and neglecting small terms, we obtain

$$r \frac{\mathrm{d}\sigma_r}{\mathrm{d}r} = (\sigma_\theta - \sigma_r) \tag{3.66}$$

for yielding $(\sigma_\theta - \sigma_r) = \sigma_Y$ and the equilibrium equation can be

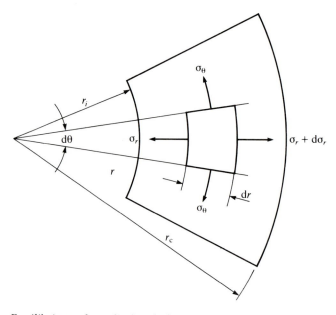

Figure 3.20   Equilibrium of a cylindrical element

integrated between $r = r_i$ (internal radius), where $\sigma_r = p$, and $r = r_e$ (external radius), where $\sigma_r = 0$, to give as the condition for overall yielding,

$$p = \sigma_Y \ln \frac{r_e}{r_i}$$

It will be noted that the longitudinal stress has not been taken into account since it does not affect the equilibrium condition. In practice, the longitudinal strain must be the same for all the elements such as the one shown in Fig. 3.20, since otherwise we would violate compatibility of displacements or, in other words, if we were to cut the cylinder into a number of slices of a given thickness, the slices would only fit together if their faces remained perfectly flat. This implies that the longitudinal strain must be independent of the radial position, being either zero (plane strain) or constant (generalized plane strain). The flow condition then says that

$$\sigma_x = \frac{\sigma_\theta + \sigma_r}{2} + \text{constant}$$

which does not introduce any additional restriction on yielding. What happens when the material strain-hardens? Equation (3.66) still applies, and, under plane strain condition (zero longitudinal extension),

$$\epsilon_\theta = -\epsilon_r$$

but, from geometrical considerations, taking the radial displacement to be u,

$$\epsilon_\theta = \frac{u}{r} \quad \epsilon_r = \frac{du}{dr}$$

It follows that $(ur) = $ constant; therefore we can write that

$$\epsilon_\theta = \epsilon_F \left(\frac{r_i}{r} \right)^2 \tag{3.67}$$

We can now combine Eqns (3.66), (3.67), the yield condition and the constitutive equation to obtain the bursting pressure. For example, assume that the constitutive equation is (3.57). Then,

$$r \frac{d\sigma_r}{dr} = k \left[\frac{4}{3} \epsilon_F \left(\frac{r_i}{r} \right)^2 \right]^n$$

which, upon integration, gives

$$p = \left(\frac{4}{3} \epsilon_F \right)^n \frac{k}{2n} \left[1 - \left(\frac{r_i}{r_e} \right)^{2n} \right] \tag{3.68}$$

Alternatively, if the material is rigid−linear strain-hardening,

$$\sigma = \sigma_Y + \left(\frac{\sigma_F - \sigma_Y}{\epsilon_F}\right)\epsilon$$

and the solution is

$$p = \sigma_Y \ln\frac{r_e}{r_i} + \frac{2}{3}(\sigma_F - \sigma_Y)\left[1 - \left(\frac{r_i}{r_e}\right)^2\right] \tag{3.69}$$

As an illustration, take the case of a vessel with $r_e = 1.5\,r_i$ made of mild steel for which:

$$\sigma_Y = 400\text{ MPa} \quad \sigma_u = 800\text{ MPa} \quad \epsilon_F = 0.4 \quad k = 900\text{ MPa}$$
$$n = 0.13$$

We obtain, from Eqn (3.68),

$$p = 319\text{ MPa}$$

and from Eqn (3.69),

$$p = 310\text{ MPa}$$

For comparison, if we had ignored strain-hardening, the bursting pressure would have been estimated to be

$$p = 162\text{ MPa}$$

Often, it is possible to model the material as rigid–ideally plastic with *flow stress* equal to the mean between σ_Y and σ_u, in which case the estimated bursting pressure would have been

$$p = 243\text{ MPa}$$

3.6.6 BURSTING OF A SPINNING DISC

The problem of the spinning disc, as for example a turbine wheel, is very similar to the one that has just been treated. In the equilibrium equation we have to introduce a term to account for the centrifugal force.

$$F = (\rho\,\omega^2)\,r\,dr\,d\theta$$

giving

$$r\frac{d\sigma_r}{dr} = (\sigma_\theta - \sigma_r) - \rho\omega^2 r^2 \tag{3.70}$$

where ρ is the density and ω the angular velocity. The state will be one of plane stress, in which σ_θ and σ_r will both be positive or zero and, for a Tresca rigid–ideally plastic material, full plasticity will set in when

$$\sigma_\theta = \sigma_F$$

which, combined with Eqn (3.70), gives upon integration,

$$\sigma_r = \sigma_Y - \frac{1}{3}\rho\,\omega^2 r^2 + \frac{C}{r} \tag{3.71}$$

This stress must be zero at $r = r_e$. For a disc with a central hole it will also be zero at $r = r_i$; therefore,

$$\sigma_Y - \frac{1}{3}\rho\,\omega^2 r_e^2 + \frac{C}{r_e^2} = \sigma_Y - \frac{1}{3}\rho\,\omega^2 r_i^2 + \frac{C}{r_i^2}$$

$$C = -\sigma_Y r_i + \frac{1}{3}\rho\,\omega^2 r_i^3 = -\sigma_Y r_e + \frac{1}{3}\rho\,\omega^2 r_e^3$$

and the bursting angular velocity is

$$\omega^2 = \frac{3\sigma_Y}{\rho}\frac{r_e - r_i}{r_e^3 - r_i^3} \tag{3.72}$$

It may be checked that σ_r does not exceed σ_Y by substituting ω^2 and C into Eqn (3.71).

For a disc without a central hole, the state of stress at the centre is isotropic, $C = 0$ and

$$\omega^2 = \frac{3\sigma_Y}{\rho}\frac{1}{r_e^2}$$

3.7   APPLICATIONS TO PLANE STRAIN PROBLEMS

The Tresca yield criterion combined with a rigid−ideally plastic model is a simple and valuable tool to study the problem that arises when the material is forced to yield under a state of plane strain. Typical applications are a punch forced on to a half-space (Fig. 3.21(a)) and a thick plate with a V-notch under tension (Fig. 3.21(b)). To solve these problems, we start by writing the yield condition in the form

$$\tfrac{1}{2}(\sigma_1 - \sigma_2) = \tau_Y$$

This implies that, for a general biaxial state of stress, defined in Fig. 3.22(a) for which the Mohr circle is as shown in Fig. 3.22(b), the radius of the circle is fixed and equal to the critical shear stress while the centre of circle may be anywhere along the horizontal axis. We can then write that

$$\sigma_x = \sigma + \tau_Y \sin 2\zeta$$

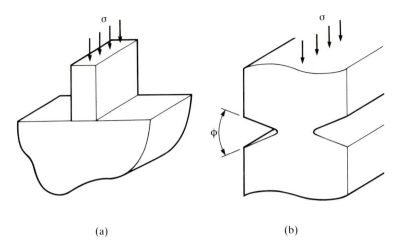

(a) (b)

Figure 3.21 Indentation by a rectangular punch (a) and tensile load applied to a thick, notched plate (b)

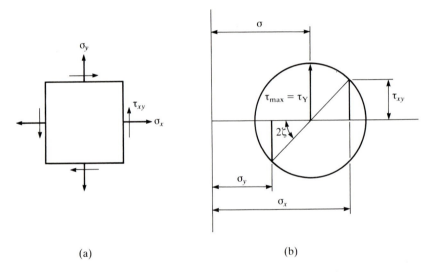

(a) (b)

Figure 3.22 State of stress corresponding to the yield condition when the maximum shear equals the critical shear τ_c

$$\sigma_y = \sigma - \tau_Y \sin 2\zeta \tag{3.73}$$

$$\tau_{xy} = \tau_Y \cos 2\zeta$$

The general equations of equilibrium for the elements are

$$\frac{\partial \sigma_x}{\partial x} + \frac{\partial \tau_{xy}}{\partial y} = 0 \tag{3.74}$$

$$\frac{\partial \sigma_y}{\partial y} + \frac{\partial \tau_{xy}}{\partial x} = 0$$

Combining Eqns (3.73) and (3.74) we obtain the following pair of equations

$$\frac{\partial \sigma}{\partial x} + 2\tau_Y \left(\frac{\partial \zeta}{\partial x} \cos 2\zeta - \frac{\partial \zeta}{\partial y} \sin 2\zeta \right) = 0 \qquad (3.75)$$

$$\frac{\partial \sigma}{\partial y} - 2\tau_Y \left(\frac{\partial \zeta}{\partial y} \sin 2\zeta + \frac{\partial \zeta}{\partial \zeta} \cos 2\zeta \right) = 0 \qquad (3.76)$$

Changing the axes (x,y) to (α,β) along the direction of maximum shear (including ζ and $\pi/2 + \zeta$ with respect to x,y) gives the modified equations,

$$\frac{\partial \sigma}{\partial \alpha} + 2\tau_Y \frac{\partial \zeta}{\partial \alpha} = 0 \qquad (3.77)$$

$$\frac{\partial \sigma}{\partial \beta} - 2\tau_Y \frac{\partial \zeta}{\partial \beta} = 0$$

i.e. the variation of σ along the directions α and β is proportional to the change in the orientation of the lines of maximum shear, increasing with positive change along α and with negative change along β (Fig. 3.23). The (α,β) lines form two orthogonal families of curves. The state of stress at any point on these curves will satisfy the Tresca yield criterion and the curves may therefore be regarded as yield or slip loci. There are many graphical solutions for problems involving these slip lines, and in all cases we start by drawing the condition at free surfaces where they will start at an inclination of $45°$ since (Fig. 3.24) the

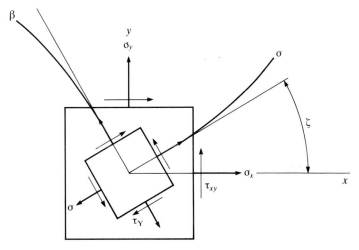

Figure 3.23   Stresses on an element tilted by ζ with respect to the element of Fig. 3.22 and slip (maximum shear) lines

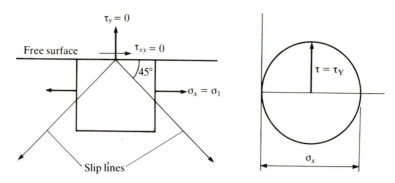

Figure 3.24 Slip lines at a free surface

principal stresses are σ_1, along the surface, and $\sigma_2 = 0$. Taking the problem of the punch, in region I the stresses do not change; hence the average stress, σ_I, is given by

$$\sigma_I = -\tfrac{1}{2}\sigma_1 = -\tau_Y$$

The slip lines start in region I at $45°$ to the surface (Fig. 3.25). In region II, if there is no friction between punch and surface in contact, the state of stress will be compression in the vertical direction and tension (or compression) horizontally, without any shear, so that the slip lines will also be at $45°$ to the surface. Linking the two sets of straight lines in regions I and II by a fan of quarter circles and radial lines completes the slip line field. The two families are mutually orthogonal. In region I σ_I remains constant; between I and II the change of orientation is $(\pi/2)$. Therefore, from Eqn (3.77), the change in average stress,

$$\sigma_{II} - \sigma_I = -2\tau_Y \Delta\zeta = -\pi\,\tau_Y$$

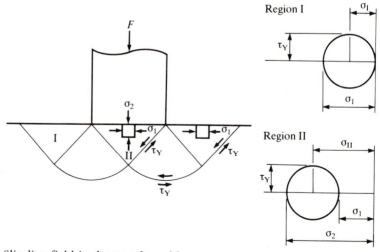

Figure 3.25 Slip line field in the punch problem

100

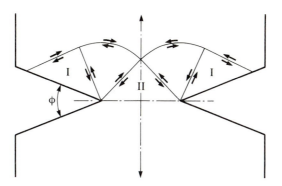

Figure 3.26   Slip line field for a notched plate in tension

and since there is no further change of orientation in region II,

$$\sigma_{II} = \sigma_I - \pi\, \tau_Y = -\tau_Y(1 + \pi)$$

The state of stress at the contact with the punch is therefore

$$\sigma_1 = \sigma_{II} - \tau_Y = -\tau_Y(2 + \pi)$$

$$\sigma_2 = \sigma_{II} + \tau_Y = -\tau_Y\, \pi$$

Therefore, the 'apparent' yield stress has increased from $2\tau_Y$ under plane stress, to $(2 + \pi)\tau_Y$ under plane strain, by a factor of about $\times\ 2.57$.

The case of a V-notch in the thick plate is very similar. It is left as an exercise to show that the apparent yield stress is equal to $\tau_Y(2 + \pi - \phi)$, with the slip line field of Fig. 3.26.

3.7.1   APPROXIMATE SOLUTIONS USING RIGID BLOCKS

It is not always easy to define the slip line field as was done in the example. An alternative is to assume that yield will occur along the boundaries of blocks that slide against each other while remaining perfectly rigid themselves. The shape of the blocks can be defined arbitrarily, the only requirement being to provide enough mobility so as to permit the overall deformation. Taking the example of a notched bar (Fig. 3.27), assume that yielding occurs by sliding of triangles such as B, C, D along lines b, c, d, e. For displacement v of B, corresponding to the crack opening by $2v$, the sliding is

B on C, (b), $v_{BC} = \dfrac{2}{\sqrt{3}} v$ C on D, (d), $v_{CD} = \dfrac{1}{\sqrt{3}} v$

C on E, (c), $v_{CE} = \dfrac{1}{\sqrt{3}} v$ D on E, (e) $v_{DE} = \dfrac{1}{\sqrt{3}} v$

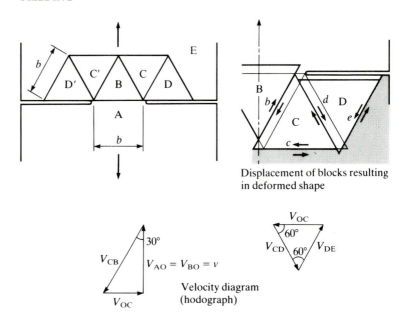

Displacement of blocks resulting
in deformed shape

Velocity diagram
(hodograph)

Figure 3.27 Rigid block approximation to the slip line field

Writing that the external work equals the energy dissipated in the yielding process

$$Fv = 2\tau_Y (bv_{BC} + bv_{CE} + bv_{CD} + bv_{DE}) = 2\,\tau_Y b\,v\,\frac{5}{\sqrt{3}}$$

$$F = \frac{10}{\sqrt{3}}\,\tau_Y b$$

and the apparent yield stress is

$$\sigma = \frac{F}{b} = \frac{10}{\sqrt{3}}\,\tau_Y \simeq 2.88\,\sigma_Y$$

This result is only slightly higher than the one obtained from the slip line solution. It can, in fact, be shown that any approximate result obtained by means of the rigid block approach will always exceed the true theoretical solution. The proof of this statement may be seen in Refs 6, 7.

It remains to be noted that the preceding treatment can be interpreted in terms of displacements per unit time, i.e. of velocities. This helps when dealing with continuous processes, as in the following example.

3.7.2 APPLICATION TO METAL FORMING

As a simple example of the application of plastic analysis to metal-forming, we consider the drawing of a sheet through a die (Fig. 3.28(a)).

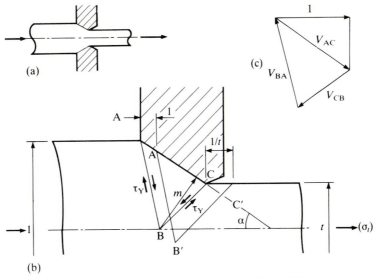

Figure 3.28   Drawing of a sheet from initial thickness 1 to final thickness t

Assume that, as the sheet passes through the die, it deforms as shown in Figure 3.28(b), the two rigid blocks sliding along the edges. The initial thickness and the entry velocity are taken as unity, the final thickness is t and, for continuity, the exit velocity is $1/t$. The relative velocities of the block with respect to sheet and die are given by the velocity diagram of Fig. 3.28(c) and, for energy balance,

$$(\sigma t)\,\frac{1}{t} = \tau_Y(l_{AB}\,v_{AB} + l_{BC}\,v_{BC}) \tag{3.78}$$

It is easily shown that

$$v_{AB} = \sqrt{\left(\frac{1}{m^2} - \frac{2\cos\alpha}{m} + 1\right)}$$

$$v_{BC} = \sqrt{\left(\frac{1}{m^2} - \frac{2\cos\alpha}{m} + \frac{1}{t^2}\right)}$$

where m is the length of the normal in the triangular block. Due to the similarity between the crosshatched triangles,

$$l_{AB} = v_{AB}\,\frac{m}{\sin\alpha}$$

$$l_{BC} = v_{BC}\,\frac{mt}{\sin\alpha}$$

Substituting in Eqn (3.78) we finally obtain

$$\sigma = \tau_Y\,\frac{1+t}{\sin\alpha}\left(\frac{1}{m} + \frac{m}{t}\right) - 2\cot\alpha$$

This is only an approximate solution that, as has been mentioned, will exceed the true one. We improve its accuracy by minimizing it with respect to the one variable that has been chosen quite arbitrarily, i.e. m. It is found that the minimum corresponds to $m = t^{1/2}$, when

$$\sigma_{min} = \frac{2\tau_Y}{\sin \alpha} (t^{1/2} + t^{-1/2}) - 2 \cot \alpha$$

EXAMPLES

3.1   A chain hoist is attached to the ceiling by three pin-jointed tie rods, one vertical and the other two at an angle θ to the vertical as shown in Fig. 3.29. Each rod is of cross-sectional area A and is made from cold-rolled steel, which is assumed to have the idealized stress–strain behaviour illustrated, with yield stress $\pm \sigma_Y$ in tension and compression.
      Draw a (vertical) load–deflection curve for the hoist.

(a)   At what load do all the tie-rods become plastic?
(b)   If this load is applied and then removed, what is the residual stress in the vertical rod?

3.2   A distinction is often made between *primary stresses*, required to maintain equilibrium with external loads, and *secondary stresses*, arising from the need to satisfy compatibility of displacements. Give typical examples of both and discuss their relative effect on structural strength.

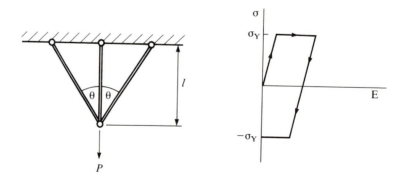

Figure 3.29

104

3.3 (a) Draw the yield loci corresponding to the Tresca and the von Mises' criteria, for a state of plane stress.

(b) A material is initially unstressed and unstrained. For the stress system defined in Fig. 3.30, find the stress σ_1 at first yield when:

    (i) $\sigma_2 = \sigma_3 = 0$ (ii) $\sigma_2 = \sigma_3 = \frac{1}{2}\sigma_1$

    (iii) $\sigma_2 = \frac{1}{2}\sigma_1, \sigma_3 = 0$ (iv) $\sigma_1 = \sigma_2 = \sigma_3$

3.4 A thin-walled, cylindrical pressure vessel has a circular cross-section of mean radius r, and a wall thickness of t. Derive expressions for the internal pressures at which failure occurs according to the failure criteria of (a) Tresca, and (b) von Mises.

3.5 Define the terms 'yield criterion' and 'flow rule' and explain how they are related for an isotropic material.

A metal tube, closed at each end, has an internal diameter equal to 20 times its wall thickness. The metal, which is isotropic, has a tensile yield stress of 69 MN m^{-2}; it obeys the von Mises' yield criterion and flow rule. The tube is subjected to an internal pressure p and an external pressure $0.8p$. Estimate (a) the value of p at yield, and (b) the ratios of the principal plastic strain rates in the wall of the tube.

3.6 A beam of rectangular cross-section, depth h and width b, yields under a moment M combined with a force F applied at the centre of the section and acting along the beam centre

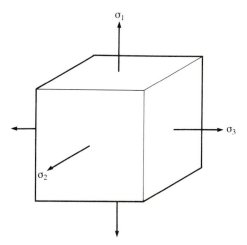

Figure 3.30

line. If the material is rigid—ideally plastic, with yield stress σ_Y, find the relationship between F and M.

3.7 A thin-walled pipe, thickness b and diameter d, yields under an axial force F combined with a torque T. If the material is rigid—ideally plastic, with yield stress σ_Y, find the relationship between F and T and describe the mode of deformation. Assume that the material obeys von Mises' criterion.

3.8 (a) Show that the equation of equilibrium for a rotating disc of density ρ may be written in the form

$$\sigma_r - \sigma_\theta + r\frac{d\sigma_r}{dr} + \rho\omega^2 r^2 = 0$$

(b) A rotating disc of uniform thickness has an outside radius b, inside radius a. It is made of a rigid—ideally plastic material with uniaxial yield stress σ_Y. If the material obeys Tresca's criterion, find the maximum angular velocity it can withstand without bursting.

3.9 Show that the bursting speed for a rotating hoop of radius c is

$$\omega = \frac{1}{c}\sqrt{\frac{\sigma_Y}{\rho}}$$

where ρ is the density and σ_Y the yield stress. Compare this result with the solution for the rotating disc of outside radius r_o and inside radius r_i.

3.10 A thick cylinder is subjected to internal pressure until it yields. Assuming that the material obeys Tresca's criterion:

(a) Find the pressure required and the tangential stress at the bore under pressure.

(b) The pressure is relieved. Find the final stress at the bore when the pressure is entirely released.

(c) Find the minimum value of the ratio between the outside and the inside radii for which reversed plasticity occurs upon unloading.

3.11 (a) As one of the final steps in fabrication, torsion bars and leaf springs used in vehicles are given a plastic deformation in the same direction as the one anticipated in service. Explain why this process can help to increase the fatigue life.

(b) After finding that a shaft has suffered a slight bowing during fabrication, it is decided to straighten it up by cold bending. Is this process likely to affect the strength of the shaft if it is to be subjected to cyclic variations of the applied torque?

3.12   Explain what is meant by *triaxiality of stress*, and *plastic constraint*. Assuming that plastic flow is governed by Tresca's criterion, and fracture by a maximum tensile stress criterion, show qualitatively how plastic constraint affects the ductility of a notched bar.

3.13   Show that for a perfectly plastic material deformed with $\dot{\epsilon}_z = 0$ (plane strain), the hydrostatic component of stress is $\frac{1}{2}(\sigma_x + \sigma_y)$. Figure 3.31 shows the slip line field (i.e. direction of maximum shear stress) near the central plane $z = 0$ of a wide tension specimen with two deep symmetrical 45° grooves. At any point in this field, the material is deforming plastically in plane strain under a shear stress τ_Y acting along the slip lines, combined with a hydrostatic tensile stress $\bar{\sigma}$; $\bar{\sigma}$ is constant in regions OAB and OCO'C' and along radii in the sector OBC. Determine (a) the value of $\bar{\sigma}$ in OAB, and (b), by taking moments about O for equilibrium of OBC, the value of $\bar{\sigma}$ along OC. Hence show that in OCO'C' the maximum tensile stress is $4.36\,\tau_Y$, and compare this value with the tensile yield stress of an unnotched specimen.

3.14   Explain why the expression

$$\overset{*}{\sigma} = (1\sqrt{2})\{(\sigma_1 - \sigma_2)^2 + (\sigma_2 - \sigma_3)^2 + (\sigma_3 - \sigma_1)^2\}^{1/2}$$

is used to define the effective stress, $\overset{*}{\sigma}$, in terms of the

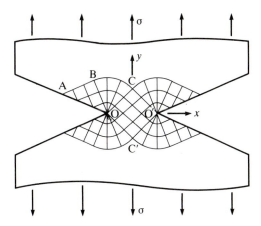

Figure 3.31

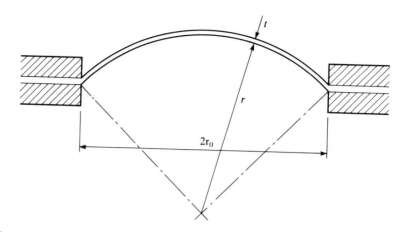

Figure 3.32

principal stresses, σ_1, σ_2 and σ_3, for a material which obeys the von Mises' criterion. Obtain the corresponding expression for the effective strain.

A flat circular plate is made of a material which obeys the von Mises' criterion and has an effective stress—effective strain curve given by

$$\overset{*}{\sigma} = k(\overset{*}{\epsilon})^n$$

where k and n are constants. The plate is rigidly clamped at the rim and a pressure, p, is applied on one side, over the unclamped region of radius r_0 (see Fig. 3.32), such that it deforms into a segment of a sphere. If the initial thickness of the plate is t_0, derive an expression relating the pressure to the current thickness, t, and given that $n = 0.5$ and $t_0 = 2$ mm, calculate the thickness of the plate at the instant of bursting.

3.15 Figure 3.33 illustrates the top half of a slip line field for plane strain frictionless direct extrusion through a square die at an extrusion ratio of 3. The field is symmetrical across the axis. The velocities of the billet and product are U and u respectively, both parallel to the axis. Region 0−1−2−3 is a centred fan consisting of straight and circular slip lines, centred on point O. The circular slip line is extended to form region 0−3−4, which is a right-angled equilateral triangle of straight slip lines meeting the die face at 45°. Region 0−3−4 undergoes simple translation at velocity v parallel to the die face. Slip lines 1−2−3−4 and 0−1 are tangential velocity discontinuities.

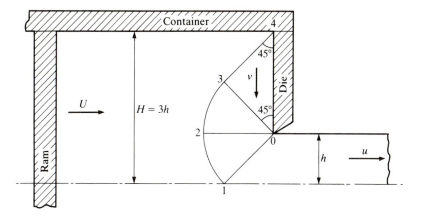

Figure 3.33

(a) Draw the stress plane and the hodograph.

(b) Determine the mean extrusion pressure as a function of τ_Y, the constant shear yield stress of the billet material.

3.16 Briefly discuss the advantages and limitations of the upper bound technique for determining the energy required and forces developed in metal-forming processes.

Asymmetric, frictionless, plane strain extrusion is carried out as shown in Fig. 3.34. Velocity discontinuities may be

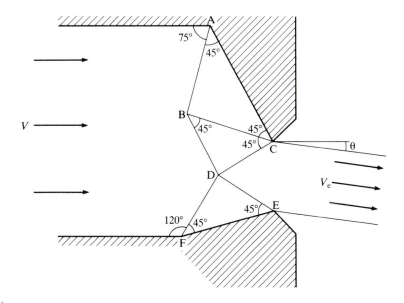

Figure 3.34

assumed along the sides of triangles ABC, BCD and DEF and the shear flow stress, τ_Y, remains constant throughout. Estimate:

(a) the exit angle θ and velocity V_e of the extruded material, and

(b) the required extrusion pressure.

REFERENCES

1 HILL R (1950) *The Mathematical Theory of Plasticity* Oxford English Science Series, Oxford University Press, Oxford

2 HOFFMAN O, SACHS G 1953 *Introduction to the Theory of Plasticity for Engineers* McGraw Hill, New York

3 BREE J 1967 Elasto-plastic behaviour of thin tubes subjected to internal pressure and intermittent high-heat fluxes with application to fast nuclear reactor fuel elements. *J. Strain Anal* **2**: 226–38

4 BREE J 1968 Incremental growth due to creep and plastic yielding of thin tubes subjected to internal pressure and cyclic thermal stresses. *J. Strain Anal* **3**: 122

5 BREE J 1989 Elastic plastic behaviour of thin tubes subjected to internal pressure and cyclic thermal stresses. In Allison I M and Ruiz C *Applied Solid Mechanics — 3* Elsevier Applied Science, London

6 PONTER A R S 1975 A general shakedown theorem for elastic/plastic bodies with work hardening. University of Leicester Report No 75–11

7 COCKS A C F, PONTER A R S 1985 The plastic behaviour of components subjected to constant primary stress and cyclic secondary strains. *J. Strain Anal* **20**: 7–14

CHAPTER FOUR

Brittle Fracture

4.1   INTRODUCTION

The two elementary ways in which a material may fail at the microstructural level are in shear (which results in plastic flow) and in cleavage (which results in brittle fracture). The simpler of the two is, without doubt, the former. Indeed, the repeatability of the yield point of engineering materials under uniaxial loading, and the precision with which pressure-independent yield criteria may be established, mean that the prediction of yielding under a multiaxial stress state is both easy to establish and reliable. All that is required is a knowledge of the state of stress at every point; the corresponding yield parameter is then readily assembled. Engineers may therefore design against first yield by carrying out an elastic analysis and ascertaining the material's uniaxial (or shear) yield strength. The question arises: 'Is it equally simple to design against brittle fracture?' In anticipation of our later results it can be stated unambiguously that it is not.

4.2   BULK BRITTLE FRACTURE

Early attempts to predict brittle fracture were based on an examination of the stress state at fracture, by analogy with experimental work on yielding, which was already well advanced by the early 20th century. Some guidelines may be drawn from this approach: in particular, under a multiaxial state of stress, fracture almost invariably occurs perpendicular to the most positive principal stress. This observation led to the conclusion that the most positive direct stress present was responsible for cleavage. Apart from the ordinary tensile test, where a true brittle fracture occurs perpendicular to the axis of the specimen

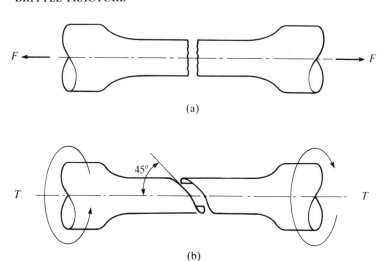

(a)

(b)

Figure 4.1   Failure of a brittle round bar: (a) in tension; (b) in torsion

(Fig. 4.1(a)), the other example commonly met in the laboratory is in a torsion test of, for example, cast iron (Fig. 4.1(b)). In the latter, failure is along a spiral whose angle is approximately $45°$ to the axis of the specimen. The reason for this is clearly seen if the Mohr's circle for stress is drawn (Fig. 4.2). Points A and B correspond to the stress state expressed in the natural cylindrical coordinate set, i.e. $\sigma_{zz} = \sigma_{\theta\theta} = 0$, $\tau_{z\theta} = 2T/\pi R^3$, from elementary torsion theory. Point C represents the more positive principal stress, and is inclined at $90°/2$ with respect to the axial (z) direction.

There are, however, two deficiencies in this theory. First there is always a considerable spread of experimental results when attempting to establish a brittle fracture strength, even under conditions of uniaxial

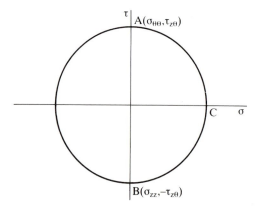

Figure 4.2   Mohr's circle for the state of stress in the bar shown in Fig. 4.1(b). There is a maximum tension at point C, at $45°$ to the axis

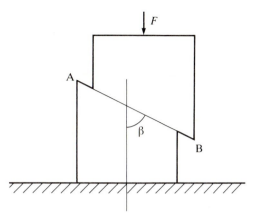

Figure 4.3   Stone column failed in compression

loading. Thus, the stress state *alone* would seem to be insufficient to predict failure. Secondly, fracture occurs by a rather different process if all the principal stresses are negative or zero, e.g. in a column. Brittle fracture of a cast iron column is rare, and normally preceded by yielding, but it can occur in the case of other brittle materials such as stone. In this case, although the micromechanics of the fracture process are different, fracture *does* occur in shear (Fig. 4.3), with a fracture angle β less than $45°$. A comprehensive treatment of this is beyond the scope of this text.[1] It will be noted that the shear strength τ_0 along the plane AB may be thought of as arising from two sources, i.e.

$$\tau_y = c + \mu p \tag{4.1}$$

where c is the cohesive strength of the material, p is the compressive stress on the plane and μ physically has the role of a coefficient of friction. Coulomb showed that with this hypothesis failure would occur at an angle β given by

$$\beta = 45° - \psi/2 \tag{4.2}$$

where $\tan \psi = \mu$. This type of bulk failure criterion is appropriate for rocks and soils, but not for the brittle fracture of metals.

Returning to the question of cleavage of engineering materials, a reasonable question to ask is 'Why is a knowledge of the stress state alone sufficient to quantify yielding, but not brittle fracture?' To answer that we must return briefly to micromechanical considerations. It has been shown (Chapter 3) that the theoretical shear strength of a solid is many times the value realized in practice. This discrepancy led to the concept of a dislocation, which permits two halves of a crystal to glide, one past the other, by the 'failure' of only one or two rows of interatomic bonds (perpendicular to the glide direction) at a time.

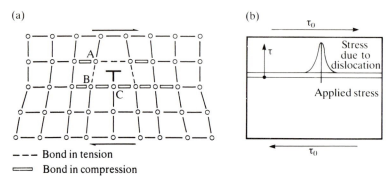

--- Bond in tension
▭▭ Bond in compression

Figure 4.4 (a) Edge dislocation in single cubic crystal showing distorted bands. (b) Self-stress superimposed on an applied shear stress

Several rows of bonds in the neighbourhood of the dislocation are distorted depending on its width, *before* an external load is applied (Fig. 4.4(a)).

As an external shear load is imposed, extra energy is put into the material near the dislocation core so that it becomes more favourable for bond AB to be replaced by a new bond AC. The dislocation has then moved one lattice spacing to the left and there has been a slight rigid body translation of the upper half of the specimen to the right. By the time the dislocation has completely passed from one side of the crystal to the other the relative displacement will be equal to the dislocation's Burger's vector. Note, however, that the process of breaking and reforming bonds is not totally conservative, but requires the input of energy. Further, there is no question of a catastrophic failure along the shear plane, since all but one row of interatomic bonds remain intact. An alternative way of looking at the role of a dislocation emerges if we view the material in a macroscopic way: the dislocation then serves as a mobile, singular, self-equilibrating entity. It is important to recognize that the stress field around the dislocation is due to the singular nature of the displacement discontinuity itself, and is not connected with any applied load (Fig. 4.4(b)). Thus, the dislocation by itself overcomes the crystal's theoretical shear strength, and the imposed load only has to cause the dislocation to move. This may be thought of as propelling the singularity through a periodic set of energy barriers at the pitch of the lattice parameter. These barriers are of characteristic height for a particular metal, and hence give rise to its particular yield strength.

An almost exactly analogous calculation to that for the theoretical shear strength may be carried out for fracture under direct stress. With the exception that the range of the interatomic bonds in tension is a little less certain than in shear (where a periodicity equal to the interatomic spacing is needed), the calculation has the same expected accuracy, and again it tells us that the theoretical fracture strength is

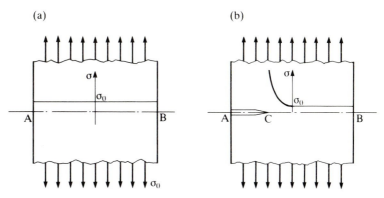

Figure 4.5   Stress distribution across prospective cleavage plane AB: (a) in perfect component; (b) in a flawed component with crack tip at C

of the order of one-tenth of the relevant modulus — in this case Young's modulus. Again, practically realized values are at least two orders less than this.

By analogy with the yielding case it is clear that something is causing a severe weakening of the crystal. In this case the weakening is caused by the presence of pre-existing, finite cracks (Fig. 4.5).

In Fig. 4.5(a) the unflawed component is shown, so that the imposed stress is uniform over the prospective cleavage plan A—B. This state of affairs can only be attained in specially prepared specimens, when strengths approaching those anticipated by the theoretical calculation can be attained. In normal glass specimens and all engineering materials when brittle fracture precedes yielding, cleavage occurs at average stress levels which are low, as stated in the previous paragraph. This is because sharp, pre-existing cracks cause a great concentration of the stress at the crack tip C, (Fig. 4.5(b)). Therefore, although the *mean* stress on the cleavage plane is very low, the peak value, although existing over only a short distance, may be two orders of magnitude greater, and hence is capable of attaining the theoretical rupture value at low applied stresses. As rupture of the crack-tip bonds occurs the tip advances, and, as long as it remains very sharp, the stress concentration will be maintained. Hence the crack will continue to extend.

These, in essence, are the micromechanics of the failure by yielding and by brittle fracture. In the case of yielding, the weakening is caused by the presence of line defects (dislocations) whose state of self-stress is sufficient to cause local bond failure. Only a lack of conservation in making and breaking bonds (the Peierls—Nabarro force) gives the material yield strength. In the case of brittle fracture, pre-existing flaws (which may be minute in the case of highly brittle materials or quite large in the case of metals) cause a severe localization of the applied stress. Thus, in the case of yielding, the applied stress, resolved, in the case of a single crystal, on to the glide plane, has to reach a critical value. This critical resolved shear stress is composed of the Peierls–Nabarro stress together with the effects of boundaries and any 'pins' such as inclusions. On the other hand, in the case of cleavage, it is the stress state near the crack tip which must reach a critical value, and, whilst this is *proportional* to the applied load, it will also be some function of the crack size and geometry. Thus, a knowledge of the stress state *alone* will be insufficient to predict the occurrence of brittle fracture, and this will make it rather difficult to design against. However, it should be noted that because the crack-tip stress level is proportional to the applied or bulk stress, a low overall design stress, as required to avoid yielding, will also lessen the chances of brittle fracture.

4.3   THE GRIFFITHS CRITERION

The first attempt at quantifying truly brittle fracture was made by Griffiths at the end of the First World War. He noticed that freshly drawn glass rods had a strength approaching the theoretical value, whilst those left for a few hours on the bench, where they suffered bombardment by dust particles and some loss of surface integrity from handling, had only the modest strength normally associated with glass. The idea that loss of strength was associated with the presence of surface flaws occurred to him, and this led to the introduction of a simple energy criterion. Although in the form to be derived it is applicable only to perfectly brittle materials such as glass, its comprehension is an essential step in developing the science of fracture mechanics, which can also explain brittle fracture of metals and other materials.

Consider a plate of glass containing some pre-existing crack of half-length *a* (Fig. 4.6(a)), loaded between the jaws of a tensile testing machine. The shape and size of the plate is arbitrary, as is the shape of the crack, although the plane in which it lies must be perpendicular to the jaw so that as the jaws are separated a tensile force tends to open the crack. As this happens, the total work done by the testing machine will be stored in the plate as elastic strain energy. Exactly how this energy is distributed over the plate will clearly depend on the precise geometry but it need not concern us at this stage. If the material remains elastic, a plot of the force–extension curve would be a straight line (Fig. 4.6(b)), and the area enclosed represents the energy stored, *U*. Therefore,

$$U = \int_0^\delta F \, . \, \mathrm{d}x \qquad\qquad (4.3)$$

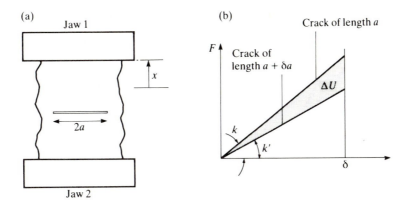

Figure 4.6 (a) A cracked plate loaded in a screw-driven test machine ('fixed grips').
(b) The corresponding force–extension curve showing the effect of crack growth

and if the gradient of the line, which represents the stiffness of the plate, is k, we could write

$$F = k.x$$

$$U = \int_0^\delta kx\mathrm{d}x = \tfrac{1}{2} k \,\delta^2 \tag{4.4}$$

Imagine now unloading the plate, extending the crack by δa (perhaps by using fine saw cuts) and reloading the plate. Because the slit is now longer, the plate will extend further for a given load, i.e. it will have a lower stiffness, and hence if the same displacement, δ, is imposed a lower force will be developed, so that the work done is

$$U' = \tfrac{1}{2} k' \,\delta^2 \tag{4.5}$$

where k' is the new stiffness.

Thus, reverting to a plate with initial crack half-length a, and stretched by an amount δ, if the crack suddenly extends by an amount $\mathrm{d}a$ the elastic energy released and therefore available to propel the crack is given (Fig. 4.6(b)) by:

$$\Delta U = U - U' = \tfrac{1}{2} \delta^2 (k - k') \tag{4.6}$$

If the plate is of thickness t, the energy released per unit area of crack is given by

$$G = \frac{1}{t} \frac{\Delta U}{\Delta a} = \frac{\delta^2}{2t} \frac{\mathrm{d}k}{\mathrm{d}a} \tag{4.7}$$

This quantity, G, is known as the strain energy release rate or generalized crack extension force.

In the case of glass, the only sink term, i.e. the only form of energy absorption, is the energy needed to create new surfaces. If the surface energy is γ, the total energy to be supplied is 2γ, since two surfaces need to be created. Thus, a *necessary* condition for crack growth in glass is

$$G = 2\gamma \tag{4.8}$$

If the crack is of sufficient length for Eqn (4.8) to be fulfilled then crack growth is energetically possible. An alternative way of viewing the above result is to consider the whole of the energy associated with the crack, i.e.

$$E = -U(a) + 2at\gamma \tag{4.9}$$

The point of instability is found by differentiating to maximize E, i.e.

$$\frac{1}{t}\frac{dE}{da} = -G + 2\gamma = 0 \tag{4.10}$$

The exercise carried out indicates the *source* of crack-propelling energy for 'fixed grips' or displacement-controlled loading. A simple example where this might occur in practice is a cam follower; the cam imposes a cyclic displacement whose maximum value is constant. If a crack grows in the follower, the stress developed will fall off as the component's stiffness decreases. The calculation will now be repeated for 'dead loading' or force-controlled loading. This might be realized by using weights to load the specimen of Fig. 4.6(a), say to a force F_0. The work done as before will be

$$U = \tfrac{1}{2}k\delta^2 = \frac{F_0^2}{2k} \tag{4.11}$$

If the crack now extends by da, the extra work performed by the external force is $F_0\,d\delta$, i.e. area ABCD in Fig. 4.7:

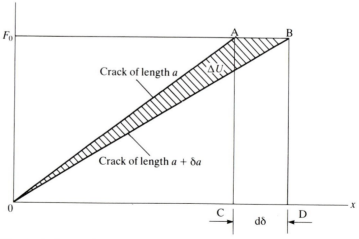

Figure 4.7   Change in internal energy under 'dead loading', i.e. force-controlled load

$$dU = k\delta \frac{d\delta}{da}. \, da$$

When the load is removed the energy recovered is area OBD so that the change in strain energy, area OAB, is one-half of ABCD. Therefore:

$$G = \frac{1}{t} \frac{dU}{da} = \frac{k\delta}{2t} \cdot \frac{d\delta}{da} \tag{4.12}$$

But

$$F = k \, \delta$$

so that

$$dF = kd\delta + \delta \, dk = 0 \tag{4.13}$$

or

$$\frac{d\delta}{da} = - \left(\frac{\delta}{k}\right) \frac{dk}{da}$$

Combining with Eqn (4.12), we find

$$G = - \frac{\delta^2}{2t} \frac{dk}{da} \tag{4.14}$$

Apart from a change of sign, this is the same as Eqn (4.7). Therefore, the expression for strain energy release rate, disregarding the sign, is the same, regardless of whether the load is applied under 'fixed grip' or 'dead loading' conditions. Note that a very crude means of determining G experimentally, for a specific geometry, has also been established. We need to measure the component's stiffness, extend the crack by a small amount, and then remeasure the stiffness. By knowing the ratio $\Delta k/\Delta a$ we can obtain an approximate value of G from Eqn (4.7).

The Griffiths criterion is a useful concept, but it has several drawbacks; the most important are that it does not have a direct relevance to the fracture of semibrittle materials (e.g. metals) and that the generalized crack extension force cannot be computed.

4.4   THE STRESS INTENSITY FACTOR

In engineering usage, it is comparatively rare to use the quantity G to characterize the crack-propelling force. An alternative parameter, the stress intensity factor, is normally used. This arises in considering the stress state of a cracked body, and an important identity relating it to

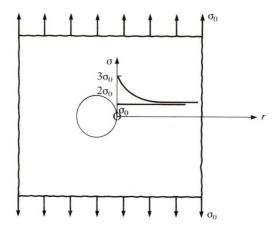

Figure 4.8   Stress concentration at a circular hold in an infinite uniaxially strained plate

the strain energy release rate will be derived later. It is important to note that the stress *concentration* factor and the stress *intensity* factor are completely different quantities: the stress concentration factor is simply a dimensionless constant, the ratio of the maximum stress in the neighbourhood of some stress-raising feature to some nominal remote stress. Thus, for example, for a large plate containing a circular hole, Fig. 4.8, under uniaxial tension the stress adjacent to the hole is three times the remote stress.

It should be noted that the higher stress depends on the geometry of the stress raiser and is not simply a feature of the reduced cross-sectional area sustaining the load; indeed, in the example cited the plate is assumed to be infinite perpendicular to the applied load. It is the curvature of the hole itself which causes the stress to be raised. The sharper the geometric discontinuity, the more severe the stress concentration. Compendia of stress concentration factors are available[2] and are extremely useful in engineering design work, since they enable the true state of stress in the neighbourhood of some stress-raising feature, such as a shoulder or keyway on shaft, screwthread or rivet hole to be estimated, but they are of no use in analysing brittle fracture, where we are concerned with the stress state near an infinitely sharp crack or slit, and for which the stress concentration factor calculated assuming elastic behaviour is infinity.

In the previous section the analysis proceeded without any attempt at understanding how the stress was localized in the neighbourhood of the crack, or how precisely the elastic strain energy may be released during crack extension. To do this the stress state near the crack tip must be determined. It would be convenient if we could do this for the case of a crack subjected to a uniform tension perpendicular to its plane *only*, but because the problem is solved by a suitable choice

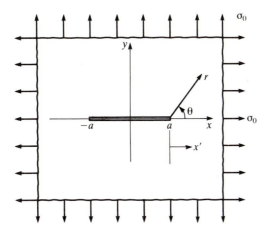

Figure 4.9   Infinite plate with a crack of half-length a under biaxial tension

of solutions to the biharmonic equation, we are somewhat restricted in the repertoire of functions available. The so-called Westergaard solution will be used, which gives rise to the following far-field tension (Fig. 4.9)

$$\sigma_{xx}, \sigma_{yy} \to \sigma_0 \quad \text{as } |x| \to \infty \text{ or } |y| \to \infty \tag{4.15}$$

$$\sigma_{yy} = \tau_{xy} = 0 \quad |x| < a, \, y = 0 \tag{4.16}$$

Details of intermediate steps are given in Appendix A at the end of this chapter. Only the results are given here. The stress components due to a complex potential $\phi(z)$ where $z = x + iy$ are

$$\begin{aligned}
\sigma_{xx} &= \text{Re } \phi(z) - y \text{ Im } \phi'(z) \\
\sigma_{yy} &= \text{Re } \phi(z) + y \text{ Im } \phi'(z) \\
\tau_{xy} &= - y \text{ Re } \phi'(z)
\end{aligned} \tag{4.17}$$

and the Westergaard solution is

$$\phi(z) = \sigma_0 / \sqrt{1 - (a/z)^2} \tag{4.18}$$

$$\frac{d\phi}{dz} = - \frac{\sigma_0 \, a^2}{[z\sqrt{1 - (a/z^2)}]^3}$$

As $|z| \to \infty$, $d\phi/dz \to 0$ and $\phi(z) \to \sigma_0$. It may therefore be seen that Eqns (4.15) are fulfilled. Now consider the $y = 0$ plane. From Eqn (4.18)

$$\phi(x) = \sigma_0 / \sqrt{1 - (a/x)^2}$$

and if $|x| < a$ the square-root has a negative argument, so that ϕ is purely imaginary. Hence Eqn (4.16) is satisfied. Now, a more detailed knowledge of the stress state in the neighbourhood of the crack tip is needed. It will be more convenient to choose a new stress function having its origin at the crack tip. Therefore, let $w = x' + iy = z - a$, i.e.

$$\phi(w) = \frac{\sigma_0}{\sqrt{1 - [a/(a + w)]^2}}$$

$$= \frac{\sigma_0 (a + w)}{\sqrt{2aw + w^2}} \tag{4.19}$$

Restricting the treatment to the case where $|w| \ll a$, Eqn (4.19) becomes approximately

$$\phi(w) = \frac{\sigma_0 a}{\sqrt{2aw}} \tag{4.20}$$

This is therefore the dominant, or singular, term of a series. From de Moivre's theorem $\phi(w)$ may be written in polar form as

$$\phi(w) = \sigma_0 \sqrt{\left(\frac{a}{2r}\right)} \exp\left(-\tfrac{1}{2} i\theta\right) \tag{4.21}$$

and the stress components, from Eqn (4.17), are:

$$\sigma_{xx} = \frac{K_I}{\sqrt{2\pi r}} \left[\cos\frac{\theta}{2}\left(1 - \sin\frac{\theta}{2}\sin\frac{3\theta}{2}\right)\right] \tag{4.22}$$

$$\sigma_{yy} = \frac{K_I}{\sqrt{2\pi r}} \left[\cos\frac{\theta}{2}\left(1 + \sin\frac{\theta}{2}\sin\frac{3\theta}{2}\right)\right] \tag{4.23}$$

$$\tau_{xy} = \frac{K_I}{\sqrt{2\pi r}} \sin\frac{\theta}{2}\cos\frac{\theta}{2}\cos\frac{3\theta}{2} \tag{4.24}$$

where

$$K_I = \sigma_0 \sqrt{\pi a} \tag{4.25}$$

The terms in square brackets are unity, but τ_{xy} vanishes on the $y = 0$ plane. This is the so-called singular solution. It must be borne in mind that it is really only the dominant term in the solution at small distances from the crack tip and that the stress components tend to Eqns (4.15), (4.16) in this case remotely. The solution is of great importance. It demonstrates clearly that the stress state is square-root singular at the crack tip, that it is proportional to the magnitude of the applied load, as might be expected, and to some function of the crack length having the dimensions of $L^{1/2}$. However, it is more than the solution to the problem originally posed. It is the solution for the crack-tip stress field for *any* remote loading, provided that (a) the value of K_I, which we now identify as the stress intensity factor, is chosen correctly, and (b) the loading is only such as to tend to open the crack, which is known as mode I. K_I has already been calculated for a crack within a large plate. Expressions for other geometries are found in Appendix B at the end of this chapter.

Very remote from the crack, the stress field is influenced mainly by the overall geometry and the crack has no effect (Fig. 4.10). As the

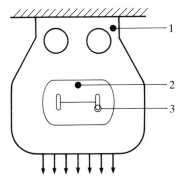

Figure 4.10   Stress fields in an arbitrarily shaped cracked plate. 1, Far field; stress strongly affected by shape, unaffected by presence of crack. 2, Singular solution dominates problems. 3, Crack-tip plasticity

crack is approached, its stress-raising influence begins to dominate the problem and the stresses are given by Eqns (4.22)–(4.24). It must be noted that the spatial variation of the stress state near the crack, and in particular the characteristics of the stress state at the crack tip, are fixed. The absolute magnitude of the stresses in this region is, however, governed by the stress intensity factor, which in turn depends on the geometry and remote load. At very small distances from the crack tip it is apparent that there must be some plasticity since, as $r \rightarrow \infty$, $\sigma_{ij} \rightarrow \infty$, but this will be discussed in more detail later. It is, however, salutary to note that the elasticity solution for brittle fracture demands at least some plasticity.

Since the nature of the crack-tip stress field is fixed, but the magnitude proportional to K_1, it might be anticipated that the stress intensity factor controls crack growth. In order to show that it is connected to the strain energy release rate the displacement field corresponding to Eqns (4.22)–(4.24) must be established. To do this, we shall first find the relevant strain field, from which the displacements may be found. It will transpire later on that for plane problems results in fracture mechanics are very sensitive to the degree of constraint in the third direction, i.e. according to whether plane stress or plane strain conditions prevail. Accordingly, we note that the two forms of Hooke's law:

$$\text{Plane stress:} \quad \epsilon_{ii} = \frac{1}{E} \{\sigma_{ii} - \nu \, \sigma_{jj}\} \quad \sigma_{zz} = 0 \tag{4.26}$$

$$\text{Plane strain:} \quad \epsilon_{ii} = \frac{1}{E} \{(1 - \nu^2) \, \sigma_{ii} - \nu(1 + \nu)\sigma_{jj}\} \tag{4.27}$$

may be combined into the single equation

$$\epsilon_{ii} = \frac{1 + \nu}{4\,E} \{(1 + \kappa)\sigma_{ii} + (\kappa - 3)\sigma_{jj}\} \tag{4.28}$$

123

where

$$\kappa = \frac{3 - \nu}{1 + \nu} \text{ in plane stress}$$

and

$$\kappa = 3 - 4\nu \text{ in plane strain}$$

Substituting Eqns (4.22)–(4.25) into Eqn (4.28) enables the singular strain field to be found. From the definition of strain the displacement may be found by direct integration, i.e.

$$u(x,y) = \int \epsilon_{xx} \, dx + f(y) \tag{4.29}$$

$$v(x,y) = \int \epsilon_{yy} \, dy + g(x) \tag{4.30}$$

where $f(y)$ and $g(x)$ are arbitrary functions. By ensuring that the displacements vanish at the crack tip (the origin) and that the shear stress is correctly represented, the arbitrary quantities may be eliminated, giving

$$u = \frac{K_I}{2E} \sqrt{\frac{r}{2\pi}} (1 + \nu) \left[(2\kappa - 1) \cos \frac{\theta}{2} - \cos \frac{3\theta}{2} \right] \tag{4.31}$$

$$v = \frac{K_I}{2E} \sqrt{\frac{r}{2\pi}} (1 + \nu) \left[(2\kappa + 1) \sin \frac{\theta}{2} - \sin \frac{3\theta}{2} \right] \tag{4.32}$$

It is important to recall that these equations only have relevance in the region of the crack. Particularly important is the displacement along the line of the crack. Outside the crack, i.e. $x > a$, $\theta = 0$ and hence

$$u = \frac{K_I}{E} \sqrt{\frac{r}{2\pi}} (1 + \nu) (\kappa - 1) \quad x > a \tag{4.33}$$

$$v = 0$$

whilst within the crack, i.e. $x < a$, $\theta = \pi$ and

$$u = 0$$

$$v = \frac{K_I}{E} \sqrt{\frac{r}{2\pi}} (1 + \nu) (\kappa + 1) \quad x < a \tag{4.34}$$

The above results may be used to establish a useful relationship between G and K_I. Figure 4.11(a) shows an infinite plate containing a crack of nominal length a, which in Fig. 4.11(b) has extended by a small amount δa. Suppose that the second configuration is chosen and tractions are applied over the portion δa to bring the faces back together again. The work expended in doing this may be found since the form of the shape of the open crack (Fig. 4.11(b)) is given by Eqn (4.34) and the final tractions present in the closed-up crack are given by Eqn (4.23).

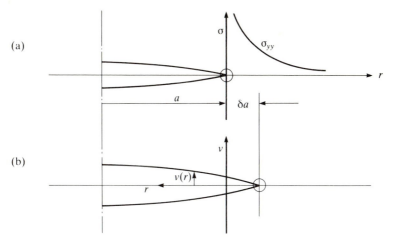

Figure 4.11   Stress and displacements along crack centre line as it extends by ∂a

Choosing the coordinate set defined by Fig. 4.11(a),

$$\sigma_{yy}(r) = \frac{K_1}{\sqrt{2\pi r}} \qquad (4.35)$$

and

$$v(r) = \frac{K_1}{E\sqrt{2\pi}} (1 + v)(\kappa + 1)\sqrt{\delta a - r}$$

The work done in closing some small segment δa is given by:

$$\delta U = \tfrac{1}{2} \int_0^{\delta a} 2v(r) \cdot \sigma_{yy}(r)\, \mathrm{d}r \qquad (4.36)$$

where the 2 within the integral arises because the total distance through which the force moves is twice $v(r)$ and the $\tfrac{1}{2}$ outside appears because this is a linear−elastic problem. Substituting Eqn (4.35) into Eqn (4.36) gives:

$$\delta U = \frac{K_1^2}{2\pi E} (1 + v)(\kappa + 1) \int_0^{\delta a} \frac{\sqrt{(1 - r/\delta a)}}{\sqrt{(r/\delta a)}}\, \mathrm{d}r \qquad (4.37)$$

which may be solved by the substitution $r/\delta a = \sin^2 \theta$ to give:

$$\delta U = \frac{K_1^2}{4E} (1 + v)(\kappa + 1)\, \delta a \qquad (4.38)$$

Since the work done in closing the crack tips must be equal in magnitude to that released upon crack extension, the strain energy release rate may now be found immediately. Substituting for κ, we obtain

$$\text{Plane strain: } G = \frac{K_1^2(1 - v^2)}{E} \qquad (4.39)$$

Plane stress: $G = \dfrac{K_I^2}{E}$ \hfill (4.40)

This establishes a simple relationship between the stress intensity factor, which governs the magnitude of the crack-tip stresses, and the strain energy release rate.

4.5 FRACTURE OF REAL BRITTLE MATERIALS

In an earlier section the Griffiths energy balance concept was used to examine the fracture of glass. In that instance, because the only energy-absorbing term is associated with the creation of new surfaces, it is readily identified as being γ. But it has just been shown that the stress field in the neighbourhood of the crack tip is singular, so that there *must*, in reality, be at least a small zone of plasticity.

As the crack grows (Fig. 4.12), the plastic zone moves. As material becomes plastic some energy is absorbed which is not recovered as relaxation to the elastic state occurs, and this represents a major expenditure of work. In most crystalline materials this energy 'sink term' far outweighs the surface energy contribution. Therefore, it is *postulated* that there is a material property, designated G_{IC}, which represents the critical value of the strain energy release rate at which unstable crack growth occurs. So, in lieu of Eqn (4.8), we write speculatively that crack growth occurs when

$G = G_{IC}$ \hfill (4.41)

It is usual, however, to write the equation down in a different form.

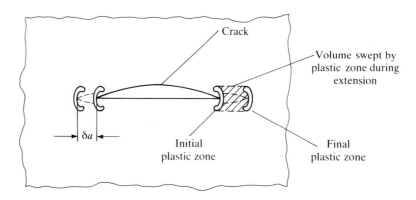

Figure 4.12 Plastic work during crack extension

Making the substitution of Eqn (4.39) and contemplating the existence of an alternative critical stress intensity, K_{IC}, we may write, in lieu of Eqn (4.41),

$$\frac{K_I^2 (1 - \nu^2)}{E} = \frac{K_{IC}^2 (1 - \nu^2)}{E}$$

That is, brittle fracture occurs when,

$$K_I = K_{IC} \text{ under plane strain} \tag{4.42}$$

This is the brittle fracture criterion as normally used by engineers. It must be emphasized that the notion of a critical stress intensity factor, or fracture toughness as it is often called, is only empirical. In order to test that it *is* a material property, it is necessary to perform a large number of experiments with plates containing pre-existing cracks or sharp notches of differing lengths, and to show that the value of the stress intensity factor at which they all fail is the same. Even more compelling evidence would be to take specimens containing cracks of widely differing geometries so that there is a considerable variation between the expressions for the stress intensity factor, and to show that Eqn (4.42) still predicts failure correctly.

It transpires that our assumption, that K_{IC} *is* a material constant, is true with some provisos. As one might expect, its value changes with temperature and metallurgical parameters such as grain size, as indeed does the yield stress. But there are slightly stronger restrictions. First, the critical stress intensity factor is found to vary with plate thickness, if the plate is thin. As the plate thickness increases and plane strain conditions are approached, the critical stress intensity factor tends to a minimum (the fracture toughness K_{IC}) for reasons that will be discussed later. Secondly, the crack must be long enough for fracture to occur under essentially elastic conditions. This point may be clarified by the following argument. For a large, centrally cracked plate the variation of brittle fracture stress with crack length is given from Eqns (4.25) and (4.42):

$$\sigma_{fail} = \frac{K_{IC}}{\sqrt{\pi a}} \tag{4.43}$$

This variation is shown by curve A in Fig. 4.13. However, gross yield in the uncracked specimen will occur at σ_Y, denoted on the diagram by line B, whilst yielding in a cracked bar will occur at a *lower* bulk stress because of the influence of the singular term (Eqns 4.22–4.24). Since the whole of the theory of the stress intensity factor is based on an *elastic* solution, this concept will therefore break down as soon as there is a significant region of plasticity. It follows that the predicted fracture stress must be much lower than the gross yield stress for a solution based on Eqn (4.42) to be completely valid.

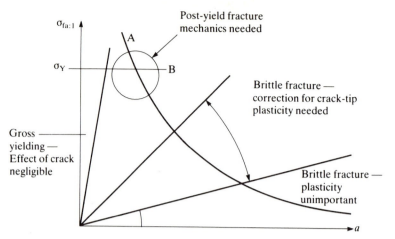

Figure 4.13 Failure in perfectly brittle, partly brittle and yielding regimes

4.6 CRACK-TIP PLASTICITY

Even in a brittle material, a plastic region will be present at the crack tip. When this region is small, as will be the case when the nominal fracture stress is much lower than the yield point, the stress state in the neighbourhood of the crack tip will be dominated by the elastic hinterland. To a good approximation, it will be possible to determine the position of the elastic–plastic boundary by substituting the stress components predicted by Eqns (4.22)–(4.24) into either Tresca's or von Mises' yield criterion and determining the function $r_p(\theta)$ at incipient yield. The expression obtained is

$$r_p(\theta) = \left(\frac{K_I}{\sigma_Y}\right)^2 \frac{1}{4\pi} \left[(1 + \cos\theta)(1 - 2\nu)^2 + \tfrac{3}{2}\sin^2\theta\right] \quad (4.44)$$

under plane strain, and the plane stress solution is recovered by setting $\nu = 0$. The form of the elastic–plastic boundary predicted is not very sensitive to the choice of yield criterion, but it does depend strongly on the degree of lateral constraint, i.e. on whether the material is under a condition of plane strain or plain stress, as shown in Fig. 4.14. For real semi-brittle materials the dominant energy sink term is not the creation of new surfaces, but the provision of work expended in causing plastic deformation and which is not recovered. As the crack extends, the crack-tip plastic zone sweeps out a strip of material (A– A′, B–B′, in Fig. 4.14 for a material under plane strain). The wider this strip is, the more energy must be fed in, or, from Eqns (4.7) and (4.41), the tougher the material is. Thus, material which is under plane stress may be seen to have a higher apparent toughness than that under

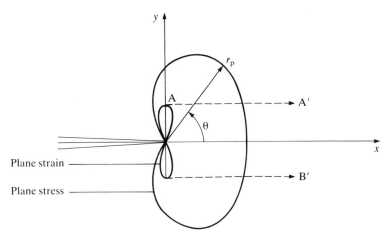

Figure 4.14   Size and shape of crack-tip plasticity zone

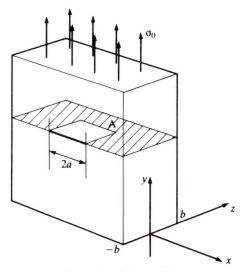

Figure 4.15   Cracked plate of thickness $2b$

plane strain. This may be seen qualitatively from Eqn (4.44) where, for $\theta = \pi/2$, we have for $\nu = 0.3$.

$$\frac{r_{\mathrm{p}|\,\text{plane stress}}}{r_{\mathrm{p}|\,\text{plane strain}}} = 1.51 \tag{4.45}$$

All real components, even those notionally 'plane' such as a sheet, are really neither under plane stress nor plane strain, which are only mathematical conveniences. Consider the question of a cracked sheet, Fig. 4.15, in more detail.

On the faces $z = \pm b$ the direct traction component of stress, σ_{zz}, must clearly vanish, so that they are in a state of plane stress. On the

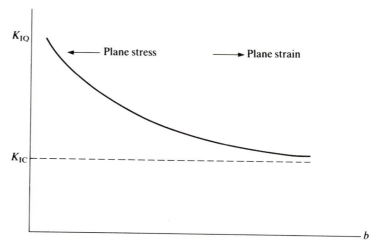

Figure 4.16   Effect of sheet thickness on measured toughness

other hand, if $b \gg a$ sections near the centre of the plate ($|z| \simeq 0$) must be in a state of plane strain. In this event, since the size of the plastic zone is small, the material's apparent toughness will be a minimum; this is the fracture toughness itself, denoted by K_{IC}. For the case of thin plates, plane stress will prevail throughout the thickness, and the toughness, denoted K_{IQ}, will be greater than K_{IC}. Unlike K_{IC}, K_{IQ} is not a material property, and varies with thickness in the manner shown in Fig. 4.16.

Some estimate is needed to enable the engineer to decide whether a plate is under plane stress, plane strain, or not dominated by either regime. Rules of thumb to carry out this estimate rely on evaluating first the plane strain plastic zone size on the plane of the crack ($\theta = 0$, Fig. 4.14). From Eqn (4.44), $r_p(0)$ is given by

$$ r_p(0) = \left(\frac{K_I}{\sigma_Y} \right)^2 \frac{1}{2\pi} (1 - 2\nu)^2 \tag{4.46} $$

and

If $b \simeq 2r_p$, plane stress may be expected
If $b > 20r_p$, plane strain may be expected

Various empirical estimates of K_{IQ} as a function of b in the intermediate range are available, but standard design practice is to assume that plane strain obtains, which is a safe lower-bound value for the fracture toughness.

There are two opposing effects of crack-tip plasticity. On the one hand, it provides the means of energy absorption. On the other, it means that the elastic singular solution *overestimates* the stress which may be carried by material adjacent to the crack tip. This means that

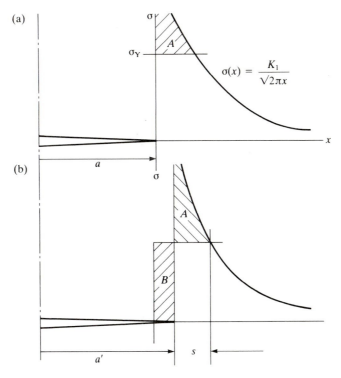

Figure 4.17   Loss of equilibrium caused by presence of plasticity zone: (a) for actual crack; (b) assumed effective size to restore equilibrium

a correction to the calculated crack-tip stress intensity factor must be made, and this will now be described.

Referring back to Fig. 4.15, it will be appreciated that in order to satisfy equilibrium, $\int \sigma_{yy} dA$ over the ligaments adjacent to the crack must equal the externally applied force. This is ensured if the material is entirely elastic, and the stress field obeys the singular solution. But now consider Fig. 4.17(a), which shows the stress state in a real material ahead of the crack tip.

The force corresponding to area A vanishes in a real material, owing to its finite yield strength. A safer estimate of the effect of the crack would be to assume that it has an apparent length a', (Fig. 4.17(b)), so that the load sustained over the yielded region B ($= \sigma_Y(a' - a)$) precisely matches the load 'lost' to yielding A. Thus

$$\sigma_Y = \frac{K_I}{\sqrt{2\pi s}} = \frac{\sigma\sqrt{\pi a'}}{\sqrt{2\pi s}}$$

$$\frac{2s}{a'} = \left(\frac{\sigma}{\sigma_Y}\right)^2 \tag{4.47}$$

and from Eqn (4.46), taking $v = 0$ for convenience,

$$r_p(0) = \left(\frac{\sigma}{\sigma_Y}\right) \left(\frac{a}{2}\right) \tag{4.48}$$

Comparing Eqns (4.47) and (4.48),

$$\frac{s}{a'} = \frac{r_p}{a}$$

or, since $s \ll a'$,

$$s \simeq r_p \tag{4.49}$$

Note, therefore, that a better estimate than Eqn (4.46) of the size of the plastic zone, allowing to a limited extent for the redistribution of stresses caused by a finite plastic zone size, is

$$r_p' = (a' - a) + r_p \tag{4.50}$$

To find $(a' - a)$ we recall that areas A and B are equal, i.e.

$$(a' - a)\sigma_Y = \int_0^s \sigma\sqrt{\frac{a'}{2r}} \, dr - \sigma_Y s \tag{4.51}$$

Since $a' \simeq a$, we find $(a' - a) \simeq r_p$, or

$$r_p' = 2r_p \tag{4.52}$$

Although this result is of fundamental interest, the most important aspect of the calculation from the engineering point of view is that the consistent crack length is a', *not* a. Thus, to estimate the stress intensity factor for a crack in a general stress state and in an arbitrary orientation we need to go through the following procedure:

(1)   Calculate $K_I(a)$ using appropriate formula.

(2)   Calculate r_p from Eqn (4.46).

(3)   Recalculate $K_I(a + r_p)$.

Note that this is the end of the procedure, and it is neither necessary nor consistent to re-estimate r_p and form an iterative sequence.

The question now arises as to the validity of the calculation carried out in any particular case. From Fig. 4.13 it should be recalled that *all* of the above calculations (indeed every one carried out so far in this chapter) are based on the solution of a classical elasticity field problem. The solution will therefore only have validity if the zone of plasticity is *small* and controlled by a large elastic hinterland. As soon as the region of plasticity becomes at all significant we move into the regime labelled 'Post-yield fracture mechanics needed' on that figure, and this is outside the scope of this book. On the other hand, if the material is really quite ductile but final separation occurs at a notch, it may be appropriate to estimate the load necessary to cause failure from a limit-state analysis, where it is assumed that all load-bearing material in the vicinity of the stress-raising feature is plastic. It is the intermediate range which is very difficult to analyse; in practice this regime occurs when testing moderately brittle materials and it is found difficult to devise a specimen of modest dimensions so that the plastic enclave is sufficiently contained by elastic material to obtain a meaningful value for K_{IC}.

4.7   EXTENSIVE PLASTICITY

In the previous section the treatment of cracks showing modest plasticity was discussed. This technique works well as a design procedure, although it becomes increasingly difficult to maintain the concept of a sharp crack tip with associated strain field as the size of the plastic zone is increased. Proceeding from the other direction, material which has a low yield strength or is ductile may form a hinge or fail by nett section yielding, so that a line across the entire component may be found where a yielded state obtains. In this case a limit-state technique, or the use of slip line fields, may enable the load at 'failure' (which we interpret as the greatest load sustainable, rather than the final breaking of the component) to be found. This is treated in Chapter 3. Much more difficult to analyse with any rigour is the case when there is significant crack-tip plasticity but at the same time no fully plastic ligament occurs.

There have been several approaches to this problem, all of them relying to a greater or lesser extent on using a fully elastic solution to characterize the *strain* field, and assuming that this is valid well into the post-yield regime. Exceptions to this are the *J* integral (which relies on using finite elements or some similar techniques to the Bilby, Cottrell and Swinden model, which although precise is mainly of interest for research use). We shall review in turn the techniques available, describing their uses and limitations.

An early attempt to extend fracture mechanics into the larger-scale yielding regime is the concept of the crack opening displacement (COD). In the elastic regime the COD is related to the stress intensity factor through the singular solution (Eqn 4.34), and therefore a knowledge of the separation of the crack faces means that K_I is known. Special clip gauges were devised to measure the COD during loading of a specimen, and the critical COD at specimen failure is recorded. In the case of significant crack-tip opening, it is argued that

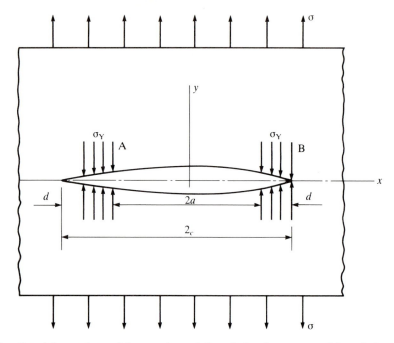

Figure 4.18   Dugdale crack model: actual crack length $2a$, but assumed length $2c$

since the crack yielding displacement at the crack tip is related to the *strains* present, a knowledge of the COD means that the amount of plastic strain at the crack tip is known. If, moreover, failure occurs at a critical plastic *strain*, there is a corresponding critical crack-tip COD which should be a material property. This criterion is not universally acknowledged, and is not currently in vogue.[3]

A second approach, and one which is fundamentally sound, but difficult to implement, is the Bilby, Cottrell and Swinden (or BCS) model. Here, the elastic solution round a crack suffering mode III loading is the starting point. Dislocations (or line singularities) are then distributed along the line of the crack to restore equilibrium in a strip where the yield criterion is exceeded. By adding the Burger's vectors of the distributed dislocations, the COD and associated strain field may be found. In principle the same technique may be used in plane strain mode I but the difficulty of predicting how plastic constraint varies ahead of the crack tip removes much of the technique's elegance.[4] Although the problem in applied mechanics is effectively satisfied, no new failure-controlling quantity is devised.

Yet another attempt at bridging the elasticity/full plasticity gap is the CEGB R6 failure analysis. This is derived from the celebrated Dugdale model in which it is assumed that the crack length is effectively the sum of the actual crack length plus the plastic zones. Within the latter, pressure is applied to force the crack faces together (Fig. 4.18).

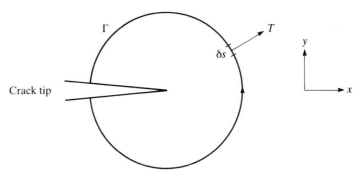

Figure 4.19   Derivation of the *J* integral

The crack-tip stress intensity is then found as the sum of that due to the remote stress σ and the imaginary stress σ_Y in the plastic zone. This nett stress intensity must *vanish* since there can no longer be a stress infinity at the crack tip. This may be used to obtain a good estimate of the plastic zone size. It must be realized, however, that ultimately it is reliant on elasticity solutions, and hence in the limit of large-scale plasticity its validity is questionable.[5]

A further contribution to this study is the so-called HRR (or Hutchinson, Rice and Rosengren) model. This is a fundamental formulation of the problem of a plate containing a slit, made from a Ramberg–Osgood (see Chapter 3) or power-law hardening material, but assuming deformation plasticity. It is, therefore, correct only for a *non-linear elastic* material having the same form of stress–strain curve.[6]

Lastly, we mention the well-known *J* integral, again due to Rice. It is one of a number of conservative, i.e. path-independent, integrals, which may be found for contours enclosing the crack tip. Essentially, along a contour we define (Fig. 4.19)

$$J = \int_\Gamma \left(W\mathrm{d}y - T\frac{\partial u}{\partial x} \right) \mathrm{d}s \tag{4.53}$$

Here $\mathrm{d}s$ is an element of the contour, T is the traction on $\mathrm{d}s$ perpendicular to Γ and W is the strain energy per unit volume, i.e.

$$W(x, y) = \int_0^\epsilon \sigma_{ij}\,\mathrm{d}\epsilon_{ij} \tag{4.54}$$

and $\mathrm{d}y$ is the projection of $\mathrm{d}s$ on to the y axis. In the case of a linear elastic material $J = G$, and it is therefore tentatively assumed that since there is a critical value of G at which fracture occurs, there might be an equivalent property which is a critical value of J, i.e. J_{IC}. For the cracked component, J (the crack-propelling term) may be found by recourse to a purely numerical technique (usually finite elements). Within certain limits J_{IC} does indeed seem to be a material property, but a full understanding of this subject is still the topic of current research.

4.8   OTHER MODES OF LOADING

All the fracture analysis of previous sections has concentrated on a plate containing a crack whose faces are perpendicular to the applied tension. Thus, material in the neighbourhood of the crack tip is being cleaved. As was stated in our discussion of bulk approaches to brittle fracture, fracture almost always occurs by catastrophic growth of a flaw perpendicular to the most positive direct stress, and therefore the case already discussed is of the greatest practical importance. However, two other modes of loading exist and these may be important when a crack grows under unusual stress states (for example within a gear tooth) or when it is channelled by a zone of weakness such as a weld. All three modes of loading are depicted in Fig. 4.20.

Each type of loading results in a stress field which shows a $1/\sqrt{r}$ singularity as the crack tip is approached, and in each case a parameter K (with subscripts I, II or III added) gives the magnitude of the stress state in terms of the remote loading. Thus, along the plane $y = 0$, Fig. 4.20, we find

$$\sigma_{yy} = \frac{K_{\mathrm{I}}}{\sqrt{2\pi x}}; \quad \tau_{xy} = \frac{K_{\mathrm{II}}}{\sqrt{2\pi x}}; \quad \tau_{yz} = \frac{K_{\mathrm{III}}}{\sqrt{2\pi x}} \tag{4.55}$$

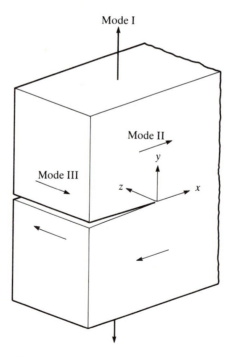

Figure 4.20   The three modes of crack loading

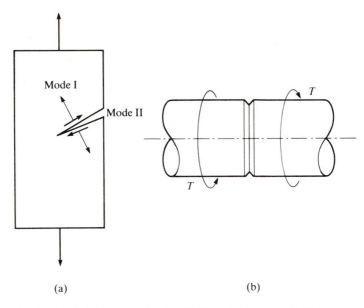

(a)   (b)

Figure 4.21   Cracks loaded (a) in modes I and II, and (b) in mode III

The full stress field for mode II loading may be found in an analogous fashion to that for mode I, using complex variable theory, whilst mode III loading requires a different treatment. The most commonly occurring case of mode II loading is a sloping crack in what is usually a tension member (Fig. 4.21(a)), whilst the most common case of mode III loading conceivable is a circumferential notch in a shaft (Fig. 4.21(b)). It is very rare for mode II loading to occur in isolation, and it is almost invariably associated with a mode I component.

Many of the arguments pursued in earlier sections for mode I loading can be extended to modes II and III — the inevitable crack-tip plasticity, the existence of theoretical toughness values K_{IIC} and K_{IIIC} etc. However, cracks never grow in a stable manner in either mode II or mode III, and these provisions are therefore not needed.

It is also possible to evaluate the work done in extending the crack in modes II and III, and therefore to establish relations between K_{II}, K_{III} and the generalized crack extension force G, analogous to Eqns 4.39 and 4.40. These are:

Plane stress: $G = [K_{II}^{2}/E] + [K_{III}^{2} (1 + \nu)/E]$ (4.56)

Plane strain: $G = [K_{II}^{2} (1 - \nu^{2})/E] + [K_{III}^{2} (1 + \nu)/E]$ (4.57)

and can, since they express energy release, be added in a scalar fashion. Alas, there is as yet no satisfactory criterion for crack extension under combined-mode monotonic loading.

4.9 DESIGN AND FRACTURE MECHANICS IN PRACTICE

Fracture mechanics does not readily provide design criteria; it is qualitatively true that the components should be as defect-free as possible, and that the most positive principal stress should be the minimum achievable value. But how do we implement these axioms in a quantitative manner? The first rule is to try to predict probable sources of crack initiation and to focus attention there. These may be positions where analysis predicts that the stress state will be severe — for instance at geometric stress raisers such as holes, fillets or keyways, which in all probability will be under scrutiny anyway for the avoidance of yielding. Large pre-existing defects are extremely unlikely in, for example, bright drawn steel bar, and therefore immediate brittle fracture of bolted components made from this type of material will be rare. It is much more likely that defects will be incorporated during welding or fabrication, or in components made by a forging or casting process. These are therefore the areas where a critical review of the tolerable flaw size is needed. Failure under a single, monotonic loading is, of course, extremely rare in practice and may be averted completely by proof loading at a value well in excess of the service load. It is much more likely that under cyclic loading fatigue cracks will grow to the critical size, leading to catastrophic failure.

Having highlighted areas of potential difficulties, the designer can establish precisely what the stress state is in the absence of a crack, perhaps by some classical technique, but more probably for most components by the finite element method. The principal stress at each point is then readily found. Most flaws are surface-breaking, and therefore attention is concentrated on the surface, although significant buried defects may be present, for instance slag inclusion in a casting. One then has to speculate on the probable crack shape in order to determine the stress intensity factor. The simplest estimate is a two-dimensional edge crack, which has a K_I value of

$$K_I = 1.12 \, \sigma \, \sqrt{\pi a} \tag{4.58}$$

Other possible geometries include so-called 'thumbnail' (semicircular) or semi-elliptical shapes. Stress intensity factors for these and other shapes are given in Appendix B. However, although the values of K_I do vary considerably, it usually suffices to write the expression as

$$K_I = Y \, \sigma \, \sqrt{\pi a} \tag{4.59}$$

where a is the larger 'diameter' of the crack, and Y is a geometric factor which is a function of crack shape. Usually Y is in the range $1-1.5$, which is quite adequate for design purposes. Using the known value of fracture toughness in conjunction with Eqn (4.59) the critical flaw size may be found.

4.10   THE TWO-CRITERIA APPROACH

In engineering practice, most 'brittle' fractures are found to be preceded or followed by a substantial amount of plastic deformation, as a glance at Fig. 1.14 will show. In a normally ductile material, a truly brittle fracture surface is confined to a central region, where plane strain conditions exist, while the edges fail in shear. Without going into a detailed study of the exceedingly complex phenomenon of quasi-brittle fracture, it is possible to provide some useful guidelines to engineers by considering two extreme situations that embrace the real one. In this approach, we establish bounds by applying the traditional methods of linear elastic fracture mechanics on the one hand and those of limit analysis on the other. Consider, as an example, a beam under three-point bending (Fig. 4.22). K_I is approximately given by

$$K_I \simeq \sigma\sqrt{\pi a} = \frac{PS}{4}\frac{6}{B(W-a)^2} \qquad (4.60)$$

The load at failure is then

$$(P)_{\text{brittle}} = \frac{2}{3}\frac{BW^2}{S}\frac{(1-\alpha)^2}{\sqrt{\alpha}}\frac{K_I}{\sqrt{\pi W}} \qquad (4.61)$$

where α is the ratio (a/W). This is represented in Fig. 4.23, in non-dimensional form for various values of the non-dimensional fracture toughness, $K_I/\sigma_F\sqrt{\pi W}$ where σ_F is the flow stress, taken arbitrarily between the yield point and the tensile strength. It is common practice to define

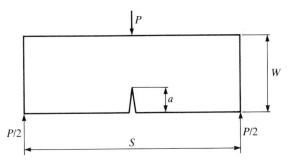

Figure 4.22   General geometry of notched beam under three-point loading. The notch is extended by fatigue precracking before the test

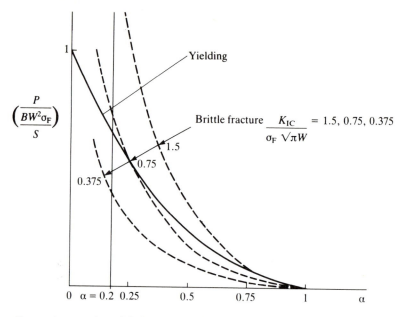

Figure 4.23 Alternative modes of failure depending on crack size, yield point and fracture toughness

$$\sigma_F = \frac{\sigma_Y + \sigma_u}{2} \tag{4.62}$$

An alternative mode of failure consists in yielding, when the bulk stress at the centre of the beam equals σ_F in tension or compression. The maximum bending moment corresponding to the limit state is then

$$(P)_{\text{yielding}} \frac{S}{4} = \frac{\sigma_F B (W - a)^2}{4} \tag{4.63}$$

The variation of $(P)_{\text{yielding}}$, expressed in non-dimensional form, against α, is also presented in Fig. 4.23. Imagine that $K_{IC} = 0.75 \, \sigma_F \sqrt{\pi W}$. A crack $a < 0.25 \, W$ cannot grow in a brittle manner since extensive yielding causes the beam to fail forming a plastic hinge first. On the other hand, larger cracks will grow before plastic deformation sets in. The maximum load that the beam can support may be determined by conducting the two extreme analyses, being intermediate between their predicted results. This can be represented by the diagram of Fig. 4.24, in which the region above the line $(P/P_{\text{brittle}}) = 1$ and to the right of $(P/P_{\text{yield}}) = 1$ may be regarded as unsafe, although it does not follow that the rest of the quarter-plane is safe since interaction between brittle and yielding modes, particularly

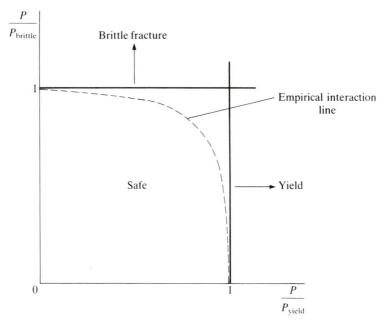

Figure 4.24   Schematic of the two-criteria diagram

near the corner $(1,1)$, may weaken the material. A safety factor is clearly necessary. As an example, take $a = 0.2\,W$ in two materials:

A (mild steel): $\sigma_F = 500$ MPa $K_{IC} = 150$ MPa m$^{1/2}$

B (low-alloy steel): $\sigma_F = 1000$ MPa $K_{IC} = 100$ MPa m$^{1/2}$

The width is 10 mm in the specimen but the thickness in the full-scale structure for which the materials are proposed is 250 mm. For material A,

$P_{brittle}/P_{yield} = 2.52$ in the specimen (yielding failure)

but

$P_{brittle}/P_{yield} = 0.50$ in the structure (fractures before yielding)

It is obvious that the laboratory test has not helped at all to assess the structural strength. For material B,

$P_{brittle}/P_{yield} = 0.84$ in the specimen

and

$P_{brittle}/P_{yield} = 0.17$ in the structure

Both fracture before yielding. Although the ratio is different, we can clearly learn more from the laboratory test than before.

EXAMPLES

4.1 Derive an expression for the fracture strength under tension of a brittle solid containing circular internal flaws of radius a oriented such that their plane is normal to the stress axis.

Glass rods of length 3 mm are found to fracture under a tensile load of 483 N. Estimate the size of circular flaws believed to occur within the rods. E for glass = 70 GN m^{-2}, $\gamma = 0.3$ J m^{-2}.

4.2 The stress intensity factor K_I for a plate containing two symmetrically placed edge cracks (Fig. 4.25(a)) is

$$K_I = \bar{\sigma} w^{1/2} \left(\tan \frac{\pi a}{w} + 0.1 \sin \frac{2\pi a}{w} \right)^{1/2}$$

where a is the crack length, w is the width of the plate and $\bar{\sigma}$ the average stress applied normal to the plane of the cracks at sections remote from that plane.

In tests on a maraging steel, specimens of this geometry were used with $a = 0.015$ m and $w = 0.15$ m. The load P at which a crack propagated was found to depend on plate thickness t as follows:

P(MN)	0.332	0.450	0.569	0.692	0.830
t(m)	0.004	0.006	0.008	0.010	0.012

(a) Calculate K_{IC} for each value of t (in units of MPa m$^{1/2}$). Discuss the significance of the results. The yield stress of the maraging steel is 1.7 GPa.

(b) Estimate the critical length of the crack below which brittle fracture would not occur in a plate of infinite width stressed as in Fig. 4.25(b) when $\bar{\sigma} = 1.0$ GN m^{-2}.

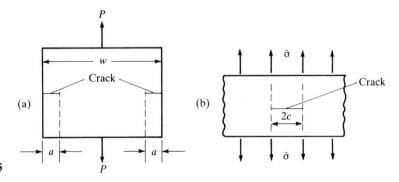

Figure 4.25

4.3 Local yielding near the crack tip alters the stress distribution in the elastic zone further ahead of the crack; discuss how the effect of this on K_I may be estimated.

(a) A high-strength aluminium alloy has a tensile yield stress σ_Y = 470 MPa; the critical stress intensity factor K_{IQ} is 98 MPa m$^{1/2}$ and 41 MPa m$^{1/2}$ for plates of 0.8 mm and 19 mm thickness, respectively. Allowing for the plastic zone correction, estimate the greatest length of crack which can be present in plates of both thicknesses, if neither is to fracture at a working stress of $2\sigma_Y/3$.

(b) Estimate the applied stress at fracture, and the extent of the plastic zone just prior to fracture, in a 19 mm thick plate with a 4 mm deep edge crack in it.

4.4 The principal stresses at a small distance from the tip of a crack in a plate are given by

$$\sigma_1 = \sigma \sqrt{\frac{a}{2r}} \cos \frac{\theta}{2} \left(1 + \sin \frac{\theta}{2} \right)$$

$$\sigma_2 = \sigma \sqrt{\frac{a}{2r}} \cos \frac{\theta}{2} \left(1 - \sin \frac{\theta}{2} \right)$$

with the notation of Fig. 4.26. The third principal stress σ_3 depends on whether the plate can be considered to be under plane stress or plane strain.

(a) Assuming that the material obeys von Mises' yield condition, draw the approximate shape of the yield locus around the crack tip for the plane stress and for the plane strain

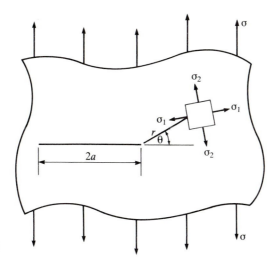

Figure 4.26

conditions, taking $\nu = 0.5$. Can you suggest a reason for the observed variation of fracture toughness with plate thickness?

(b) The following data have been obtained for two types of steel at room temperature:

Mild steel (boiler quality): $\sigma_Y = 275$ N mm^{-2},
$$K_{IC} = 3.4 \text{ kN mm}^{-3/2}$$

Ni−Mo−V low-alloy steel: $\sigma_Y = 750$ N mm^{-2},
$$K_{IC} = 4.5 \text{ kN mm}^{-3/2}$$

For each steel estimate the thickness required for the development of a full plastic constraint and the maximum allowable crack length if the design stress is $\frac{2}{3}\sigma_Y$.

4.5 A large plate of alloy steel is to be loaded in tension. The design stress is taken as $0.5 \times \sigma_u$ (σ_u = tensile strength), and the minimum detectable flaw size is 2 mm. In order to reduce the weight of the structure it is proposed to alter the heat treatment of steel in order to raise its tensile strength from 1500 MPa to 2000 MPa. Concomitant with this increase is a reduction in K_{IC} from 70 MPa m$^{1/2}$ to 35 MPa m$^{1/2}$. Can you recommend this proposal? (You may assume plane strain conditions.)

4.6 (a) A plate containing a through-thickness crack of half-length a is subjected to a tensile stress σ, normal to the plane of the crack. Show that with limited plasticity at the crack tip, the stress intensity factor for mode I opening, K_I, is given by:

$$K_I \simeq \sigma(\pi a)^{1/2} \left[1 + 1/2 \left(\frac{\sigma}{\sigma_Y} \right)^2 \right]$$

where σ_Y is the yield stress for the material of the plate.

(b) A pressure vessel of 2 m mean diameter is to contain a maximum pressure of 70 MPa. The pressure vessel safety requires that the mean section stress in the wall of the vessel must not exceed the lesser of $2\sigma_Y/3$ or $0.35\,\sigma_u$ (tensile strength), and the average stress should not exceed $0.9\sigma_f$, where σ_f is the fracture stress as defined by the material's critical stress intensity factor K_{IC}. Available inspection methods cannot detect a crack of depth less than one-tenth of the wall thickness of the pressure vessel.

Three steels, the properties of which are given below, are under consideration for adoption for the material from which the vessel is to be constructed. Advise the manufacturer as to the appropriate choice of material. If constructed of the chosen steel, would the vessel fail by general yield or fast fracture if an accidental uncontrolled rise in pressure occurred? How would this influence the choice?

	Steel	σ_Y(MPa)	σ_u(MPa)	K_{IC} (MPa m$^{1/2}$)
A	Mild steel	280	550	142
B	Low-alloy steel	625	950	47
C	Maraging steel	1700	2000	120

4.7 A cylindrical steel pressure vessel with closed ends is to contain a pressure (p) of 15 MPa and the steel is to be chosen to give minimum mass for the whole vessel. The vessel is fabricated by edge-welding of steel plates, and is to have a volume of 100 m^3. The maximum stress must not exceed $\sigma_Y/2$, and the fracture stress must exceed the working stress by a safety margin of 10%. Defects may be present in the form of thumbnail cracks at the inner surface of the welds, having an aspect ratio $a/2c = 0.2$, and their stress intensity factor is given by

$$K_I = 1.12\sigma(\pi a/Q)^{1/2}$$

where Q, the flaw shape parameter, may be taken as ~1. The largest defect which may be present without detection can have a length ($2c$) of 20 mm.

Determine which of the following three steels, available in the thicknesses given, is the most appropriate for the construction of the vessel.

Steel	Thickness (m)	σ_Y(MPa)	K_{IC} (MPa m$^{1/2}$)
A	0.08	965	280
B	0.06	1310	66
C	0.04	1700	40

4.8 (a) It is desired to obtain the fracture toughness of an adhesive in shear. A novel experiment, Fig. 4.27, is designed to

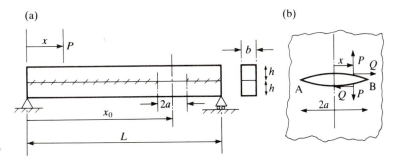

Figure 4.27

achieve this. Two rectangular beams, each of height b and breadth b, are glued together over their entire length, save a small strip of width $2a$. The applied load P is gradually moved along the beam until failure occurs at $x_c(<x_0)$. Derive an expression for K_{IIc}.

The stress intensity factors for a crack, length $2a$, in an infinite plate, loaded by self-equilibrating forces P,Q as shown, are

$$K_I^B = \frac{P}{\sqrt{\pi a}} \left(\frac{a + x}{a - x} \right)^{1/2}$$

$$K_{II}^B = \frac{Q}{\sqrt{\pi a}} \left(\frac{a + x}{a - x} \right)^{1/2}$$

If these equations are utilized in solving the crack problem of Fig. 4.27, what assumptions must be made?

(b) Suppose that the crack is positioned centrally, i.e. $x_0 = L/2$, and that failure occurs when the load is positioned somewhere over the crack, i.e. $x_0 - a < x < x_0 + a$. Determine the stress intensity factor in this case.

4.9 Explain qualitatively why a material's toughness depends on its thickness (or equivalently, the degree of lateral constraint).

Figure 4.28 shows an axial section along a flawed gas pipeline, pressurized to internal pressure p_0. Approximate values of the stress intensity factors at points A and B are given by:

$$K_A = 1.12 \frac{p_0 R}{b} \left(\pi a \sec \frac{\pi a}{2b} \right)^{1/2}$$

$$K_B = \frac{p_0 R}{b} \left[\pi c \left(1 + 1.61 \frac{c^2}{Rb} \right) \right]^{1/2}$$

and the pipeline material has fracture toughness values of $K_{IC} = 77$ MN m$^{-3/2}$ (in plane strain) and $K_c = 85$ MN m$^{-3/2}$ (in plane stress). Further details are as follows:

$R = 280$ mm $b = 14$ mm

(a) Derive expressions for growth of the crack in radial and axial directions (i.e. at points A and B).

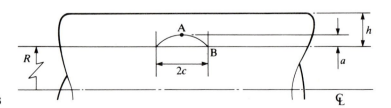

Figure 4.28

(b) If the minimum depth of crack which can be detected is 3.5 mm, what is the maximum axial crack length where radial growth may be expected to precede axial growth?

(c) Recommend a safe working pressure for the pipeline.

REFERENCES

1 JAEGER J C (1969) *Elasticity, Fracture and Flow with Engineering and Geological Application* Methuen, London

2 PETERSON R E (1974) *Stress Concentration Factors* Wiley Interscience, New York

3 BROEK D (1982) *Elementary Engineering Fracture Mechanics* 3rd edn, Martinus Nijhoff

4 KNOTT J F (1973) *Fundamentals of Fracture Mechanics* Butterworths, Guildford

5 EWALD H L, WANHILL R J H (1984) *Fracture Mechanics* Edward Arnold/Delftse Uitgereres Maatschappej

6 HUTCHINSON J W (1968) Plastic stress and strain fields at a crack tip. *J Mech Phys Solids* **16**: 337−47

Appendix A   THE WESTERGAARD SOLUTION

This appendix provides further details of the derivation of the stress field at the tip of a crack, and is included for the reader who wishes to understand the whole calculation.

First we will review some of the basic field equations which apply in two-dimensional elastic problems. For a component free of body forces, i.e. where no stresses are induced within the body by, for instance, gravitational or centrifugal effects, equilibrium considerations of an element of material give:

$$\frac{\partial \sigma_{xx}}{\partial x} + \frac{\partial \tau_{xy}}{\partial y} = 0$$

$$\frac{\partial \tau_{xy}}{\partial x} + \frac{\partial \sigma_{yy}}{\partial y} = 0$$

(A.1)

In continuum mechanics, strains are conveniently defined in terms of displacement gradients. Thus

$$\epsilon_{xx} = \frac{\partial u}{\partial x}$$

$$\epsilon_{yy} = \frac{\partial v}{\partial y} \tag{A.2}$$

$$\gamma_{xy} = \frac{\partial u}{\partial y} + \frac{\partial v}{\partial x}$$

Differentiating the first equation twice with respect to y, the second with respect to x and the third once each with respect to x and y enables us to deduce the compatibility condition, viz.

$$\frac{\partial^2 \epsilon_{xx}}{\partial y^2} + \frac{\partial^2 \epsilon_{yy}}{\partial x^2} = \frac{\partial^2 \gamma_{xy}}{\partial x \partial y} \tag{A.3}$$

Although the compatibility condition is concerned, by definition, with strains or displacements, invoking Hooke's Law, i.e. in plane stress;

$$\epsilon_{xx} = \frac{1}{E}(\sigma_{xx} - \nu\sigma_{yy})$$

$$\epsilon_{yy} = \frac{1}{E}(\sigma_{yy} - \nu\sigma_{xx}) \tag{A.4}$$

$$\gamma_{xy} = \frac{\tau_{xy}}{G}$$

enables us to cast Eqn (A.3) in terms of stress components, i.e.

$$\left(\frac{\partial^2}{\partial x^2} + \frac{\partial^2}{\partial y^2}\right)(\sigma_{xx} + \sigma_{yy}) = 0 \tag{A.5}$$

We seek a function (the stress function) which simultaneously satisfies Eqns (A.1), (A.5) together with the boundary conditions on the surface of the solid. If we define

$$\sigma_{xx} = \frac{\partial^2 \phi}{\partial y^2}, \quad \sigma_{yy} = \frac{\partial^2 \phi}{\partial x^2}, \quad \tau_{xy} = \frac{-\partial^2 \phi}{\partial x \partial y} \tag{A.6}$$

then it may be seen that the equilibrium conditions (A.1) are automatically satisfied. Expanding (A.5) we obtain

$$\frac{\partial^2 \sigma_{xx}}{\partial x^2} + \frac{\partial^2 \sigma_{yy}}{\partial x^2} + \frac{\partial^2 \sigma_{xx}}{\partial y^2} + \frac{\partial^2 \sigma_{yy}}{\partial y^2} = 0$$

i.e.

$$\frac{\partial^4 \phi}{\partial x^4} + 2\frac{\partial^4 \phi}{\partial x^2 \partial y^2} + \frac{\partial^4 \phi}{\partial y^4} = 0 \tag{A.7}$$

Thus, any function $\phi(x,y)$ which satisfies Eqn (A.7) is a feasible stress function, and the particular choice of ϕ depends on the boundary conditions to be satisfied.

Usually only real solutions to Eqn (A.7) are met in an undergraduate course — the so-called Airy stress functions, usually written as polynomials. However, there is another much wider class of functions which are of particular interest in the examination of stress distribution in the neighbourhood of concentrating features such as holes and cracks.

Let

$$z = x + iy \tag{A.8}$$

and take a complex function $\phi(z)$, whose real and imaginary parts are $a(x,y)$ and $b(x,y)$ respectively, i.e.

$$\phi(z) = a(x,y) + ib(x,y) \tag{A.9}$$

We shall need the derivative of ϕ, defined by

$$\phi'(z) = \frac{d\phi}{dz} = \underset{\Delta z \to 0}{\text{Lt}} \left[\frac{\phi(z + \Delta z) - \phi(z)}{\Delta z} \right] \tag{A.10}$$

and we shall require the derivative to be unique at a point; that is, its value shall be constant regardless of the direction in which it is approached. Such a function is said to be analytic, and uniqueness of its derivative means that

$$\phi'(z) = \underset{\substack{\Delta x \to 0 \\ y = 0}}{\text{Lt}} \left\{ \frac{a(x + \Delta x, y) - a(x,y) + i\,[b(x + \Delta x, y) - b(x,y)]}{\Delta x} \right\}$$

$$= \frac{\partial a}{\partial x} + i\,\frac{\partial b}{\partial x}$$

$$= \underset{\substack{\Delta y \to 0 \\ x = 0}}{\text{Lt}} \left\{ \frac{a(x, y + \Delta y) - a(x,y) + i\,[b(x + \Delta y) - b(x,y)]}{i\Delta y} \right\}$$

$$= -i\,\frac{\partial a}{\partial y} + \frac{\partial b}{\partial y}$$

Thus, the requirement that a complex function be analytic leads to the following relations, known as the Cauchy–Riemann conditions:

$$\frac{\partial a}{\partial x} = \frac{\partial b}{\partial y} \qquad \frac{\partial a}{\partial y} = - \frac{\partial b}{\partial x} \tag{A.11}$$

If we rewrite Eqn (A.11) as

$$\frac{\partial \mathrm{Re}(\phi)}{\partial x} = \frac{\partial \mathrm{Im}(\phi)}{\partial y} \qquad \frac{\partial \mathrm{Re}(\phi)}{\partial y} = - \frac{\partial \mathrm{Im}(\phi)}{\partial x}$$

it is apparent that this means that $\text{Re}(\phi)$ and $\text{Im}(\phi)$ must both be harmonic:

$$\left(\frac{\partial^2}{\partial x^2} + \frac{\partial^2}{\partial y^2}\right) \text{Re}(\phi(z)) = \frac{\partial^2 \text{Re}(\phi)}{\partial x^2} + \frac{\partial^2 \text{Re}(\phi)}{\partial y^2}$$

$$= \frac{\partial^2 \text{Re}(\phi)}{\partial x^2} + \frac{\partial}{\partial y}\left(-\frac{\partial \text{Im}(\phi)}{\partial x}\right)$$

$$= \frac{\partial^2 \text{Re}(\phi)}{\partial x^2} + \frac{\partial}{\partial x}\left(-\frac{\partial \text{Im}(\phi)}{\partial y}\right)$$

$$= \frac{\partial^2 \text{Re}(\phi)}{\partial x^2} + \frac{\partial}{\partial x}\left(-\frac{\partial \text{Re}(\phi)}{\partial x}\right) = 0$$

Similarly,

$$\left(\frac{\partial^2}{\partial x^2} + \frac{\partial^2}{\partial y^2}\right) \text{Im}(\phi(z)) = 0$$

Stress functions may therefore be constructed from the real and imaginary parts of $f(z)$.

If $\phi(z)$ is harmonic, so are $x\phi(z)$ and $y\phi(z)$. Further, if the Cauchy–Riemann conditions apply for the first derivative of a function $\phi(z)$, then they apply to all subsequent derivatives. To investigate the stress field near a crack, we consider a plate under uniform biaxial tension σ, and containing a slit of length $2a$ along the x axis, with the origin at the midpoint of the slit.

Note that to a good approximation the far-field tension σ_{xx} is not influenced by the crack (at least the crack has no effect when closed). We are really interested only in the effect of σ_{yy}; σ_{xx} is merely a by-product of the stress function we shall choose, and we shall tolerate its presence. Choose a complex function ψ defined by

$$\psi = \text{Re}(\phi) + y\,\text{Im}(\phi') \tag{A.12}$$

From Eqns (A.6) and (A.12),

$$\frac{\partial \psi}{\partial y} = \frac{\partial}{\partial y}\text{Re}(\phi) + \text{Im}(\phi') + y\frac{\partial}{\partial y}\text{Im}(\phi')$$

$$\sigma_{xx} = \frac{\partial^2 \psi}{\partial y^2} = \frac{\partial^2}{\partial y^2}\text{Re}(\phi) + 2\frac{\partial}{\partial y}\text{Im}(\phi') + y\frac{\partial^2}{\partial y^2}\text{Im}(\phi')$$

$$\tag{A.13}$$

$$\sigma_{yy} = \frac{\partial^2}{\partial x^2}\text{Re}(\phi) + y\frac{\partial^2}{\partial x}\text{Im}(\phi') \tag{A.14}$$

$$-\tau_{xy} = \frac{\partial^2}{\partial x \partial y}\text{Re}(\phi) + \frac{\partial}{\partial x}\text{Im}(\phi') + y\frac{\partial^2}{\partial x \partial y}\text{Im}(\phi') \tag{A.15}$$

150

Using the Cauchy−Riemann equations,

$$\sigma_{yy} = \mathrm{Re}(\phi'') + y\,\mathrm{Im}(\phi''') \tag{A.16}$$

$$\tau_{xy} = -y\,\mathrm{Re}\phi''' + \frac{\partial}{\partial x}\left[\frac{\delta}{\partial y}\,\mathrm{Re}(\phi)\right] + \frac{\partial}{\partial y}\,\mathrm{Im}(\phi')$$

$$= -y\,\mathrm{Re}\phi''' - \frac{\partial}{\partial x}\,(\mathrm{Im}(\phi)) + \frac{\partial}{\partial x}\,\mathrm{Im}(\phi')$$

$$= -y\,\mathrm{Re}\phi''' \tag{A.17}$$

$$\sigma_{xx} = \mathrm{Re}(\phi'') - y\,\mathrm{Im}(\phi''') \tag{A.18}$$

Choose a function $\phi(z)$ such that

$$\phi''(z) = \frac{\sigma z}{\sqrt{z^2 - a^2}} \tag{A.19a}$$

$$\phi'''(z) = -\frac{\sigma a^2}{(z^2 - a^2)^{3/2}} \tag{A.19b}$$

Remote from the crack, $z \to \infty$.

$$\phi''(z) \to \sigma$$

$$\phi'''(z) \to 0$$

Hence σ_{xx}, $\sigma_{yy} \to \sigma$, $\tau_{xy} \to 0$ remote from the crack. On $y = 0$, if $|z| < a$, ϕ'' and ϕ''' become entirely imaginary, so that $\sigma_{xx} = \sigma_{yy} = 0$. Thus, the crack faces are cleared of all tractions.

In the neighbourhood of the crack tip, i.e. $z \to a$,

$$\phi''(z) = \frac{\sigma z}{\sqrt{(z-a)(z+a)}} \to \frac{\sigma}{\sqrt{z-a}}\sqrt{\frac{a}{2}} \tag{A.20}$$

Now,

$$z - a = x - a + iy$$
$$= re^{i\phi}$$

Where r is a radial coordinate from the crack tip. The corresponding stresses are

$$\sigma_{xx} = \sigma\sqrt{\frac{a}{2r}}\left[\mathrm{Re}(e^{-i\theta/2}) + \frac{y}{2r}\,\mathrm{Im}(e^{i3\theta/2})\right] \tag{A.21}$$

$$= \sigma\sqrt{\frac{a}{2r}}\left(\cos\frac{\theta}{2} - \frac{y}{2r}\sin\frac{3\theta}{2}\right)$$

$$= \sigma\sqrt{\frac{a}{2r}}\left(\cos\frac{\theta}{2} - \sin\frac{\theta}{2}\sin\frac{3\theta}{2}\right)$$

$$\sigma_{yy} = \sigma \sqrt{\frac{a}{2r}} \cos \frac{\theta}{2} \left(1 + \sin \frac{\theta}{2} \sin \frac{3\theta}{2} \right) \qquad (A.22)$$

$$\sigma_{zz} = \sigma \sqrt{\frac{a}{2r}} \cos \frac{\theta}{2} \sin \frac{\theta}{2} \cos \frac{3\theta}{2} \qquad (A.23)$$

Appendix B SOME USEFUL STRESS INTENSITY FACTORS

Crack of half-length *a* in infinite sheet subject to uniform tension *σ*

$$K_I = \sigma \sqrt{\pi a}$$

May be used for a small crack well away from any edge

Edge crack of length *a* in semi-infinite sheet subjected to uniform tension *σ*

$$K_I = 1.12 \, \sigma \sqrt{\pi a}$$

May be used for an edge crack with no other edges nearby.

Semi-elliptical edge crack, of depth *a* and diameter 2*c* (see Fig. B.1)

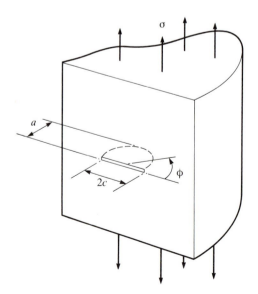

Figure B.1

$$K_I \cong 1.12 \frac{\sigma}{\Phi} \sqrt{\pi a} \left[\sin^2\phi + \left(\frac{a}{c}\right)^2 \cos^2\phi \right]^{1/4}$$

where

$$\Phi\left(\frac{a}{c}\right) = \int_0^{\pi/2} \left\{ 1 - [1 - \left(\frac{a}{c}\right)^2 \sin^2\theta] \right\}^{1/2} d\theta$$

This is an elliptic integral of the second kind, of argument (a/c). Note that the crack-tip stress intensity varies along the crack front.

Forces on crack faces (see Fig. B.2)

$$\frac{K_{IA}}{P} = \frac{K_{IIA}}{Q} = \frac{1}{\sqrt{\pi a}} \sqrt{\frac{a + \zeta}{a - \zeta}}$$

$$\frac{K_{IB}}{P} = \frac{K_{IIB}}{Q} = \frac{1}{\sqrt{\pi a}} \sqrt{\frac{a - \zeta}{a + \zeta}}$$

Centrally cracked plate of finite width, w

$$K_I = \sigma\sqrt{\pi a} \left(\sec \frac{\pi a}{w} \right)^{1/2}$$

Circular (penny-shaped) crack in an infinite body subject to uniform tension (Fig. B.3)

$$K_I = \frac{2}{\pi} \sigma\sqrt{\pi a}$$

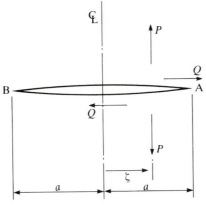

Figure B.2

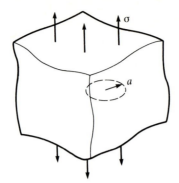

Figure B.3

CHAPTER FIVE
Fatigue

5.1 INTRODUCTION

Very few structures suffer a constant load. Even those which are static (such as bridges or steel-framed buildings), which have to sustain a constant self-weight, are subject to varying forces: in the first case these arise from the vehicles, trains or pedestrians being transported, and in both cases from windage forces, which are often considerable. A pressure vessel supports a fluctuating internal pressure as it is filled and emptied, and this gives a low-frequency fluctuating stress. But the pump which fills it may be of the reciprocating kind, in which case a high-frequency ripple will be present, and in addition there will certainly be vibrations which will induce additional high-frequency effects. In some cases (such as the low-frequency cycling of the pressure vessel, or the loading seen by, say, the wings of a civil aeroplane), the stress state may vary very considerably during one 'cycle' of load, but the 'period' of the cycle is anything but constant, as in the last example cited when the duration of flights may vary enormously. However, although the exact shape of the waveform, i.e. the stress versus time curve, may not be regular or conform to a simple mathematical description, the magnitudes of the stresses during each phase of the cycle are quite well defined, and the number of cycles of safe life is much more critically dependent on this parameter.

The components and structures referred to above all suffer complex states of stress, but fatigue cracks nearly always grow perpendicular to the highest tensile stress, which approximately determines its trajectory. An exception is cracking due to rolling-contact fatigue, where the shear stress is much more influential. There are essentially two ways of analysing fatigue: by the classical way, which requires only a knowledge of the notional state of stress in the unflawed components, and by fracture mechanics, which can give additional information such as the expected crack growth rate. This may be used to determine safe inspection intervals for critical components.

5.2 MICROMECHANICS OF FATIGUE

Fatigue occurs when the majority of a component is in an elastic state. One might therefore expect all loading to be reversible, and for there to be no damage. Macroscopically this is true, but cracks grow because of some irreversibility and this is actually the exhaustion of ductility within the tiny plastic zone always found at the tip of a crack. In order to understand fatigue in detail, it is therefore necessary to understand the crack-tip process fully. This is inevitably rather qualitative as the micromechanics of void formation ahead of the crack tip and the coalescence of the voids with the growing crack is statistical, depending on grain size and orientation and on the degree of prior work-hardening. Several models based on the interaction of dislocations have been put forward but, from the engineering point of view, once a crack is well developed the Paris or Forman laws predict the growth rate rather well when a description of the crack-tip mechanism is unnecessary. What *is* important is an understanding of the initiation process for fatigue cracks.

Some cracks initiate from well-defined pre-existing defects, such as welds or imperfectly bonded, brazed, or glued components, and there is no mystery about growth, since there is a finite stress intensity already present. Much more difficult to understand are those cracks which start from an apparently smooth and flawless surface. It is rare for cracks to start if completely unprovoked, and the most common feature from which they start is a stress raiser, such as a notch, fillet, keyway, screwthread or indeed any change of profile (including, for example, stiffening ribs in sheet components). These features cause a local increase in stress above the mean. In addition, all surfaces are irregular, with marks produced on a microscopic scale by machining, filing, extrusion or any mechanical process, as well as those produced on a slightly larger scale by inadvertent scratching or handling damage. It is possible that at least one feature of this type will lie in the region of the stress concentration, and the two effects together may be sufficient to cause an extremely localized plastic zone. Dislocation slip will be concentrated on the most favourably orientated glide plane, and even though the average strain is modest, large dislocation excursions will occur on this plane. Complete reversibility of dislocation motion will not be possible, as some interaction between dislocation gliding on intersecting planes may result in locking.

Eventually a notch (intrusion) or bulge (extrusion) will form which is now geometrically quite sharp, and a finite stress intensity factor will develop. Some very recent work, still at the research state,[1] is attempting to quantify the development of a crack from dislocation motion, and may eventually lead to a useful initiation criterion. A study of short cracks is still also the subject of present research, but we can draw some simple design guidelines from the above remarks.

First, it is important that all stress-raising features are made as mild as possible, so that the state of stress is as homogeneous as possible. This is, of course, not easily achievable. However, it is also important to ensure that the microscopic stress raisers are as small as possible, which in practice means achieving the best possible surface finish, and if necessary polishing the final surface. Also, processes which involve 'ironing' (such as the expansion of rivet holes with a tapered mandrel, or blanking) are preferable to those involving cutting. The induction of surface compressive residual stresses is beneficial, as any cracks which do tend to form are unable to grow. Similarly, tensile residual stresses have a bad effect as they tend to keep embryo cracks open. Shot peening, bead blasting, ion implantation or the driving of oversize mandrels through holes as cited above, all tend to leave significant compressive stresses in the surface, although the effect of ion implantation is very short-lived. Electrodeposition often leaves significant tensile stresses in the surface (particularly from chromium plating) and various bath additives must be used or other measures taken to annul this effect. Corrosion is also a serious problem, and can cause cracks to grow even under a static load. Certainly the rate of growth under cyclic loading is increased.

Mechanical damage to contacting surfaces may occur during service, by a process known as fretting. When two components are clamped together (for example by a ring of bolts) and an external oscillatory force applied, perhaps simply vibration, it is customary to find that there are areas over the contacting surfaces where complete adhesion has not occurred. Slight relative motion results, which causes a deterioration of the surface finish, often the production of oxide, and the concentration of interfacial surface tractions near the stick/slip interface. A combination of the above provides ideal conditions for crack initiation.

5.3   THE BULK APPROACH TO FATIGUE

The traditional approach to fatigue, where all that is necessary to quantify the rate of growth of a crack is the nominal stress present,

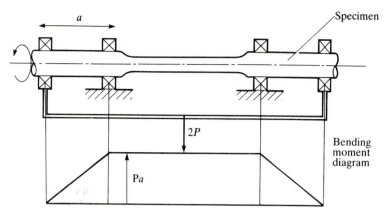

Figure 5.1   Geometry and bending moment diagram for four-point bend test

might be called the 'bulk' approach. Although it is no longer studied extensively it still has its uses where there are many fatigue data available, but it is limited in the way it can both handle multi-axial states of stress, and estimate the residual life of a flawed component.

The origins of the technique lie in the extensive tests of Wohler, at the end of the 19th century, on axles of railway vehicles. These were carried out by using a rotating beam bend specimen as shown schematically in Fig. 5.1. Since the bending moment is constant over the central portion, each cross-section is in the same state of stress, and a fatigue crack is equally likely to initiate at any point on the surface. Note that the stress history seen by any point on the specimen is fully reversed, i.e. it experiences an alternating stress (where extreme values are easily found from elementary bending theory) shown schematically in Fig. 5.2(a). Many tests are carried out on apparatus of this type with a range of values of P, and the number of cycles to failure of the specimen found. Although the apparatus is limited insofar as it is not possible to vary the mean stress, it is easy to carry out tests at a comparatively high speed, and to obtain results quickly. Tests are often stopped at either 10^7 or 10^8 cycles, at which point the specimen is considered to have an 'infinite' life. Typical results of the magnitude of the alternating stress (S) against the number of cycles to failure (N) are shown in Fig. 5.3. These are frequently called $S-N$ curves as S was often used to represent stress in the early days.

Two types of response emerge. Many metals, such as aluminium, copper and their alloys, exhibit a continuously decreasing curve, i.e. it would appear that no matter how low the stress, the specimen will fail at some finite life, as in Fig. 5.3, curve (a). On the other hand, low-carbon steels, titanium alloys and a few aluminium alloys show some cut-off stress (known as the fatigue limit or endurance limit) below

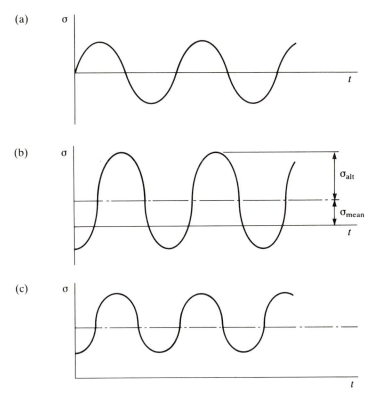

Figure 5.2 Types of stress history: (a) fully reversing; (b) fluctuating; (c) pulsating

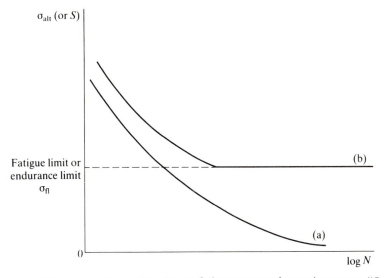

Figure 5.3 Logarithm of number of cycles to failure versus alternating stress ('$S-N$ curves'): (a) no fatigue limit; (b) fatigue limit

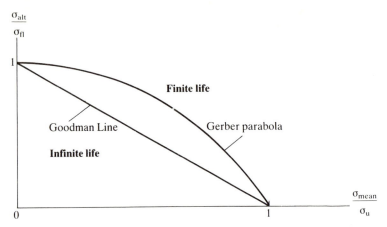

Figure 5.4   One form of the Goodman diagram

which the specimen has an infinite life — Fig. 5.3, curve (b). For materials showing no distinct plateau the fatigue limit is sometimes artificially defined as the stress at which the life expected is 10^7 cycles. There is always a spread of lives found from carrying out a number of fatigue tests under notionally identical conditions. This statistical nature of fatigue is accounted for by the uncertainties in crack initiation, to which we shall refer later.

Although the $S-N$ diagram can be used to design against failure (or give a probable life of a given number of cycles), its use is limited to uniaxial fully reversing stress of constant amplitude. The assumption that the mean stress is zero may be readily relaxed by using the empirical Goodman diagram. Stress waveforms of the types shown in Fig. 5.2(b,c) are catered for, although it should be stated that few experimental tests have been carried out when the mean stress is compressive. One version of the diagram (sometimes associated with the names Haig–Soderberg) is shown in Fig. 5.4. On one axis the alternating component of stress, normalized with respect to the fatigue limit σ_{fl}, is plotted. In the absence of a mean stress this ratio must be less than unity if the component is to have an infinite life. On the other axis the mean stress, normalized with respect to the ultimate tensile strength, σ_u, is plotted. If there is not to be immediate failure, this ratio must be less than unity. Goodman postulated that the line delineating infinite life from finite life would be a straight one between the two extremes, and suggested that, for an infinite life,

$$\frac{\sigma_{alt}}{\sigma_{fl}} + \frac{\sigma_{mean}}{\sigma_u} \leq 1 \tag{5.1}$$

Testing performed under combined static reversing loading tended to show an infinite life at higher stresses, however, and Gerber suggested the parabola shown in Fig. 5.4, whose equation is

$$\frac{\sigma_{alt}}{\sigma_{fl}} + \left(\frac{\sigma_{mean}}{\sigma_u}\right)^2 \leq 1 \tag{5.2}$$

However, only a rash engineer would wish to admit static stresses as high as the tensile strength and a more conservative rule, due to Soderberg, is to use Eqn (5.1) but with σ_{TS} replaced by σ_Y, the yield stress. There is no underlying physical reason why any one of these rules should be preferred over the others, although for practical design the last is the most conservative.

The Goodman diagram may be used in another way. Above we used it to determine conditions for infinite life, assuming that the material exhibits a fatigue limit. But we may also use it to predict the finite life of a component when there is a non-zero mean stress, using $S-N$ data obtained under fully reversing stress conditions. To do this we use Eqn (5.1) and take the equality. Values of the mean stress and alternating stress present are then substituted in, together with the material's tensile strength. The equation may now be solved for σ_{fl}, i.e. the quantity which is normally the fatigue limit of the material. In this instance, however, the variable takes on a different role and is, in effect, a concreted alternating stress which allows for the influence of the mean stress. It may be used in conjunction with an $S-N$ curve to determine the component's predicted life.

It has been stated that there have been few tests carried out with a negative mean stress, and hence, although the substitution of negative values of σ_{mean} into Eqns (5.1) and (5.2) is perfectly possible, great caution should be exercised in using the predicted fatigue strengths, as the mode of crack growth is now at least partly in shear, whereas in a tensile environment it is invariably in an opening mode (see Chapter 4).

Although fatigue tests are normally carried out with a sinusoidal stress–time curve, the exact form of the curve does not influence the component's life significantly, nor does the frequency of loading. However, it is frequently the case that the amplitude of loading may vary slowly during the life of a component. For example, if Fig. 5.1 represents the loading diagram for a rail-vehicle axle, adding the payload that the vehicle carries will increase the bending moment, and hence the magnitude of the stresses. The question arises, 'What is the expected life of such a component?' Figure 5.5 shows a typical stress history curve, with the number of cycles at each load level considerably reduced for clarity. If, at stress σ_{alt}^i the number of cycles to failure is N_f^i from the $S-N$ curve, and the number of cycles actually sustained is n_f^i, then the fraction n_f^i/N_f^i of the life of the specimen is considered to have been used up, and when the sum of the fractions reaches unity failure will occur, i.e. when

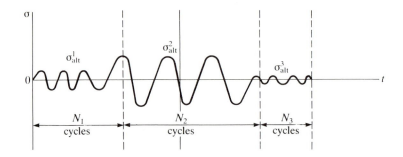

Figure 5.5 Stress history when the stress amplitude varies — use of Miner's law

$$\sum_i \frac{n_f^i}{N_f^i} = 1 \qquad (5.3)$$

In cases where the mean stress is non-zero, Eqn (5.3), which is known as Miner's law, may be used in conjunction with the Goodman diagram.

For materials not exhibiting a clear fatigue limit the form of the curve in the $S-N$ diagram (Fig. 5.3) may sometimes be represented by

$$\sigma_{alt}^a N_f = \text{constant} \qquad (5.4)$$

where a lies in the range $8 < a < 20$, found from a curve-fit. This empirical fatigue law is known as Basquin's law, and is usefully used in conjunction with the Miner law. Basquin's law might be interpreted as stating that two sets of cycles produce the same amount of damage if Eqn (5.4) holds. Hence, if we have cycled for n_1 cycles at stress σ_{alt}^1, the remaining number of cycles n_2 that we expect will cause failure at a stress σ_{alt}^2 is given by

$$N_2^{\text{remaining}} = N_2 - N_1 \left(\frac{\sigma_{alt}^1}{\sigma_{alt}^2} \right)^a \qquad (5.5)$$

where N_2 is the expected life if all cycling had been at a level σ_{alt}^2.

Similarly, if the component had undergone n_1 cycles at σ_{alt}^1, n_2 cycles at σ_{alt}^2 and was now stressed to σ_{alt}^3 we might anticipate the remaining life to be

$$N_3^{\text{remaining}} = N_3 - N_2 \left(\frac{\sigma_{alt}^2}{\sigma_{alt}^3} \right)^a - N_1 \left(\frac{\sigma_{alt}^1}{\sigma_{alt}^3} \right)^a \qquad (5.6)$$

Although very useful, the above results relate to uniaxial loading. Usually this is not restrictive as, even under a multi-axial state of stress, a crack will grow perpendicular to the most positive principal stress, and be governed by that stress. However, the aircraft industry (in particular) has been anxious to understand bulk fatigue under multiaxial conditions. Two sets of results are worth pointing out. Gough carried out two sets of fatigue tests — one in pure bending and the other in pure torsion. He established fatigue limits under these conditions, σ_{fl}^b and τ_{fl}^t respectively. He then carried out combined loading tests with stress components σ_{alt}^b, τ_{alt}^t. The failure envelope was of the form

$$\left(\frac{\sigma_{alt}^b}{\sigma_{fl}^b}\right)^2 + \left(\frac{\tau_{alt}^t}{\sigma_{fl}^t}\right)^2 = 1 \qquad\qquad (5.7)$$

A more ambitious set of tests was carried out by Sines, and resulted in a generalized form of the Goodman law which might be written in the form

$$\frac{\sqrt{3 J_2^{alt}}}{\sigma_{fl}} + \frac{I_i^{mean}}{\sigma_u} \leq 1 \qquad\qquad (5.8)$$

Here J_2 is the second invariant of the stress deviator referred to the alternating stress component and I is the first invariant of the complete stress referred to the mean stress component. For a detailed explanation of these terms refer to Chapter 3.

5.4   FRACTURE MECHANICS APPROACH

A more modern method of tackling the fatigue problem, although still, ultimately, relying on experimental data for determination of life, is to apply fracture mechanics. In order to appreciate this section fully it is essential that Chapter 4 is studied first, so that the concept of the stress intensity factor is fully understood. 'Bulk' fatigue has its limitations: it is not possible, if a crack is found during inspection, to predict how long it will take to grow to a dangerous state, and hence how long the component may safely be left in service before the next inspection. Also, a fundamental problem with the classical approach is that it is not possible to separate out the phases of crack initiation and crack propagation. In some cases the first of these may occupy the majority of the life of the component, whilst in other cases it is the latter which is the longer process.

The starting point of the application of fracture mechanics is the result of a large series of tests on differing materials (but particularly those used in the aircraft industry) where specimens have a pre-existing crack installed, usually by growing a crack from a saw-cut, and are then loaded cyclically in tension in a servo-hydraulic test machine. The length of the crack is then monitored continuously during the test,

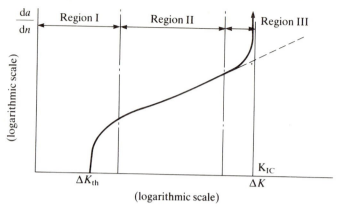

Figure 5.6 Typical da/dN versus ΔK curve

and a plot obtained of the crack growth rate (da/dN) against the instantaneous stress intensity factor, ΔK, obtaining as the test proceeds. We use the symbol ΔK to denote the amplitude of the varying stress intensity. For example, for an edge crack in a semi-infinite plate, subject to a sinusoidally varying stress of amplitude $\Delta\sigma$, the stress intensity factor would be

$$\Delta K = 1.12 \, \Delta\sigma \, \sqrt{\pi a} \tag{5.9}$$

A typical graph of crack growth rate is shown in Fig. 5.6. Note that many tests have to be performed at a range of values of crack length and stress level, but resulting in the same ΔK, to ensure that it *is* the stress intensity factor alone which correlates with crack growth rate. This is found to be the case except for extremely short cracks (with correspondingly high stresses) where initiation has not been completed and conventional elastic fracture mechanics does not apply.

It may be seen that the typical growth rate versus stress intensity figure has three regimes, although the transition from one to the next may be blurred and there is always some scatter. At low stress intensities, where initiation is being completed, or pre-existing cracks such as weld defects are beginning to grow, the microstructure of the material, together with environmental effects, are very important. During this phase the crack may grow in shear, but as it moves into the second region it will turn to grow perpendicular to the most positive principal stress (so that K_{II} is also zero). The influence of microstructure is now very small, and this regime may continue for a significant time. Eventually the crack becomes so large that crack-tip plasticity begins to have a significant effect (region III), and the rate of crack growth accelerates. Eventually, as $K \rightarrow K_{\mathrm{IC}}$, the growth rate becomes infinite and final fracture occurs.

Experiments have been performed with a spread of values of stress ratio $R = \sigma_{min}/\sigma_{max}$, but with the same range of stress intensity, i.e. the same ΔK. Some materials show a weak dependence of the growth rate on R, whilst in others it is much more marked. In the latter case a new da/dN versus ΔK plot must be established for each stress ratio. We wish to be able to put data from these plots in a numerical form for rapid estimation of a crack's life. The most widely used equation is that due to Paris, and is

$$\frac{da}{dN} = C\Delta K^m \tag{5.10}$$

where C and m are constants, chosen so as to fit the linear portion (region II) of the growth rate, which is linear on a log−log plot. The difficulty with using this equation is that once region III is entered it underestimates the rate of growth, and it is therefore not conservative. A second equation, widely recognized in the research literature but for which comparatively few data are available, is one due to Forman, i.e.

$$\frac{da}{dN} = \frac{C\Delta K^m}{(1-R)K_c - \Delta K} \tag{5.11}$$

and this expression is intended to cater for regions II and III. Data for the Paris equation have been obtained for many common metals and alloys, and values of C and m, together with ΔK_{th}, the threshold for a finite rate of crack growth (Fig. 5.6) are given in Tables 5.1, 5.2 and 5.3. ΔK_{th} corresponds to the concept of a fatigue limit in bulk fatigue. More specialized data for aircraft alloys are available.

Equation (5.10) may readily be used to find the number of cycles of loading needed to cause the crack to increase in size, say from a_1 to a_2. Often the amplitude of the applied stress is constant, so that the magnitude of the stress intensity factor depends only on the crack length, i.e.

$$\Delta K = \Delta K(a;\sigma) \tag{5.12}$$

Combining Eqn (5.10) with Eqn (5.12) we see that the number of cycles N required to cause the crack to grow from length a to a_2 is

$$N = \int_{a_1}^{a_2} \frac{da}{C\,\Delta K^m(a;\sigma)} \tag{5.13}$$

As an example, suppose we are concerned with an edge crack, for which

$$\Delta K = 1.12\,\Delta\sigma\,\sqrt{\pi a} \tag{5.14}$$

Equation (5.13) then becomes

Table 5.1 Fatigue crack growth data for various materials for use with Paris equation (5.10)[2] (Crown copyright)

Material	Tensile strength (MN m^{-2})	0.1 or 0.2% Proof stress (MN m^{-2})	R	m	ΔK for da/dN = 10^{-6} mm (cycle^{-1}) (MN m$^{-3/2}$)
Mild steel	325	230	0.06−0.74	3.3.	6.2
Mild steel in brine*	435	—	0.65	3.3	6.2
Cold-rolled mild steel	695	655	0.07−0.43	4.2	7.2
			0.54−0.76	5.5	6.4
			0.75−0.92	6.4	5.2
Low-alloy steel*	680		0−0.75	3.3	5.1
Maraging steel*	2010		0.67	3.0	3.5
18/8 austenitic steel	665	195−255	0.33−0.43	3.1	6.3
Aluminium	125−155	95−125	0.14−0.87	2.9	2.9
5% Mg−aluminium alloy	310	180	0.20−0.69	2.7	1.6
HS30W aluminium alloy (1% Mg, 1% Si, 0.7% Mn)	265	180	0.20−0.71	2.6	1.9
HS30WP aluminium alloy (1% Mg, 1% Si, 0.7% Mn)	310	245−280	0.25−0.80	3.9	2.6
			0.50−0.78	4.1	2.15
L71 aluminium alloy (4.5% Cu)	480	415	0.14−0.46	3.7	2.4
L73 aluminium alloy (4.5% Cu)	435	370	0.50−0.88	4.4	2.1
DTD 687A aluminium alloy (5.5% Zn)	540	495	0.20−0.45	3.7	1.75
			0.50−0.78	4.2	1.8
			0.82−0.94	4.8	1.45
ZW1 magnesium alloy (0.5% Zr)	250	165	0	3.35	0.94
AM503 magnesium alloy (1.5% Mn)	200	107	0.5	3.35	0.69
			0.67	3.35	0.65
			0.78	3.35	0.57
Copper	215−310	26−513	0.07−0.82	3.9	4.3
Phosphor bronze*	370		0.33−0.74	3.9	4.3
60/40 brass*	325		0−0.33	4.0	6.3
			0.51−0.72	3.9	4.3
Titanium	555	440	0.08−0.94	4.4	3.1
5% Al−titanium alloy	835	735	0.17−0.86	3.8	3.4

Table 5.1
(continued)

Material	Tensile strength (MN m^{-2})	0.1 or 0.2% Proof stress (MN m^{-2})	R	m	ΔK for da/dN = 10^{-6} mm cycle^{-6} (MN m$^{-3/2}$)
15% Mo–titanium alloy	1160	995	0.28–0.71	3.5	3.0
			0.81–0.94	4.4	2.75
Nickel*	430		0–0.71	4.0	8.8
Monel*	525		0–0.67	4.0	6.2
Inconel*	650		0–0.71	4.0	8.2

* Data of limited accuracy obtained by an indirect method

$$N = \frac{1}{1.12\ C(\Delta\sigma\ \sqrt{\pi}\,)^m} \int_{a_1}^{a_2} \frac{da}{a^{m/2}}$$

$$= \frac{1}{1.12\ C(1-m/2)(\Delta\sigma\sqrt{\pi}\,)^m}\ (a_2^{\,1-m/2} - a_1^{\,1-m/2}) \qquad (5.15)$$

The usual application of this equation is to estimate how long it will take a pre-existing defect to grow to the critical size at which brittle fracture occurs. In the above example, for instance, we could let a_2 be given by the crack length at fracture, i.e.

$$a_2 = a_c = \frac{1}{\pi}\left(\frac{K_{IC}}{1.12\ \sigma}\right)^2 \qquad (5.16)$$

where σ is the maximum stress in a cycle.

Note, however, that this is not a conservative estimate of life as in region III (Fig. 5.6) the rate of growth is underestimated. It is, however, extremely useful in establishing inspection intervals for scheduling non-destructive testing. Non-destructive tests are performed on critical pieces of equipment where failure will have disastrous consequences, for example colliery winding gear, aeroplane structural frames and pressure reservoirs. For each of these items the engineer must establish the critical and most highly stressed components, choose a non-destructive test method and establish the frequency with which these components must be inspected. Different non-destructing testing (NDT) techniques have different resolutions, which depend at least partly on the orientation and shape of the crack being sought, but typical values might be in the region of 0.3–1.0 mm. We imagine that a crack *just* smaller than that which can be detected was present at the time of an inspection. If we substitute the size of this not-quite-detected flaw into Eqn (5.15) for a_1, the expected time for it to grow to the critical size will be given. Bearing in mind that our calculation for the number of cycles is an overestimate, we might halve the value of N suggested to determine the inspection interval.

Pre-existing flaws may be of any shape or orientation: in the case of welded steel structures, weld connections which may incorporate slag or lack penetration are frequent origins of cracks, whilst in aluminium alloy structures rivet holes for riveted connections cause stress concentrations which may give rise to cracks. Because of their arbitrary orientation the initial crack may experience any combination of modes I, II and III loading. It might therefore be thought that crack

Table 5.2 Fatigue crack growth thresholds for some common metals[2] (Crown copyright)

Ferrous materials	Tensile strength (MN m^{-2})	R	ΔK_{th} (MN m$^{-3/2}$)
Mild steel	430	-1	6.4
		0.13	6.6
		0.35	5.2
		0.49	4.3
		0.64	3.2
		0.75	3.8
Mild steel at 300 °C	480	-1	7.1
		0.23	6.0
		0.33	5.8
Mild steel in brine	430	-1	~ 2.0
		0.64	1.15
Mild steel in brine with cathodic protection	430	0.64	3.9
Mild steel in tap water or SAE30 oil	430	-1	7.3
Low-alloy steel	835	-1	6.3
	680	0	6.6
		0.33	5.1
		0.50	4.4
		0.64	3.3
		0.75	2.5
NiCrMoV steel at 300 °C	560	-1	7.1
		0.23	5.0
		0.33	5.4
		0.64	4.9
Maraging steel	2010	0.67	2.7
18/8 Austenitic steel	685	-1	6.0
	665	0	6.0
		0.33	5.9
		0.62	4.6
		0.74	4.1
Grey cast iron	255	0	7.0
		0.50	4.5

(a)

Table 5.2
(continued)

Non-ferrous materials	Tensile strength $(MN\ m^{-2})$	R	ΔK_{th} $(MN\ m^{-3/2})$
Aluminium	77	−1	1.0
		0	1.7
		0.33	1.4
		0.53	1.2
L65 aluminium alloy	450	−1	2.1
(4.5% Cu)	495	0	2.1
		0.33	1.7
		0.50	1.5
		0.67	1.2
		0.50	1.15
L65 aluminium alloy			
(4.5% Cu) in brine†			
ZW1 magnesium alloy	250	0	0.83
(0.6% Zr)		0.67	0.66
AM503 magnesium	165	0	0.99
alloy (1.6% Mn)		0.67	0.77
Copper	225	−1	2.7
	215	0	2.5
		0.33	1.8
		0.56	1.5
		0.69	1.4
		0.80	1.3
Phosphor bronze	325	−1	3.7
	370	0.33	4.1
		0.50	3.2
		0.74	2.4
60/40 brass	330	−1	3.1
	325	0	3.5
		0.33	3.1
		0.51	2.6
		0.72	2.6
Titanium	540	0.60	2.2
Nickel	455	−1	5.9
	430	0	7.9
		0.33	6.5
		0.57	5.2
		0.71	3.6
Monel	525	−1	5.6
		0	7.0
		0.33	6.5
		0.50	5.2
		0.67	3.6
Inconel	655	−1	6.4
	650	0	7.1
		0.57	4.7
		0.71	4.0

(b)

† Unpublished data.

Table 5.3 Some fracture toughness and tolerable flaw sizes for selection of aircraft alloys

Material	σ_u (N mm^{-2})	σ_Y (N mm^{-2})	K_{IC} (N mm$^{-3/2}$)	$2a^*$ (mm)
7075-T6 Al	550	490	1020	6−1
2014-T6 Al	460	580	1030	8−8
D6AC steel	1990	1710	1540	1.05
4340 steel	1780	1440	1470	1.02
Maraging (300) steel	1810	1690	2840	4.4
6Al−4V titanium	1200	1100	1200	1.8
4 Al−4Mo−2Sn−0.5Si titanium	1050	960	2450	9−6

* This column is tolerable flaw size if the maximum permitted design stress is 60% of σ_u and a central crack is assumed.

growth laws analogous to the Paris equation but for the other two modes, and indeed combined modes, might be needed. However, experience has shown that, unless they are guided in some contrived way, almost all cracks rapidly turn so that they grow perpendicular to the most positive principal stress, i.e. they are under pure mode I loading. Laws corresponding to the shear modes of growth are the subject of current research, but are rarely needed.

5.5 APPLICATIONS TO DESIGN

It is essential that careful inspection of all critical components which are liable to brittle fracture is carried out prior to their entering service. NDT is an attempt to ensure that all pre-existing flaws (normally associated with welding, although conceivably produced by casting) are well below the critical size. Many engineering components see some sort of fluctuating load, and there is therefore every probability of crack development which may ultimately be catastrophic. In such cases a regular inspection (usually by ultrasonic detection or dye penetrant) is carried out and fracture mechanics enable the inspection interval to be found. This is chosen so that any small crack 'missed' in a particular inspection cannot grow to the critical size before the next routine test. Even assemblies not fabricated by welding (and therefore not containing significant pre-existing defects) may subsequently develop structurally important cracks initiating from stress raisers such as keyways, rivet holes, machining errors, screwthreads and so on.

5.6   NON-DESTRUCTIVE TESTING

The simplest form of non-destructive testing is inspection by eye—either by the naked eye or with the help of a low-power magnifier. First, the size of fillets may be checked and the overall quality of welds verified — sometimes penetration may be checked at the end of a weld-run if the end is visible. Also, the evenness of fillets may easily be seen. Sometimes inclusions or slag may be visible, but usually only in very bad cases. These defects may be made more noticeable by using a dye penetrant. This is sprayed on, and is attracted into crevices or cracks. The excess is then wiped off and the component illuminated by ultra violet light, whereupon the entrained dye fluoresces. More sophisticated testing techniques include the use of ultrasonics; here, a sound source and detector are moved steadily over the surface, the detector registering reflections from the back of the plate. The presence of a significant flaw will also produce a reflection which is recorded. However, this will only occur if the plane of the crack is correctly orientated, and any crack which lies parallel to the trajectory of the ultrasonic wave will be missed. A magnetic circuit between two surface points is greatly weakened by slag or non-magnetic material where the flux will spread over the surface, and this principle is also used for flaw detection.

All the above techniques may be carried out 'on site' with quite portable equipment. Small components may be inspected by radio-graphic techniques, the source being X- or γ-rays. Since irradiation from such short-wavelength sources is very dangerous, tests must be carried out in well-protected chambers, although pipeline is also inspected by radiographic techniques.

The size of flaw which may be detected depends on many factors, including the morphology and orientation of the crack, as well as the test method. In ideal cases cracks or flaws of sub-millimetre dimensions may be observed, but in most cases the smallest flaws which can be reliably found are at least 5 mm across and may be rather more.

5.7   INSPECTION INTERVALS

Using the techniques described previously it is straightforward, for a crack of given shape, to establish the growth of a crack with the number of cycles. For example, for the edge crack of initial length a_0 the length $a(n)$ after n cycles is, from Eqn (5.15),

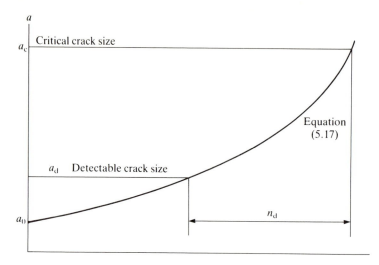

Figure 5.7 Growth of crack with number of cycles

$$a(n) = [a_0^{1-m/2} + 1.12\ Cn\ (1-m/2)(\Delta\sigma\sqrt{\pi})^m]1/1-m/2 \qquad (5.17)$$

This curve is shown schematically in Fig. 5.7. Suppose that the crack size at which failure occurs, from Eqn (5.16), is a_c, and that the minimum size of crack which can reliably be detected is a_d. The number of cycles it will take a crack to grow from a_d to a_c is n_d cycles. Hence, if, at a given inspection, the largest crack present is *just* smaller than a_d, i.e. it cannot quite be found, it is essential to have another inspection before n_d cycles have elapsed, i.e. before the crack approaches the critical size.

The difficulty with the above procedure is that it is not known into what shape the crack will develop. Providing the stress gradients near the crack are slight this may not matter, as the stress intensity factor may be written as

$$K_I = Y\sigma\sqrt{\pi a} \qquad (5.18)$$

and $1 < Y < 1.5$, approximately.

5.8   LEAK-BEFORE-BREAK

Where, despite the engineer's best efforts, the possibility of a crack developing to a critical size cannot be ruled out (for instance in a buried pipeline), it is important to be able to predict just how the final rupture will manifest itself. Returning to the pipeline example, many pipes have a longitudinal seam which is a source of weakness, and in the presence of a fluctuating circumferential stress a crack may develop and grow in both the radial and axial directions. When the crack breaks through

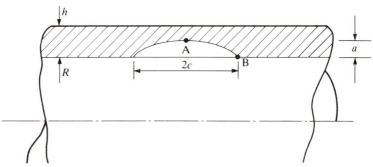

Figure 5.8   Crack of approximately semi-elliptical form in the wall of a pipe

the wall thickness, escape of whatever is within the pipe will occur, i.e. it will 'leak'. Undesirable though this is, it may not be catastrophic. However, if the crack has already grown longitudinally to a significant extent, breakthrough may be accompanied by high-speed running of the crack along its length, i.e. the seam may 'unzip' or 'break' along a great distance, causing complete release of all the fluid or gas contained. This outcome is catastrophic, and much effort has been expended in trying to predict if 'leak-before-break' will occur in any particular instance. Leakage is detectable, and hence the pipeline may be decommissioned before the crack runs.

Of the two principal stresses in a thin-walled pipe, the σ_θ one is greater by a factor of $2(\sigma_\theta = P(R/t)$ where R is the mean radius and t the wall thickness). Imagine that the crack has grown from some point both radially and longitudinally, and that its present front is approximately semi-elliptical, as shown in Fig. 5.8. The stress intensity factor at point A is approximately given by

$$K_I^A = 1.12 p_0 \, (R/b) \left(\pi a \sec \frac{\pi a}{2b} \right)^{1/2} \tag{5.19}$$

whilst at point B it is given by

$$K_I^B = p_0 \left[\pi c \left(1 + 1.61 \frac{c^2}{Rb} \right) \right]^{1/2} \tag{5.20}$$

The crack will continue to grow in both directions until the stress intensity becomes critical. This normally occurs at point A first, and breakthrough to the outside of the pipe or 'leak' will occur when

$$K_I^A = K_{IC} \tag{5.21}$$

The corresponding pressure, p_0, at which this occurs will be the working pressure within the pipe. The fracture process may result in a partial or total loss of pressure if the pipe is gas-filled, but if it is liquid-filled hydraulic-hammer or other inertia effects may result which actually transiently *increase* the pressure. In any event the stress intensity at B may become critical, i.e.

$$K_I^B = K_C \tag{5.22}$$

so that the crack runs longitudinally, producing potentially catastrophic 'break' conditions as the tube peels open, potentially for great distances if the stress intensity is maintained. Note that in Eqn (5.22) we have used the symbol for the critical stress intensity under non-plane strain conditions. As the tube wall is unlikely to be thick enough to enable full plane strain to be attained, this point actually works in the favour of safety, as $K_C > K_{IC}$.

If we make the simplifying assumption that the pressure neither rises nor falls during the leak event, break will not ensue providing

$$\left(\frac{K_I^B}{K_I^A}\right) < \left(\frac{K_C}{K_{IC}}\right) \tag{5.23}$$

This may be used to establish a critical ratio for (c/a), i.e. the crack aspect ratio. We see that the longer the crack is in the longitudinal direction the more prone it will be to cause 'break'. Similarly, a series of small cracks all lying on the same plane (perhaps a weld line) may link up and produce break.

5.9   FATIGUE IN NON-METALLIC MATERIALS

The mechanisms of fatigue detailed earlier in this chapter are essentially plastic, i.e. they involve local dislocation flow because of the presence of stress raisers. In polymeric materials, yield or a phenomenon similar to yield in metals occurs and, as in metals, slow crack growth occurs under cyclic loading. Experiments have shown that a relation identical to the Paris equation (5.10) describes crack growth in polymers. However, there is also a threshold below which no crack growth occurs. In common with metals, the application of a constant stress around which the cyclic loading takes place influences the fatigue life of a polymer, but there is no 'universal' relation similar to the Goodman relation (Eqns 5.1 and 5.2) which is applicable to polymers. Indeed, some polymers show an *increase* in life if they are statically loaded. This is thought to be caused by the static load blunting the crack tip.

In ceramics and other new brittle engineering materials, e.g. intermetallics, the Peierl's stress for dislocation motion is very high (see Chapter 7). The plastic zone at the tip of the crack is extremely small and virtually no dislocation otion occurs in uncracked specimens under cyclic loading. Therefore, for most practical engineering purposes, conventional fatigue can be ignored in these materials. However, strength degradation can occur under certain conditions. In Chapter 7 it will be shown that very high stresses can be generated on ceramic surfaces which lead to the nucleation of cracks. Repeated loading of ceramics can then lead to crack formation at the point of loading followed by a decrease in strength that can eventually result in failure. There are no fixed rules that the designer can apply to deal with this problem, which is inherently random in nature. Statistical techniques applicable to ceramic materials are disucssed in detail in Chapter 7.

Composite materials are not immune to fatigue, although because they are normally reinforced with ceramic fibres, the reinforcement is, at least, expected to be highly fatigue-resistant. Composites are discussed in detail in Chapter 8 and here the discussion will be confined to the phenomenon of fatigue in fibre-reinforced polymers. When a composite is subjected to a cyclic stress in the direction of its fibres, cracks grow in the matrix at a reduced rate compared with an unreinforced polymer because a large proportion of the load is taken by the fibres and the crack-tip stress intensity is reduced (the fibres are said to 'shield' the crack). In many cases these cracks can grow completely across the composite without total fracture, with the fibres now taking the load. Such fatigued composites are still intact but have a much reduced elastic stiffness and may be regarded as totally inadequate in many applications. If the composite is loaded normal to the direction of the fibres, cracks can grow without hindrance across the matrix and it will behave in this orientation much like an unreinforced polymer. It is common for composites to have fibres aligned in many different directions and, thus, parts of a component made of a composite may show rapid crack growth in fatigue while other points will remain intact. Catastrophic cracking may be prevented by selecting the fibre orientation, although a reduction of stiffness will still be experienced. The development of design techniques to predict this reduction is an area of current interest and readers are referred to the references at the end of Chapter 8 for further information.

EXAMPLES

5.1   A clevis supporting a lift car is fabricated from structural steel to BS 4360. The nominal design stress is 155 N mm^{-2} and this is attained whenever the lift accelerates from a floor fully loaded. If the smallest crack which can be detected by periodic NDT inspection is 5 mm, determine what the minimum inspection interval should be. The weld steel may be assumed to have the following properties:

Fracture toughness: K_{IC} = 66 MPa m$^{1/2}$
Yield stress: σ_Y = 250 MPa

Paris equation: $\dfrac{da}{dN} = 5.01 \times 10^{-12} (\Delta K)^{3.1}$

where da/dN is in m cycle^{-1} and ΔK in MPa m$^{1/2}$.

Assume that the most probable type of crack which may initiate is an edge type with stress intensity factor $K_I = 1.12 \, \sigma\sqrt{\pi a}$, and the lift makes 400 movements per day.

If it is desired to extend the inspection interval to one year, what sensitivity is required of the NDT technique?

5.2 (a) Sketch the form of the da/dN versus ΔK_I relationship found for most metals, labelling the threshold and other salient features, and explain its relationship to the Paris fatigue law.

(b) This part-question is concerned with the design of a spherical pressure vessel, to be used as an air accumulator. The pressure range is to be maintained in the range 50 N mm^{-2} to 150 N mm^{-2}, the radius of the vessel is 1.2 m, and the material from which it is made has the following properties:

$E \quad = 205 \times 10^9$ N m^{-2}
$\nu \quad = 0.31$
$\sigma_Y \quad = 560 \times 10^6$ N mm^{-2}
$K_{IC} = 94 \times 10^6$ N m$^{-3/2}$

$\dfrac{da}{aN} = 5.1 \times 10^{-12} (\Delta K)^{3.2}$ m cycle^{-1}

where ΔK has units of MN m$^{-3/2}$

Carry out the following steps:

(i) Assuming a factor of safety of 1.3, determine a suitable wall thickness, using classical bulk elastic design.

(ii) Determine the crack length at which a crack will spontaneously propagate through-thickness.

(iii) If the minimum crack which can be detected by NDT is one 2 mm deep, and the pressure vessel goes through a cycle once every five minutes, what would be a suitable inspection interval?

(iv) Conversely, if the inspection interval were to be every two months, what would be the minimum acceptable sensitivity of the test, i.e. what should be the smallest detectable crack?

5.3 Define the design concept 'leak-before-break', applied to pressure vessels and pipelines. Explain the relationship between the initial aspect ratios of a crack, and the required characteristics of the NDT techniques.

Figure 5.8 shows a length of pipeline, which may be assumed to be thin-walled, containing a semi-elliptical crack.

Stress intensity factors for growth in the r and z directions are, respectively,

$$K_1 = \frac{pR}{b} \, \sigma_{\theta\theta} \, 1.12\sqrt{\pi a}$$

$$K_1 = \frac{pR}{b} \sqrt{\pi c} \left(1 + 1.61 \, \frac{c^2}{Rb} \right)^{1/2}$$

where in the second equation it is assumed that the crack has grown through-thickness, i.e. $a \to b$.

In a test with cyclically varying pressure, post-failure sectioning showed that through-thickness failure has occurred when $a = 0.5b$, for a pressure vessel with $R = 10b$. Find the maximum value of c to avoid longitudinal failure at this pressure.

Under what conditions may the pressure available to propel axial cracks be (a) less than and (b) greater than, that producing through-thickness failure?

5.4 One type of specimen for measuring fracture toughness is the double cantilever beam specimen, shown in Fig. 5.9.

The calibration for the stress intensity factor under plane strain condition is

$$K_1 = \frac{2\sqrt{3}}{1 - \nu^2} \, \frac{Pa}{b^{3/2}}$$

where ν, Poisson's ratio, is 0.3. The Paris law for the material concerned is

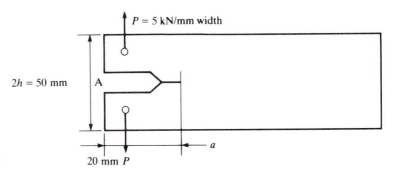

Figure 5.9

$2h = 50$ mm

A

$P = 5$ kN/mm width

20 mm P

a

$$\frac{da}{dN} = 5.01 \times 10^{-12} \, (\Delta K_\mathrm{I})^{3.1} \text{ m cycle}^{-1}$$

where ΔK has units of MPa m$^{1/2}$, and its fracture toughness is $K_{\mathrm{IC}} = 132$ MPa m$^{1/2}$.

(a) Find the crack length at which fracture will occur at the load indicated.

(b) Find the number of cycles to failure if the initial crack length is 0.1 mm and P is cycled to the same value.

(c) Suppose that loading is performed by driving a wedge into the crack instead of applying P. If $a = 20$ mm and the specimen is made from steel, estimate the separation needed at point A to produce failure.

REFERENCES

1 VENKATARAN G, CHANG Y-W, NAKASONE Y, MURA T (1990) Free energy formulation of fatigue crack initiation along persistent slip bands: calculation of $S-N$ curves and crack depths. *Acta Metall Mater* **38**: 31−40

2 POOK L P (1978) Analysis and application of fatigue crack growth data. In *A General Introduction to Fracture Mechanics, MEP*, HMSO, London

Time-Dependent Behaviour

6.1 INTRODUCTION

The best-known form of time-dependent deformation and ultimate fracture of materials under stress is creep, already mentioned in Chapter 1. In general, the state of strain at a given instant depends not only on the stress at that instant, but also on how long the stress has been maintained on the one hand and on the rate of loading or of straining on the other. Time must therefore be considered as another factor, together with temperature, influencing the response of a material to a given load as represented by the stress—strain curve. In structures maintained under load for a long time and exposed to a sufficiently high temperature, *creep* may arise as a potential source of failure. At the other extreme, a sudden *shock loading* results in stress waves within the structure and the material response depends on the straining rate. Ideally, a constitutive equation relating all four variables — time, temperature, stress and strain — could be constructed and used for design, but the practical difficulties are formidable, given the vast variety of materials in common use and the wide range of values to be covered for each variable. Indeed, creep and shock loading are entirely different problems that have no real physical link. It is convenient to combine strain and time to obtain expressions relating stress and strain to temperature and strain rate. Indeed, the strain rate characterizes several regimes of interest, as shown in Table 6.1.

Each regime is of interest in particular fields. The very low strain rates characteristic of high-temperature creep apply to boilers, pressure vessels and high-temperature pipework as well as to steam and gas turbines whilst, at the other end of the spectrum, high strain rates are associated with high-speed metal shaping and ballistic impact. To a greater or lesser extent, all materials are strain-rate sensitive. Since the

Table 6.1 Regimes associated with different strain rates

Strain rate(s^{-1})	Type of loading	Test condition	Inertial forces	Thermal regime
$10^{-8}-10^{-4}$	Constant	Creep	No	Isothermal
$10^{-4}-10^{-1}$	Slow change	Tensile compressive	No	Isothermal
$10^{-1}-10$	High frequency	Fatigue (resonance)	Yes	Isothermal
$10^{4}-10^{6}$	Hypervelocity impact	Shock wave	Yes	Adiabatic

vast majority of structural elements operate at relatively low temperature under static or quasi-static loading, strain-rate sensitivity may be neglected and the design may be based on a straightforward — although not always simple — elastic stress analysis. A linear stress–strain relation obtained from a conventional strain- or load-controlled uniaxial tensile test at a strain rate of around $10^{-1}\,s^{-1}$ and at the service temperature provides all the information that is needed. At higher service temperatures, creep may acquire some importance and tests under constant load or extension have to be specified. Equally, in components subjected to shock-loading and hence to strain rates well in excess of those reached in the tensile test, the stress analyst must know how this affects the mechanical properties of the material.

Empirical constitutive equations — albeit with some theoretical justification — have been proposed for specific materials under a given strain rate. Their application to the stress analysis of any but the simplest shapes requires advanced numerical techniques, forming a whole area of research.

6.2   CREEP OF METALS

The typical shape of the creep curve is shown in Fig. 1.17. Creep occurs at any temperature provided that the stress level is sufficiently high, but it is steady loading at temperatures above one-half of the melting point which is industrially important, and which we will now discuss.

When a metal is subjected to a stress slightly higher than σ_Y, it deforms rapidly until work-hardening increases the yield point to the level of the applied stress. In contrast, below yield, dislocations remain locked and the strain remains constant. The presence of flaws and small notches may raise the stress above the yield point over a small region but will not affect the behaviour of ductile materials since blunting of the crack tip or notch radius will relieve the stress and stop the deformation. The same will not be true in brittle materials, where progressive growth takes place. Limiting our discussion to ductile materials, the strain rate will be extremely high immediately upon loading and it will rapidly decrease as the amount of work-hardening increases. A simple equation that models this observed behaviour is:

$$\dot{\epsilon} = \alpha \, t^{-1} \tag{6.1}$$

where α is a function of temperature and of $(\sigma - \sigma_Y)$, the initial plastic component of the applied stress, and t is the time. This equation may also be expressed in the form

$$\epsilon = \epsilon_0 + \alpha \ln t \tag{6.2}$$

The balance between the driving and resisting forces on the dislocations is easily altered by thermal activation. At low temperatures the applied stress has to remain close to the self-stresses due to the dislocations themselves, but as the temperature increases, thermal softening occurs as a result of dislocation climb. This is a process relatively insensitive to the applied stress level, and causes the forces resisting dislocation glide to relax at the same rate as they tend to build up as a result of work-hardening. Assuming a configuration in which the applied force is developed by a 'dead load', the nominal stress remains constant, and hence

$$d\sigma = \frac{\partial \sigma}{\partial \epsilon} \, d\epsilon + \frac{\partial \sigma}{\partial t} \, dt = 0 \tag{6.3}$$

where $\partial \sigma / \partial \epsilon = h$ is the work-hardening coefficient (tangent modulus) and $-\partial \sigma / \partial t = r$ is the coefficient of thermal softening. From Eqn (6.3) it follows that, during this second stage,

$$\dot{\epsilon} = \frac{r}{h} \tag{6.4}$$

As the extensional deformation increases, the cross-sectional area decreases. Since most tests are conducted under constant load, the resulting gradual increase in the stress is responsible for the accelerating strain rate observed in the final stage of creep which leads, eventually, to rupture. This is not, however, the only mechanism that explains tertiary creep, nor is it the most important. Acceleration is mainly due to metallurgical softening, to the migration of inclusions towards the intercrystalline bonds where they form brittle components or, in general, to the formation of small intercrystalline cracks. To explain this type of rupture, it is customary to define a creep damage factor q, which can be visualized as a measure of the increase in the effective stress due to the growth of voids,

$$\sigma_{\text{eff}} = \frac{\sigma}{1-q} \tag{6.5}$$

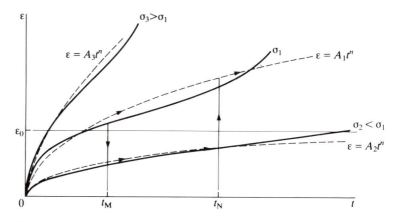

Figure 6.1 Typical creep curves and approximate constitutive equations

and at rupture, $q = 1$. The damage factor at a given temperature grows with time and with the effective stress,

$$\dot{q} = C \left(\frac{\sigma_{\text{eff}}}{1-q} \right)^n \tag{6.6}$$

The three equations, (6.2), (6.4) and (6.6), define the three stages of creep. In these equations, the values of α, r and C depend strongly on the temperature. For design, a single equation incorporating all the variables is clearly needed, even if its use results in a loss in the power of the equation to describe accurately each stage of the process. The simplest equation is an exponential (Fig. 6.1),

$$\epsilon = A\,t^n \quad n < 1 \tag{6.7}$$

A is a function of the temperature and of the stress:

$$A = B \exp \left(\frac{-Q}{RT} \right) \sigma^m = \zeta\,(T)\,\sigma^m \tag{6.8}$$

where Q is the activation energy for self-diffusion and R is the universal gas constant; m is found to be approximately equal to 4. Taking the derivative with respect to time in Eqn (6.7) and eliminating the time between the strain and the strain rate, we obtain the following expression:

$$\sigma = \left[\frac{1}{\zeta(T)} \left(\frac{\dot{\epsilon}}{n} \right)^n \right]^{1/m} \epsilon^{(1-n)/m} \tag{6.9}$$

giving the stress—strain relationship for a given strain rate; $(1-n)/m$ is the strain-hardening exponent. In many polycrystalline materials at high temperatures, n is approximately equal to $\frac{1}{3}$, giving a strain-hardening exponent of about $\frac{1}{6}$. At the early stages of creep and low temperatures, n is almost zero, giving a strain-hardening exponent of about $\frac{1}{4}$. The $n = 0$ case corresponds to the initial stage (Eqn 6.2), as can be seen by comparing Eqn (6.1) with the expression of the strain rate obtained by taking the derivative of ϵ from Eqn (6.7) and setting $(An) = \alpha$ and $n = 0$:

$$\dot{\epsilon} = (An)\, t^{n-1} = \alpha\, t^{-1}$$

We may then obtain the logarithmic expression, Eqn (6.2).

Many other equations have been proposed,[1-3] amongst them Wilshire and Evans',

$$\epsilon = \theta_1[1 - \exp(-\theta_2 t)] + \theta_3[\exp(\theta_4 t) - 1] \qquad (6.10)$$

where θ_1, θ_2, θ_3 and θ_4 are stress- and temperature-dependent coefficients obtained experimentally. The minimum creep rate corresponds to a time,

$$t_{min} = \frac{1}{\theta_2 + \theta_4} \ln \frac{\theta_1 \theta_2^2}{\theta_3 \theta_4^2}$$

6.3   IRRECOVERABLE DEFORMATION AND TIME TO RUPTURE

The major weakness of the proposed equations is that they imply that, should the stress be reduced (for example, from σ_2 to σ_1 at t_M in Fig. 6.1), all the deformation is immediately recovered and, alternatively, that when the stress increases, the strain jumps to a higher value instantly (σ_2 to σ_1 at t_N in Fig. 6.1). An alternative formulation consists in splitting the strain into an elastic, time-independent and fully recoverable component, a plastic component, also time-independent but irrecoverable, and a time-dependent component that can be partly recovered, i.e.

$$\epsilon = \epsilon_0 + \epsilon_p + \epsilon_c$$

where

$$\epsilon_0 = \frac{\sigma}{E}, \quad \epsilon_p = \beta \left(\frac{\sigma}{\sigma_Y}\right) p, \quad \epsilon_c = \zeta(T) \sigma^m t^n$$

Another weakness of Eqn (6.7) and of this improved formulation is that they both fail to provide a means to estimate the time to rupture, which is often the main design criterion. The problem here consists in the extrapolation of short-duration tests to the long service life that is specified by the designers. Typically, a creep test may take of the order of a thousand hours, while the service life may be one or two hundred thousand hours. The extrapolation follows semi-empirical rules based on the linear representation of the relationship between the test variables by means of suitable changes in coordinates and the definition of auxiliary functions. The most widely used parametric methods consist in plotting the stress in ordinates against a variable f that depends, in turn, on the time to rupture t_r and the temperature T. Larson and Miller have proposed:

$$f = T (\log t_r + L)$$

where L is an empirical constant.[4] Manson and Haferd proposed a different variable,

$$f = \frac{T_0 - T}{\log t_r - \log t_{r,0}}$$

where subscript 0 denotes a reference set of data. More simply, ($\log \sigma$) may be plotted against ($\log t_r$) for a given temperature and ($1/T$) against ($\log t_r$) for a given stress. Extrapolation is then possible, but may not be reliable!

6.4   CREEP-RESISTANT ALLOYS

A high melting point is the first feature of creep-resistant metals and alloys. This must be coupled with high-temperature oxidation resistance, low-temperature toughness, machinability, weldability, etc. The prevention of dislocation glide by setting obstacles stable at high temperatures is essential to the development of the so-called refractory alloys. Face-centred cubic (FCC) metals are preferred to body-centred cubic (BCC) metals because the extended dislocations that exist in the former find it more difficult to overcome the obstacles present in the structure by cutting through, cross slip or climbing. Similarly, structures with long-range orders are preferable. The most obvious and effective way to block the dislocations is by dispersed particles that form a stable dispersion-hardened structure or provide a work-hardened state through strain-ageing. The dispersion can be achieved by precipitation during the manufacturing process or even in service. In the design of creep-resistant alloys, a compromise must be reached between the conflicting requirements of good solubility at high temperature for manufacture, and long-term stability in service. Most of the alloys in common use are nickel-based, with cobalt, tungsten and chromium in solid solution and stable, hard precipitates such as Ni_3Al, Ni_3Ti and MoC. Protection against oxidation is provided by chromium, which forms a tough adherent skin of Cr_2O_3. In steel, chromium and nickel are the main alloy elements. Carbides form to provide stable obstacles to the dislocations.

Grain size is an important factor. A large grain size is essential to defeat creep deformation and prevent grain boundary sliding. Whenever possible, the microstructure should be such as to ensure that the grain boundaries are parallel to the principal tensile stress. This is achieved in castings, e.g. turbine blades, by directional solidification. Single-crystal turbine blades with excellent creep resistance are now widely used.

6.5   POLYMERS

The structure of composites consists of long molecules with a backbone of carbon atoms linked by covalent bonds. In non-crystalline or amorphous polymers the molecular chains have an entirely random orientation and are cross-linked occasionally by a few strong covalent bonds and numerous but weaker van der Waals bonds. These weaker bonds break as the temperature reaches a value known as the glass transition temperature, characteristic for each polymer. Below the glass transition temperature, T_g, the polymer behaves as a linear elastic solid. Creep becomes increasingly significant as the temperature increases and, above T_g, the polymer deforms in a viscous manner under load. Typical values of T_g are given in Chapter 2 (Table 2.1). In crystalline polymers the molecules are oriented along preferred directions, bringing with them optical and mechanical anisotropy. It is quite normal for a polymer to consist of a mix of amorphous and crystalline regions. Figure 6.2 describes these various structures. Polymers are divided into *thermoplastics* and *thermosets*.

In thermoplastics, particularly in those with low T_g, there are very few covalent bonds between molecular chains. It is then relatively easy to cause the molecules to follow the direction of the principal tensile stress, even at very low values of that stress. Once orientated, they deform first by straightening up and then, as the stress increases, by stretching. The mechanical behaviour that results from these various modes of deformation combined with the progressive breaking of the cross-links — covalent and van der Waals — and of the primary links within the molecular backbone is extremely complex. It cannot be described by the simpler constitutive equations that have been proposed for the metals.

Thermosets differ from thermoplastics in that there are more covalent cross-links. Indeed, the glassy transition may be preceded by the breakdown of the whole structure through oxidation, combustion or sublimation of some form or another. In a typical epoxide, for example, the monomer polymerizes into long chains at temperatures

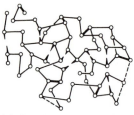

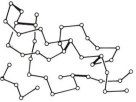

(a) Covalent cross links (—) 6
     Van der Waals links (- -) 8

(b) Covalent cross links (—) 5
     Van der Waals links (- -) 0

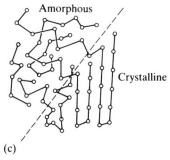

(c)

Figure 6.2 Polymer structures: (a) amorphous polymer below T_g; (b) amorphous polymer above T_g; (c) crystalline/amorphous structure

between room temperature and 200 °C. Once polymerized, the chains are tightly interlocked and the material retains the stiffness well above the polymerization temperature. Above T_g, sufficient covalent cross-linking remains to prevent viscous flow. Indeed, the ability of such thermosets to regenerate the secondary links between molecules and thus to 'freeze' the molecular chains into preferred orientation upon cooling makes them particularly useful as photoelastic materials, since the optical birefringence that they acquire can be related to the state of stress (Fig. 6.3).[4]

6.6 MODELS OF MECHANICAL BEHAVIOUR OF POLYMERS

The time-dependence of the deformation is reflected in Fig. 6.4. Under a constant stress, σ, the strain increases with time from an initial elastic value ϵ_0. Upon removal of the stress, recovery takes place also with time. Mechanical analogues, consisting of linear elastic springs and viscous dampers, have been devised to model this behaviour. The first is the Maxwell model, Fig. 6.5. When a fixed stress σ is applied we can write that

$$\sigma = E_m \epsilon_e = \eta_m \dot{\epsilon}_m \tag{6.11}$$

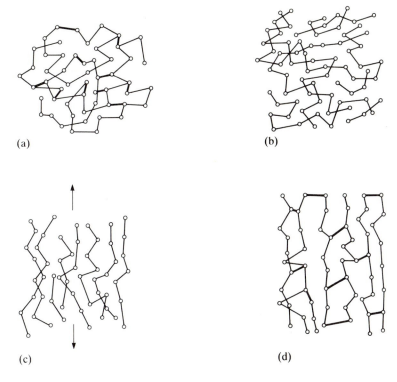

Figure 6.3   Principle of molecular orientation in a photoelastic epoxy: (a) structure at room temperature; (b) structure (unloaded) at 150 °C; (c) structure (loaded) at 150 °C; (d) structure obtained upon cooling and unloading

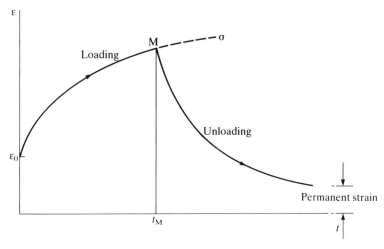

Figure 6.4   Typical strain−time curves obtained upon loading to a stress σ and unloading

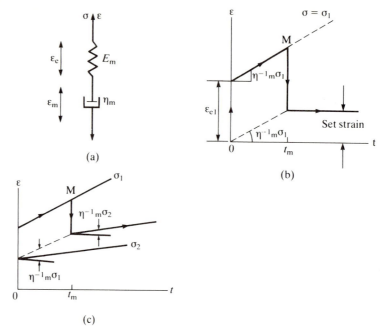

Figure 6.5 Maxwell analogue: (a) model; (b) response to a constant stress, unloading at t_M; (c) response to a change in stress from σ_1 to σ_2

where ϵ_e is the elastic component of the strain and $\dot{\epsilon}_m$ is the viscous strain rate. The total strain is

$$\epsilon = \epsilon_e + \epsilon_m = \frac{\sigma}{E_m} + \int_0^t \frac{\sigma}{\eta_m}\, dt \qquad (6.12)$$

The response is illustrated in Fig. 6.5(b). Upon the sudden application of a stress σ_1 the strain becomes ϵ_{e1}. The strain increases linearly with time while σ_1 is maintained. Upon unloading, ϵ_{e1} is recovered but the viscous strain is not. A reduction from σ_1 to σ_2 has the effect shown in Fig. 6.5(c).

In the Voigt model the spring and the damper are in parallel, as in Fig. 6.6(a). We can then write

$$\sigma = E_v \epsilon + \eta_v \dot{\epsilon} \qquad (6.13)$$

At constant stress this equation has the following solution:

$$\epsilon = \frac{\sigma}{E_v} + A \exp\left(-\frac{E_v}{\eta_v} t\right) \qquad (6.14)$$

where A is an integration constant. Taking, for $t = 0$, $\epsilon = 0$, gives

$$\epsilon = \frac{\sigma}{E_v}\left[1 - \exp\left(-\frac{E_v}{\eta_v} t\right)\right] \qquad (6.15)$$

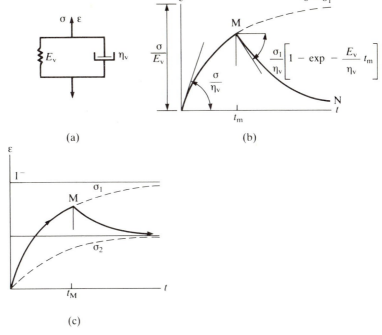

Figure 6.6 Voigt analogue: (a) model; (b) response to a constant stress, unloading at
t_M; (c) response to a change in stress from σ_1 to σ_2

represented in Fig. 6.6(b). Unloading completely at M results eventually
in total recovery following the curve MN, given by the equation

$$\epsilon = \frac{\sigma_1}{E_v}\left[1 - \exp\left(-\frac{E_v}{\eta_v}t_m\right)\right]\exp\left[-\frac{E_v}{\eta_v}(t - t_m)\right]$$

(6.16)

Maintaining σ_1 would eventually result in a strain (σ/E_v), comparable
with the elastic component in the Maxwell model. A reduction to σ_2
is represented by

$$\epsilon = \frac{\sigma}{E_v}\left[1 - \exp\left(\frac{E_v}{\eta_v}t_m\right)\right] - \frac{\sigma_2}{E}\left\{1 - \exp\left[-\frac{E_v}{\eta_v}(t - t_m)\right]\right\}$$

(6.17)

and shown in Fig. 6.6(c).

It is evident that neither of these two elementary models is an
accurate analogue of the observed behaviour. The initial elastic strain
and the permanent set strain are properly modelled by the Maxwell
model, whilst the gradual change in strain rate together with more
realistic recovery pattern are reproduced by the Voigt model. It
therefore follows that the two could be combined, as shown in Fig.
6.7, to provide an accurate analogue. For this Maxwell–Voigt model,

189

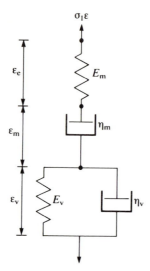

Figure 6.7   Maxwell–Voigt model

$$\sigma = E_m \, \epsilon_e = \eta_m \, \dot{\epsilon}_m = E_v \, \epsilon_v + \eta_v \, \dot{\epsilon}_v \tag{6.18}$$

$$\epsilon = \epsilon_e + \epsilon_m + \epsilon_v \quad \dot{\epsilon} = \dot{\epsilon}_e + \dot{\epsilon}_m + \dot{\epsilon}_v \tag{6.19}$$

The three components of the total strain can be eliminated between these equations to give

$$E_v \epsilon + \eta_v \dot{\epsilon} = \sigma \left(\frac{E_v}{E_m} + \frac{\eta_v}{\eta_m} + 1 \right) + \frac{E_v}{\eta_m} \int \sigma \, \mathrm{d}t + \frac{\eta_v}{E_m} \, \dot{\sigma} \tag{6.20}$$

This is a complicated equation that may be simplified without any great loss by assuming that $\eta_m = \infty$, i.e. that there is no Maxwell damping,

$$E_v \, \epsilon + \eta_v \, \dot{\epsilon} = \sigma \left(1 + \frac{E_v}{E_m} \right) + \dot{\sigma} \, \frac{\eta_v}{E_m} \tag{6.21}$$

Just after a stress σ has been applied, the strain is σ/E_m. As time passes, the strain increases following the same pattern of Fig. 6.6. All the deformation is eventually recovered when the material is unloaded.

Polymers and rubbers are used to damp out mechanical vibrations. There are innumerable examples to be found in machines and structures: vibration isolation pads, wheels and tyres, shaft couplings, belts, etc. In these cases it is important to find the response of the material to a harmonic excitation of frequency ω,

$$\sigma = \sigma_0 \exp i\omega t \tag{6.22}$$

The strain will be given by an equation of the form

$$\epsilon = \epsilon_0 \exp i(\omega t + \phi) \tag{6.23}$$

where ϕ is a phase shift that can be obtained, together with the strain

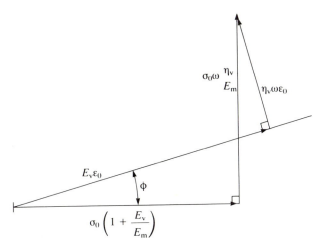

Figure 6.8   Phasor diagram applied to the determination of the vibration response of a Maxwell−Voigt material

amplitude, by solving Eqn (6.21). The simplest method consists in using the phasor diagram,[5] Fig. 6.8, from which we obtain

$$\epsilon_0 = \sigma_0 \frac{E_m + E_v}{E_m E_v} \sqrt{\frac{1 + \left(\dfrac{\eta_v \omega}{E_m + E_v}\right)^2}{\left(\dfrac{\eta_v \omega}{E_v}\right)^2}}$$

$$\tan \phi = \frac{\eta_v E_m \omega}{(E_m + E_v)E_v + \eta_v^2 \omega^2}$$

The energy dissipated per cycle in a cube of unit volume is obtained by integration of the product (σdE):

$$U = \pi \omega^2 \sigma_0 \epsilon_0 \sin \phi$$

This is the fundamental equation when designing an elastomeric suspension with damping. It provides the information needed to calculate the size of the suspension block for a given set of material properties.

6.7   STRESS RELAXATION

So far it has been assumed implicitly that the load was controlled in the test or during the service life of the component. There are cases where the controlling variable is, in fact, the strain, for example, in bolts pretightened to hold together two flanges, prestressing tendons in a concrete beam, etc. The effect of creep will be to relieve the stress in bolt or tendon, thus reducing the contact pressure between the flanges or the state of compression in the beam. The relaxation by creep of the residual stresses that are found after welding, machining or forming of metals has a beneficial rather than harmful influence.

Referring to Fig. 6.1, stress relaxation at a constant strain ϵ_0 is illustrated by drawing a line parallel to the time axis. The stress drops very rapidly at first, slowing down as time increases. Eventually, it may reach a stationary, elastic value proportional to the strain. Equation (6.9) is not applicable since the strain is constant and the strain rate is zero. Combining Eqns (6.7) and (6.8), we obtain

$$\sigma = \left[\frac{\epsilon}{\zeta(T)} \right]^{1/m} t^{-n/m} \tag{6.24}$$

In the Maxwell model, Eqn (6.12) gives

$$\sigma = \epsilon E_m \exp \left(- \frac{E_m}{\eta_m} t \right) \tag{6.25}$$

whilst in the Voigt model only an elastic, time-independent component appears. In the combined Maxwell–Voigt model of Eqn (6.21), the stress relaxation is given by

$$\sigma = \frac{E_m}{E_m + E_v} \epsilon \left[E_v + E_m \exp \left(- \frac{E_v + E_m}{\eta_v} t \right) \right] \tag{6.26}$$

The initial stress is $E_m\epsilon$, relaxing eventually to a minimum value equal to $E_m\epsilon \times (E_v/(E_m + E_v))$ without ever reaching zero. Equation (6.26) can be expressed in a slightly different form to highlight the relative amount of stress relaxation:

$$\frac{\Delta\sigma}{\sigma_{\text{initial}}} = \frac{\sigma_{\text{initial}} - \sigma_{\text{residual}}}{\sigma_{\text{initial}}}$$

$$= \frac{E_m}{E_m + E_v} \left\{ 1 - \exp \left(- \frac{E_v + E_m}{\eta_v} t \right) \right\} \tag{6.27}$$

This expression helps us to understand the mechanism of stress relaxation in, for example, a bolted joint (Fig. 6.9). The total cross-sectional area of the bolts, A_b, will be only a small fraction of the contact area A_c. The initial bolt stress σ_i, which may approach the yield point, results in a contact pressure $\sigma_c = -\sigma_i A_b/A_c$. With time, the rate of stress relaxation, from Eqn (6.27), is

$$(\Delta\dot{\sigma})_{\text{bolts}} = \sigma_i \frac{E_m}{\eta_v} \exp \left(- \frac{E_v + E_m}{\eta_v} t \right)$$

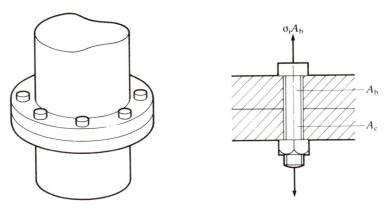

Figure 6.9   Stress relaxation in a bolted joint

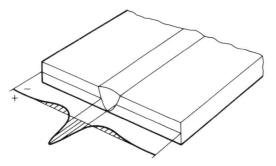

Figure 6.10   Residual stress distribution in a welded joint between two plates

for the bolts and

$$(\Delta\dot{\sigma})_{\text{joint}} = (\Delta\dot{\sigma})_{\text{bolts}} \frac{A_b}{A_c}$$

for the joint. The bolts are thus seen to relax much faster than the two halves of the joint. A similar situation is found in a welded seam between two plates (Fig. 6.10). Upon cooling, the weld run tends to shrink but, being restrained by the two cooler plates, it becomes subjected to a tensile residual stress of the order of the yield point while a low compressive residual stress distributed over a wide zone in the plate appears. The system of tensile and compressive residual stress is self-balanced. A stress-relieving treatment consists in heating the assembly to a temperature sufficiently high to allow creep stress relaxation to occur but not so high as to cause a metallurgical transformation. Typically, for a mild steel this is around 500 °C.

Stress relaxation plays an important role in limiting the design life of bolted joints and prestressed concrete. On the other hand, it is an essential step in the fabrication of welded structures such as pressure vessels.

193

6.8 DESIGN STRESSES IN THE PRESENCE OF CREEP

Two criteria are normally specified when designing against creep, one of rupture and one of deformation. The design stress is taken to be between 0.6 and 1.0 times the minimum or average stress for rupture after 100 000 hours at the design temperature. As an additional check, the design stress may be required to remain below the stress for rupture after 100 000 hours at a temperature 15 °C above design temperature. The reason for this safeguard is that even such a small rise in temperature, well within the likely experimental errors and service control limits, can cause a significant drop in the creep strength. In accordance with the deformation criterion, the design stress is maintained below the stress required to produce a creep strain of 1% after 100 000 hours or a minimum creep strain rate of 10^{-7} h^{-1}. In general, the limitation of the nominal stress is sufficient to ensure satisfactory designs. Most failures are due to unforeseen metallurgical factors or to excessively high temperatures or stresses. Designs are usually based on an assumed elastic behaviour which overestimates the time to rupture and the creep strains, as can be shown by a simple example.

Consider the three-bar system of Fig. 6.11. Due to the difference in length between the central and the outer bars,

$$\epsilon_I = K \epsilon_{II} \qquad \sigma_I q + \sigma_{II} = P \qquad (6.28)$$

For an elastic system,

$$\sigma = E\epsilon$$

and hence we obtain

$$(\sigma_I)_e = P \frac{K}{1 + qK} \qquad (\sigma_{II}) = \frac{1}{K}(\sigma_I)_e \qquad (6.29)$$

where e refers to the elastic regime.

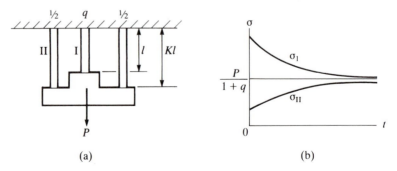

(a)   (b)

Figure 6.11   Three-bar system subjected to creep: (a) system; (b) stress history

On the other hand, in the presence of creep with P held constant, we can write

$$\dot{\epsilon}_I = K\dot{\epsilon}_{II} \qquad \dot{\sigma}_I q + \dot{\sigma}_{II} = 0 \qquad\qquad\qquad (6.30)$$

For a Maxwell–Voigt material it can be shown by solving Eqn (6.21) with the preceding equilibrium and compatibility conditions that

$$\sigma_I = \frac{P}{1+q}\left[1 + \frac{K-1}{1+qK}\exp\left(-\frac{E_v + E_m}{\eta_v}t\right)\right]$$

$$\sigma_{II} = \frac{P}{1+q}\left[1 - \frac{(K-1)q}{1-qK}\exp\left(-\frac{E_v + E_m}{\eta_v}t\right)\right]$$

$$(6.31)$$

On application of the load, $t = 0$ and σ_I and σ_{II} have respectively the initial elastic values of Eqn (6.29). As $t \to \infty$, the exponential term decays, and eventually σ_I and σ_{II} tend asymptotically to the common value $P/(1+q)$ equal to the average or primary stress, as explained in Chapter 1. The stress concentration factor K is therefore seen to decrease with time due to creep deformation in which the shorter bar will deform more rapidly than the longer bars. In a material with limited ductility, fracture may well arrest the whole process before the stresses have had time to even out. Nevertheless, it is clear that an elastic analysis will overestimate the strain rate as much as it overestimates the true maximum stress.

6.9   MATERIAL BEHAVIOUR AT STRAIN RATES ABOVE 10 s^{-1}

The effect of the strain rate on the mechanical properties of materials is of some importance in the assessment of the integrity of structures subjected to shock loading, or high-speed forming and cutting operations, etc. As a crude approximation, the strain rate is equal to the impact velocity divided by the size of the impact zone. Thus, in the collision between two ships sailing at 10–20 knots, the strain rate will be of the order of 10 s^{-1}, in the case of a head-on collision between two cars travelling at 50 miles h^{-1}, it will increase to 50 s^{-1}, whilst a bird hitting the leading edge of a fan blade in an aero-engine might result in a strain rate of 1000 s^{-1}. Even higher strain rates are possible. For example, in metal cutting (Fig. 6.12), the peripheral speed may be 50 m s^{-1}, the shear band spreads over a depth of 0.5 mm and the shear strain rate is therefore 10^5 s^{-1}. Typically, the effect of strain rate is to increase the yield point, as seen in Fig. 6.13. On the other hand, as the strain rate increases, there is tendency for the temperature in the process zone to rise, the thermodynamic process changing from isothermal to adiabatic, and for thermal softening to counteract the

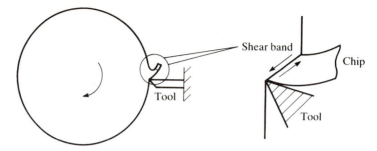

Figure 6.12   Schematic of metal cutting in a lathe

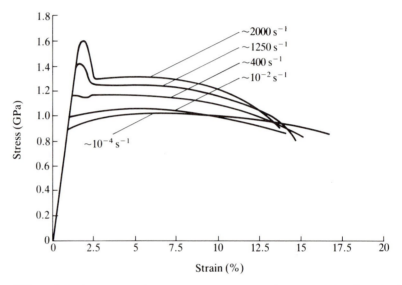

Figure 6.13   Effect of strain rate on the stress–strain curve of annealed Ti–6% Al–4% V alloy[6]

strain rate hardening. Many semi-empirical equations have been proposed to model the strain rate sensitivity of the yield point of metals and the rate-dependence detected in the tensile test after the onset of yielding. One of the first (Ludwik, 1909) was

$$\sigma = \sigma_1\,(\epsilon)\; +\; A(\epsilon)\,\ln\left(\frac{\dot{\epsilon}_p}{\dot{\epsilon}_{1p}}\right)$$

where $\epsilon_p = \epsilon - \sigma/E$ and the subscript 1 denotes a reference state. Two other classical equations are:

$$\dot{\epsilon} = C\,\sinh\,\alpha\sigma \qquad\qquad\text{(Prandtl, 1928)}$$

$$\sigma = \sigma_1\,\dot{\epsilon}^n \qquad\qquad\text{(Sokolov, 1946)}$$

In these equations, A, C, α and n are determined experimentally. A more general expression (Perzyna, 1966) is

$$\dot{\epsilon}_p = \Upsilon \left[\phi \left(F \right) \right]$$

where $F = (\sigma/\sigma_Y - 1)$ and the function ϕ takes different possible forms:

$$\phi [F] = F^\partial \qquad\qquad\qquad \text{(Symonds and Cowper, 1957)}$$

$$\phi [F] = \exp (F) - 1$$

$$\phi [F] = \Sigma A_n [\exp (F^n) - 1]$$

$$\phi [F] = \Sigma B_n F^n$$

These expressions only give the strain rate when a stress exceeding yield is applied to the material. The Lindholm–Johnson constitutive equation provides a relationship between stress and strain,

$$\sigma = [A + B\epsilon^n] \left[1 + C \ln \left(\frac{\dot{\epsilon}}{\dot{\epsilon}_0} \right) \right] \theta(T) \qquad (6.32)$$

This equation combines a static term (similar to the one in the Ramberg–Osgood model), a strain-rate term, referred to a strain rate $\dot{\epsilon}_0$ that may be taken as 1 s^{-1}, and a temperature term, $\theta(T)$, that should be 1 at room temperature (T_0) and 0 at melting point (T_m). A simple expression that satisfies both extremes is

$$\theta(T) = \frac{T_m - T}{T_m - T_0} \qquad (6.33)$$

Some typical values of the constants are given in Table 6.2.

In any mechanical test, a certain proportion of the external work is converted into heat. For a material of density ρ and specific heat c, the temperature rise is

$$\Delta T = \frac{\beta}{\rho c} \int_0^\epsilon \sigma \, d\epsilon \qquad (6.34)$$

where β is a proportionality constant. The integral is the strain energy

Table 6.2 Typical values of the parameters in the Lindholm–Johnson equation (after [7])

	A (MPa)	B (MPa)	n	C
Cu	69	106	0.32	0.027
Al	193	157	0.41	0.010
Fe	76	196	0.25	0.028
Tool steel	883	248	0.18	0.012

Table 6.3 Typical values of the parameters in Eqn (6.36)

	σ_0 (MPa)	σ_u (MPa)	λ_Y (MPa)	λ_u (MPa)
Mild steel	375	400	50	30
Stainless steel	370	713	26	18
Al alloy (Dural)	400	480	4	4

λ_Y refers to yield stress, λ_u to ultimate tensile stress

per unit volume. Consider now a test at uniform strain rate $\dot{\epsilon}$. The maximum stress reached is found by differentiating Eqn (6.32),

$$d\sigma = \left(\frac{\partial \sigma}{\partial \epsilon}\right)_T d\epsilon + \left(\frac{\partial \sigma}{\partial T}\right)_\epsilon dT = 0$$

which, expanded and combined with Eqns (6.33) and (6.34), gives the following condition at instability:

$$\epsilon^{n-1} = D(A + B\epsilon^n)^2 \tag{6.35}$$

where $D = \dfrac{\beta}{\rho C}\left[1 + C\ln\left(\dfrac{\dot{\epsilon}}{\dot{\epsilon}_0}\right)\right]\dfrac{1}{T_m - T_0}$

With ϵ from Eqn (6.35), it is then possible to obtain σ (maximum) for a given temperature from Eqn (6.32).

There are many other forms of constitutive questions that have been proposed. For example, taking the material as elastic–ideally plastic, we can write that

$$\sigma = E\epsilon \text{ for } \sigma < \sigma_Y \text{ or } \sigma = \sigma_Y = \sigma_0 + \lambda_Y \log_{10}\left(\frac{\dot{\epsilon}}{\dot{\epsilon}_0}\right) \tag{6.36}$$

where σ_0 is the yield stress at quasi-static rate $\dot{\epsilon}_0$ (1 s^{-1}) and σ_Y is the yield stress at $\dot{\epsilon}$. A similar expression may be applied to the fracture stress σ_u or to the flow stress σ_F, intermediate between σ_Y and σ_u, replacing λ_Y by λ_u. Typical values are given in Table 6.3.

6.10   STRAIN RATE AND FRACTURE TOUGHNESS

Plastic constraint was seen in Chapter 3 to be instrumental in raising the stress required to produce yielding in an element in a state of plane strain above the yield point detected in the conventional tensile test. The same effect was found when comparing the size of the plastic zones in the plane strain and in the plane stress cases in Chapter 4. The Charpy impact test has been for many years, and indeed still is, the main

(a)   Specimen

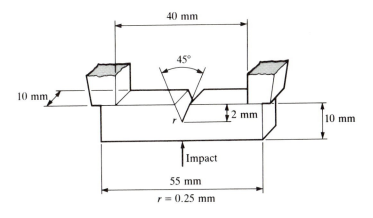

(b)

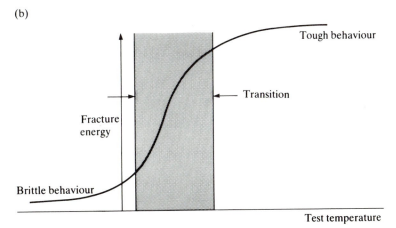

Figure 6.14   Charpy V-notch test

technological test for notch toughness of materials. In the test, a notched beam of standardized dimensions (shown in Fig. 6.14) is fractured under the impact of a heavy pendulum travelling at a velocity of around 5 m s^{-1}. The energy absorbed by the fracture is easily found by measuring the swing of the pendulum before and after the point of impact. Alternatively, a dynamometer fixed to the pendulum provides a force–time plot from which, assuming constant velocity, the energy and the maximum force just prior to fracture can be detected. Even in a normally ductile material such as mild steel, a transition from a high-energy, tough behaviour to a low-energy, brittle behaviour is observed as the test temperature falls below a transition region (Fig. 6.14(b)). Such transition is not found in an unnotched specimen nor is it easily obtained when the beam is loaded under quasi-static conditions. The presence of the notch, the temperature and the

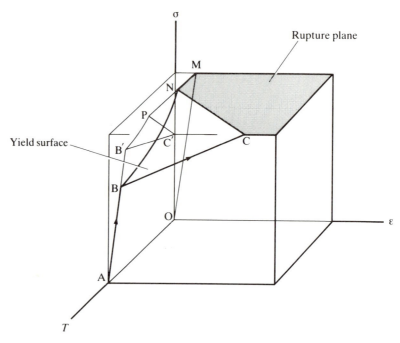

Figure 6.15   Effect of the test temperature on the yielding and fracture of a specimen under tension

straining rate are therefore found to be instrumental in causing the material to fracture in a brittle manner. The process whereby this happens may be visualized by assuming that yielding and fracture are independent and that, whilst the yielding stress varies with the testing temperature, with the state of stress (plane stress or plane strain) and with the straining rate, the fracture stress remains approximately constant. As the temperature drops below the transition region in a notched specimen in a state approaching plane strain and the strain rate is sufficiently high, yielding ceases to precede fracture. This can be seen by plotting stress–strain–temperature diagrams as in Fig. 6.15. At a temperature corresponding to A, the uniaxial stress–strain curve is ABC, from A to B the material is elastic, yielding takes place at B and, from there on, the material deforms plastically until it fractures at C. At lower temperatures the yield point increases and, at a sufficiently low temperature, it is so high that fracture occurs without any plastic deformation at all. This temperature corresponds to point N in the yield surface BNC. The effect of plastic constraint — as found at the root of a notch or in plane strain conditions — is to raise the yield surface to B′P′C′. The same effect is achieved by increasing the strain rate, as evidenced by Fig. 6.13. An even more conclusive argument can be built on the constitutive equation (6.36). From this equation and with the data of Table 6.3, we construct the diagrams

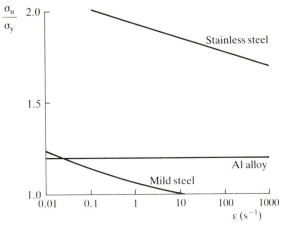

Figure 6.16   Variation of yield point and fracture stress with strain rate from Eqn (6.36)

of Fig. 6.16. In the stainless steel and in the aluminium alloy the yield point is always lower than the fracture stress, but in the mild steel there is a cross-over point at a strain rate of 10 s^{-1}. A mild steel beam, with a notch, can therefore be expected to behave in a brittle manner when tested in a Charpy pendulum, even though its behaviour in a quasi-static test is ductile, while the aluminium alloy and the stainless steel would not be affected by the testing conditions. This strain rate sensitivity of mild steel (ferritic, face-centred structure) has been known for a long time to be among the main reasons for the brittle fracture of large welded structures.

6.11   DESIGN AGAINST CREEP

The preceding discussion has emphasized the effect of stress, temperature and time on the creep of the material. To design against creep it would appear that it is sufficient to maintain the stress levels below those specified by the various design codes, control the service temperature so that it does not exceed the design value and withdraw the structure or component from service once its design life has been reached. In practice, of course, the stresses are calculated assuming linear—elastic behaviour in the belief that this gives an inaccurate but safe assessment of the time-dependent situation; the estimated service life, based on this incorrect analysis and, possibly, inadequate test data, can only be a rough approximation. There is, however, no fundamental reason why this simple design philosophy should not be implemented.

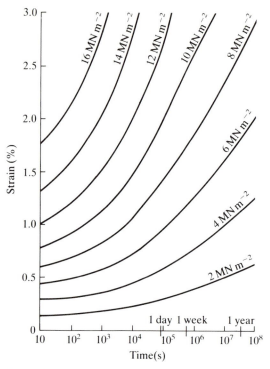

Figure 6.17   Typical creep curves for polypropylene at 20 °C (from ICI Technical Service note PP 110)

Indeed, the introduction of numerical analysis techniques and the widespread use of powerful computers have made it possible to model mathematically complex, non-linear material behaviour; the only limitation arises from the lack of reliable test data rather than from any weakness in the analysis.

In most applications, a detailed study is not warranted. An elastic analysis, in which Young's modulus is replaced by a time-dependent creep modulus, is generally regarded as adequate. The only information needed is a set of creep curves such as those of Fig. 6.1. The approach is the same for metals and for polymers, for which creep is often the main mode of failure that must be considered in design. Typical curves for a polypropylene are shown in Fig. 6.17. From them, we deduce the following approximate values of the apparent elastic modulus at 20 °C and 6 MPa:

Time(s):	10	10^3	10^5	10^7
Average modulus (MPa):	1200	730	600	370

It must also be recognized that structural components will not be under full load throughout their whole life. A linear criterion, similar to Miner's criterion for cumulative fatigue damage, has been proposed to take this fact into consideration. Assuming that σ_1, σ_2, ... σ_i, ... are possible stress levels during operation, active during the times t_1, t_2, ... t_i, ... at the temperatures T_1, T_2, ... T_i, ... and that, at these temperatures, they would cause rupture after t_{r1}, t_{r2}, ... t_{ri}, ..., the criterion is expressed as

$$\sum \frac{t_i}{t_{ri}} < 1$$

The values of t_{ri} can be obtained from conventional constant-load creep tests.

A similar crtierion can be used when deformation is the governing factor. If $t_i{}'$ is the time required for the stress σ_i at a temperature T_i to cause a specified creep strain or strain rate and t_i is the time during which σ_i is applied,

$$\sum \frac{t_i}{t_i'} < 1$$

Although such a simplistic approach to a very complex problem is not always justified, it still provides some guidance as to the relative magnitude of the design stresses on the basis of the design life and the required time between inspections. Practical experience, embodied in design codes, manufacturing standards etc., plays an important role in designing against creep.

Nothing has been said about the effect of the environment on unstressed components. Oxidation, corrosion and erosion are only some forms in which the environment can cause a degradation in the properties of the material and weaken the component. Prolonged exposure at elevated temperature can be more harmful than might be anticipated from an accelerated creep test. Numerous unexpected failures in boilers, pressure vessels and pipes are witness to the unduly optimistic attitude of metallurgists and engineers in the past.

Yet another complication arises when the material is fatigued at an elevated temperature. It has already been mentioned that at temperatures within the creep range, some relaxation in the secondary stresses induced by structural constraints takes place. Taking, for instance, the case of a straight run of pipe fixed to rigid end supports, heating will cause a compressive stress assumed to be below yield. This compressive stress will eventually be relaxed by creep, so that upon cooling the same stress will appear as tensile if we assume for simplicity that the modulus of elasticity and the yield point do not change. In subsequent cycles the stress will vary between a positive value, when the pipe is cold, and zero when it is at the operating temperature. The

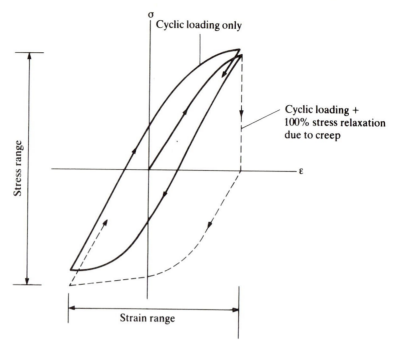

Figure 6.18   Combined cyclic loading and creep

effect of creep in this simple example is, therefore, beneficial except when the material is brittle and might fail under tension. When the initial compressive stress exceeds yield and the total strain range is $\Delta\epsilon$, as shown in Fig. 6.18, it is apparent that creep stress relaxation broadens the hysteresis cycle. A material subjected to creep in combination with fatigue suffers more work-hardening than one undergoing simple fatigue cycles. On the other hand, the observations concerning the nature of creep at the beginning of the chapter imply that work-hardening at elevated temperature may be less serious than might be expected.

6.12   CONCLUDING REMARKS

There are relatively few occasions when it is necessary to ascertain the time-dependence of the mechanical properties of materials. In structures or machines subjected to impact loading, test data obtained at an appropriate strain or loading rate are needed but the analysis follows conventional practice. The material is modelled as linear–elastic or non-linear, with a Young's modulus, yield stress etc., modified by the strain rate. Creep presents a very different problem in that 'time' must appear as one of the main variables in any accurate mathematical model. Since this is not always possible it follows that designers have to rely heavily on their own experience and on the application of high safety factors on stress while maintaining the temperature as low as possible. Sometimes, as in power or process plant, this is possible by using thermal shields. In gas turbines, it may be necessary to cool the parts exposed to hot gases, such as blades and discs, by blowing cold air through internal ducts or channels. This practice, although necessary from the point of view of thermodynamic efficiency, still results in mechanical losses and increases the cost of the components by making them more difficult to manufacture. There is a continuous requirement for better creep-resistant materials that has led to the development of the superalloys already mentioned and, currently, of the high-strength structural ceramics that will be treated in Chapter 7.

EXAMPLES

6.1 The following rupture times were obtained for stress rupture tests on a steel alloy:

Stress (MPa)	Temperature (°C)	Rupture time (h)
550	580	0.43
550	554	6.1
550	538	22.4
550	524	90.8
69	760	1.95
69	732	6.9
69	704	26.3
69	675	84.7

Examine the validity of the Larson−Miller parameter and suggest a value for the constant L.

6.2 Creep tests carried out on a given alloy gave the following results:

Stress (MPa)	Temperature (°C)	Minimum creep rate (s^{-1})
25	620	3.1×10^{-12}
25	650	6.25×10^{-12}
30	620	7.71×10^{-12}

Assuming that a single power-law creep equation describes the deformation process, obtain a relation between stress, temperature and minimum creep rate for this material.

A tie bar in a chemical plant is made from this alloy and is designed to withstand a stress of 25 MPa at 620 °C. In service it was found that, for 30% of the running time, the stress and temperature increased to 30 MPa and 650 °C. Calculate the average creep rate under service conditions.

6.3   From a series of tests at a constant temperature of 600 °C of an austenitic stainless steel, the following creep data are obtained:

Stress (MPa)	Minimum creep rate (% h^{-1})
65	2×10^{-5}
97.5	1.5×10^{-3}
130	3×10^{-2}
162.5	3.3×10^{-1}
260	5

The minimum creep rate (during the secondary or steady-state stage) is given by

$$\dot{\epsilon} = B \exp\left(- Q/RT\right) \sigma^{m} = C \, \sigma^{m}$$

based on Eqns (6.7) and (6.8) with $n = 1$. Obtain C and m.

6.4   The material of the previous example is used to make a cylindrical vessel with a diameter of 1 m, subjected to an internal pressure of 50 bar. The allowable creep rate is 1% per 10 000 h. If the material has a yield strength of 200 MPa and a stress for rupture in 100 000 h of 80 MPa, find the minimum wall thickness, taking the following safety factors:

(a)   on creep deformation, 1.2;

(b)   on creep rupture, 1.5;

(c)   on yield, 2.

The design life is 100 000 h. State any assumptions made.

6.5   The strain−time curve of a viscoelastic polymer under a constant tensile stress of 50 MPa is given in Fig. 6.19. Assuming that the behaviour can be modelled by Eqn (6.20) (Maxwell−Voigt model), suggest values for E_m, E_v, η_m and η_v.

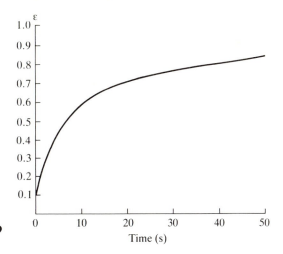

Figure 6.19

Time (s)

6.6 A rubber pad 10 mm thick, used as a vibration absorber, is subjected to a cyclic load with an amplitude of 0.2 MPa and a frequency of 50 Hz. The material behaves as a simplified Maxwell–Voigt model (Eqn 6.21) with E_m = 100 MPa, E_v = 20 MPa, η_v = 500 MPa s. How much energy is stored in a 1 m^2 pad per cycle?

6.7 The minimum creep rate for a steel bolt (Young's modulus 200 GN m^{-2}) at a given temperature varies with stress in a manner shown:

Stress (MN m^{-2}):	21.6	43	107.7	216	430
Creep rate (% per 1000 hr):	0.0037	0.0388	0.45	3.60	28.8

The bolt is used to clamp together two rigid plates at this temperature. The initial stress in the bolt is 15 MN m^{-2} and relaxation follows the empirical law $\dot{\epsilon}_p \propto \sigma^n$. Estimate the stress after two years. Note that the strain in the bolt (elastic strain and creep strain) remains constant.

REFERENCES

1 DORN J E (1961) *Mechanical Behaviour of Materials at Elevated Temperature* McGraw-Hill, New York

2 COTTRELL A H (1961) *Dislocation and Plastic Flow in Crystals* OUP, Oxford

3 WILSHIRE B, EVANS R W (1987) *Creep and Fracture of Engineering Materials and Structure* Inst Metals, London

4 DUGDALE D S, RUIZ C (1971) *Elasticity for Engineers* McGraw-Hill, London

5 THOMPSON W T (1966) *Vibration Theory and Applications* George Allen & Unwin, London

6 HARDING J (1986) The effect of high strain rate on material properties. In Blazynski T Z (ed) *Materials at High Strain Rates* Applied Science Publishers, London

7 ZUKAS J A, NICHOLAS T, SWIFT H F, GRESZCZUK L B, CURRAN D R (1982) *Impact Dynamics* John Wiley, New York

CHAPTER SEVEN

Ceramics and Brittle Materials

7.1 INTRODUCTION TO PROPERTIES

Mechanical engineers have traditionally worked with metals. There are many good reasons which explain why this is so and why engineers are often resistant to changes in materials. There is a wealth of experience of working with metals which extends back over thousands of years. Many metals possess an important degree of ductility which allows them to be machined and worked. The presence of ductility and plasticity can introduce a degree of safety into a design where failure by yield results in an energy-absorbing deformation before final catastrophic failure occurs. However, metals have their limits, with lower melting points and elastic moduli than some other materials. Hence for certain applications at high temperatures, in corrosive environments and in regions of high frictional or erosive wear, stiffer and stronger ceramic materials are sometimes preferred. All metals also conduct electricity and a ceramic part is sometimes used where an electrical insulator of high structural integrity is required, e.g. within a spark plug for an internal combustion engine. Some examples of new ceramic components are shown in Fig. 7.1.

Ceramics have a wide range of engineering applications, both potential and actual, some of which are listed in Table 7.1. Clearly there is not only a range of applications but also a range of materials which can be described as ceramics. A more comprehensive definition of a ceramic is best left to more specialist books.[1-3] However, a convenient simple definition for practical purposes is that of solids which possess an ordered arrangement of atoms bonded together by covalent or ionic forces. This definition includes inorganic glasses because even they possess an order at very short ranges within their structure. The bonding in most useful ceramics is usually of a hybrid

209

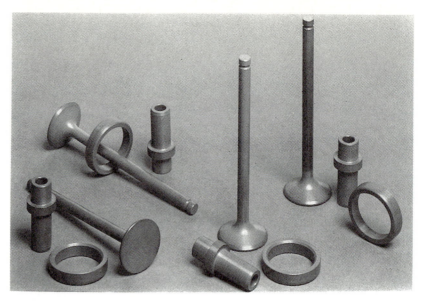

Figure 7.1 A selection of experimental ceramic components developed for use in internal combustion engines (courtesy of Cookson Group plc)

nature with the extremes represented by silicon carbide, which is practically 100% covalent, and by magnesium oxide, which is almost completely ionic. The strong ionic and covalent nature of the bonding in most ceramics leads to a stable crystal structure with a high melting point and high stiffness. Figure 7.2 illustrates the range of properties found in ceramics as compared with other materials.

The physical nature of covalent or ionic bonding will impose its own constraints on ceramic crystal structures. Unlike metal alloys, ceramics are often inorganic compounds of rigidly defined chemical composition which govern the short-range order of the different atomic species within the lattice. Ionic bonding occurs with the exchange of charge between atoms to create a structure of positively and negatively charged ions. These ions are held together in the crystal lattice by electrostatic forces. The crystal structure will be defined by conditions of minimum electrostatic potential energy, global charge neutrality, and local ionic arrangement separating like-charged ions. A typical ionic structure is shown for MgO in Fig. 7.3. Covalent bonding occurs when atoms share electrons in specific bond orbitals between the bonded atoms. Carbon and silicon atoms can exist in a state with four potential bond orbitals radiating out from the atomic centres along the vertices of a tetrahedron. A very stable bond can form when two such orbitals overlap and allow electron sharing. The conditions of quantum mechanics constrain these bond orbitals to the tetrahedral directions and a very open, yet rigid, structure occurs when these elements combine to form β-SiC (Fig. 7.4).

Table 7.1
Engineering
application of
ceramics

	Ceramic	Applications	Critical properties
(1)	Clayware	Bricks, floor tiles, pipework	Cheap mouldable raw material
(2)	Pottery, e.g. earthenware, porcelain	Tableware, sanitary whiteware, wall tiles, electrical insulators	Cheap material producing a stronger product than (1)
(3)	Refractories based on alumina, silica, magnesia, zircon, graphite	Furnace linings, moulds, molten metal handing	Very stable at high temperatures in aggressive environments
(4)	Glasses (including enamels and glass-ceramics)	Windows, containers, decorative ware	Easily mouldable in viscous glassy state. Glass-ceramics are crystallized to give better mechanical properties. Many glasses are transparent to large ranges of the spectrum.
(5)	Cement and concrete	Structural engineering, floors, roadways	Cheap bulk material which solidifies in place at ambient temperature
(6)	Abrasives, e.g. Al_2O_3, SiC, diamond	Grinding wheels, polishing papers, cutting tools	Very high wear resistance
(7)	Engineering ceramics, e.g. Al_2O_3, SiC, Si_3N_4, ZrO_2	Bearings, seals, dies, high-temperature structural applications	Higher toughnesses and reliability than conventional ceramics

Both the ionic and covalent types of structure can contain dislocations whose nature is similar to those responsible for plastic flow in metals. However, new constraints imposed by the ceramic crystal structure severely restrict dislocation motion in most cases. In an ionic crystal dislocations cannot violate the condition of global charge neutrality, nor can they create a shear which brings like charges into contact. To satisfy these conditions it is common for ionic materials to have dislocations not on their closest packed planes of ions but on some less well packed plane more resistant to shear. In covalent materials the local lattice distortion is highly energetic because of the highly direction-specific covalent bond. In addition there will be an unbonded electron orbital at the dislocation core, a so-called dangling bond. For both ionic and covalent crystals, the passage of a dislocation must not disrupt local chemical order. The net result of all these chemical and crystallographic conditions is to increase greatly the stress required to cause dislocation motion. In many ceramics dislocation

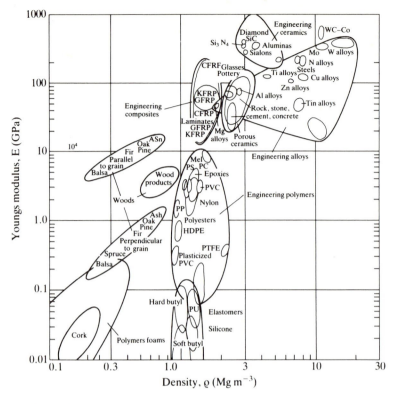

Figure 7.2 A plot of the Young's modulus versus density for a number of engineering materials. Ceramics tend to have very high elastic moduli coupled with lower densities than many metals.

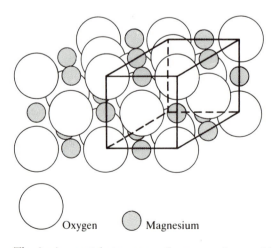

Figure 7.3 The ionic crystal structure of a magnesium oxide ceramic

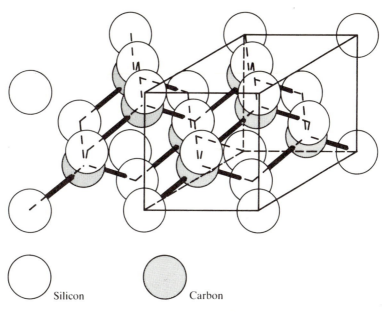

Silicon   Carbon

Figure 7.4   The covalent structure of a silicon carbide ceramic

motion is only possible at temperatures close to the material's melting point or else requires very high initiation stresses close to the material's ideal shear strength.

In most ceramics the bonding is intermediate in character between pure ionic and pure covalent forces and so a combination of the properties discussed above applies. Ceramic crystal structures are often of lower symmetry than the cubic examples of MgO and SiC discussed earlier. In such crystals there are significant differences in elastic properties along different crystal directions. This obviously leads to severe design and stress analysis complications. However, except for a few highly specialized applications, structural ceramics are normally used in polycrystalline form and so over a conventionally sized component a mean isotropic behaviour is observed. Many ceramics, especially traditional clay-based ceramics and whitewares, are polycrystalline multiphase materials. Again, these materials will be close to isotropic in their bulk behaviour but will show local property differences between the grains. Although all these materials are globally isotropic, grain-scale inhomogeneities and anisotropies can have secondary effects on bulk behaviour.

7.2   BRITTLENESS, DEFECTS AND FLAWS

It is very difficult to initiate dislocation flow in ceramics unless they are near their melting points. Thus they might be expected to show a very high strength. However, this is not so. Instead, lower strengths are measured and the material fails in a brittle manner. Brittle behaviour is discussed in Chapter 4 and is known to be governed by the presence of cracks, defects or flaws in the material. In order to design satisfactorily with brittle materials, the source of these flaws must be understood.

7.2.1   DEFECTS FROM FABRICATION

Engineering ceramics, because of their high melting point and brittleness, are usually manufactured by techniques not commonly used for metals. Most ceramics start off in the form of a very fine powder. Even the common clay-based materials used in traditional ceramics are really mixtures of finely divided particles bonded together by water. These powders are converted into solid ceramic bodies by a process known as sintering. At high temperatures (above about half the material's melting point) atoms move by diffusion. A collection of compacted powder contains a large quantity of free surface in the form of interconnected voids. Such free surfaces have an associated energy which is released if the voids shrink and close. This void shrinkage occurs by diffusional processes and is known as sintering. If a ceramic is held for insufficient time at the temperature necessary for sintering, incomplete closure of the voids can occur. Such a microstructure results in defects or *flaws* in the ceramic. Interested readers are referred to Kingery et al.[1] and Richerson[3] for more detailed accounts of sintering and ceramic microstructures.

Inorganic glasses are generally easier to fabricate than ceramics. A glass undergoes a transition at a critical temperature into a highly viscous liquid. The viscosity of the liquid decreases with further increases in temperature but at all temperatures above the *glass transition temperature* it behaves as a *Newtonian fluid*, i.e. its strain rate varies linearly with stress. Materials which obey this law deform without necking and can be easily moulded into complex shapes. This property explains many of the reasons why glass is preferred to ceramics for a large number of applications despite being more brittle than most ceramics. Defects can also form in glasses during manufacture, the most common being gas bubbles which, because of the isotropy of glass, will be spherical in shape.

One technique used to make ceramics easier to fabricate is to incorporate a small amount of glass into their structure. This glass occurs along the boundaries between the particles being sintered and, because it becomes liquid at comparatively low temperatures, will flow into any pores between the particles, filling them up and reducing the number of defects present. In these microstructures defects can still be present from regions where poor packing of the particles prior to

Table 7.2 Physical properties of some ceramics and metal alloys

Material	Density (kg m$^{-3} \times 10^3$)	Melting point (K)	Young's modulus (GPa)	Coefficient of thermal expansion (K$^{-1} \times 10^{-6}$)	Mean strength (MPa)	Thermal conductivity (W m^{-1} K^{-1})
Alumina (Al$_2$O$_3$)	3.8	2300	345	7.8	400	20
Silicon carbide (SiC)	3.2	2970	400	4.7	600	100
Silicon nitride (Si$_3$N$_4$)	2.6	2170*	400	3.7	600	20
Zirconia TZP (ZrO$_2$)	5.8	2960	200	8.4	1000	2.3
Soda glass	2.5	600†	74	9	65	1.4
Mild steel	7.9	1700	210	16	300	63
2024 aluminium alloy	2.8	800	71	23	550	201

* Decomposes.
† Softening temperature.

sintering leads to large pores which could not be filled by the glassy phase.

All ceramics and glasses are manufactured at elevated temperatures and will contract in size on cooling to room temperature. Some ceramics with an anisotropic crystal structure have different values for their coefficient of thermal expansion (CTE) along different crystal directions. Some examples of the CTEs of different materials are shown in Table 7.2, along with other physical properties. Note how in general ceramics change their dimensions with temperature much less readily than metals. This property can lead to severe problems if ceramics are to be used in conjunction with metals at elevated temperatures. Alumina (Al$_2$O$_3$) has a maximum difference in its CTE ($\Delta\alpha$) along its crystal directions of about 10^{-6} K^{-1}. In a polycrystalline ceramic made of alumina with many different grains in different orientations it is possible to generate differential strains between adjacent, misoriented, grains of:

$$\epsilon_T = \frac{\Delta\alpha\Delta T}{(1-\nu)} \tag{7.1}$$

where ΔT is the change in temperature and ν is Poisson's ratio. Most engineering ceramics are fabricated at very high temperatures. For alumina these are usually above 1000 °C and the thermal strain generated can be in excess of 10^{-3}. A similar condition for thermal strains exists in multiphase ceramics, where now $\Delta\alpha$ is the difference in CTE between the phases present. These strains will generate high internal stresses which can, in certain circumstances, nucleate internal fracture events. For such a fracture to occur the same conditions as apply to bulk fracture must be satisfied, i.e. the same conditions as apply in the Griffiths equation (Chapter 4). In order to analyse this properly we must have some knowledge of the nature of other cracks and flaws in the neighbourhood of the stressed grain boundary. Without this exact information it is still possible to make some estimates of likely behaviour. The total elastic energy stored in the grains will be related to the elastic energy density (U^*):

$$U^* = \frac{\epsilon_T^2 E}{2} \tag{7.2}$$

The total stored energy in each grain (U_v) will be determined by the grain volume:

$$U_v \propto \epsilon_T^2 E D^3 \tag{7.3}$$

where D is a linear dimension of the grain. If a crack is nucleated and crosses a grain or separates two grains, the total surface energy requirement (Γ_s) will be determined by the grain dimensions.

$$\Gamma_s \propto \gamma D^2 \tag{7.4}$$

where γ is the material's surface energy. Thus the stored energy available for internal fracture increases with D^3 and the energy required for crack advance only with D^2. Hence we would expect a large-grained ceramic to be more susceptible to thermal cracking than a small-grained material.

Defects generated by local differences in thermal expansion do not generally extend beyond one grain dimension. The stored energy is generated by the constraint of contacting grains on shape change from local CTE differences. Once a fracture is nucleated, either across one grain or between two, the constraint is removed and a local shape change occurs.

7.2.2 CONTACT DAMAGE

If brittle materials are indented by other hard materials, cracks are seen to nucleate from the indentation (Fig. 7.5). It is observed that the extent of cracking from these indents is of the same order of size as the indent itself. Tests on brittle materials show that in many cases fracture initiates from surface defects a few microns in size. Hence we might

Figure 7.5   Cracking in a brittle material has been nucleated from the corners of an indentation made during hardness testing

expect the damage to be initiated by the contact of micron-sized hard particles. This is not an impossible speculation because dust contains a large fraction of quartz (SiO_2) particles of the appropriate dimensions, and the impact of such particles carried by air currents or resting on the surface of other materials can be quite common. Quartz is hard enough to indent all but super-hard materials such as diamond.

The problem of investigating stress around regions of contact was first analysed (elastically) by Hertz in 1881. He considered an ideal case of a spherical ball in contact with an infinite surface. If E and ν are the elastic modulus and Poisson's ratio of the surface and E' and ν' the corresponding values for the ball, Hertz's analysis shows a relation between the radius of the contact area (a) and the contact load (P) with:

$$a = \sqrt[3]{\tfrac{3}{4} Pr \left[\frac{(1-\nu^2)}{E} + \frac{(1-\nu'^2)}{E'} \right]} \qquad (7.5)$$

217

Figure 7.6   Side view of a Hertzian cone crack induced by the action of a steel ball on a glass surface

where r is the radius of the sphere. The precise stress distribution generated under the surface by the contact is too complicated to derive fully. However, the maximum tensile stress is found at the edge of the contact region acting in a radial direction, and is of magnitude:

$$\sigma_{max} = \frac{P(1-2\nu)}{2\pi a^2} \tag{7.6}$$

Within the circle of contact the radial stress becomes compressive; outside the circle it rapidly falls away from its maximum value.

$$\sigma_r = \sigma_{max} \left(\frac{a}{x}\right)^2 \quad \text{for } x > a \tag{7.7}$$

where x is the distance from the centre of the contact. In the absence of defects, the maximum radial stress increases linearly with contact load until the ultimate fracture strength of the surface is reached. Once this level is reached a crack propagates rapidly around the contact and descends into the material at an angle of about 68° to the vertical axis, following an approximately conical path (Fig. 7.6). This is the Hertzian cone crack. The tensile stress field beneath the contact falls off rapidly with increasing depth and the crack arrests when there is insufficient stress to propagate it further. The exact relation between cone crack length (c) and load is not easily soluble. Approximate analysis gives:

$$c = \sqrt[3]{\frac{C_1 P^2}{E\Gamma}} \tag{7.8}$$

where Γ is the crack propagation surface energy and C_1 is a material constant related to the elastic properties of the sphere and surface.

The critical load at which fracture initiates can be estimated if we use the assumption that the ideal fracture strength of a brittle material (Chapter 4) is related to its elastic modulus by:

$$\sigma_{ideal} = \frac{E}{10} \tag{7.9}$$

Taking the case of a quartz particle of radius 1 μm in contact with a glass surface, with $E \simeq E' \simeq 70$ GPa and $\nu \simeq \nu' \simeq 0.3$, Eqns (7.5) and (7.6) give a critical contact load of approximately 0.5 N. This is much larger than the weight of such a particle but is easily attained if two surfaces are in contact with dust particles trapped between them. Also, if a dust particle is moving at any speed when it hits the surface, after being carried by air currents or by Brownian motion, the impulse loading experienced may be sufficient to nucleate a crack. Dust particles tend to be angular in shape; thus the contacting radius may be much smaller than the radius of the particle with corresponding increases in contact stress.

If a very small-radius indentation contact is made, a different cracking behaviour is seen. Instead of a circumferential or cone crack around the contact region, a pattern of radial cracks is seen to be nucleated, intersecting the surface with subsurface lateral cracks forming on *unloading* (Fig. 7.7). These lateral cracks follow an ascending curve which may intersect the surface allowing a piece of material to *spall* or be removed. The exact analysis of these small radius or sharp contacts is complicated because a small region of plastic flow occurs beneath the indentor. Plastic flow can occur even in the most brittle materials in this region because the loading generates very high loads in *compression* which do not have a crack-opening contribution. Although dislocation motion is difficult in ceramics, it will still be possible above some critical (though rather high) stress. The lateral cracks are seen to nucleate from the edge of the plastic zone and are believed to be generated by the formation of residual tensions on unloading. The radial cracks which run normal to the surface are the most important in terms of stress degradation since they will lie across likely stress distributions. However, lateral cracking and surface spalling is an important mechanism of material removal during ceramic machining.

7.2.3   MACHINING DAMAGE

Engineers must have precise control over the final shape and dimensions of a component. This is easily achieved with metals, where highly precise cutting and machining is possible. Such techniques may

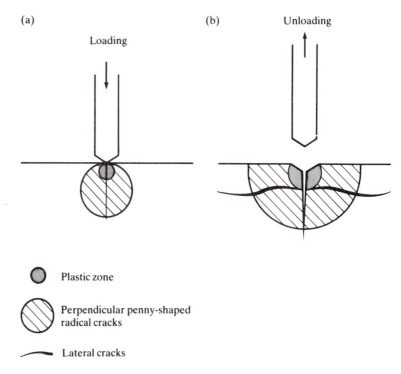

(a) Loading

(b) Unloading

● Plastic zone

◯ Perpendicular penny-shaped radical cracks

— Lateral cracks

Figure 7.7 Schematic view of the process of crack nucleation beneath a sharp indenting point: (a) loading nucleates a penny crack and generates a plastic zone beneath the indentor; (b) on unloading the penny crack extends to intersect the surface and a residual tensile stress nucleates lateral cracks

not be used with ceramics because most ceramics are harder than many cutting tool materials and also because most cutting configurations develop significant tensile stresses in the component. For these reasons most ceramic machining is carried out by abrasive techniques. For a more comprehensive review of alternative machining techniques see Richerson.[3]

In abrasive machining, small hard particles are used to remove ceramic material from the surface. The particles can be incorporated into a resin or metal grinding wheel or used in lapping as a suspension in a liquid slurry. Abrasive media tend to be hard or super-hard ceramics themselves, typically Al_2O_3, CeO_2, SiC and diamond. To machine efficiently, a medium harder than the ceramic to be machined is used.

The machining mechanism is similar to that of sharp indentation. A collection of small angular particles from the grinding medium are dragged along the surface of the ceramic and a long scratch or groove is formed. Beneath the scratch is a region which has been through plastic deformation and beneath this is a long median crack running parallel to the abrasion direction (Fig. 7.8), which penetrates deep into

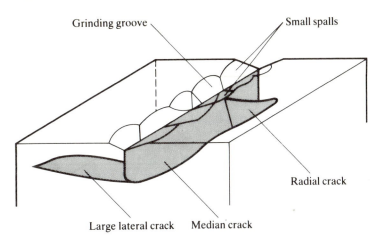

Figure 7.8   Schematic view of damage nucleated below a machining groove, showing its similarities to the damage beneath a sharp indentor

the ceramic. At 90° to this direction are occasional radial cracks. Beneath, and running parallel to the surface, there are lateral cracks. These can lead to spalling and material removal when they intersect the surface. Of these cracks, the median crack is largest and leads to the most severe reduction in strength. All the crack lengths are related to the size of abrasive particle and machining conditions such as normal load and grinding speed. Thus, the conditions which lead to rapid stock removal by encouraging large spalls, will also introduce the most severe machining damage by inducing large median cracks. To minimize strength degradation, a machining sequence must be devised which consists of a series of abrasive processes using smaller particle sizes such that each stage removes the machining damage of the previous stage and introduces less severe damage itself. By such techniques, and finishing with lapping using very fine particles, extremely smooth and almost defect-free ceramic surfaces can be achieved, albeit after much time, effort and expense. Machining also introduces a certain degree of strength anisotropy. Of the various damage types introduced during machining, the median crack weakens the material most effectively. Hence strengths measured perpendicular to this crack will be much lower than in other directions (Fig. 7.9). Therefore, if grinding is the final surface treatment, care should be taken to ensure that it is not carried out normal to the direction of maximum tensile stress in the component.

For a complete understanding of intrinsic flaws generated by ceramic processing, handling, and machining we must be able to include the probability of there being small flaws from imperfect sintering and the chance that they may become the initiation sites for larger flaws under the action of internal and external stresses. Many ceramic microstructures are complicated and multiphase; the interactions

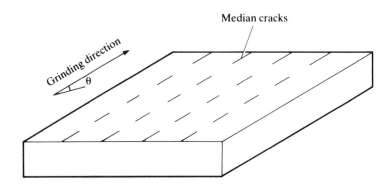

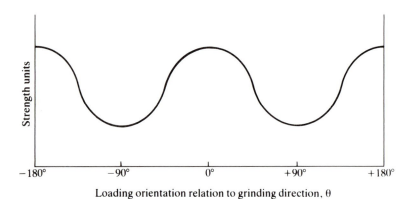

Loading orientation relation to grinding direction, θ

Figure 7.9   Anisotropic strength distribution of a machined ceramic block. The strongest direction is parallel to the direction of grinding

between the different mechanical properties of the phases in contact can lead to further mechanisms of flaw nucleation. The effect of all these mechanisms of flaw nucleation is to make it very difficult to create completely flaw-free material. Instead, most ceramics contain many flaws of a range of sizes.

7.3   DESIGNING WITH BRITTLE MATERIALS

7.3.1   THE STRENGTH OF A CERAMIC

In order to design a component for some mechanical application, the engineer must have an adequate measure of the material's physical properties, in particular its strength. Ceramics are usually brittle materials under the conditions in which they are used and so the strength is normally taken to mean the fracture strength. All physical measurements of any material property will have some uncertainty associated with them. Either there is some limit to the accuracy of the technique used to measure the property in question, or there is some intrinsic variation within the material as supplied. Engineers are most accustomed to working with metals and are usually reliant on datasheets or books to provide the required design information. Any uncertainty in yield strength or fracture strength is typically only of a few percentage points and is customarily dealt with by designing with a comfortable degree of safety by allowing maximum stresses well below the likely range of failure stresses. Ceramics, however, are highly brittle materials and strengths are limited by a combination of the toughness or susceptibility to fracture and the size of flaw present. These are combined to produce a fracture stress of:

$$\sigma_F = \frac{K_{IC}}{\sqrt{(\pi c)}} \tag{7.10}$$

Given that a typical value of K_{IC} for an advanced engineering ceramic is about 4 MPa m$^{-1/2}$, if the ceramic contains a crack of length 5 μm its failure strength will be in the region of 1000 MPa, but if the crack were of length 50 μm its strength would be about 320 MPa and a 500 μm crack would produce a strength of only 100 MPa. Hence the failure strength of a piece of ceramic is strongly dependent on the size of defects present in it; or, to be more accurate, the strength will be determined by the size of the *largest* flaw present. Thus, as the fracture toughnesses of ceramics are generally much lower than those of the more familiar engineering alloys, the size of defect which adversely affects a ceramic's strength is also much smaller than those important in metals. In real engineering ceramics it is very difficult to eliminate all cracks below about 100 μm in size. Typical flaw sizes will be distributed between about 50 and 500 μm. The strong dependence of strength on flaws also generates a strong variation in the measured fracture strength of a batch of ceramic parts; a scatter of ±50% may be found in a large sample size.

7.3.2   DESIGN TECHNIQUES

This high degree of variability in the fracture strength of a ceramic is a challenge for the designer. Because the failure of a highly brittle part is sudden and often catastrophic in nature, it is an event which must be avoided. There are a number of approaches to problems of this nature. The traditional approach to brittle materials is to design by trial and error in an empirical fashion. This technique is epitomized by manufacturing parts, or close analogues to the final part, and testing them to destruction in likely service conditions. This is the way many ceramic parts have been designed successfully in the past using traditional clay-based ceramics. It may be the only way to design a component which is to be used in a highly aggressive environment, or else under conditions where the material's properties are poorly known and the expense of accurately determining them is prohibitive for the part in question. However, with ceramics being proposed for critical applications such as parts of gas turbines and compressors, a more exacting design methodology is required.

The traditional way of dealing with unquantifiable uncertainties in design with metals is to 'overdesign' or to introduce a factor of safety where the fracture stress is considerably greater than the greatest design stress in the component. Stresses within the component are either determined by analytical techniques or, more commonly these days, by computer-aided methods such as the *finite element method*. The safety factor is usually chosen by experience and so there is still an empirical component to this technique. With a safety factor and a knowledge of the stresses present an appropriate material will be chosen, or if none is available the design will be altered to lower the stresses in the critical areas. This technique works very well with metals and is used with brittle materials such as concrete in civil engineering practice. However, these applications are either with materials of well-defined mechanical properties, or in the case of concrete when they are loaded in compression. Ceramics do not have a well-characterized fracture strength and their strength is known to be a function of the size of test specimen tested. Traditional design techniques developed for metals cannot take these effects into account and, if a highly stressed component is considered, extremely large safety factors become necessary. Such design solutions soon become uneconomic as a large safety factor increases cost and reduces any advantage a ceramic material has over competing materials.

A final traditional design technique which must be mentioned is the *proof-test*. This is where a component is tested after fabrication up to stress levels exceeding those likely to be experienced in service. This method is long established in industries manufacturing brittle components such as grinding-wheel manufacture where a proof 'burst test' is carried out on all wheels prior to despatch. Although this test is perhaps valuable as an ultimate safety test, it assumes the major design problems have been ironed out. It would otherwise be highly wasteful of expensive ceramic material.

Traditional design approaches were adequate for typical ceramic applications in the past but are limited in critical applications or where large and complex stress distributions occur. In these situations design techniques must be capable of considering the distributions of both stress and flaws within a material. Statistical techniques must be used to account for likely flaw populations in the material.

7.3.3   WEIBULL STATISTICS

Statistical techniques are now commonly used in design with ceramics. A statistical distribution first proposed by Weibull in 1951 is currently the most favoured. This approach can be thought of as modelling the behaviour of a ceramic as that of a multipart component whose strength is determined by the weakest part. Alternatively, in terms of a ceramic, the strength of a ceramic object is limited by the size of the largest flaw present. A simple mechanical analogy is that of a long chain, the strength of which will be determined by the weakest link. If the strength of individual links follow some probability distribution we can clearly see that the longer the chain, the more links are present, and hence the greater the likelihood of a weak link. This type of behaviour is seen in brittle materials, where a large specimen usually fails at a lower strength than a smaller one.

In order to model the behaviour of these materials, consider a chain of length λ under a load F with a probability of failure $P_f(\lambda)$. There will be a complementary probability that the chain survives of $P_s(\lambda)$ $= 1 - P_f(\lambda)$. If the length of chain is now doubled to 2λ, then the probability of survival will be that of two chains of length λ joined together, i.e.

$$P_s(2\lambda) = P_s(\lambda).P_s(\lambda) \tag{7.11}$$

If instead of a chain we had a piece of ceramic of unit volume V under a stress σ, we can define a similar survival probability of $P_s(V)$. Using the same argument as we used to calculate the probability of failure of the chain, the probability that a piece of ceramic of volume xV will fail at the same stress is:

$$P_s(xV) = [P_s(V)]^x \tag{7.12}$$

For ease of manipulation later, we can rewrite the above equation as:

$$P_s(xV) = \exp \{x \ln [P_s(V)]\} \tag{7.13}$$

In Weibull's terminology the risk of fracture is defined as:

$$R = -x \ln [P_s(V)] \tag{7.14}$$

and at infinitesimally small volumes of ceramic this is taken to depend solely on the stress experienced by that volume of ceramic, i.e.

$$dR = f(\sigma)dV \tag{7.15}$$

Weibull then proposed a simple three-parameter expression for $f(\sigma)$ of:

$$f(\sigma) = \left(\frac{\sigma - \sigma_c}{\sigma_0}\right)^m \tag{7.16}$$

where σ_c is a critical stress below which no failure is observed and σ_0 is a normalizing reference stress of no physical significance, to which we shall return later. Combining the above equations, we can calculate a failure probability as a function of stress and volume by:

$$P_f(V) = 1 - P_s(V) = 1 - \exp\left[-\int_V \left(\frac{\sigma(V) - \sigma_c}{\sigma_0}\right)^m \frac{dV}{V_0}\right] \tag{7.17}$$

where $\sigma(V)$ is the stress of an element dV as a function of its position in the stressed volume and V_0 is some normalizing reference volume. If the volume of material tested is stressed uniformly, Eqn (7.17) becomes:

$$P_f(V) = 1 - \exp\left[-\left(\frac{V}{V_0}\right)\left(\frac{\sigma - \sigma_c}{\sigma_0}\right)^m\right] \tag{7.18}$$

The magnitude of m, known as the *Weibull modulus*, determines the width of this probability distribution in σ. If m is large the distribution becomes narrow and the material behaves as if there is a very narrow range of σ over which failure might occur. Conversely, if m is small there will be a large range of possible failure stresses. By statistical manipulation, it is possible to relate the Weibull modulus to the covariance (C_v) of the strength distribution, in which case the following approximate relation is found:

$$C_v \approx \frac{1.2}{m} \tag{7.19}$$

Equation (7.18) is not easy to visualize in terms of probabilities and stresses. In order to determine the Weibull modulus of a material a large number of specimens must be tested in as near a uniform way as possible (30 is often taken as a minimum to give a meaningful result). The failure stresses are measured and the survival probability at each stress determined. In terms of a *survival* probability, Eqn (7.18) gives:

$$P_s(V) = \exp\left[-\left(\frac{V}{V_0}\right)\left(\frac{\sigma - \sigma_c}{\sigma_0}\right)^m\right] \tag{7.20}$$

which can be rewritten:

$$\ln\{P_s(V)\} = -\left(\frac{V}{V_0}\right)\left(\frac{\sigma - \sigma_c}{\sigma_0}\right)^m \tag{7.21}$$

or in a form more convenient for plotting:

$$\ln\{-\ln[P_s(V)]\} = \ln\left(\frac{V}{V_0}\right) + m\ln\left(\frac{\sigma - \sigma_c}{\sigma_0}\right) \tag{7.22}$$

(a)

Ranked data value (*n*)	Strength (MPa)	Estimated probability of failure $P_f \neq n/(N+n)$
1	178	0.1
2	210	0.2
3	235	0.3
4	248	0.4
5	262	0.5
6	276	0.6
7	296	0.7
8	318	0.8
9	345	0.9

(b)

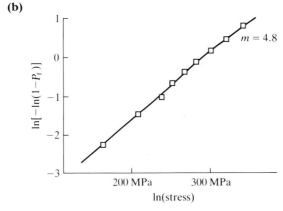

Figure 7.10   Weibull plots of tests on glass rods: (a) data from 10 tests ranked in order; (b) plot of the data from (a) to generate the Weibull modulus

The physical meaning of the term σ_c is a minimum stress below which no fracture can occur. This implies a maximum defect size. In practice the properties of ceramics cover such a range of strengths that there appears to be no evidence for a clearly defined maximum defect size and σ_c is taken as zero. Hence, Eqn (7.22) can be plotted graphically with $\ln\{-\ln[P_s(V)]\}$ against $\ln(\sigma)$ giving a gradient of m and an intercept of $\ln(V/V_0) - m\ln(\sigma_0)$. This is shown in Fig. 7.10 to give a measure of the Weibull modulus of a glass rod. The value obtained of $m = 4.8$ is low and the scatter that this implies is clearly shown by the range of failure stresses measured.

From the Weibull plot of Fig. 7.10 let us consider the effect of changing the value of m. As m increases, the gradient of the plot also increases until at $m = \infty$ the line sits vertically through a single failure stress (Fig. 7.11). Thus, a Weibull modulus of infinity implies a uniquely

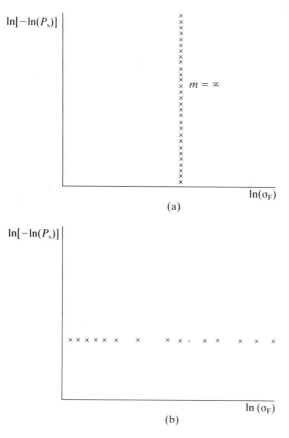

Figure 7.11 Examples of failure strength/probability plots for: (a) infinite and (b) zero Weibull moduli

defined failure stress. Conversely, if $m = 0$ the plot represents all stresses having the same failure probability. Thus $m = \infty$ is the ultimate precise definition of a unique failure stress and $m = 0$ defines a totally random fracture behaviour. Clearly all materials lie between these limits, with most ceramics having values in the range $5 < m < 25$. In contrast, metals would have m values in the 100s.

A final caution concerns determining the probability of failure from the list of failure stresses. If a total of N specimens were tested, the mean probability of survival of the nth strongest specimen can be defined as $P_s = 1 - n/(N+1)$. However, the median probability of survival of that specimen will be given by $P_s = 1 - (n-0.3)/(N+0.4)$. There are a number of slightly different statistical derivations of the survival probability from a ranked list of failure stresses and the method chosen can significantly alter the probabilities at the tails of the distributions where the probability of large flaws is considered.

7.3.4   THE EFFECT OF STRESS DISTRIBUTION

In most engineering applications the stress within a component will not be constant and approximate solutions such as Eqn (7.18) will not be valid. Instead, the complete volume integral of the stress distribution in Eqn (7.17) must be considered. It is easier to consider the evaluation of this volume integral if some measurable reference stress is used to define the stress distribution, i.e. $\sigma(V) = f(V)\sigma_r$ where $f(V)$ is some function of the position of a volume element dV. This stress σ_r might be, for example, the maximum tensile stress observed on the surface of a specimen tested in bending. This can now be included into Eqn (7.17) to give the probability distribution function for a volume V which contains some stress distribution $\sigma(V)$:

$$P_f(V) = 1 - P_s(V)$$

$$= 1 - \exp\left[-\left(\frac{V}{V_0}\right)\left(\frac{\sigma_r}{\sigma_0}\right)^m \int_V \left(\frac{\sigma(V)}{\sigma_r}\right)^m \frac{dV}{V}\right] \quad (7.23)$$

The volume integral in the above equation has the normalizing constants σ_0 and V_0 removed and is known as the stress–volume integral. Because $\sigma(V)$ will be a linear relation with the reference stress, the stress–volume integral will be independent of the load on the specimen. This integral can be determined analytically or numerically but we still need to replace the normalizing constants by physically meaningful terms.

Let us consider the constant-stress probability distribution function with $\sigma_c = 0$ (Eqn 7.18). This probability is a cumulative probability of failure and from it the mean failure stress $\bar{\sigma}$ may be determined:

$$\bar{\sigma} = \int_0^\infty \exp\left[-\left(\frac{V}{V_0}\right)\left(\frac{\sigma}{\sigma_0}\right)^m\right] d\sigma$$

$$= \Gamma\left(\frac{1}{m} + 1\right)\left(\frac{V_0}{V}\right)^{1/m} \sigma_0 \quad (7.24)$$

where $\Gamma(1/m + 1)$ is the numerically evaluated gamma function, given by:

$$\Gamma(q) = \int_0^\infty x^{q-1} e^{-x} dx \quad (7.25)$$

which can be referred to in tables of mathematical constants. Note that the overbar here denotes a mean value not, as in Chapter 3, the hydrostatic stress. From this we can replace σ_0 by a measured average stress $\bar{\sigma}$ to give:

$$P_f(V) = 1 - \exp\left[-\left(\frac{1}{m}!\right)^m \left(\frac{\sigma}{\bar{\sigma}}\right)^m\right] \quad (7.26)$$

where $1/m!$ is conventionally used to represent the gamma function in Eqn (7.24).

Using the above equation, we now consider the behaviour of some *real* reference volume V^* which is uniformly stressed (σ^*) and has a mean failure stress $\bar{\sigma}^*$. This mean failure stress is given by:

$$\bar{\sigma}^* = \int_0^\infty \exp \left[-\left(\frac{V^*}{V_0}\right) \left(\frac{\sigma_r}{\sigma_0}\right)^m \int_V \left(\frac{\sigma^*(V)}{\sigma_r}\right)^m \frac{dV}{V^*} \right] d\sigma_r \tag{7.27}$$

Note that the asterisk here denotes a reference value. In the case of a uniformly stressed volume the stress distribution will be constant and must be equal to the system's reference stress, i.e. $\sigma^*(V) = \sigma_r$ and the stress−volume integral in Eqn (7.27) will then be equal to 1. Thus the above equation gives:

$$\bar{\sigma}^* = \left(\frac{V_0}{V^*}\right)^{1/m} \left(\frac{1}{m}!\right) \sigma_0 \tag{7.28}$$

If this is now substituted into Eqn (7.23),

$$P_f(V) = 1 - P_s(V) =$$

$$1 - \exp \left[-\left(\frac{V}{V^*}\right) \left(\frac{\sigma_r}{\bar{\sigma}^*}\right)^m \left(\frac{1}{m}!\right)^m \int_V \left(\frac{\sigma(V)}{\sigma_r}\right)^m \frac{dV}{V} \right] \tag{7.29}$$

Now all the undefined normalizing constants have been replaced by well-defined or measurable stresses and volumes. The average failure stress for any specimen is then given by:

$$\bar{\sigma} = \int_0^\infty \exp \left[-\left(\frac{V}{V^*}\right) \left(\frac{\sigma_r}{\bar{\sigma}^*}\right)^m \left(\frac{1}{m}!\right)^m \int_V \left(\frac{\sigma(V)}{\sigma_r}\right)^m \frac{dV}{V} \right] d\sigma_r$$

$$= \bar{\sigma}^* \left[\frac{V^*}{V \, I(V)} \right]^{1/m} \tag{7.30}$$

where $I(V)$ represents the stress−volume integral.

The use of a stress−volume integral assumes that all the strength-limiting defects are distributed uniformly throughout the brittle body. Earlier we discussed the nucleation of these defects within ceramics and identified two basic types of defect. Defects can be nucleated within the body of the material by internal stresses or by flaws intrinsic to the fabrication process. A second family of defects is generated solely on the surface of the material by handling and contact damage. In transparent glasses and some very fine-grained ceramics, the uniformly distributed defects are very much smaller in size than the surface defects. Thus the strength−volume integral will produce an overestimate of the failure probability because it is considering stresses in regions where no fracture initiation is possible. Instead the stresses should be integrated over the surface of the body and a *stress−area* integral used instead with

$$P_f(S) = 1 - \exp\left[-\left(\frac{A}{A^*}\right)\left(\frac{\sigma_r}{\bar{\sigma}^*}\right)^m\left(\frac{1}{m}!\right)^m\int_S\left(\frac{\sigma(A)}{\sigma_r}\right)^m\frac{dA}{A}\right]$$

(7.31)

and

$$\bar{\sigma} = \bar{\sigma}^*\left[\frac{A^*}{A\ I(A)}\right]^{1/m}$$

(7.32)

where S is the surface over which the integration is carried out, A is the surface area, A^* the surface area over which the measured reference stress σ^* was determined and $\sigma(A)$ is the stress as a function of position on the surface.

The measurement of σ_r for a given reference volume is difficult because it would be necessary to stress uniformly a volume of material in tension without the introduction of further defects during manufacture. However, it is possible to use the stress–volume integral approach to determine σ_r from a series of measurements on any test geometry as long as the stress distribution is accurately known. A common testing geometry is shown in Fig. 7.12(a), where the strength of a rectangular-section ceramic bar is tested in three-point bend. The maximum tensile stress occurs on the top surface opposite the central loading point and is given by:

$$\sigma_{max} = \frac{3PL}{2wt^2}$$

(7.33)

where the beam dimensions are as given in Fig. 7.12(a). If the reference stress is taken as this maximum stress (i.e. $\sigma_{max} = \sigma_r$), the stress distribution in the bar is given by:

$$\sigma(V) = \frac{4x(L/2 - |y|)}{Lt}\sigma_r$$

(7.34)

The stress–volume integral for the bar under three-point bend is then:

$$\int_V\left(\frac{\sigma(V)}{\sigma_r}\right)^m\frac{dV}{V} = 2\int_0^{L/2}\int_0^{t/2}\left[\frac{4x(L/2 - |y|)}{Lt}\right]^m\frac{dx}{L}\frac{dy}{t}$$

$$= \left[\frac{1}{2(m + 1)^2}\right]$$

(7.35)

From Eqn (7.30) the relationship between the three-point bend mean fracture stress and the reference stress σ^* will be:

$$\frac{\bar{\sigma}^*}{\bar{\sigma}_{bend}} = \left[\frac{V_{bend}}{2V^*(m + 1)^2}\right]^{1/m}$$

(7.36)

For the case when the reference stress and bend stress are equal, Eqn (7.36) will predict the ratio between the mean tensile strength and mean three-point bend strength for a given ceramic specimen. If $m = 10$, a reasonably high value of the Weibull modulus typical of many

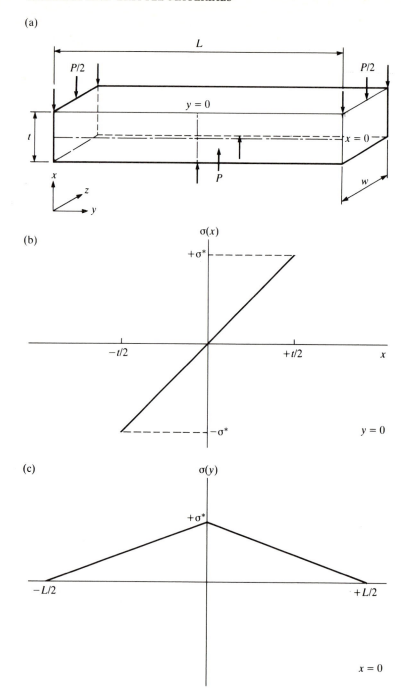

Figure 7.12 Schematic of three-point bend specimen showing stress distribution within the bar: (a) the dimensions of the bar and orientation with respect to the axes; (b) the stress distribution along the x direction of the bar at $y = 0$; (c) the stress distribution along the y direction of the bar at $x = 0$

engineering ceramics, this equation predicts $\bar{\sigma}^*/\bar{\sigma}_{bend} = 0.577$. This analysis of the stress—volume integral indicates that strengths measured in three-point bend will always overestimate the mean strength of a ceramic in tension.

7.3.5 TIME-DEPENDENT STRENGTH

So far we have assumed that a ceramic component has its strength limited by the presence of a collection of flaws introduced during fabrication or subsequent handling. These flaws, once created, remain of fixed length unless they become unstable under load and grow catastrophically as a fracture. However, the strengths of many brittle ceramics, in particular silica glass, are found to decrease with increasing time under load. This phenomenon, known as *stress rupture* or *static fatigue*, is often noted during a simple laboratory test when the strength of a glass rod is assessed by hanging weights from its centre while it is suspended in a 3-point bend from its ends. In this experiment the load on the rod is gradually increased by adding more weights. Sometimes the rod fails not when the load is increased but a short while afterwards and before the next load increment is added. If the Griffiths criterion for fracture initiation is still satisfied, tis implies that an existing flaw has increased in size under load. To a design engineer this phenomenon can be likened to creep in metals, i.e. time-dependent plastic flow below the yield stress. The information required is similar too — a suitable method of assessing the life of a component under a given load is needed. Such an analysis, we will find, requires a statistical treatment for brittle materials. However, before this is achieved, the mechanisms of crack growth under load must be understood.

In all cases where static fatigue is observed, the mechanism is believed to be the subcritical growth of existing cracks or flaws under the action of the applied stress. This effect is seen under two environmental regimes. At high temperatures, above about half the absolute melting temperature, all ceramics experience crack growth under stress. The precise mechanisms are unclear, and there may well be several acting in parallel; however, thermally activated dislocation motion in the grains and void growth in glassy grain-boundary phases have both been suggested. There is a general relation between the velocity of crack advance (v) and the stress intensity (K_I) such that:

$$v = C_1 K_I^n \tag{7.37}$$

where C_1 is a constant and the exponent n is typically about 10.

At low temperatures there is also evidence for slow, subcritical crack growth in some brittle materials at ambient temperatures. This is apparent in silica glass which, in common with some other oxide

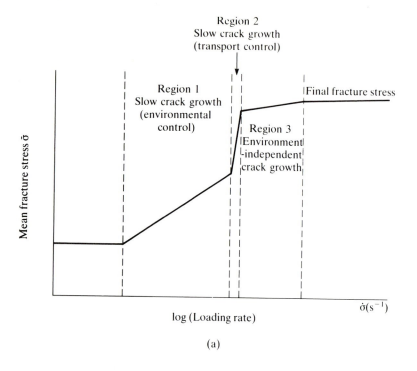

(a)

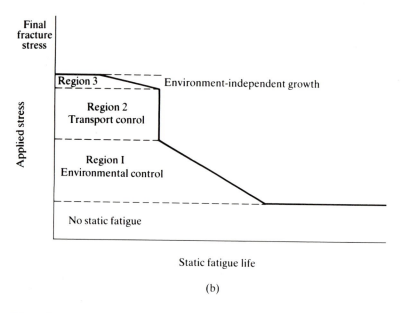

(b)

Figure 7.13 Mean fracture stress of glass rods as a function of (a) loading rate, and (b) time under load

ceramics, is particularly sensitive to water, even in the form of atmospheric moisture. Water is believed to break the interatomic bonds in a corrosive reaction at the highly stressed crack tip. Experimental measurements of dynamic fracture stress as a function of loading rate and time to fracture under static loading are shown in Fig. 7.13. The mean fracture stress is seen to be independent of loading rate below some critical loading rate, indicating a possible critical stress (low) below which no crack growth occurs. Above this critical stress there is a region (region 1) where the time to fracture falls steadily with increasing stress during static tests and the fracture stress rises rapidly with loading rate during dynamic tests. This region is determined by environmental conditions and is an environment-controlled regime of slow crack growth. Above a further critical stress the time to failure is independent of stress. This second fracture regime is believed to be controlled by the rate of transport of the environment to the growing crack tip and is thus independent of the local stress. At high stresses and loading rates the applied stress again controls crack propagation and the behaviour in region 3 is similar to that observed in region 1. In region 3 the growth mechanisms are independent of the environment but are at stresses sufficiently close to the ultimate defect-controlled fast fracture to be of little design importance. The low-stress regime of region 1 on the diagram is of most design importance because within this regime of slow crack growth the majority of the specimen's life will occur. Here it is found that the crack growth rate can be modelled by an equation functionally equivalent to Eqn (7.37):

$$v = C_2 K_I^n \tag{7.38}$$

In this case n is measured in the range $10-20$ for most oxide ceramics and glasses. Other non-oxide ceramics, such as silicon nitride and silicon carbide, do not show this effect to any degree. Because both mechanisms of crack advance can be described by similar equations, it is possible to develop a general treatment of the influence of subcritical crack growth on ceramic strength.

If a ceramic is loaded to a stress σ, it will fail when any flaw present grows to a critical length a_f, defined by the Griffiths equation. Under conditions of constant stress this will be achieved in a time t_f defined by the crack velocity such that:

$$t_f = \int_{a_i}^{a_f} \frac{da}{v} \tag{7.39}$$

where a_i is the initial length of the flaw and v its velocity. The stress intensity factor of a crack in a thin plate under plane stress conditions is given by:

$$K_I = \sigma\sqrt{\pi a} \tag{7.40}$$

and hence from Eqn (7.38)

$$\sigma^n t_f = \int_{a_i}^{a_f} \frac{da}{C_2(\pi a)^{n/2}}$$

$$= \left[\frac{2a^{(1-n/2)}}{\pi^{n/2} C_2(2-n)} \right]_{a_i}^{a_f} \qquad (7.41)$$

Given that the crack velocity exponent n is typically about 10 and that for most applications of interest to designers $a_i \ll a_f$, Eqn (7.41) can be rewritten:

$$t_f = \frac{2a_i^{(1-n/2)}}{C_2(2-n)(\sqrt{\pi}\sigma)^n} \qquad (7.42)$$

Hence, for any specimens of the same initial maximum flaw size, the relation

$$t_f \sigma^n = \text{constant} = C_3 \qquad (7.43)$$

must apply. Thus in a particular specimen, with the initial crack length set constant, Eqn (7.43) can be rewritten to give a relation between lifetimes t_1, t_2, at given stresses σ_1, σ_2:

$$\left(\frac{\sigma_1}{\sigma_2} \right)^n = \left(\frac{t_1}{t_2} \right) \qquad (7.44)$$

This equation can be used to define lines of gradient n on a Weibull strength plot (Fig. 7.14) which represent equivalent times to failure at different stresses. These composite plots are known as *strength–probability–time* (SPT) diagrams and were first proposed by Davidge.

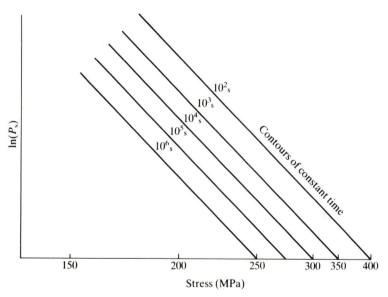

Figure 7.14   Strength–probability–time (SPT) diagram for an alumina ceramic

In order to generate these SPT diagrams a number of tests should be performed to measure the probability of failure at given strengths after given times under load. This can be carried out more conveniently by testing the material under a range of strain rates. Normally specimens are tested at a constant strain rate in a tensile testing machine, i.e. the specimen is gradually loaded from zero to its failure stress (σ_f) at a constant rate. SPT diagrams can then be constructed from the measured mean strengths of materials tested at different strain rates. If a material is tested at a constant strain rate (and hence a constant stress rate $\dot{\sigma}$), the stress at any time t will be given by

$$\sigma = \sigma(t) = \dot{\sigma}t \tag{7.45}$$

This can now be substituted into Eqn (7.41) to give:

$$\int_0^{t_f} \sigma^n \, dt = \dot{\sigma}^n \int_0^{t_f} t^n \, dt = \int_{a_i}^{a_f} \frac{da}{C_2(\pi a)^{n/2}} \tag{7.46}$$

which, assuming $a_i \ll a_f$, integrates to

$$\frac{\dot{\sigma}^n}{(n+1)} t_f^{n+1} = \frac{2a_i^{(1-n/2)}}{C_2(2-n) \, \pi^{n/2}} \tag{7.47}$$

The fracture stress at time t_f will be given by $\sigma_f = \dot{\sigma} \, t_f$, assuming $\dot{\sigma}$ to be a true constant, and so Eqn (7.47) can be rewritten

$$\sigma_f^{n+1} = \frac{2(n+1) \, \dot{\sigma} \, a_i^{(1-n/2)}}{C_2(2-n) \, \pi^{n/2}} \tag{7.48}$$

which reduces to

$$\sigma_f = C_4 \, \dot{\sigma}^{1/(n+1)} \tag{7.49}$$

and an analogous relation to Eqn (7.44) can be used to define the gradient n of the SPT diagram

$$\frac{\sigma_{f1}}{\sigma_{f2}} = \left(\frac{\dot{\sigma}_1}{\dot{\sigma}_2} \right)^{1/(n+1)} \tag{7.50}$$

The SPT diagram can be used as a design tool allowing us to estimate component lifetimes in terms of failure probabilities after a time at a given stress level. Note that the diagram contains an implicit crack growth law of $\dot{a} = C_1 K^n$ with a given C_1 and n. If different environmental conditions were to apply (e.g. a change in temperature) then a new SPT plot must be determined, or possibly a modification made to the crack growth law to take this into account.

An important implication of the SPT diagram is that there is no intrinsically safe design stress for a ceramic component. Instead we must satisfy ourselves with an understanding of the probability of failure under various conditions of loading. Thus an important contribution can be made by proof-testing to remove the 'tail' of any flaw distribution. However, proof-testing cannot stop flaws below the proof level from growing subcritically. Indeed, the very stress level used to proof-test could be itself responsible for some subcritical crack growth. Thus proof conditions must be chosen carefully to eliminate unsatisfactory components without weakening the survivors excessively.

In both the time-independent and time-dependent treatment of ceramic strength, the most important parameter which determines strength is clearly the distribution of cracks and related flaws in the material. The precise treatment of the Weibull approach is affected by the position of the flaws with different integrals required for predominantly surface flaws and volume flaws. In all these cases the flaw distribution is definitely *not* a simple material parameter but is a combination of the material's intrinsic resistance to crack nucleation and the precise sequence of manufacturing events leading to the final fabricated brittle component. Thus great care must be exercised in extrapolating from test data to design strengths. In practice, any test pieces used to provide design data should be fabricated in the same way as the component being designed in order to ensure a similar population of defects. This often requires a test piece to be of similar size to the final component because the techniques used to manufacture glass and ceramic parts are often very different for different-sized components.

7.4 THERMAL STRESSES

Earlier, we showed how changes in temperature can result in the generation of flaws by an internal stress driven by local differences in the coefficient of thermal expansion. Thermal stresses of similar magnitude can also be generated within a large component if various regions are held at different temperatures, as might happen if there was a heat flow down a temperature gradient. This can occur on cooling from fabrication temperatures if a specimen is thick, or if it is subjected to local heating. These effects are grouped under the generic title of thermal shock and can be particularly severe if the ceramic material has a large coefficient of thermal expansion or a low thermal conductivity.

If a large, solid ceramic body is cooled slowly it will develop a temperature differential (ΔT) between its outer surface and its interior. The thermal expansion of a ceramic can be taken as linear with temperature and hence this temperature distribution will generate a strain distribution within the specimen. The peak strain will be given by an equation analogous to Eqn (7.1). Such strain distributions will always occur if there are differences in temperature across parts of a material. Consider a ceramic part which acts as a divider between two different temperature environments, e.g. a transparent view-port on the side of a furnace. During operation a steady heat flux will occur through the port, setting up a temperature gradient across it. The temperature gradient will cause differential thermal expansion across the port and introduce a bend distortion. As a very simple example, consider the port to act as a beam, i.e. we reduce the problem to two dimensions. If the beam is thin and of thickness t there will be a constant temperature gradient $\Delta T/t$ across its length and the different thermal expansion induced by this gradient will introduce a bend in the beam of radius

$$r = \frac{t}{\Delta T \alpha} \tag{7.51}$$

If the beam is unconstrained this will generate no stresses within it. However, if the beam is constrained from bending, a moment must be generated to keep the beam straight and this will introduce a stress distribution across the beam with a maximum top-surface tensile stress of

$$\sigma_{max} = \frac{E \alpha \Delta T}{(1 - \nu)} \tag{7.52}$$

which is identical to the stress predicted by Eqn (7.1) for a fixed temperature and a variation in CTE. Thus thermal stresses are generated by a combination of differences in temperature and mechanical constraints resisting the resultant thermal deformation.

The mechanical response of brittle materials to differences or changes in temperature is a very important design consideration since many potential applications for ceramics are at high temperatures. For realistic design purposes, thermal stresses will be more accurately determined using numerical stress analysis techniques such as the finite element method. Nonetheless, certain materials parameters relevant to mechanical and thermal response are important in determining thermal shock resistance of materials as an aid to materials selection prior to complex design. To quantify thermal shock resistance fully, the likely thermal and mechanical loads on a component need to be assessed; however, these can be simplified to idealized cases for comparison.

For an infinite rate of heat transfer the surface of the ceramic instantly acquires a new temperature T_1 and the difference between this and the original temperature T_0 is $\Delta T = T_1 - T_0$. The resulting thermal stress will be tensile if ΔT is negative, and it will be given by Eqn (7.42). The thermal shock resistance parameter R_1 is given by the maximum temperature change the ceramic can experience before failure occurs, i.e.

$$R_1 = \frac{\bar{\sigma}_F(1-\nu)}{E\alpha} \qquad (7.53)$$

where $\bar{\sigma}_F$ is the mean fracture stress of the component.

If the rate of temperature change at the specimen surface is low or if the material has a high thermal conductivity, there will be a conduction of heat into or out of the ceramic, in which case the surface temperature changes more slowly, reduced thermal stress levels are experienced and R_1 is no longer suitable as a thermal shock parameter. The surface temperature will now depend on both the relative rate of heat transfer into the material and the rate of conduction of the heat into the interior of the specimen. These are in turn dependent on both the method of surface heating and the material. If the heat transfer coefficient H_T is given in units of $W\ m^{-2}\ K^{-1}$ and the thermal conductivity of the material (k) is given by $W\ m^{-1}\ K^{-1}$, then it is possible to define a dimensionless number representative of the relative heating efficiencies of materials — the Biot modulus (β) — with:

$$\beta = \frac{H_T\, x}{k} \qquad (7.54)$$

where x is a characteristic linear dimension of the material. On heating a material the surface maximum stress is given by:

$$\sigma_{max} = \frac{E\alpha\Delta T}{(1-\nu)}\, f(\beta) \qquad (7.55)$$

The function $f(\beta)$ is difficult to determine universally but when β is small $f(\beta)$ is linearly related to β, giving:

$$\sigma_{max} = B_1 \frac{E\alpha\Delta T}{(1-\nu)} \left(\frac{H_T\, x}{k} \right) \qquad (7.56)$$

where B_1 is a constant. From Eqn (7.56) a second thermal shock parameter can be defined:

$$R_2 = \frac{\bar{\sigma}(1-\nu)}{E\alpha}\, k \qquad (7.57)$$

but now the maximum temperature drop which can be withstood is given approximately by $\Delta T = R_2/B_1\, xH_T$ which is dependent on a specimen dimension.

If the ceramic surface is experiencing a constant temperature change $\dot{T}(K\ s^{-1})$, the surface stress is found to depend on the thermal diffusivity $D_T = k/(\rho C_p)$, where ρ is the materials density and C_p its heat capacity, such that:

$$\sigma_{max} = \frac{B_2 E\alpha}{(1-\nu)} \left(\frac{\dot{T}}{D_T} \right) \qquad (7.58)$$

Table 7.3 Thermal
shock resistance
parameters for some
ceramics

Property at 500 °C	Soda glass	Pyrex glass	Alumina	Hot-pressed silicon carbide	Hot-pressed silicon nitride	TZP zirconia
Young's modulus (GPa)	70	70	345	400	400	200
Mean fracture strength (MPa)	65	70	400	600	600	1000
Poisson's ratio	0.20	1.13	0.22	0.19	0.23	0.22
Mean expansion coefficient ($K^{-1} \times 10^{-6}$)	9.0	3.2	7.7	4.5	2.6	8.4
Thermal conductivity ($W\ m^{-1}\ K^{-1}$)	1.70	1.70	11.0	80	5.0	2.3
Density ($kg\ m^{-3} \times 10^3$)	2.5	2.4	3.8	3.2	2.6	5.8
Heat capacity ($J\ kg^{-1} \times 10^3$)	1.13	1.13	1.16	1.12	1.05	3.98
R_1 (K)	80	250	120	270	440	460
R_2 ($W\ m^{-1}$)	140	430	1290	21600	2220	1070
R_3 ($m^2\ K\ s^{-1} \times 10^{-6}$)	50	160	290	6030	810	4.6

where B_2 is a constant. This gives a third thermal shock parameter related to the maximum allowable rate of surface heating:

$$R_3 = \frac{\bar{\sigma}_F(1-\nu)D_T}{E\alpha} = \frac{\bar{\sigma}_F(1-\nu)}{E\alpha}\left(\frac{k}{\rho C_p}\right) \qquad (7.59)$$

These different thermal shock parameters are tabulated for a number of brittle materials in Table 7.3. On comparison with the materials' data in Table 7.2 it can be seen that high strength, high thermal conductivity and low expansion coefficient lead to the best general thermal shock properties for these materials.

7.5   GENERAL DESIGN PRACTICE

With highly brittle materials there can be no absolute certainty as to fracture strength. A probabilistic approach combined with pre-service proof-testing can be used to produce adequate results from a design exercise. The Weibull approach assumes a very high sensitivity of failure probability to applied stress and hence any local stress raisers such as sharp corners or notches should be eliminated. Residual stresses from fabrication or thermal stresses under service conditions can also significantly increase the risk of failure. However, if defects can be kept to a minimum, the high stiffness and low densities of many ceramics (Table 7.2) result in a material with many potential applications, especially at high temperatures where the high melting points of most ceramics allow them to be used in conditions unsuitable for many engineering alloys. However, their properties of high stiffness and low thermal expansion coefficient make them very difficult to use in contact with metals because these differences can generate further service stresses at the metal/ceramic boundary.

Ceramics cannot be machined by the techniques developed for metals, and very expensive diamond grinding is utilised. Thus it is often necessary to manufacture the ceramic component to as near the final shape as possible in the initial fabrication stages to reduce subsequent machining. Ceramics are highly sensitive to surface defects, which are easily introduced during manufacture. In practice, engineers have found ways of modifying the surfaces of brittle materials to reduce the influence of these defects. One such technique is to introduce a residual compression into the surface which acts as a closure force on surface defects. This can be done in a number of ways but one of the most widespread is the technique used to toughen glass for automobile use. In this case the reduction in volume on solidification of the glass is used to introduce compressive surface stresses. A sheet of glass is held at a temperature where it is just plastic and air jets are blown on to it, cooling the surface and solidifying the top layers of the material. The central regions solidify later and their contraction draws the already-solid surface into compression. This results in a stress distribution with the outer layers in residual compression and the inner region in tension to preserve stress equilibrium. The residual surface compressive stress increases the external stress required to initiate the fracture, but now if a crack propagates the stored elastic energy in these residual stresses leads to massive multiple cracking on failure. Similar surface stresses can be introduced chemically by substituting larger ions in the glass or ceramic surface structure, either by an ion-exchange mechanism or by using high-energy ion beams to implant them physically into the crystal lattice.

This chapter is intended to give the reader an introductioin to the problems faced by the engineer when he has to work with highly brittle materials. More comprehensive studies are to be found in specialist books on ceramic engineering,[3,4] which also cover the problems inherent in manufacturing ceramic components.

7.1   Silicon carbide has a fracture toughness of approximately
       5 MPa m$^{1/2}$. Calculate the typical fracture-initiating defect size
       in:

       (a)   a sintered silicon carbide rod tested in three-point bend
              with a mean fracture stress of 500 MPa;

       (b)   a silicon carbide fibre tested in tension with a strength
              of 4 GPa.

7.2   Discuss the likely mechanisms responsible for the high
       density of fracture-inducing flaws found in most engineering
       ceramics.

7.3   A series of tests show that a ceramic component has a mean
       failure probability of 0.25 when stressed by 207 MPa and a
       probability of 0.1 when stressed 176 MPa. Determine the
       Weibull modulus and the maximum allowable stress if a
       failure of only one component in a thousand can be
       tolerated.

7.4   The following table shows the measured strengths of a
       number of 3 mm square-section ceramic bars. These tests
       were carried out in four-point bend over a 60 mm length of
       specimen with the outer loading points spanning 55 mm and
       the inner ones 25 mm. Calculate the expected mean fracture
       strength of a 50 mm length of the bar tested in tension.

Specimen number	Bend strength (MPa)	Specimen number	Bend strength (MPa)
1	73.2	11	65.3
2	65.3	12	70.2
3	50.7	13	64.5
4	72.3	14	54.8
5	49.7	15	57.9
6	59.0	16	52.0
7	59.1	17	57.1
8	49.1	18	61.2
9	49.2	19	54.4
10	65.8	20	54.4

7.5   What materials properties are most important for the
       following cases where ceramics are subjected to heat:

(a)   the inner lining of a furnace designed to reach 1600 °C;

(b)   the ceramic hotplate on an infrared heating cooker?

REFERENCES

1   KINGERY W D, BOWEN H K, UHLMANN D R (1975) *Introduction to Ceramics* 2nd edn, John Wiley, New York

2   DAVIDGE R W (1981) *Mechanical Behaviour of Ceramics* Cambridge University Press, Cambridge

3   RICHERSON F W (1985) *Modern Engineering Ceramics* Marcel Dekker, New York

4   CREYKE W E C, SAINSBURY I E J, MORRELL R (1982) *Design with Non-ductile Materials* Applied Science Publishers, London

CHAPTER EIGHT

Composites

8.1 INTRODUCTION AND DEFINITION

In the preceding chapter some of the properties of ceramics were discussed. Many ceramic materials have very high elastic moduli and ideal strengths, but the advantages these properties bestow are often outweighed by their highly brittle nature, which leads to low and unpredictable failure stresses from the presence of flaws. Therefore, in order to improve the strength of a ceramic component, two approaches are followed, in which either extreme care is taken to minimize the presence of flaws, or the ceramic is toughened and made more resistant to cracks. At present (1991), the toughest (and most expensive!) ceramics have a fracture toughness no better than cast iron.

Another approach to the utilization of brittle materials is to impede the crack propagation stage rather than concentrating on the condition for fracture initiation as defined by the Griffiths criterion. Catastrophic failure occurs during brittle fracture when the crack travels without interruption across the material, resulting in separation of the broken halves. If instead the stressed component consists of a number of discrete pieces, the fracture of one piece will not necessarily lead to complete failure. For example, let us consider the case illustrated in Fig. 8.1. Here we have a weight suspended from a fixing by a large ceramic rod, or by an array of smaller rods of the same composition as the first rod and of identical total cross-sectional area so that both arrangements are stressed to the same extent. As both configurations are identical they will have the same mean flaw size distribution and thus the same failure probability, as can be defined using Weibull statistics (see Chapter 7). If the largest flaw present in each example is such as to exceed the critical flaw size for fracture, there will be catastrophic failure in the first case but the failure of only one of the rods in the second case. With an array of N rods loaded to a stress σ, a single failure will result in an increase in the mean stress on the

(a)   (b)

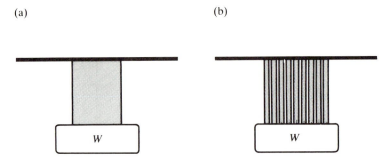

Figure 8.1   Weight suspended by a brittle material. (a) One large column gives a high risk of complete and catastrophic failure. (b) If several small columns are used there is a higher probability that one column will break but, because there are many separated pieces, there is a lower probability of complete failure

Figure 8.2   The core and sheath construction of an abseiling rope reduces the chance of catastrophic failure under load (rope courtesy of Marlow Ropes)

surviving rods of $\sigma N/(N-1)$. Another rod will fail only if the new stress exceeds its failure stress as defined by the next-largest flaw. Thus we can see that decreasing the size of the rods, and hence increasing their number, decreases the chance of catastrophic failure. It is not surprising to discover that climbing ropes are designed in this way with many parallel polymer fibres held together by a sheath (Fig. 8.2).

Reducing the size and increasing the number of brittle components in a critical system is the principle on which many engineering composite materials are based. Strong yet brittle fibres of glass, carbon

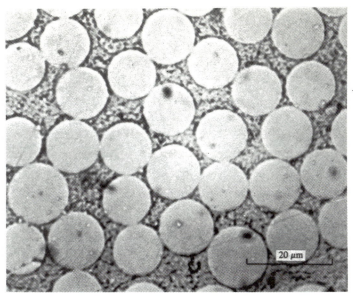

Figure 8.3   Transverse cross-section through a glass fibre reinforced polyester resin showing the high density of fibres present (from Hull 1981)

or ceramic are incorporated into weaker polymer matrices, the fibres giving great strength and stiffness with the matrix function confined mainly to fixing the bundles of fibres together. A section through a typical composite material of glass fibres and polyester resin is shown in Fig. 8.3. A large fraction of the cross-sectional area is taken up by the fibres present and the mechanical properties of the composite approach those of the fibre alone. In this chapter we shall review the properties of composites and show that some can be thought of as the result of a simple mixing of the properties of the constituent components of the composite, whilst other properties are determined by mechanisms unique to the composite material's structure.

8.2   ELASTIC PROPERTIES OF COMPOSITES

8.2.1   THE RULE OF MIXTURES

Consider the simplest possible composite structure consisting of alternate layers of constant-thickness material as illustrated in Fig. 8.4. If this material is stressed in tension parallel to the layers, all the layers must deform elastically to the same extension in order to preserve

(a)

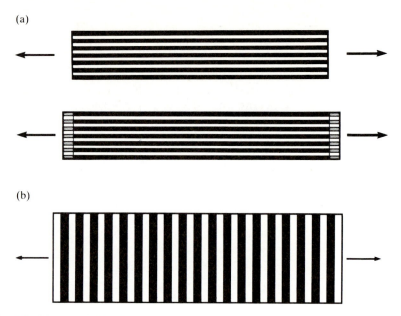

(b)

Figure 8.4 Ideal limits of the elastic behaviour of a laminated structure. (a) Stress applied parallel to the axis of the laminate; continuity ensures both components must be strained to the same extent (Voigt limit). (b) Stress applied perpendicular to the laminate; equilibrium requires the same stress in each component (Reuss limit)

material continuity, or strain compatibility (Fig. 8.4(a)). If the two materials are material a and material b and if they both undergo a strain ϵ_c because of an applied stress σ_c, each individual layer in the composite will experience a stress as defined by:

$$\epsilon_c = \epsilon_a = \epsilon_b = \frac{\sigma_a}{E_a} = \frac{\sigma_b}{E_b} = \frac{\sigma_c}{E_c} \tag{8.1}$$

where E is the Young's modulus and the subscripts a, b and c refer to the properties of the constituents a, b, and the resultant composite respectively. The force (F) can be calculated in each layer with

$$\sigma_a = \frac{F_a}{A_a} = \frac{\epsilon_c E_a}{A_a}$$

$$\sigma_b = \frac{F_b}{A_b} = \frac{\epsilon_c E_b}{A_b} \tag{8.2}$$

The mean stress can be calculated from the total force per area of the composite and the Young's modulus of the composite defined thus:

$$E_l = \frac{\sigma_c}{\epsilon_c} = \frac{E_a A_a}{A} + \frac{E_b A_b}{A} \tag{8.3}$$

where the total area of the composite A is $A = A_a + A_b$ and where the subscript l is used to identify the longitudinal Young's modulus parallel to the layers of the composite. In this case, because the components run continuously along the direction perpendicular to the area A, the area fractions are equivalent to the volume fractions V_a, V_b of each component, i.e. $V_a = A_a/A$ and $V_b = A_b/A$ with $V_a + V_b = 1$, and Eqn (8.3) can be rewritten:

$$E_l = E_a V_a + E_b V_b \tag{8.4}$$

If the composite is loaded perpendicular to its constituent layers (Fig. 8.4(b)), each layer must sustain the same stress in order to preserve equilibrium, i.e.

$$\sigma_c' = \sigma_a' = \sigma_b' = \epsilon_a' E_a = \epsilon_b' E_b \tag{8.5}$$

The composite will extend by an amount δ_c given by the sum of the extension of each component, δ_a and δ_b. These will be determined by the total thickness of each component L_a and L_b perpendicular to the layers, i.e.

$$\delta_c = \delta_a + \delta_b = L_a \epsilon_a' + L_b \epsilon_a' = \frac{L_a \sigma_a'}{E_a} + \frac{L_b \sigma_b'}{E_b} \tag{8.6}$$

so that the mean 'composite' strain is given by $\epsilon_c' = \delta_c/L$. The composite strain ϵ_c' is related to the stress transverse to the layers by the composite modulus in this direction. Young's modulus is then given by:

$$\frac{1}{E_t} = \frac{\epsilon_c'}{\sigma_c'} = \left(\frac{L_a}{LE_a} + \frac{L_b}{LE_b} \right) \tag{8.7}$$

where the subscript t indicates the modulus transverse to the layers. With the composite of Fig. 8.4, the length fractions of each component are equivalent to the volume fractions (as were the area fractions earlier), i.e. $V_a = L_a/L$ and $V_b = L_b/L$, and Eqn (8.7) can be rewritten:

$$\frac{1}{E_t} = \frac{V_a}{E_a} + \frac{V_b}{E_b}$$

or

$$E_t = \frac{E_a E_b}{V_a E_b + V_b E_a} \tag{8.8}$$

The above simple derivations of the two orthogonal elastic constants are usually known as the rule of mixtures. They both assume very simplified, ideal geometries of the components in a composite. Equation (8.4) is known as the Voigt approximation and it assumes that both components of the composite undergo the same strain. This maximizes the stiffness of the composite and is an upper bound of the possible composite elastic moduli. Equation (8.8) is the Reuss approximation. In this geometry both components hold the same stress

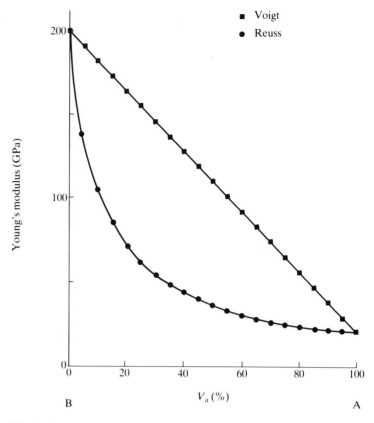

Figure 8.5 The Voigt and Reuss equations act as upper and lower bounds for the prediction of the elastic modulus of a fibre-reinforced composite structure

and thus for a given stress both can extend as if they were in isolation. This gives the minimum possible stiffness and is a lower bound on the composite elastic modulus. The variation in E_l and E_t with V_a and V_b predicted by these equations is illustrated in Fig. 8.5. In both simple models we have neglected Poisson's ratio, which would be expected to constrain the deformation in both of the geometries discussed above.

The majority of composites are manufactured from stiff fibres and not from plates. Thus the elasticity relations derived above will need to be modified slightly to reflect this geometry and predict the true mechanical behaviour of real composites. Equation (8.4) is found to be a good approximation to the elastic behaviour of fibre-reinforced composites when measured parallel to the fibre direction, if the fibres are sufficiently long for us to assume that they extend along the length of the composite. This is because the assumption which leads to this formula, that the strain is equal in each component, will also be approximately satisfied when the fibres are loaded along their axis. The modulus normal to the fibre direction will be smaller than this

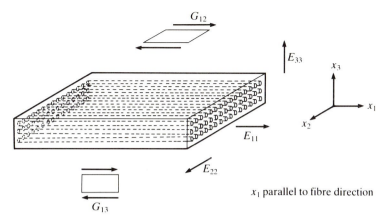

Figure 8.6   Composite materials are usually fabricated as thin sheets. A standard
coordinate system is used with the three orthogonal directions x, y and z.
These are conventionally represented by the subscripts 1, 2 and 3 when
describing stresses, strains and elastic constants of the composite, as
illustrated above

but will be slightly greater than the limit of Eqn (8.8) because the equal
stress approximation is not satisfied when an array of fibres is loaded
normal to their axes. In this geometry the loading conditions will be
intermediate between the Voigt and Reuss limits.

The calculation of all the elastic properties of a single sheet of a fibre-
reinforced composite material is clearly a very complicated elastic
analysis, well beyond the scope of this book. However, there are some
simple empirical extensions to the general rule of mixtures approach
that can be used to give adequate approximations to many composite
elastic properties. Using the coordinate system of Fig. 8.6 to define
the axes and principal elastic constants of a composite laminate, a series
of approximate equations were put forward by Halpin and Tsai, with
the subscripts f and m used to identify the fibre and matrix components
of a composite. They considered the rule of mixtures as an adequate
representation of the Young's modulus in the direction parallel to the
fibres, identified by the subscript 11 in Fig. 8.6:

$$E_{11} \approx E_f V_f + E_m(1 - V_f) \qquad (8.9a)$$

The rule of mixtures was also used to calculate the Poisson's ratio of
the composite when strained parallel to the fibres. Referring to Fig. 8.6
there will be two Poisson's ratios describing the strain perpendicular
to the fibres either within the plane of the composite sheet or
perpendicular to it. These two Poisson's ratios must be identical
because of the homogeneous distribution of fibres in each sheet (see
Fig. 8.3).

$$\nu_{12} = \nu_{13} \approx \nu_f V_f + \nu_m(1 - V_f) \qquad (8.9b)$$

For all other elastic properties, e.g. the Young's moduli perpendicular to the fibre directions E_{22}, E_{33} or the shear moduli G_{12}, G_{13}, G_{23}, and the remaining Poisson's ratio ν_{23}, Halpin and Tsai proposed the relation

$$\frac{M}{M_m} = \frac{(1 + \alpha\chi V_f)}{(1 - \chi V_f)} \tag{8.9c}$$

where

$$\chi = \frac{\left(\dfrac{M_f}{M_m} - 1\right)}{\left(\dfrac{M_f}{M_m} + \alpha\right)} \tag{8.9d}$$

In the above equations, M is a composite elastic constant (e.g. E_{22}, G_{12}, ν_{23}) for which the rule of mixtures does not give an adequate result, M_f and M_m are the corresponding values of this elastic constant for the fibre and matrix respectively, and α is a geometrical or shape factor which is altered to get the best fit for different reinforcement arrangements. For a simple rectangular array of long circular-section fibres, the constant $\alpha = 2$.

Many composite materials are fabricated from long stiff fibres of graphite (carbon fibres), glass or high-modulus polymers combined with a light, low-modulus polymer matrix. The mechanical properties of several of these materials are listed in Table 8.1. From these values and Eqns (8.4) and (8.8), we can see that there will be a considerable difference between the values of E_i and E_t for many typical composites. Composite materials can be used to exploit the very high Young's modulus and low densities of brittle reinforcing fibres. If a composite is manufactured from 40% by volume of carbon fibres and the balance epoxy resin (materials' data from Table 8.1), it will have a density of 1500 kg m^{-3} and a Young's modulus parallel to the fibres of 158 GPa. This produces a material of high modulus and low density which is said to have a high *specific stiffness* (E/ρ). The specific stiffness of this composite is compared with that of a number of common engineering materials in Table 8.2. Composites, especially those manufactured using carbon fibres in polymer matrices, have many potential applications where low weight is a premium.

8.2.2 ELASTIC PROPERTIES OF A UNIDIRECTIONAL COMPOSITE LAMINATE

The large difference between the elastic moduli of a composite measured in the directions parallel and perpendicular to the fibre direction leads to severe problems of anisotropy. The composite discussed in the previous section had a modulus parallel to the fibres

Table 8.1
Mechanical
properties of some
reinforcing fibres and
light matrices

Material	Density $\rho(\text{kg m}^{-3} \times 10^3)$	Young's modulus $E(\text{GPa})$	Tensile strength σ^* (GPa)	Fibre radius $r(\mu\text{m})$
E-Glass fibres	2.56	76	1.4–2.5	10
Carbon fibres (high-modulus)	1.75	390	2.2	8.0
Carbon fibres (high-strength)	1.95	250	2.7	8.0
Kevlar fibres	1.45	125	3.2	12
Silicon carbide (monofilament)	3.00	410	8.6	140
Silicon carbide (Nicalon)	2.50	180	5.9	14
Alumina (Saffil)	2.80	100	1.0	3
Epoxy resin	1.2–1.4	2.1–5.5	0.04–0.08	—
Polyester resin	1.2–1.4	1.3–4.5	0.04–0.08	—
Aluminium (strong alloy)	2.70	71	0.6	—

Table 8.2 Specific
properties of some
materials

Material	Young's modulus E (GPa)	Tensile strength σ^* (MPa)	Density ρ (kg m^{-3} × 10^3)	E/ρ (Nm kg^{-1} × 10^6)	σ^*/ρ (Nm kg^{-1} × 10^3)
Aluminium (strong alloy)	71	600	2.71	26	221
Titanium alloy	110	500	4.54	24	110
Nylon	4	40	1.15	3	35
Mild steel	210	460	7.86	27	58
40% Carbon fibre-reinforced epoxy	158	880	1.50	105	590
40% Glass fibre-reinforced polyester	32	800	1.74	18	460

of 158 GPa but if the Halpin–Tsai equations are used (Eqns 8.9), the modulus perpendicular to this direction is only 9 GPa. Clearly there is expected to be a smooth change in the measured Young's modulus as the testing direction is varied between these two extreme axes. The Halpin–Tsai equations allow calculation of the composite elastic properties in a coordinate system defined by Fig. 8.6. In order to determine the general elastic behaviour of a unidirectional composite laminate, the change in these properties with a change in loading direction is required. In elasticity theory the properties of such anisotropic materials are usually treated using a tensor notation and such an approach can be followed in specialist texts on composite behaviour.[1–3] Here we will use a simpler approach based on the principles of two-dimensional axes rotation using the Mohr's circle construction.

Composite structures fabricated from arrays of long fibres are usually constrained by production problems to thin sheet shapes. These sheets can be moulded at a later stage into more complex shapes but for the purposes of this analysis thin plane sheets with the z axis much thinner than the x or y directions (from Fig. 8.6) will be considered. Such a sheet will be loaded in a plane stress configuration and thus only in-plane elastic properties need be considered: E_{11}, E_{22}, v_{12}, v_{21}, and G_{12}. The properties of these materials, known as *orthotropic*, will be investigated now.

Consider a composite lamina with fibres oriented as shown in Fig. 8.6. When a stress is applied along the direction of the fibres, longitudinal and transverse strains result, expressed by the equations:

$$\epsilon_{11} = \frac{1}{E_{11}} \sigma_{11} \tag{8.10a}$$

$$\epsilon_{22} = -\frac{v_{12}}{E_{11}} \sigma_{11} \tag{8.10b}$$

Similarly, when a transversal stress is applied, the strains are

$$\epsilon_{11} = -\frac{v_{21}}{E_{22}} \sigma_{22} \tag{8.10c}$$

$$\epsilon_{22} = \frac{1}{E_{22}} \sigma_{22} \tag{8.10d}$$

Finally, applying a shear stress τ_{12} results in a shear strain γ_{12} defined in

$$\gamma_{12} = \frac{1}{2G_{12}} \tau_{12} \tag{8.10e}$$

The factor of 2 is introduced because this is a true shear strain and not the engineering strain. The deformation of a composite laminate oriented in this particular manner can be combined using matrix notation:

$$
\begin{bmatrix} \epsilon_{11} \\ \epsilon_{22} \\ \gamma_{12} \end{bmatrix} =
\begin{bmatrix}
\dfrac{1}{E_{11}} & \dfrac{-\nu_{12}}{E_{11}} & 0 \\[2ex]
\dfrac{-\nu_{21}}{E_{22}} & \dfrac{1}{E_{22}} & 0 \\[2ex]
0 & 0 & \dfrac{1}{2G_{12}}
\end{bmatrix}
\begin{bmatrix} \sigma_{11} \\ \sigma_{22} \\ \tau_{12} \end{bmatrix}
= [S] \begin{bmatrix} \sigma_{11} \\ \sigma_{22} \\ \tau_{12} \end{bmatrix}
\quad (8.11)
$$

where $[S]$ is the compliance matrix whose inverse $[C]$, the stiffness matrix, can easily be shown to be:

$$
[C] =
\begin{bmatrix}
\dfrac{E_{11}}{1-\nu_{12}\nu_{21}} & \dfrac{\nu_{12}E_{22}}{1-\nu_{12}\nu_{21}} & 0 \\[2ex]
\dfrac{\nu_{21}E_{11}}{1-\nu_{12}\nu_{21}} & \dfrac{E_{22}}{1-\nu_{12}\nu_{21}} & 0 \\[2ex]
0 & 0 & 2G_{12}
\end{bmatrix}
\quad (8.12)
$$

and

$$
\begin{bmatrix} \sigma_{11} \\ \sigma_{22} \\ \tau_{12} \end{bmatrix}
= [C]
\begin{bmatrix} \epsilon_{11} \\ \epsilon_{22} \\ \gamma_{12} \end{bmatrix}
\quad (8.13)
$$

The two terms in both the stiffness and compliance matrices relating strains in the x direction to stresses along the y axis, and vice versa, must be identical from considerations of the fundamental symmetry of elasticity. This implies $\dfrac{\nu_{12}}{E_{11}} \equiv \dfrac{\nu_{21}}{E_{22}}$, thus the two matrices can be written:

$$
[S] =
\begin{bmatrix}
\dfrac{1}{E_{11}} & \dfrac{-\nu_{12}}{E_{11}} & 0 \\[2ex]
\dfrac{-\nu_{12}}{E_{11}} & \dfrac{1}{E_{22}} & 0 \\[2ex]
0 & 0 & \dfrac{1}{2G_{12}}
\end{bmatrix}
\quad (8.14a)
$$

and

$$
[C] =
\begin{bmatrix}
\dfrac{E_{11}}{1-\nu_{12}^2(E_{22}/E_{11})} & \dfrac{\nu_{12}E_{22}}{1-\nu_{12}^2(E_{22}/E_{11})} & 0 \\[2ex]
\dfrac{\nu_{12}E_{22}}{1-\nu_{12}^2(E_{22}/E_{11})} & \dfrac{E_{22}}{1-\nu_{12}^2(E_{22}/E_{11})} & 0 \\[2ex]
0 & 0 & 2G_{12}
\end{bmatrix}
\quad (8.14b)
$$

All the above elastic stiffness or compliance terms can be calculated using the Halpin–Tsai equations. These matrices are also known as the reduced stiffness and compliance tensors of the composite. In both cases the matrices show that an applied tensile or compressive strain either parallel or normal to the fibre axes does not introduce any shear strains. Materials showing these properties are known as orthotropic.

In order to calculate the elastic displacements induced by stresses applied along other orientations it will be necessary to transform these stresses along the orthotropic directions. Once the stresses are oriented thus, the elastic strains can be calculated for this orientation using the compliance matrix above (Eqns 8.11 or 8.14a). These strains can then be transformed back along the axes of the original deformation to predict the change in shape of the laminate. The transformation of axes with respect to fibre and loading direction is conventionally done by tensor notation. However, it is also possible to carry out these transformations in this comparatively simple two-dimensional problem using Mohr's circle construction.

For example, a composite is tested in a pure tensile stress along a direction at an angle θ to the fibre direction. This tensile stress is given by σ_{11} in the axis system $x-y$ (Fig. 8.7). In order to calculate the strain induced by this stress it must be resolved into components parallel and perpendicular to the fibre direction in the coordinate system $x'-y'$. Using Mohr's circle (Fig. 8.8) the stresses in the fibre axes' coordinates become:

$$\sigma'_{11} = \tfrac{1}{2}\sigma_{11}(1 + \cos 2\theta) = \sigma_{11}\cos^2\theta$$

$$\sigma'_{22} = \tfrac{1}{2}\sigma_{11}(1 - \cos 2\theta) = \sigma_{11}\sin^2\theta \qquad (8.15)$$

$$\tau'_{12} = \tfrac{1}{2}\sigma_{11} \sin 2\theta = \sigma_{11}\cos\theta\sin\theta$$

These stresses are now used in Eqns (8.14) to calculate the strains in the composite relative to the fibre direction.

$$\epsilon'_{11} = \sigma_{11}\left(\frac{\cos^2\theta}{E_{11}} - \frac{\nu_{12}\sin^2\theta}{E_{11}}\right)$$

$$\epsilon'_{22} = \sigma_{11}\left(\frac{\sin^2\theta}{E_{22}} - \frac{\nu_{12}\cos^2\theta}{E_{11}}\right) \qquad (8.16)$$

$$\gamma'_{12} = \gamma'_{21} = \sigma_{11} \frac{1}{2G_{12}} \cos\theta\sin\theta$$

Mohr's circle is now used again to transform these strains back to the original $x-y$ axes. In order to determine the Young's modulus in this orientation (E^θ_{11}) only the strain ϵ_{11} need be evaluated, and from Fig. 8.9:

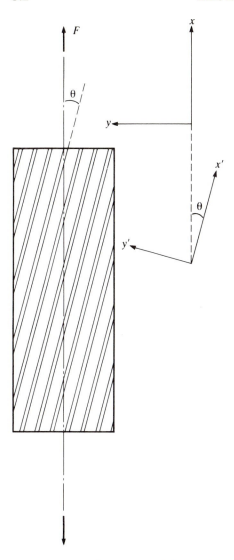

Figure 8.7   A unidirectionally reinforced long-fibre composite loaded at some angle θ from the mean fibre direction. Two axes sets are used, x–y parallel to the loading force and x'–y' aligned parallel to the fibre direction

$$\epsilon_{11} = \tfrac{1}{2}(\epsilon'_{11} + \epsilon'_{22}) + R\cos(2\alpha - 2\theta) \tag{8.17}$$

where R is the radius of the Mohr's circle of strain and α is the angle between the direction of the non-shear components and the direction of principal strains. Combining Eqns (8.16) and (8.17), and using the relations $R\cos2\alpha = \tfrac{1}{2}(\epsilon'_{11} - \epsilon'_{22})$ and $R\sin2\alpha = \gamma'_{12}$ gives:

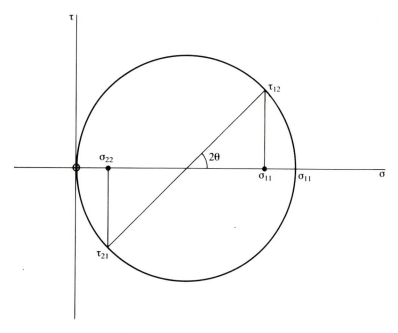

Figure 8.8 Mohr's circle construction is used to resolve the applied force on to the system of axes parallel to the fibre direction

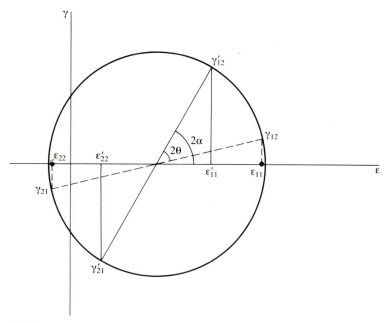

Figure 8.9 Mohr's circle is used to resolve the composite strains determined along the orthotropic directions back to the original set of axes parallel to the loading direction. Note that there is no factor of 2 on the shear strain because the tensorial definition is used

$$\epsilon_{11} = \frac{1}{2}\sigma_{11}\left[\frac{\cos^2\theta}{E_{11}} + \frac{\sin^2\theta}{E_{22}} - \frac{\nu_{12}(\cos^2\theta + \sin^2\theta)}{E_{11}}\right]$$

$$+ \frac{1}{2}\sigma_{11}\left[\frac{\cos^2\theta}{E_{11}} - \frac{\sin^2\theta}{E_{22}} - \frac{\nu_{12}(\sin^2\theta - \cos^2\theta)}{E_{11}}\right]$$

$$(\cos^2\theta - \sin^2\theta) + \sigma_{11}\frac{1}{G_{12}}\cos^2\theta\sin^2\theta \qquad (8.18)$$

Hence the Young's modulus is given by:

$$\frac{\epsilon_{11}}{\sigma_{11}} = \frac{1}{E_{11}^{\theta}} = \frac{\cos^4\theta}{E_{11}} + \left(\frac{1}{G_{12}} - \frac{2\nu_{12}}{E_{11}}\right)\sin^2\theta\cos^2\theta + \frac{\sin^4\theta}{E_{22}}$$

$$(8.19)$$

In Fig. 8.9 it can be seen that the strains induced by the tension σ_{11} in the $x-y$ axes occur in both the x and y directions, and there is also a shear strain.

The new compliance matrix will become:

$$\begin{bmatrix} \epsilon_{11} \\ \epsilon_{22} \\ \gamma_{12} \end{bmatrix} = \begin{bmatrix} S_{11} & S_{12} & S_{16} \\ S_{12} & S_{22} & S_{26} \\ S_{16} & S_{26} & S_{66} \end{bmatrix} \begin{bmatrix} \sigma_{11} \\ \sigma_{22} \\ \tau_{12} \end{bmatrix} \qquad (8.20)$$

where S are the compliance terms of the composite oriented at angle θ to its fibre directions; the subscript 6 relates to the shear terms, i.e. S_{16} is used to calculate the shear induced by the tension σ_{11}. S_{11} is given by $1/E_{11}^{\theta}$ and the other terms can be determined by transforming the stresses and strains as illustrated above to give:

$$S_{11} = \frac{1}{E_{11}^{\theta}} = \frac{\cos^4\theta}{E_{11}} + \left(\frac{1}{G_{12}} - \frac{2\nu_{12}}{E_{11}}\right)\sin^2\theta\cos^2\theta + \frac{\sin^4\theta}{E_{22}}$$

$$(8.21a)$$

$$S_{12} = \frac{\nu_{12}}{E_{11}}(\cos^4\theta + \sin^4\theta) - \left(\frac{1}{E_{11}} + \frac{1}{E_{22}} - \frac{1}{G_{12}}\right)\sin^2\theta\cos^2\theta$$

$$(8.21b)$$

$$S_{22} = \frac{1}{E_{22}^{\theta}} = \frac{\cos^4\theta}{E_{22}} + \left(\frac{1}{G_{12}} - \frac{2\nu_{12}}{E_{11}}\right)\sin^2\theta\cos^2\theta + \frac{\sin^4\theta}{E_{11}}$$

$$(8.21c)$$

$$S_{66} = \frac{1}{2G_{12}^{\theta}} = 2\left(\frac{2}{E_{11}} + \frac{2}{E_{22}} + \frac{4\nu_{12}}{E_{11}} - \frac{1}{G_{12}}\right)\sin^2\theta\cos^2\theta$$

$$+ \frac{1}{G_{12}}(\sin^4\theta + \cos^4\theta) \qquad (8.21d)$$

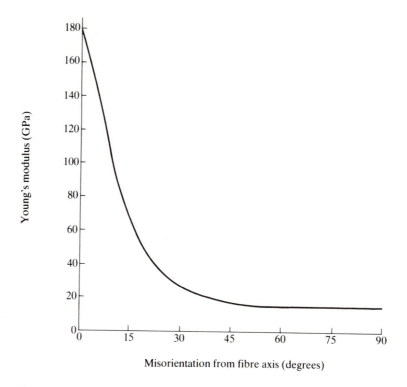

Figure 8.10 The angular dependence of the Young's modulus of a unidirectionally reinforced composite of 50% carbon-fibre-reinforced epoxy resin

$$S_{16} = \left(\frac{2}{E_{11}} + \frac{2\nu_{12}}{E_{11}} - \frac{1}{G_{12}} \right) \sin\theta \, \cos^3\theta$$

$$- \left(\frac{2}{E_{22}} + \frac{2\nu_{12}}{E_{11}} - \frac{1}{G_{12}} \right) \sin^3\theta \, \cos\theta \qquad (8.21e)$$

$$S_{26} = \left(\frac{2}{E_{11}} + \frac{2\nu_{12}}{E_{11}} - \frac{1}{G_{12}} \right) \sin^3\theta \, \cos\theta$$

$$- \left(\frac{2}{E_{22}} + \frac{2\nu_{12}}{E_{11}} - \frac{1}{G_{12}} \right) \sin\theta \, \cos^3\theta \qquad (8.21f)$$

Equations (8.21a–f) illustrate two problems which this anisotropy of elastic properties introduces into mechanical design. First, there is a comparatively rapid change in elastic modulus with small angular deviations from the fibre axis. This variation in Young's modulus with angle in an example composite system is illustrated in Fig. 8.10. Second, the presence of non-zero S_{16} and S_{26} compliance terms means that tensile strains in these directions will induce substantial shears and possibly complex shape changes may occur.

Although there may be applications where an extremely high modulus is required in only one direction, isotropic properties are normally preferred by the designer. To obtain a completely isotropic elastic behaviour within the plane of the lamina, the fibres would need to be completely randomly oriented. This is not practical and the cylindrical nature of the fibres requires them to be parallel to maximize V_f and hence lamina stiffness. The technique which is normally used to reduce elastic anisotropy is to make a composite sheet from a number of different laminae each oriented at a different angle to the previous one. Fabrication practicalities limit the thickness of each lamina to a fraction of a millimetre, hence only a few different orientations are practical within one laminated composite sheet.

8.2.3   THE ELASTIC PROPERTIES OF LAMINATES

Consider a composite laminate of thickness $2t$ made from two identical laminae, a and b in Fig. 8.11, oriented at 90° to each other, loaded in P in the x direction along one of the lamina's fibre directions. For equilibrium,

$$\sigma_{xx,a} + \sigma_{xx,b} = \frac{P}{t} \tag{8.22a}$$

$$\sigma_{yy,a} + \sigma_{yy,b} = 0 \tag{8.22b}$$

We assume that the ends are constrained in such a way that the strain is the same for both laminae:

$$\epsilon_{xx,a} = \epsilon_{xx,b} = \epsilon \tag{8.22c}$$

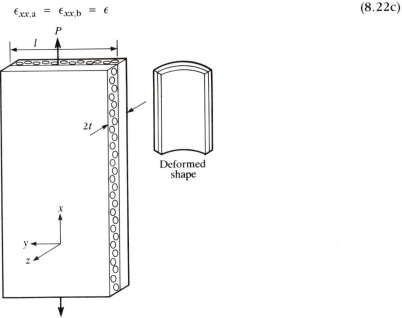

Figure 8.11   Laminated composite structure constructed from two laminae at 90° to each other

The strain at the interface between the laminae must also be the same for both laminae in the y direction but the laminae will be free to bend, so that their strain at the mid-points will be:

$$\epsilon_{yy,a} = (\epsilon_{yy})_o + \rho \frac{t}{2} \tag{8.22d}$$

$$\epsilon_{yy,b} = (\epsilon_{yy})_o - \rho \frac{t}{2} \tag{8.22e}$$

where ρ is the curvature and the subscript o denotes the interface. Combining Eqns (8.22) with the stress–strain equations (8.10) gives

$$\epsilon = \frac{\sigma_{xx,a}}{E_{11}} = \frac{\sigma_{xx,b}}{E_{22}} = \frac{1}{E_{11} + E_{22}} \frac{P}{t} \tag{8.23a}$$

$$\epsilon_{yy,o} = -\frac{1}{2} \frac{\nu_{12} + \nu_{21}}{E_{11} + E_{22}} \frac{P}{t} \tag{8.23b}$$

$$\rho' = \frac{2}{t} \frac{\nu_{12} + \nu_{21}}{E_{11} - E_{22}} \frac{P}{t} \tag{8.23c}$$

Bending results in a deformation of the composite that may be regarded as unacceptable. It could be avoided by fixing the composite to a rigid frame. This solution could make things worse by introducing a stress system in the y direction that could weaken the laminae. The alternative is to manufacture the composite by symmetrical lay-ups of material either side of the centre of the composite plate.

In the case of symmetrical laminate lay-ups, the strain is uniform throughout the thickness of the plate and we can write that, for each layer,

$$\begin{bmatrix} \sigma_{11} \\ \sigma_{22} \\ \tau_{12} \end{bmatrix}_k = [C]_k \begin{bmatrix} \epsilon_{11} \\ \epsilon_{22} \\ \gamma_{12} \end{bmatrix}_k \tag{8.24}$$

where $[C]$ represents the reduced stiffness matrix of the lamina. Adding up the contributions of each lamina, and observing that

$$\begin{bmatrix} \sigma_{11} \\ \sigma_{22} \\ \tau_{12} \end{bmatrix}_{average} = \sum_{k=1}^{N} V_k \begin{bmatrix} \sigma_{11} \\ \sigma_{22} \\ \tau_{12} \end{bmatrix}_k \tag{8.25}$$

where V_k is the volume fraction of the kth layer, then

$$\begin{bmatrix} \sigma_{11} \\ \sigma_{22} \\ \tau_{12} \end{bmatrix}_{average} = \sum_{k=1}^{N} [C]_k V_k \begin{bmatrix} \epsilon_{11} \\ \epsilon_{22} \\ \gamma_{12} \end{bmatrix}_k \tag{8.26}$$

This expression defines the stiffness matrix of the composite laminate. Taking, for example, a typical cross-ply laminate where alternate layers are aligned with fibres at $0°$ and $90°$ to the $x-y$ axes, symmetrically about its central plane, it follows from the expression of the stiffness matrix (Eqns 8.12 and 8.14) taking the fibre directions 11 along the x or y axes alternately, that

$$\sum_{k=1}^{N} [C_{xy}]_k \, V_k = \frac{1}{2} \begin{bmatrix} \dfrac{E_{11}+E_{22}}{1-\nu_{12}^2(E_{22}/E_{11})} & \dfrac{\nu_{12}E_{22}}{1-\nu_{12}^2(E_{22}/E_{11})} & 0 \\[3ex] \dfrac{\nu_{21}E_{11}}{1-\nu_{12}^2(E_{22}/E_{11})} & \dfrac{E_{11}+E_{22}}{1-\nu_{12}^2(E_{22}/E_{11})} & 0 \\[3ex] 0 & 0 & 4G_{12} \end{bmatrix} \quad (8.27)$$

taking the thickness of each lamina as unity (Fig. 8.12). The stiffness matrix can also be rotated to give the angular dependence of the tensile modulus of such a laminate. The effect of such a rotation is shown in Fig. 8.13 for a carbon-fibre-reinforced epoxy resin composite and can be compared with that of a unidirectional laminate. Although the $0°-90°$ laminate has a reduced anisotropy, it is by no means isotropic in its elastic behaviour. Other fibre orientation combinations have been used to reduce further anisotropies; however, with materials of widely different elastic properties such as carbon fibre and plastic matrices, some degree of anisotropy will always be encountered in practical laminate constructions.

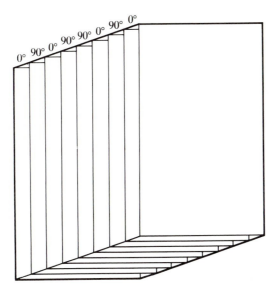

Figure 8.12   Schematic representation of an eight-ply symmetrically laminated structure with fibre axes at $90°$ to each other

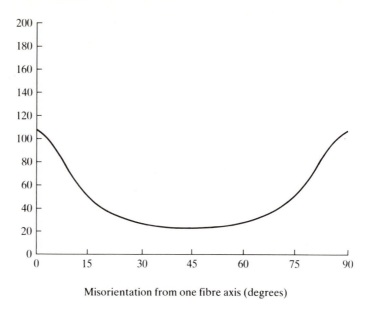

Misorientation from one fibre axis (degrees)

Figure 8.13 The angular dependence of the Young's modulus of a symmetrical 0°−90° reinforced composite of 50% carbon-fibre-reinforced epoxy resin

8.2.4 ELASTIC PROPERTIES OF SHORT-FIBRE COMPOSITES

Long-fibre composites have very good specific properties parallel to their fibre orientations, as was shown in Table 8.2. Their elastic properties off-axis are less good and this, coupled with the expense of manufacturing complex multi-orientation laminated structures, has tended to limit their use to specific high-cost applications where a great premium is placed on weight reduction. It is found that composites made using short fibres or particles can be manufactured significantly more cheaply than those made from long or continuous fibres. Techniques such as injection moulding can be used to produce complex shapes from plastics reinforced by short fibres.

The compliance of the matrix in most composites is much greater than that of the reinforcing fibres and hence, from the rule of mixtures, the compliance of the composite will also be greater than its fibres. The interatomic bonding across the interface between fibre and matrix in fibre-reinforced plastics is usually very weak. Thus, if a composite is manufactured from short fibres, an important aspect of their behaviour will be determined by how load is transferred from the matrix to fibres buried in the structure during straining. If the strength of the interface is low we would not expect fibres to be loaded from the ends in tension but instead by an increasing shear force parallel

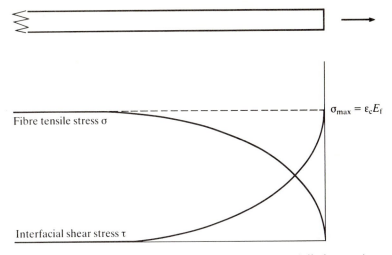

Figure 8.14   The distribution of stress at the ends of fibres is modelled assuming a transfer by shear across the fibre/matrix interface

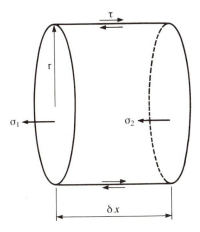

Figure 8.15   The force increase $\pi r^2 \delta\sigma$ in a small element δx of a short fibre of radius r is generated by the action of a shear stress τ such that $\pi r^2 \delta\sigma = 2\pi r \tau$

to the fibre/matrix interface, as shown in Fig. 8.14. This shear traction will gradually transfer load from the matrix to the fibre away from the ends. If a fibre of radius r and length L is divided into small units of length δx, Fig. 8.15 shows that the shear stress τ at a location x must balance the increase in fibre longitudinal stress $\delta\sigma$ such that:

$$\pi r^2 \, \delta\sigma \ = \ 2\pi r \tau \, \delta x \tag{8.28}$$

From symmetry, the interfacial shear stress must be zero at the fibre mid-point and the fibre stress will be related to the mean composite strain by $\sigma_f = \epsilon_c E_f$. At the fibre ends the fibre stress will be zero and the shear stress a maximum. This equation can be solved exactly: the solution is known as the shear lag equation, which relates the fibre tensile stress to interfacial shear stress by

$$\sigma_f = \epsilon E_f \left\{ 1 - \frac{\cosh[\beta(L/2 - x)]}{\cosh(\beta L/2)} \right\} \tag{8.29a}$$

with

$$\beta = \left[\frac{2G_m}{E_f r^2 \ln(R/r)} \right]^{1/2} \tag{8.29b}$$

where the interfibre spacing is given by $2R$. The interfacial shear stress is then given by

$$\tau_i = \left[\frac{2G_m}{E_f \ln(R/r)} \right]^{1/2} \left\{ \frac{\sinh[\beta(L/2 - x)]}{\cosh(\beta L/2)} \right\} \tag{8.30}$$

The maximum fibre tensile stress is given by $\sigma_{max} = E_f \epsilon$ at $x = L/2$ and the maximum interfacial shear stress at $x = 0$ and $x = L$ is given by

$$\tau_{max} = \left[\frac{2G_m}{E_f \ln(R/r)} \right]^{1/2} \tanh(\beta L/2) \tag{8.31}$$

These relations are shown schematically in Fig. 8.14.

The nett effect of this shear lag at the fibre ends is to reduce the load held by each fibre when the composite is strained. This load is now transferred to the matrix, which is of lower modulus than the fibres, and hence the composite will extend to a greater length than would occur in a similar volume-fraction, continuous-fibre composite. The simple rule of mixtures (Eqn 8.4) is now replaced by:

$$E_{11} = E_f V_f \Omega + (1 - V_f)E_m \tag{8.32}$$

where Ω is a reinforcement efficiency factor which will be less than 1. From further consideration of the shear lag theory, Ω is shown to be given by:

$$\Omega = 1 - \left[\frac{\tanh(\beta L/2)}{\beta L/2} \right] \tag{8.33}$$

where β is given in Eqn (8.29b).

Equation (8.33) shows a reduction in the Young's modulus of a composite with decreasing fibre length. In fact it is found that the reduction in modulus is very small for most composites where the aspect ratio of the reinforcement (i.e. the ratio of its mean length to diameter $L/2r$) is greater than about 25. This relation is demonstrated in Fig. 8.16.

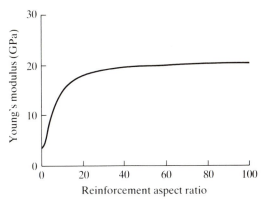

Figure 8.16 The Young's modulus of a unidirectional short fibre composite (made with 40% V_f glass fibres in an epoxy resin matrix) rises with increasing aspect ratio to an asymptote of the long-fibre Voigt limit. However, the composite has attained about 90% of its maximum possible modulus with an aspect ratio above about 25

8.3 STRENGTH OF FIBRE COMPOSITES

8.3.1 TENSILE STRENGTH OF LONG-FIBRE COMPOSITES

It is reasonable to expect that the strength of a composite material will be determined by the strength of both components present. These components may have quite different failure properties and it is found that the composite's strength will depend on the relative behaviour of both fibre and matrix. If a composite is strained parallel to the fibre direction in a unidirectional laminate, both components will be strained by the same extent. As the strain is increased, the stress in each component will increase until the failure stress is reached in one component. Subsequent fracture behaviour will depend on which component fails first. Thus the important parameter to consider is the *strain* at which fracture occurs in each component (ϵ_f^* and ϵ_m^*). There are significant differences in composite behaviour depending on whether the matrix or fibre fails first, and these cases will be considered in some detail.

If the matrix fails before the fibres, the matrix failure strain must be smaller than the fibre failure strain, i.e. $\epsilon_m^* < \epsilon_f^*$. Consider a unidirectional laminate which is being progressively strained parallel to its fibres under an increasing load. Both components of the composite will be under the same strain at any instant but the load will be shared between the two components in proportion to their volume fractions and Young's moduli. The load and strain in both components will increase until the critical failure strain of the matrix is reached. At this point the matrix fails and we assume that all the load previously sustained by the matrix is now transferred to the fibres. However, if the volume fraction of fibres present is small, this sudden transfer of load will cause the fibres' failure stress to be exceeded and complete failure of the composite occurs. In this case the longitudinal failure stress of the composite will then be given by the total stress held by fibres and matrix at the moment the matrix failure strain is reached, i.e.

$$\sigma_{11}^* = \epsilon_m^* E_{11} = \epsilon_m^* [E_f V_f + E_m(1 - V_f)] \qquad (8.34)$$

The matrix failure strain can be related to the matrix failure stress if the matrix is assumed to behave elastically up to failure with $\sigma_m^* = \epsilon_m^* E_m$. In this case the failure stress can be related to the matrix failure stress with:

$$\sigma_{11}^* = \sigma_m^* \left[(1 - V_f) + \frac{E_f V_f}{E_m} \right] \qquad (8.35)$$

Thus at low volume fractions of reinforcing fibres the longitudinal composite failure stress is controlled by the matrix failure stress.

If the volume fraction of reinforcing fibres is large, the failure of the matrix may not lead to the immediate failure of the composite. Instead, the fibres are able to withstand the sudden increase in load and will only fail at a later extension when the fibre failure strain is exceeded. At this point the load is only being held by the fibres and the failure stress will be given by:

$$\sigma_{11}^* = \epsilon_f^* E_f V_f \qquad (8.36a)$$

or in terms of the fibre strength,

$$\sigma_{11}^* = \sigma_f^* V_f \qquad (8.36b)$$

Thus, above some critical volume fraction V_c, the composite strength will be controlled by the fibre's properties. The transition in fracture behaviour from matrix to fibre control will occur when the stresses given by Eqns (8.35) and (8.36) are equivalent at the above critical volume fraction (Fig. 8.17a) with:

$$V_c = \frac{\sigma_m^*}{\sigma_f^* - \sigma_m^*(1 - E_f/E_m)} \qquad (8.37)$$

The other extreme of behaviour is the case of a composite where $\epsilon_f^* < \epsilon_m^*$. As before, under increasing load the composite's strain will increase with the load shared between matrix and fibre. In this case the fibres fail first when the critical fibre failure stress is exceeded. All load carried by the fibres is now transferred on to the matrix. If there is only a small volume fraction of fibres present in the matrix, this increase in load on the matrix will not exceed the matrix failure stress and the composite will take a further increase in load until its final failure strain is reached at a stress given by:

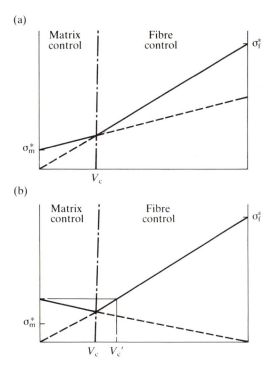

Figure 8.17   The variation in strength of a unidirectional fibre composite with increasing volume fraction of fibres in the case when: (a) the matrix failure strain is lower than the fibre failure strain; (b) the fibre failure strain is lower than the matrix failure strain

$$\sigma_{11}^{*} = (1 - V_{f})\sigma_{m}^{*} \tag{8.38}$$

Thus, at low-fibre volume fractions, the composite failure stress is clearly under matrix control. However, note that in this case the composite failure stress is predicted to *decrease* with increasing fibre volume fraction. This is because the fibres play no direct part in determining the composite failure strength; instead they act after fracture to reduce the area of matrix capable of sustaining the applied load.

If the fibres are present as a large volume fraction, their breaking will result in a sufficiently large transfer of load to the matrix for the matrix failure stress also to be exceeded, in which case the longitudinal strength will be given by

$$\sigma_{11}^{*} = \sigma_{f}^{*}V_{f} + \sigma_{m}(1 - V_{f}) = \sigma_{f}^{*}\left[V_{f} + \frac{(1 - V_{f})E_{m}}{E_{f}}\right] \tag{8.39}$$

and is now under fibre control. As the fibre volume fraction is increased, behaviour will transfer from one regime to the other at a critical volume fraction of reinforcement V_c when the stresses given by Eqns (8.38) and (8.39) are equal:

$$V_c = \frac{\sigma_m^* - \sigma_f^*(E_m/E_f)}{\sigma_m^* + \sigma_f^*(1 - E_m/E_f)} \qquad (8.40)$$

In this case V_c defines a composite failure stress lower than the matrix failure stress σ_m^*. Thus there will be a second critical volume fraction of reinforcement V_c', which is the minimum fraction of reinforcement required to ensure strengthening of the matrix in the longitudinal direction. This will be defined by the volume fraction required for σ_{11}^* (from Eqn 8.39) to exceed σ_m^*. Hence:

$$\sigma_m^* = \sigma_f^* \left[V_c' + \frac{(1 - V_c')E_m}{E_f} \right] \qquad (8.41)$$

and

$$V_c' = \frac{\sigma_m^* - \sigma_f^*(E_m/E_f)}{\sigma_f^*(1 - E_m/E_f)} \qquad (8.42)$$

This behaviour is illustrated in Fig. 8.17(b).

In order to exploit fully the benefits of strength as well as stiffness of reinforcing fibres, the fibre volume fraction should be sufficient to be in the region of fibre control. Therefore it is necessary to ensure an adequate volume fraction of reinforcement in both of the above cases of relative failure strain.

8.3.2 STRENGTH OF SHORT-FIBRE COMPOSITES

Earlier we discussed the reduction in composite modulus which occurs because the fibre ends present in short-fibre composites reduce the efficiency of load transfer to the fibres. We would naturally expect such effects to alter the efficiency of strengthening in short-fibre composites also. If the composite strength is in the regime of fibre control, the ultimate strength of the composite will be defined by the fibre failure strength σ_f^*. From our earlier considerations of the shear lag theory we would expect this stress to be reached at the middle of a stressed fibre. For the purposes of this discussion let us consider a simplified shear lag. If the matrix is very weak in shear, the shear transfer stress will be limited by its shear yield stress and can be taken as a constant at this value. Similarly, if the adhesion between fibre and matrix is low, there will be a maximum interfacial shear stress τ^* above which the interfacial bond fails, and subsequent interface shear lag occurs at a constant shear stress governed by friction across the unbonded interface. If this value of τ^* is sufficiently low, we can assume that the shear transfer stress will be exceeded very close to

(a)

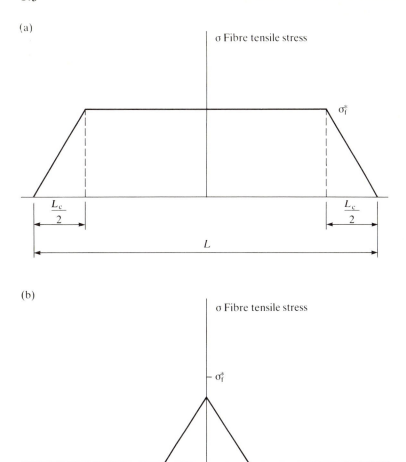

(b)

Figure 8.18 In a short-fibre composite where the fibres are stronger than the matrix, the stress transferred to the fibres builds up in a linear manner from zero at the fibre ends: (a) to its failure stress in the central region if the fibres are longer than a critical length $L > L_c$; (b) to a maximum below the fibre failure stress if the fibres are shorter than this length $L < L_c$.

the fibre ends and shear load transfer can be considered at a constant stress τ^*, such that:

$$r \, d\sigma_f = 2\tau^* dx \tag{8.43}$$

This implies a linear increase in stress from the fibre ends over some finite distance $L_c/2$ (Fig. 8.18) until the stress in the fibre reaches that predicted by the rule of mixtures for a long-fibre composite, i.e. $\sigma_f = \epsilon E_f$, where ϵ is the strain in the composite. If the fibre is to fail by

271

fracture, the mid-point stress must reach σ_f^*, and thus from equation (8.43):

$$\sigma_f^* = \int_0^{L_c/2} \frac{2\tau^* \, dx}{r} = \frac{\tau^* L_c}{r} \qquad (8.44)$$

Hence in short-fibre composites the fibres must exceed some critical length L_c if they are to be loaded to their failure stress. L_c can be defined by rearranging Eqn (8.44) to give

$$L_c = \frac{r\sigma_f^*}{\tau^*} \qquad (8.45)$$

The interfacial shear strength will be defined either by the strength of the fibre/matrix interface or by the shear yield strength of the matrix.

Consider a short-fibre composite with fibre length L greater than L_c. At the moment of fibre fracture the stress distribution within each fibre is shown in Fig. 8.18(a). The mean stress within each fibre will be given by:

$$\bar{\sigma}_f = \left[\left(\frac{L - L_c}{L} \right) + \left(\frac{L_c}{2L} \right) \right] \sigma_f^* \qquad (8.46)$$

and the longitudinal composite failure stress given by using $\bar{\sigma}_f^*$ in place of σ_f^* in Eqn (8.36).

$$\sigma_{11}^* = \left(\frac{2L - L_c}{2L} \right) \sigma_f^* V_f \qquad (8.47)$$

If the fibre length is smaller than L_c then the fibre stress distribution will never exceed σ_f^* (Fig. 8.18(b)); the mean stress will be given by:

$$\bar{\sigma}_f = \left(\frac{L}{2L_c} \right) \sigma_f^* \qquad (8.48)$$

and the composite failure stress by:

$$\sigma_{11}^* = \left(\frac{L}{2L_c} \right) \sigma_f^* V_f + \sigma_m^* (1 - V_f) \qquad (8.49)$$

8.8.3 MATRIX CRACKING IN COMPOSITES

A high proportion of composite materials contain matrices which have a failure strain below that of their reinforcing fibres. This is commonly the case in both polymer and ceramic matrix composites where the matrix is brittle. If the composite failure is in the regime of fibre control, then when a load parallel to the fibres is steadily increased the matrix fracture strain will be exceeded prior to composite failure and a matrix crack will propagate across the specimen. The fractured matrix must

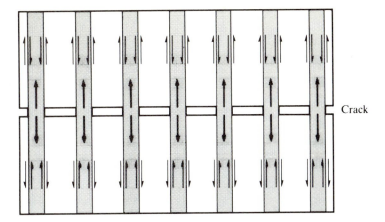

Crack

Figure 8.19   Around each matrix crack the crack surface will be traction-free with the excess load carried by the fibres. Away from the crack, load is transferred back to the matrix by a fibre/matrix interfacial shear

have its crack surfaces free of stress and the fibres bridging the crack will take all the composite load. This is effectively the inverse of the loading conditions which exist at the ends of a fibre in a short-fibre composite as discussed previously. Thus there will be a shear transfer of the excess fibre load back into the matrix over a distance away from the matrix crack. If there is a sufficient difference between fibre and matrix fracture strains, the composite will continue straining under the increasing load until a second matrix crack is nucleated. In practice, repeated cracking events occur during increasing load until a condition of multiple parallel matrix cracks occurs prior to the final fibre failure which leads to complete rupture of the composite. Thus, in brittle matrix composites a regime of continued subcritical crack growth generates substantial damage in the matrix before final failure.

Thus, provided there are sufficient fibres to bear the increased load during matrix cracking, the material will develop a series of parallel cracks as the stress increases. These cracks will be bridged by unbroken fibres. The load, which is held in its entirety by the fibres as they bridge a crack, will be transferred back into the matrix by an interfacial shear acting along the fibre/matrix interface (Fig. 8.19). If the fibre/matrix interface is weak it will transfer load at a constant low shear stress τ^*; the load transferred from a single fibre into the matrix over a distance x will be given by $2\pi r \tau x$ where r is the fibre radius. The number of fibres per unit area of the composite is given by $N_f = V_f/\pi r^2$. Hence the stress in the matrix will vary linearly with distance x from the crack with:

$$V_m \sigma_m^* = 2\pi r N_f \tau x = \frac{2V_f \tau^* x}{r} \tag{8.50}$$

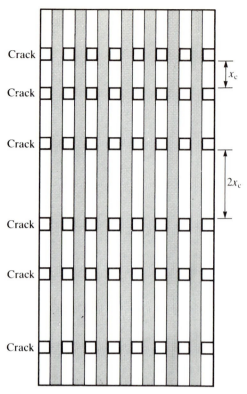

Figure 8.20 The final state of a microcracked matrix consists of a matrix saturated with cracks spaced between x_c and $2x_c$

up to a distance x_c given by:

$$x_c = \frac{V_f}{V_m} \cdot \frac{\sigma_m^* r}{2\tau^*} \tag{8.51}$$

Beyond this distance no further stress transfer into the matrix occurs. Thus, with increasing load, further cracking occurs in the matrix at distances greater than x_c from existing cracks. The matrix gradually breaks up into small blocks connected by unbroken fibres, and within each block the stress level builds up over a distance x_c; hence in any block of size greater than $2x_c$ a fracture will occur. Finally the matrix consists entirely of blocks between x_c and $2x_c$ in size (Fig. 8.20).

Once the matrix has broken into an array of blocks, further extension of the fibres will occur under increasing load until failure. No further extension of the matrix blocks is possible during the continued extension because of the fixed shear transfer length in each block. Hence, at the limit of multiple matrix cracking, the composite modulus will be reduced to:

$$E_{c(\text{cracked})} = E_f V_f + 3E_m(1 - V_f)/8 \tag{8.52}$$

The onset of multiple microcracking must then be accompanied by a decrease in composite modulus which continues to the limit of Eqn (8.52) when the microcracking is saturated. The introduction of these stable matrix cracks, although not catastrophic, can still be considered as damage because of the consequent reduction in material stiffness. The presence of the cracks also provides an easy path for environmental access and possible attack of the fibres.

The reduction in modulus observed after matrix cracking is caused by the transfer in stress from the matrix close to a crack into the crack bridging fibres. The unloaded matrix will relax from the crack. In the case assumed here where fibre/matrix adhesion is weak, this relaxation results in a relative translation at the fibre/matrix interface. This movement occurs under the action of the interfacial shear stress τ^* and mechanical work is done. The mechanism of matrix cracking therefore requires the input of energy by the external loading system. The implication of this mechanism on the behaviour of crack growth and energy absorption during composite fracture will be discussed in more detail later.

8.3.4   TRANSVERSE TENSILE STRENGTH

When a composite is loaded at 90° to the fibre direction, its strength relative to that measured parallel to the fibre direction will be considerably reduced. If the fibres are bonded very weakly to the matrix the transverse tensile strength can be approximated by considering the effect of the fibres to be identical to the presence of long cylindrical voids in the matrix. Thus the transverse strength of the composite will be dominated by the matrix failure strength as determined by the reduced area of matrix carrying the load. If a simple square array of fibres is assumed (Fig. 8.21) and the composite is loaded parallel to the array axis, the reduced area fraction will be approximately:

$$A_f = \frac{R}{R+r} = 1 - \sqrt{\frac{V_f}{\pi}} \tag{8.53}$$

with the transverse tensile strength given by:

$$\sigma_{22}^* = \sigma_m^* A_f \tag{8.54}$$

This does not take into account any stress concentration factors from the presence of the fibres which might further increase the matrix stress on loading. When fibre volume fractions become high or where irregular fibre distributions lead to regions of locally high V_f, further non-linear stress concentrations become important.

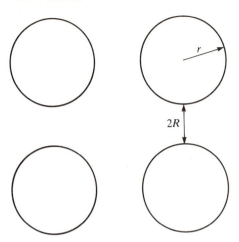

Figure 8.21   If composites are loaded perpendicular to the fibre directions the weak fibre matrix interface carries no load. The behaviour of the composite can then be modelled as a block containing holes in the positions of the fibres

8.3.5   COMPOSITE SHEAR STRENGTH

As with the transverse tensile strength, this strength will be determined chiefly by the matrix shear strength. From Fig. 8.22 it is clearly possible to envisage matrix deformation independent of the fibres on shear loading, giving a composite shear strength in the plane of the fibres of

$$\tau_{12}^* = \tau_m^* \tag{8.55}$$

Again, as with the transverse strength, there are further stress concentration factors to be taken into account which model the effect of the presence of the fibres, and the strength predicted by Eqn (8.55) will be further reduced when V_f becomes large.

8.3.6   COMPRESSIVE STRENGTH

Under low compressive loads, composites behave elastically with their moduli obeying the rule of mixtures and similar laws derived earlier for loading in tension. However, after further stressing in compression parallel to the fibre axes, the composites often fail at stresses much lower than the equivalent tension failure. The most common failure mode in compression is by a kinking of fibres accommodated by a local shear deformation of the matrix, as shown schematically in Fig. 8.23. The failure process is initiated by local buckling of the fibres in defective matrix-rich regions of the composite. This nucleates a crease or kink band which propagates across the specimen with fibre fracture at an

(a)

(b)

Figure 8.22 (a) Consider a fibre reinforced composite loaded in shear. (b) If the matrix is considerably weaker than the fibres any deformation will be confined to the matrix

angle of approximately 30° to the fibre direction. An upper bound for the compressive strength of fibre-reinforced composites has been proposed which assumes the initial buckling to be resisted by matrix shear. The compressive failure strength can then be derived as

$$\sigma_c^* = \frac{G_m}{(1 - V_f)} \tag{8.56}$$

This equation gives an overestimate of the compressive strength and does not predict the observed compressive behaviour of a fibre-reinforced composite which shows an increasing strength with increasing volume fraction up to a maximum at about $V_f = 50\%$.

Other failure mechanisms observed in composites loaded in compression include shear fracture at 45° to the applied load and longitudinal splitting parallel to the fibres allowing individual blocks of composite to buckle. When loaded in compression transverse to the fibre direction, composites fail by a 45° shear which does not require fibre failure, which is similar to the shear failure mechanism discussed earlier. However, in this case the compressive failure strength is insensitive to the fibre volume fraction.

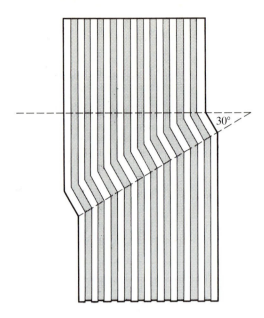

Figure 8.23 Failure in compression occurs by the compressive buckling of the fibres which is accommodated by matrix shear to form bands of deformation, kink bands, running at about 30° to the fibre axis

8.3.7 STRENGTH OF A UNIDIRECTIONAL COMPOSITE LAMINA LOADED IN TENSION

There are three possible modes of failure that have been identified for fibre-reinforced materials loaded in tension: (i) fracture by the sequential failure of matrix and fibres loaded parallel to the fibres; (ii) fracture in the transverse direction by the failure and kinking of fibre/matrix interfaces; and (iii) a shear failure of the matrix or fibre/matrix interface. If a unidirectional lamina is loaded in tension at some angle θ to the fibre axis, the stresses can be resolved into components parallel and perpendicular to the fibres and a shear component, thus:

Longitudinal stress: $\quad \sigma_{11} = \sigma^{\theta}\cos^2\theta$ $\qquad$ (8.57a)

Transverse stress: $\quad \sigma_{22} = \sigma^{\theta}\sin^2\theta$ $\qquad$ (8.57b)

Shear stress: $\quad \tau_{12} = \sigma^{\theta}\cos\theta\sin\theta$ $\qquad$ (8.57c)

where σ^{θ} is the applied stress at some angle θ to the fibre direction. In the previous sections the failure stress expected with a simple loading in each of these stresses has been calculated. A *maximum stress criterion* of lamina failure assumes that failure occurs when any of the resolved stresses exceeds the critical value defined earlier, in which case fracture occurs by the mode in which the critical stress is first

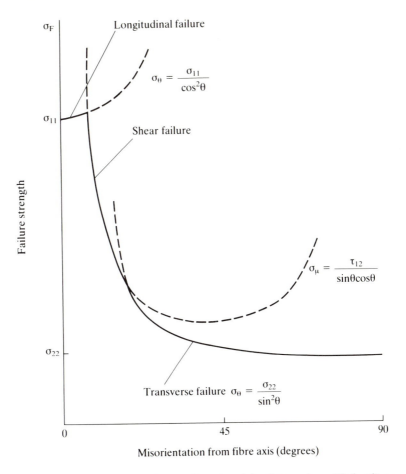

Figure 8.24   The failure mechanism of a unidirectional lamina varies with loading direction. A simple maximum stress criterion can be invoked where the stress is resolved on to each failure direction (longitudinal, transverse and shear), with the first exceeded determining the mechanism

exceeded. This is illustrated in Fig. 8.24, which shows the predicted variation in failure strength with angle of loading and the corresponding failure mode.

8.3.8   STRENGTH OF MULTIDIRECTIONAL LAMINATES

Earlier in this chapter the use of multi-ply laminates was described as a method of reducing the considerable elastic anisotropy found in unidirectional composite materials. However, if there are now many different orientations of ply present in the composite, the strength of the laminate will depend on the stresses resolved on to each individual

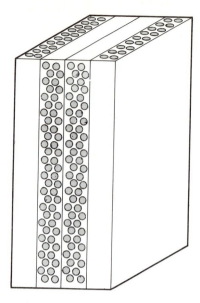

Figure 8.25 A laminate containing four plies of 0° and 90° orientation, arranged symmetrically about the centre axis to eliminate bending

ply. In a multi-ply laminate of many orientations the stress in each ply must be determined and compared with the failure criteria in order to predict the likely failure mode in any orientation.

Here we will consider the behaviour of a simple symmetrical 0°–90° laminate loaded along one of its fibre directions such as shown in Fig. 8.25. In this simple case the inner 90° plies will have all their stress resolved in the 22 or transverse direction and the outer 0° plies in the 11 or longitudinal direction; there will be no resolved shear stresses in either orientation. If we assume each layer has strengths typical for a glass-fibre-reinforced plastic then we can take: $\sigma_l^* = 500$ MPa, $\sigma_t^* = 20$ MPa and $\tau_f^* = 50$ MPa. The material has measured elastic moduli of $E_{11} = 30$ GPa and $E_{22} = 7$ GPa for each ply. If the laminate is strained parallel to the outer fibres failure will occur in the transverse tensile mode in the central 90° plies when they reach their failure strain of $\epsilon = 0.29\%$. At this point, just prior to matrix fracture, the mean stress on the laminate will be 54 MPa and the outer 0° plies will be stressed approximately 86 MPa. The load transferred to the 0° plies can be calculated because the laminate has equal thicknesses of 0° and 90° orientations. After fracture the load held by the 0° ply adjacent to the transverse crack will be 126 MPa, which is still below the failure stress of this ply loaded in the longitudinal direction. Therefore the laminate can be subjected to an increased load without further fracture occurring immediately.

The fracture event in the 90° ply will probably have been nucleated at some flaw or defect in the matrix because of the brittle nature of

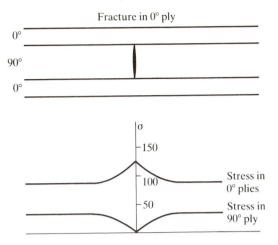

Figure 8.26   In a 0°−90° composite laminate, cracks in the transversely oriented plies
must have traction-free surfaces. Thus around each crack the stress which
was in the uncracked ply must be transferred across the interlaminar
interface by shear into the longitudinal plies. A shear lag calculation can
be used to determine the stress distribution in each ply after failure

most polymer matrices when loaded in the transverse direction. Thus
we would expect there to be a range of such defects in the ply which
may initiate further transverse cracking on increased loading. The crack
surface in the 90° ply must be stress-free and all load formerly held
by the uncracked matrix in this region will be transferred into the
adjacent 0° ply. However, because the two plies are still bonded along
their length, the 90° ply will transfer its load over some distance away
from the crack by interlaminar shear much in the way that load was
described as being transferred to the ends of short fibres. The stress
in the 90° ply will decrease towards zero as it approaches the crack,
while the stress in the 0° ply will increase from 86 to 126 MPa adjacent
to the crack (Fig. 8.26). As the laminate is loaded further, another
transverse-ply crack will nucleate away from the first crack and the
stress distribution will change to reflect the load transfer from the two
cracks. On increased loading, a third and further cracks will nucleate
between the initial cracks. This continues until cracking in the 90°ply
has reached a state in which effectively all the load is held by the 0°
plies and failure occurs at a composite mean stress of 250 MPa when
the longitudinal failure stress is exceeded. Between the initial fracture
event at 56 MPa and the final fracture at 250 MPa repeated cracking
will have occurred in the 90° ply and a stress−strain curve such as
that illustrated in Fig. 8.27 will be seen.

    In Fig. 8.27 there is a clearly defined discontinuity indicating the
onset of cracking and a reduced slope on the stress−strain curve. If
the test is arrested in the region intermediate to the initial and ultimate

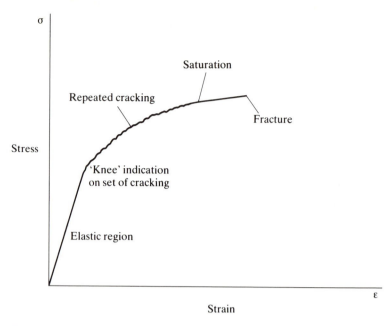

Figure 8.27 Typical stress-strain relation for a 0°–90° composite laminate. A linear elastic region is followed, after a discontinuity, by a region of decreasing tangent modulus. This represents the onset of transverse ply cracking which continues until a saturated crack regime is attained prior to failure

failure events a decreased elastic modulus is observed, reflecting the increased load now held by the 0° plies and the extra extension this implies. After testing many cracks can be seen in the 90° plies (Fig. 8.28). This is an important observation because it shows that composite laminates have two failure criteria. There is a lower stress which indicates the onset of matrix damage and initial cracking, and an upper ultimate failure stress representative of laminate fracture. Between these two limits is a region of increasing laminate damage accompanied by extensive laminate stiffness reduction as the crack density increases. The behaviour of more complex laminate orientations will be broadly similar, but all failure modes must be considered.

8.4 FRACTURE AND TOUGHNESS OF COMPOSITES

When a unidirectional composite laminate is tested to final fracture in the longitudinal direction, the fracture energy might be expected to be that required for the fracture of the matrix and fibres, i.e. we might expect a simple rule of mixtures with:

Figure 8.28   Regularly spaced transverse ply cracks in glass-fibre/epoxy resin 0°–90° symmetric laminate loaded in tension to about 0.8% strain along one of the fibre directions (courtesy of Dr L Boniface, University of Surrey)

Figure 8.29 Fracture surface of a fibre-reinforced composite showing the characteristic 'shaving brush' morphology with numerous fibres pulled out of the matrix from the opposite fracture surface (from Hull 1981)

$$G_{composite} = G_f V_f + G_m (1 - V_f) \tag{8.58}$$

where G is the critical strain energy release rate for each component. However, for a typical glass-fibre-reinforced epoxy composite, $G_f \simeq 10\ \text{J m}^{-2}$, $G_m \simeq 50\ \text{J m}^{-2}$ and $G_{composite} \simeq 10^5\ \text{J m}^{-2}$. This difference is much greater than can be explained by simple deviations from the rule of mixtures and suggests that, during the fracture of a fibre composite, new and powerful energy-absorbing mechanisms must exist. Indeed, the fracture surfaces of such composites (Fig. 8.29) are completely unlike the fracture surfaces of monolithic materials, with extensive debonding between fibre and matrix and pull-out of the debonded fibres resulting in a highly characteristic brush morphology.

In order to explain the extremely high fracture energies, let us consider the fracture processes which lead to the fracture surfaces observed. In the scheme illustrated in Fig. 8.30 a crack, loaded in simple tension, progresses through the matrix and is arrested by a fibre. We assume that the fibre is much stronger than the matrix and the crack-tip stress field is insufficient to cause fracture. For further crack propagation to occur, the matrix stress must increase. At the crack tip there is a complex stress field with stress components in all loading

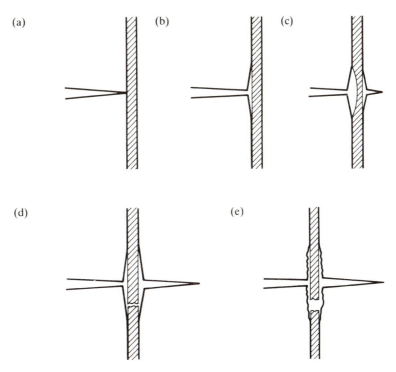

Figure 8.30   Schematic view of the progress of a crack through the matrix of a
composite normal to the fibre orientation. In (a)–(e) the crack is at first
arrested by the fibre, then a debond crack nucleates around the fibre
before fibre fracture occurs away from the matrix crack plane

orientations; immediately ahead of the crack there is a dilatational
stress, i.e. there are tensile components in all directions. If the
fibre/matrix interface has little strength, the crack-tip stress will initiate
a debonding crack between fibre and matrix. Once the debond occurs
the crack can by-pass the fibre, leaving it intact and bridging the crack
surfaces. As the crack advances it meets new fibres and similar
debonding and by-passing operations occur, leading to a region of fibres
bridging the crack behind the crack tip. Behind the crack the bridging
zone of debonded fibres is strained further with increasing load. Within
each individual fibre the load increases with increasing strain until it
reaches the fibre fracture stress. Most reinforcing fibres have brittle
failure characteristics and hence fracture will occur at a location in
the debonded region governed by the largest flaw present, and not
necessarily between the two matrix fracture surfaces. The fractured
fibre will now relax and be pulled out of the matrix by the separating
fracture surfaces. Within this sequence of events we can identify three
potential energy-absorbing mechanisms absent from the failure of
monolithic materials: crack bridging; fibre/matrix debonding; and fibre

pull-out. The energy required to produce a new fracture surface will then be given by:

$$\Gamma_f = 2\gamma + \Gamma_{bridge} + \Gamma_{debond} + \Gamma_{pull\text{-}out} \tag{8.59}$$

where γ is the mean surface energy of the composite and the Γ terms are those responsible for the enhanced composite toughening. These will now be considered in further detail.

8.4.1 CRACK BRIDGING AND DEBONDING

After a crack has by-passed a fibre which has debonded from the matrix, the fibre acts as a bridge across the opening crack. In Fig. 8.31 a hypothetical crack is seen with a number of fibres bridging the crack behind the matrix crack tip. The precise influence of these bridging fibres on crack propagation is difficult to quantify but they must reduce the matrix crack stress intensity by bearing some of the load which would otherwise act as the driving force for crack advance. The Griffiths criterion for brittle fracture (Chapter 4) assumes that as a crack advances all the elastic energy in the region surrounding the crack tip is released, providing the driving force for crack propagation. In the case of a fibre-reinforced composite, the elastic energy stored in the bridging fibres is not released. Thus, in order to determine the energy balance relevant for crack propagation in composites, this energy will not be available. The stored elastic energy in a fibre is given by $\frac{1}{2}\sigma_f\epsilon_f$ per volume of fibre. The stress in fibres within the debonded zone will be greater than in equivalent fibres in the rest of the composite. This increase in stress is transferred from the matrix, which relaxes away

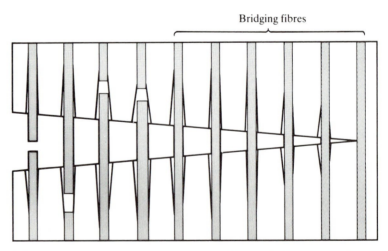

Bridging fibres

Figure 8.31 Behind a growing crack in a composite there will be a zone where unbroken fibres bridge the crack

from the crack surface; this stress transfer is assumed to occur by a shear lag mechanism similar to that assumed earlier for the transfer of stress to the ends of short fibres. If the fibres at the crack surface are assumed to be at the point of fracture, the stress in the fibre at a distance x from the surface will be approximately given by a simple linear relation:

$$\sigma_f = \sigma_f^* \left(1 - \frac{x}{d_f}\right) \tag{8.60}$$

where d_f is the length of debonded fibre/matrix interface either side of the crack. The elastic energy stored in the fibre will then be given by:

$$U_{el} = 2 \left(\frac{\pi r^2}{2E_f}\right) \int_0^{d_f} \sigma_f^{*2} \left(1 - \frac{x}{d_f}\right)^2 \mathrm{d}x = \frac{\pi r^2 \sigma_f^{*2} d_f}{3E_f} \tag{8.61}$$

where the factor of 2 is included to account for the debonding on either side of the fracture plane. To a first approximation the characteristic debonding length d_f is taken to be the same as the critical length for stress transfer at the end of a short fibre, i.e. $d_f = L_c/2 = r\sigma_f^*/2\tau_f$ where τ_f is the constant value of the shear transfer stress between fibre and matrix. Inserting this value for d_f into Eqn (8.61) gives:

$$U_{el} = \frac{\pi r^3 \sigma_f^{*3}}{6E_f\tau_f} \tag{8.62}$$

and for the total energy stored in the fibres bridging the crack,

$$\Gamma_{bridge} = \frac{N\pi r^3 \sigma_f^{*3}}{6E_f\tau_f} \tag{8.63}$$

where N is the number of fibres per unit area and r is the mean fibre radius. In practice, the debond length will be shorter than L_c because the stress in the fibres at the edge of the debonded region is not zero but instead it is given by the stress in the fibres when the matrix reaches its failure stress σ_m^*, i.e. $\sigma_f = \sigma_m^*(E_f/E_m)$. The modified debond length will be given by:

$$d_f = \frac{r(\sigma_f^* - \sigma_m^* E_f/E_m)}{2\tau_f} \tag{8.64}$$

8.4.2 FRICTIONAL RELAXATION AFTER DEBONDING

When the debond grows along the fibre/matrix interface, the matrix alongside the fibre is partially relaxed. This relaxation will result in a relative motion of the matrix with respect to the fibre (Fig. 8.32). This motion occurs against the interfacial friction stress τ_f, and thus work is done. At the crack surface the matrix relaxes fully and the matrix strain is zero, whereas the strain in the fibre is the strain just

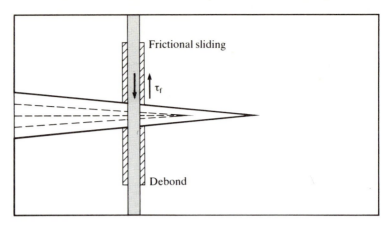

Figure 8.32 As the matrix crack opens around a bridging fibre there will be frictional energy losses from fibre/matrix sliding in the debonded region

prior to failure, i.e. $\epsilon_f^* = \sigma_f^*/E_f$. The shear transfer stress τ_f is constant over the debond; hence the difference in strain between fibre and matrix will change linearly from ϵ_f^* at $x = 0$ (the crack surface) to zero at $x = d_f$. Thus the displacement of the fibre relative to the matrix at a point x during relaxation will be given by:

$$\delta x = \int_x^{d_f} \epsilon_f^*(1 - x/d_f)\mathrm{d}x = \epsilon_f^*\left(\frac{d_f}{2} - x + \frac{x^2}{2d_f}\right) \tag{8.65}$$

The total frictional work per fibre on relaxation will be given by this displacement working against the interfacial friction. Hence:

$$W = 2\int_0^{d_f} 2\pi r\tau_f \delta x\mathrm{d}x = 4\pi r\tau_f\epsilon_f^*\left(\frac{xd_f}{2} - \frac{x^2}{2} + \frac{x^3}{6d_f}\right)_0^{d_f}$$

$$= \frac{2\pi r\tau_f\epsilon_f^* d_f^2}{3} = \frac{\pi r^3\sigma_f^{*3}}{6\tau_f E_f} \tag{8.66}$$

The increase in fracture energy is calculated for N fibres per unit area bridging the crack, to give

$$\Gamma_{\text{debond}} = \frac{N\pi r^3\sigma_f^{*3}}{6\tau_f E_f} \tag{8.67}$$

8.4.3 FIBRE PULL-OUT

The final stage of the fracture process occurs after the fibres break in the debonded region. The reinforcing fibres most commonly used in composite materials are brittle in nature and therefore fail when the stress exceeds the fracture stress as determined by a local defect. Thus fracture can occur anywhere within the debonded zone and not

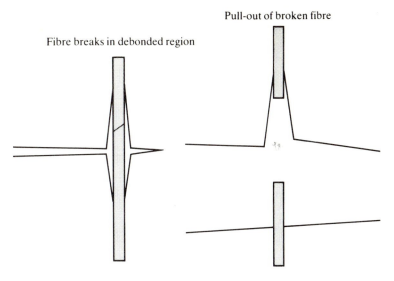

Fibre breaks in debonded region

Pull-out of broken fibre

Figure 8.33 Final fibre failure occurs at some critical defect in the debonded region. The broken fibre end is pulled out during fracture surface separation, requiring more work to be done against friction

necessarily at the crack plane where the fibre stress is greatest. If a fibre fails away from the crack plane, it must be pulled out of the matrix to allow final fracture (Fig. 8.33). This final pull-out will be resisted by the fibre/matrix friction stress and again further work must be done. If a length l of a fibre is left embedded in the matrix, the work done in pulling it out against the interfacial friction will be given by:

$$W = \int_0^l 2\pi r \tau_f x \, dx = \pi r \tau_f l^2 \tag{8.68}$$

This contribution is difficult to calculate for a fibre composite because it requires a value for the mean fibre pull-out length, which in turn requires a model to predict the likely site of fibre fracture. However, if it is assumed that fracture is equally likely anywhere within the debonded region, the average work per fibre will be given by:

$$W = \int_0^{d_f} \frac{\pi r \tau_f x^2}{d_f} \, dx = \frac{\pi r \tau_f d_f^2}{3} = \frac{\pi r^3 \sigma_f^{*2}}{12\tau_f} \tag{8.69}$$

and for N fibres per unit area

$$\Gamma_{\text{pull-out}} = \frac{N\pi r^3 \sigma_f^{*2}}{12\tau_f} \tag{8.70}$$

Hence the total work of fracture of the composite is obtained by combining Eqns (8.63), (8.67) and (8.70):

$$\Gamma = N\pi r^2 \left(\frac{r\sigma_f^{*3}}{6\tau_f E_f} + \frac{r\sigma_f^{*3}}{6\tau_f E_f} + \frac{r\sigma_f^{*2}}{12\tau_f} \right) \tag{8.71}$$

or after using $V_f = N\pi r^2$ and $d_f \approx r\sigma_f^*/2\tau_f$,

$$\Gamma = \frac{V_f r \sigma_f^{*2}}{12\tau_f} \left(\frac{4\sigma_f^*}{E_f} + 1 \right) = \Gamma_{\text{pull-out}} \left(\frac{4\sigma_f^*}{E_f} + 1 \right) \qquad (8.72)$$

from which it can be seen that the fibre pull-out energy is the most important term in composite fracture. However, in the cases where a high-strength but low-elastic-modulus reinforcement is used the other terms become more significant. In all cases it can be seen that, for all other variables remaining constant, the fracture energy will increase with increasing volume fraction of reinforcement, increasing fibre strength, increasing mean fibre radius, and with decreasing interfacial friction stress.

8.5 METAL AND CERAMIC MATRIX COMPOSITES

Recently there has been much interest in reinforcing matrices of metal or ceramic with ceramic fibres and other reinforcements. Before discussing these two new forms of composite we must identify the potential advantages a composite architecture offers these materials. With polymer matrix composites it is almost true to say that the properties of the composite are essentially those of the fibres, with little contribution from the properties of the matrix. This is certainly the case with properties parallel to the fibre directions, where the large differences in stiffness and strength between fibre and matrix are emphasized. However, perpendicular to the fibre direction, the weak fibre/matrix interface leads to properties being controlled chiefly by the matrix. This has led to the construction of complex multidirectional laminated structures to provide some isotropy. There are also properties of composites, such as their toughening mechanisms, which are uniquely determined by the fibre-reinforced structure of the material. Both metal and ceramic matrices have properties closer to those of likely reinforcements and this leads to a different choice of properties for which these composite systems are optimized.

Metal matrix composites typically comprise a light metallic alloy matrix, usually based on aluminium, magnesium or titanium alloys. A number of reinforcements have been used, including fibres of carbon, silicon carbide and alumina. The fabrication of these materials usually requires temperatures close to the matrix melting point and this precludes the use of glass and polymer fibres. The driving force for most of the applications of metal matrix composites is the potential for an increase in stiffness over the matrix alloy with little or no increase in density. Such composites fabricated from long continuous ceramic fibres behave in a manner similar to polymer matrix composites with their strengths and stiffnesses predicted as discussed earlier in this chapter. They have a different fracture behaviour for two major reasons. First, there is generally a much stronger fibre/matrix bond than is found with polymer matrix composites. This is believed to be because these composites are usually fabricated at high temperatures and at these temperatures chemical reactions can occur between matrix and reinforcement, so promoting an increased adhesion. These reactions can sometimes form brittle intermetallic phases at the fibre/matrix interface which can crack very easily and, by the introduction of a sharp crack next to the reinforcing fibre, nucleate premature fibre fracture. Second, when the metal matrix fractures in these materials it will only do so after considerable plastic work, and therefore another term must be considered in calculating the total work of fracture.

In order to reduce the costs of expensive ceramic fibres, reinforcement has been carried out with extremely short-aspect-ratio fibres and platelike reinforcements. These can have aspect ratios as low as 2, but still lead to an increase in stiffness. Curiously, these short-fibre materials do not obey the simple rule-of-mixtures arguments for composite strength as derived earlier for the case of short fibres. It is believed that there is some interaction between the reinforcement and matrix microstructure in these cases which cannot be easily explained using simple theory. The toughnesses of these very short-reinforcement composites are also a lot lower than might be expected from the simple theories expounded here. The reader is referred to more specialized books on this topic for information in these areas.[4]

Despite these limitations, there are a number of potential applications for either high specific stiffness or high wear resistance for which metal matrix composites are being considered.

Ceramic matrix composites are being developed for reasons other than increases in specific stiffness. Ceramic matrices already have large values of Young's modulus, and in any case the reinforcements used are likely to be very similar materials to the matrices. The major difficulty with the use of ceramic components in design applications is their low toughness and high variability in strength which require techniques such as Weibull statistics for their use (Chapter 7). Polymer matrices, such as epoxy resins, have very low fracture toughness, but when combined with brittle fibres an increase of many orders of magnitude of fracture energy occurs. Hence, it is hoped to achieve similar increases in the fracture toughness of ceramic matrices by the incorporation of long ceramic fibres. This has been achieved recently, with carbon and silicon carbide fibres being used to reinforce glass and glass—ceramic matrices. In this case the composite achieves toughnesses many times greater than those found in monolithic ceramics by fibre bridging, frictional delamination and pull-out, much as is found with polymer matrix composites. The reinforcement of ceramics by short fibres or 'whiskers' (short single-crystal fibres of radius $<1\ \mu m$) has also been investigated, but toughening is less efficient than has been found with long fibres. For whiskers (usually of silicon carbide) the chief interest so far has been to increase the wear resistance of ceramics.

EXAMPLES

8.1 A composite is made from glass fibres of Young's modulus 70 GPa and a polymer matrix of modulus 3 GPa. Estimate the transverse and longitudinal elastic moduli if it contains 33% V_f fibres using the rule of mixtures and the Halpin–Tsai equations.

8.2 A composite is manufactured from 50% V_f high-modulus carbon fibres with Young's modulus 390 GPa, shear modulus 160 GPa and Poisson's ratio 0.25 embedded in an epoxy resin with equivalent elastic constants of 4 GPa, 1.4 GPa and 0.39 respectively. In order to reduce elastic anisotropy it is proposed to use a laminated structure of six plies, each running in the directions $0°$, $+60°$ or $-60°$. Suggest a suitable construction for such a laminate and calculate its maximum and minimum tensile moduli.

8.3 (a) Glass fibres of tensile strength 1.5 GPa and elastic modulus 70 GPa are used to reinforce a polymer resin matrix of strength 50 MPa and elastic modulus 4 GPa.

(b) Silicon carbide fibres of tensile strength 6 GPa and elastic modulus 410 GPa are used to reinforce a metal matrix of yield strength 300 MPa and elastic modulus 70 GPa.

Plot the variation in strength of the composite with increasing volume fraction of fibres in each case.

8.4 Short ceramic fibres of strength 2 GPa and Young's modulus 200 GPa are used to strengthen pure aluminium of yield strength 20 MPa and Young's modulus 70 GPa. The fibres are approximately 5 μm in diameter. What is the shortest fibre length which is practical for this use?

8.5 Using the data of Table 8.1 calculate the expected energy absorbed on fracture of a 40% V_f glass-fibre-reinforced epoxy resin which has a fibre/matrix interfacial shear stress of 20 MPa. In order to increase the stiffness of the component and reduce its weight the glass fibres can be replaced by carbon or Kevlar fibres. If the interfacial stress remains unchanged, which of the proposed alternative fibres retains the highest energy-absorbing capability?

REFERENCES

1 HULL D 1981 *Introduction to Composite Materials* Cambridge University Press, Cambridge

2 HARRIS B 1987 *Engineering Composite Materials* Institute of Metals, London

3 MORLEY J G 1987 *High Performance Fibre Composites* Academic Press, London

4 TAYA M, ARSENAULT R J 1989 *Metal Matrix Composites: Thermomechanical Behaviour* Pergamon, Oxford

ANSWERS TO SELECTED EXAMPLES

CHAPTER 1

1.1 700 MN m^{-2}

1.2 Straight line, $\Delta = Fl/E(1 + 2\cos^3\alpha)$ up to $F = \sigma_Y(1 + 2\cos^3\alpha)$; plastic deformation given by $F = \sigma_Y(1 + 2\cos\alpha)$; elastic–plastic deformation by $\Delta = 1(F - \sigma_Y)/2E\cos^3\alpha$ in the range $\sigma_Y(1 + 2\cos^3\alpha) < F < \sigma_Y(1 + 2\cos\sigma)$

1.3 Similar plot to Example 1.2, with straight elastic–plastic section replaced by curved section tending asymptotically to $M = 312$ kN m

CHAPTER 2

2.3 $q = \dfrac{T}{a^2}\dfrac{2}{\sqrt{3}}$ $\qquad \theta = \dfrac{4T}{ta^3G}$

$q_1 = \dfrac{6}{3.4\sqrt{3}}\dfrac{T}{a^2}$ $q_2 = 1.6q_1$, $\theta = 0.47\dfrac{T}{ta^3G}$

2.4 Maximum shear flow, $\dfrac{T}{b^2} \times 0.2443$, 0.1974 and 0.1652

respectively, $\theta = \dfrac{T}{Gtb^3} \times 1.955$, 1.579 and 1.322 respectively

2.5 0.8

2.8 3.44 kN m, linear distribution from -35 MN m^{-2} at the surface to 105 MN m^{-2} at the centre

CHAPTER 3

3.1 (a) $A\sigma_Y(1 + 2\cos\theta)$ (b) $2\sigma_Y(\cos^3\theta - \cos\theta)/(1 + 2\cos^3\theta)$

3.3 (b)

	Tresca	von Mises
(i)	σ_Y	σ_Y
(ii)	$2\sigma_Y$	$2\sigma_Y$
(iii)	σ_Y	$(2/\sqrt{3})\sigma_Y$
(iv)	∞	∞

3.4 (a) $p = \sigma_Y t/R$ (b) $p = 2\sigma_Y t/\sqrt{3}R$

3.5 (a) 26 MPa (b) $\dot{\epsilon}_\theta : \dot{\epsilon}_z : \dot{\epsilon}_r = 4 : 1 : -5$

3.6 $f^2 + m = 1$ where $f = \dfrac{F}{bh\sigma_Y}$ $m = \dfrac{4M}{bh^2\sigma_Y}$

3.7 $f^2 + t^2 = 1$ where $f = \dfrac{F}{\pi bh\sigma_Y}$ $t = \dfrac{2\sqrt{3}\,T}{\pi d^2 h\sigma_Y}$

3.8 (b) $\omega^2 = \dfrac{3\sigma_Y}{\rho}\dfrac{1}{(b^2 + ab + a^2)}$

3.10 (a) $\sigma_\theta = \sigma_Y[1 - \log(r_o/r_i)]$

(b) $\sigma_{residual} = \sigma_Y\left(1 - \dfrac{2r_o^2}{(r_o^2 - r_i^2)}\log(r_o/r_i)\right)$

(c) $\dfrac{r_o}{r_i} = 2.22$

CHAPTER 4

4.1 Flaw size $a \cong \mu$m

4.2 (a) $K_{IC} = 110$ MPa$\sqrt{m}$
(b) Critical crack length $a = 3.85$ mm

4.3 (a) Plane stress (thin plate) $a = 24.2$ mm
Plane strain (thick plate) $a = 5.05$ mm

4.4 (b) Required thickness: (taken as 20 × plastic zone radius)
Mild steel 16.22 mm
Alloy 3.82 mm

4.5 Before changes: $a_{crit} = 2.77$ mm
After changes: $a_{crit} = 0.39$ mm

Latter cannot be detected and hence potentially catastrophic brittle fracture may ensue

4.6 Based on bursting criteria materials B or C look attractive but in each case the detectable crack size is well above the critical level. Hence material A is safest as the critical flaw size is about 140 mm and the detectable value is about 40 mm. It is, however, very heavy on material

4.8 (a) $K_{\text{IIC}} = \dfrac{3Px_c\sqrt{\pi a}}{4Lbh}$

(b) $K_{\text{IIc}} = \dfrac{3P}{4bh}\sqrt{\dfrac{a}{\pi}}\left[\pi\left(\dfrac{1}{2}-\dfrac{2}{L}\right) - \sqrt{1-\left(\dfrac{s}{a}\right)^2} - 2\sin^{-1}\sqrt{\dfrac{1-s/a}{2}}\right]$

where s is the distance measured from the centre of the beam

4.9 (b) Critical maximum length $c = 5.67$ mm
(c) Max. working pressure $p_0 = 20$ MPa (allowing a factor of safety of about 1.5)

CHAPTER 5

5.1 Minimum inspection interval is 165 days. If inspection interval is to be annual, NDT must be capable of detecting 2.0 mm crack

5.2 (i) Thickness 209 mm
(ii) Crack length leading to fracture 10.6 mm
(iii) Inspection interval is only every 45 days
(iv) Two-monthly inspection would require a crack detection ability of about 1.4 mm

5.3 $c = 0.85a$

5.4 (a) 0.5 mm
(b) 3.6×10^6 cycles
(c) Separation of 0.28 mm

CHAPTER 6

6.1 $L = 33.5$ to 19

6.2 6.84×10^{-12} s^{-1}

6.3 $m = 10.5$

6.4 50 mm

6.5 E_m 500 MPa, $E_v = 100$ MPa, $\eta_m = 10,000$ MPa s, $\eta_v = 526.3$ MPa s

6.6 79 J

6.7 5.8 MPa

CHAPTER 7

7.1 (a) >32 μm; (b) 0.5 μm

7.3 Weibull modulus = 6.2, maximum stress = 83 MPa

7.4 Weibull modulus = 8.4, mean tensile strength = 32 MPa

CHAPTER 8

8.1 $E_1 = 25$ GPa, $E_t = 4.4$ GPa (rule of mixtures),
$E_t = 6.7$ (Halpin−Tsai)

8.2 The best construction will be arranged
$0°|+60°|-60°|-60°|+60°|0°$ or
$0°|-60°|+60°|+60°|-60°|0°$, in order to eliminate bending or
twisting moments. $E_{max} = 75$ GPa, $E_{min} = 21.4$ GPa

8.4 Take fibre matrix shear transfer stress as matrix *shear* yield
strength $L = L_c = 500$ μm

8.5 Γ (glass) = 7.0×10^4 J m^{-2}
Γ (high-modulus carbon) = 6.6×10^4 J m^{-2}
Γ (high-strength carbon) = 1.0×10^5 J m^{-2}
Γ (Kevlar) = 2.3×10^5 J m^{-2}

INDEX

Spying Without Spies

Spying Without Spies

Origins of America's Secret Nuclear Surveillance System

Charles A. Ziegler *and* David Jacobson

Westport, Connecticut
London

Library of Congress Cataloging-in-Publication Data

Ziegler, Charles A.
    Spying without spies : origins of America's secret
nuclear surveillance system / Charles A. Ziegler and David Jacobson.
      p. cm.
    Includes bibliographical references and index.
    ISBN 0–275–95049–2 (alk. paper)
    1. Nuclear weapons—Testing—Detection—History. 2. Military
surveillance—United States—History. I. Jacobson, David.
II. Title.
U264.3.Z54 1995
355.02′17747′44—dc20   94–33263

British Library Cataloguing in Publication Data is available.

Library of Congress Catalog Card Number: 94–33263
ISBN: 0–275–95049–2

First published in 1995

Praeger Publishers, 88 Post Road West, Westport, CT 06881
An imprint of Greenwood Publishing Group, Inc.

Printed in the United States of America

The paper used in this book complies with the
Permanent Paper Standard issued by the National
Information Standards Organization (Z39.48–1984).

10 9 8 7 6 5 4 3 2 1

CONTENTS

PREFACE

This book is based on research that was supported by the National Endowment for the Humanities (Grant RH-20893-89). We wish to thank Lloyd R. Zumwalt, Robert P. Multhauf, and Stanley Goldberg for their help and collaboration in carrying out this research. Our major sources were unpublished documents in the archives of various government agencies, published articles and books, and informant interviews. The government documents we cite are in collections at the National Archives (NA), the Herbert Hoover (HHPL) and Harry Truman (HTPL) Presidential Libraries, the Library of Congress (LC), the archives of the Los Alamos National Laboratory (LANL), the Air Force Historical Research Center (AFHRC), the Air Force Phillips Laboratory (AFPL), and the Air Force Geophysical Laboratory (AFGL). Additional documents were obtained through the Freedom of Information Act from the Departments of Defense (DOD) and of Energy (DOE) and the Central Intelligence Agency (CIA).

We would be remiss if we failed to point out that our appeals under the Freedom of Information Act for certain documents (which we could identify with precision) were rejected by the relevant government agencies. We believe that these documents contain still-classified diplomatic information that is (fortunately) peripheral to the scientific and organizational themes we have attempted to enlarge upon in this book. One such top secret document, however, will be especially useful to later historians interested in this topic. The following information will aid in its retrieval at some future date when the government decides that its contents can at last be revealed: "History of Long Range Detection, 1947–1953," by Doyle Northrup and Donald Rock, Call No. TS-HOA-79 L6, 330 pages, Archives, Air Force Technical Applications Center.

Although archival materials constitute the primary evidence on which this book is based, we also obtained supplementary information from twenty-five informants who participated in the events described and whom we interviewed as part of our research. They gave freely of their time, their memories, and, in some cases, the relevant documents and memoirs in their possession. We gratefully extend our thanks to Dana Atchley, Robert F. Bacher, William E. Barbour, Jay L. Beaufait, Ralph D. Bennett, Edward A. Doty, Herbert Friedman, William T. Golden, Frederick C. Henriques, Paul A. Humphrey, Jerome Kohl, Arnold Kramish, Gerard M. Leies, Leon Leventhal, Luther B. Lockhart, Wilfred B. Mann, Charles B. Moore, Wendell C. Peacock, Roger R. Revelle, Walter Singlevich, Roderick W. Spence, Athelstan F. Spilhaus, Julius Tabin, Albert C. Trakowski, and Lloyd R. Zumwalt.

We thank Mildred Crary and Sybil Northrup for providing us with material on their husbands' role in creating the nuclear surveillance system. We also thank Robert G. Todd for his documentation on Mogul and Alan Stockdale for his review of relevant social studies of science and technology. We are greatly indebted to Alan Needell and Paul Forman for commenting on the manuscript and to Gerard Leies, Roderick Spence, and Lloyd Zumwalt for their advice in assessing the significance of highly technical government documents, but we assume full responsibility for the interpretations that are presented here. Last but not least, we thank Virginia Ziegler and Lois Jacobson, who kindly accepted our occasional preoccupation with the research and writing of this book.

A comment on terminology: most of the sources cited in this book treated "America" as synonymous with "United States" and "Russia" as synonymous with "Soviet Union," and they generally used "atomic" instead of "nuclear" when referring to bombs and reactors. We have adopted these usages to make our text consonant with quotations from these sources.

INTRODUCTION

On August 29, 1949, the Russians exploded their first atomic bomb, dubbed "Joe-1" by American intelligence analysts, at a test site deep within the Soviet Union.[1] Joe-1 was a momentous event, signalling the end of the U.S. monopoly of such weapons and the beginning of a nuclear arms race that was to plague the world for the next forty years. Despite Russian efforts to keep it hidden, the United States detected the explosion, at a distance of several thousand miles, by means of a secret technical surveillance system that had been established to monitor certain atomic activities in the Soviet Union and elsewhere. This is an account of the genesis and institutionalization of that system.

In the years that followed Joe-1, the agency that was formed to develop and control the nuclear surveillance system continued its covert operations to collect and evaluate data crucial to the assessment of the Soviet Union's nuclear capabilities and the capabilities of other foreign powers as well. As the global scope of its activities increased, so did its size, in terms of personnel, facilities, and funding.

The existence of an effective American nuclear surveillance capability has had a pervasive, but unheralded, effect on the arms race and on crucial aspects of U.S. foreign policy, such as the test ban treaties with other nuclear powers and nonproliferation agreements to limit the spread of nuclear weapons.[2] Nevertheless, the agency responsible for this task has remained relatively unknown, despite the fact that nuclear surveillance has been well funded by the American taxpayer and has been and continues to be an important factor in policymaking. Originally designated by a different acronym, the agency is now called AFTAC (Air Force Technical Applications Center). Although brief descriptions of its current organization and goals have appeared, little has been published about its formation.[3] This story of the agency's birth and the early development of America's nuclear surveillance capability thus fills a gap in the historical record.

It is a story that has never been told before, and, as such, it throws new light on questions that have been raised in previous histories of the post–World War II era: Why were the president and government "insiders" so convinced, contrary to the belief of the public at large, that America's atomic monopoly would endure into the 1950s? How was the shadowy world of atomic intelligence influenced by the popular belief, in the years prior to Joe-1, that the Russians would soon possess the "secret" of the atomic bomb? What was the role of science and scientists during the critical post–World War II years in determining America's stance on intelligence-gathering? And were scientists "on top" or merely "on tap" in formulating government science policy?

Our account also corrects some of the misconceptions about the formation of the nuclear surveillance system that have arisen over the years. Chief among these is the notion that the system was initiated in 1947 at the instigation of one man, Lewis L. Strauss, who was then an Atomic Energy commissioner.[4] Our findings indicate that, like many complex government undertakings, the system resulted from the work of many individuals, chiefly military officers, that began in World War II and continued into the postwar period.

But this book is more than a historical narrative. It is also a study of the organization of diversity in the creation of a large and complex technological system, and of the network of individuals and groups that produced and operated it. The formation and functioning of the system was threatened by disagreements among scientists and between scientists and non-scientists holding different beliefs about its feasibility, and by the conflicting interests of the organizations involved in its development and utilization. And yet, despite the disagreements and the conflicts, the system *was* created. How this was accomplished is the central theme of this book.

NOTES

1. How the term *Joe-1* originated is not known with certainty, although it was patently derived from Soviet Premier Josef Stalin's first name. At least two individuals have separately claimed to have coined it: Arnold Kramish, who was then a member of the intelligence unit of the Atomic Energy Commission, and Dr. Anthony Turkevitch, then a scientist at Los Alamos Scientific Laboratory. A. Kramish interview, July 31, 1985, and A. Turkevitch, letter to D. Jacobson, March 27, 1990. Copy in the authors' collection.

2. See, for example, R. A. Divine, *Blowing on the Wind*, 1978.

3. For a description of AFTAC circa the 1980s, see J. Richelson, *The U.S. Intelligence Community*, 2nd ed., 1989, pp. 26, 85–89, 214–224. Published works that, in part, deal with the formation of the monitoring system are "The Decision to Detect," Chapter X in L. L. Strauss, *Men and Decisions*, 1962, pp. 201–207, and C. A.

Ziegler, "Waiting for Joe-1: Decisions Leading to the Detection of Russia's First Atomic Bomb Test," *Social Studies of Science*, Vol. 18, 1988, pp. 197–229. See also *Transcript of Hearing in the Matter of J. Robert Oppenheimer*, 1954, pp. 684, 691, 695, 801, and H. S. Truman, *Years of Trial and Hope*, 1956, p. 306.

4. For example, the official history of the Atomic Energy Commission reflects this erroneous version of how the system was formed. See R. G. Hewlett and F. Duncan, *Atomic Shield*, 1969, pp. 130–131. Later historians have echoed this view. See, for example, P. Glynn, *Closing Pandora's Box*, 1992, p. 126; J. Richelson, *American Espionage and the Soviet Target*, 1987, p. 115; L. Freedman, *U.S. Intelligence and the Soviet Strategic Threat*, 2nd ed., 1986 [1977], p. 64; G. Herken, *The Winning Weapon*, 1982 [1981], p. 300; J. W. Kunetka, *Oppenheimer, the Years of Risk*, 1982, p. 146.

CHAPTER I

WORLD WAR II ORIGINS OF RADIOLOGICAL SURVEILLANCE

In mid-1944, the war was going well for the United States and its allies. The Japanese were retreating in the Pacific, and the Germans were reeling from the combined effect of a Russian offensive in the east and the successful cross-channel invasion by Anglo-American forces in the west. In late summer of 1944, the American Ninth Air Force shifted its base of operations from England to the Continent. Following this move, a few of its Douglas A-26 medium bombers were required to make mysterious low-level reconnaissance flights over areas deep within Germany that were designated by the air intelligence staff.

These forays were "mysterious" to the flight crews because the bombardier's compartment of the aircraft was fitted with equipment consisting largely of metal tubes that bore no resemblance to the aerial cameras normally used in reconnaissance.[1] And tight-lipped intelligence officers gave the pilots who flew these missions no clue as to the purpose of the equipment or the nature of the information gleaned on these overflights of Germany.

What the pilots *did* know was that they were asked to fly at suicidally low altitudes. By the fall of 1944, when these flights were made, German fighters were no longer the threat they had posed at the beginning of the year owing to attrition and fuel shortage. But, in compensation, the Germans had deployed many more antiaircraft guns which were especially effective against low-flying planes. Thus, although the A-26s were stripped of their bombs and could therefore exceed their loaded top speed of 355 mph, these missions were considered so dangerous that some pilots recommended they be scrapped.[2]

By late fall, requests for these hazardous flights ceased and the strange equipment was removed. The crews, who returned with relief to the more familiar dangers of bombing Germany, had no inkling that they had participated in the first attempt by the United States to use a radiological method of overhead reconnaissance to monitor the atomic activities of a foreign power.

BELIEFS ABOUT THE GERMAN THREAT

To understand why and at whose order the monitoring of Germany was carried out, it is necessary to examine the thinking of the men who led the American effort to develop the atomic bomb. From the early days of 1940, when the United States first seriously considered using nuclear fission as the basis for a superbomb of unprecedented power, the fear of a parallel German effort had overshadowed the deliberations of government planners. Nor were their fears confined to the possibility of a German atomic bomb. Other types of atomic weapons, based on radioactive poisons, were known to be possible as well.[3] Indeed, in 1941 American scientists had briefly entertained the idea of making a "radioactivity bomb," after an advisory committee of the National Academy of Sciences pointed out that it might be produced more quickly and easily than a true atomic bomb.

The radioactivity bomb was based on the fact that nuclear fission within a reactor could be used to create huge quantities of radioisotopes (the radioactive form of a given element). These intensely radioactive substances could then be incorporated in a bomb together with ordinary high explosives. Dropped on a city, such radioactivity bombs would have no more blast effect than conventional aerial bombs, but they would be much more devastating since detonation would disperse radioactive poisons over a large area, making it necessary to evacuate the population and rendering the city unlivable for years to come. In other words, a city subjected to such bombs would become a kind of instant Chernobyl, but on a much larger scale.[4] Nevertheless, despite their frightfulness, such weapons lacked the war-winning potential of a fission-based atomic bomb. Thus, as initial studies revealed the growing feasibility of making an atomic bomb, the notion of using radioactive poisons was dropped.[5]

But the very success of their preliminary research increased the belief among scientists in the United States that the German atomic program was proceeding apace. The prevalence of this belief at the Manhattan Engineer District, the designation for the organization responsible for the United States' atomic bomb project, made it inevitable that Brigadier General Leslie R. Groves, after being named head of that organization in September 1942, would be thoroughly briefed on the German threat by senior Manhattan Engineer District scientists. Thus, from the beginning of his tenure as head of the Manhattan Project, Groves was keenly aware of the need to assess the progress of German atomic research.

Groves was able to build up a crude delineation of the German atomic program from the somewhat disjointed information supplied by American intelligence agencies and from the guarded overview provided by the British. This picture indicated a heartening lack of progress on the part of the Germans. Indeed, the British Secret Intelligence Service, which at that time had a more detailed picture at their disposal, reported that the threat of a German atomic bomb had, if not evaporated, become more remote. The reaction to these findings, according to the official history of British intelligence, was that "the threat from Germany's nuclear programme . . . was ceasing to be a source of grave anxiety to the Allied Governments."[6]

By the late spring of 1943, when Groves became reasonably assured that the Germans were not likely to produce an atomic bomb, he was still very much concerned that they would develop and use "an ordinary bomb containing radioactive material."[7] American scientists, who had investigated this possibility, concluded that for various technical reasons, notably the problem of safely handling the enormous quantities of radioisotopes involved, it was "unlikely that a radioactive weapon will be used."[8] But Groves believed that the safety problems that deterred U.S. scientists would not inhibit the Germans because the Nazis would use technicians and workers drawn from "inferior" groups within the populations they controlled.[9]

Groves soon had the opportunity to assuage his fears about the possible German use of reactor-produced radioactive poisons by direct action. In the fall of 1943, in addition to his primary job of managing the Manhattan Project, he was given the overall responsibility for atomic intelligence by General George C. Marshall, Army chief of staff. Groves had no previous experience in intelligence work. He had been selected to head the Manhattan Project because, as an engineering officer, he had evinced outstanding talent in carrying out large-scale military construction programs. Nevertheless, he tackled the intelligence assignment with characteristic vigor by establishing a Foreign Intelligence Section within his Manhattan Project headquarters. This Section acted as a central clearinghouse for atomic information collected by the various intelligence agencies.

In his prior dealings with these agencies, however, Groves had found that the information they produced was not sufficient to provide an adequate picture of the German atomic program, partly because their intelligence-gathering efforts were fragmented and partly because the secrecy imposed on atomic matters in the United States precluded briefing agents on the kind of information sought.[10] Realizing that the U.S. intelligence community, as it was then constituted, was unlikely to provide all the information he needed, Groves decided to create his own intelligence operation whose agents would have a knowledge of the Manhattan Engineer District's secrets and access to its technical and human resources.

Groves had no intention of setting up conventional espionage networks such as those established by Army Intelligence, the Office of Strategic Services, and the British Secret Intelligence Service. Instead, he preferred to introduce methods that would add new dimensions to the overall Allied intelligence effort. That is, Manhattan Engineer District intelligence units would, as he later recalled, "supplement but not overlap or duplicate the other agencies."[11] In complying with this self-imposed policy, Groves introduced two innovative approaches to gathering atomic intelligence: first, the Alsos concept, and second, the notion of radiological surveillance.

The first approach was based on an idea he had thought about even before he had been given responsibility for atomic intelligence. In mid-1943, he had discussed with Major General George V. Strong, head of Army Intelligence, the possibility of creating specially trained teams, made up of civilian atomic scientists and military personnel, to follow in the wake of the Allied advance in Italy. These teams would interrogate newly "liberated" Italian scientists and collect scientific reports and other items related to atomic developments. After he took on the job of intelligence gathering, Groves implemented this concept with his usual dispatch and the first such team arrived in Italy in December of 1943. It was code named Alsos. The name stuck and was later applied to the team that followed the Allied armies into France and Germany.[12] Alsos was a successful wartime expedient whose utility ended with the war. On the other hand, the second innovation introduced by Groves, the concept of radiological surveillance, was to prove a lasting contribution to the field of intelligence.

It is tempting to speculate that the notion of radiological surveillance was an idea that stemmed from Groves's engineering background. If so, it is true only in the most general sense for, although he had the technical prescience to know that something might be done in this area, he lacked the scientific knowledge to conceive of an effective methodology. Thus, he delegated this task to Dr. Luis Alvarez, a Manhattan Engineer District scientist with a demonstrated ability to turn ideas into practical hardware.

TAKING THE AIR

Alvarez, a Nobel laureate-to-be, was a recent addition to the Manhattan Engineer District staff, having spent the previous three years at the Radiation Laboratory of the Massachusetts Institute of Technology. There he had worked on projects, such as a ground-controlled aircraft landing system, that showcased his ability as an experimenter with a flair for using his knowledge of physics to produce practical engineering results.

Groves apparently knew of Alvarez's innovative work at the Massachusetts Institute of Technology and sought him out after he joined the University of Chicago's Metallurgical Laboratory, a part of the Manhattan Project commonly known as "Met Lab." In recalling their first meeting in the fall of 1943, Alvarez later wrote:

I was so far down in the Met Lab structure that I didn't appear on any organization chart; nevertheless the General knew who I was. I was summoned one day to meet him. . . . He called me in and asked me how we could find out whether the Germans were operating nuclear reactors. If they were and if we could locate them, the Air Force would destroy them or at least interrupt their operation. . . . He cautioned me to let no one else in the laboratory know what I was doing.[13]

The final admonition reflected Groves's policy of keeping information "compartmentalized" for reasons of security. This policy irritated Manhattan Project scientists, who were accustomed to exchanging ideas freely. It created practical problems as well. For example, the fragmentation of information imposed by compartmentalization made it difficult to coordinate the work of various groups.[14] Nevertheless, despite complaints and problems, Groves adhered to the rule of compartmentalization. This meant that Alvarez had to conceive of a method of radiological surveillance on his own and develop it using staff and outside subcontractors who could be given only the bare minimum of information needed to complete their assigned tasks.

In the days following their meeting, Alvarez thought about the problem Groves had posed and devised a scheme to use a specially equipped airplane to sweep the skies over Germany searching for one of the radioactive gases emitted during the operation of an atomic reactor. In implementing this scheme, he became the first scientist to develop a radiological air-sampling method of overhead reconnaissance, a form of intelligence gathering that would become the cornerstone of the United States' effort to monitor Russian atomic progress during the Cold War period.

Alvarez decided that, of the various gases produced by a reactor, radioactive xenon-133 would be the easiest to detect partly because of the readily identified characteristic radiation it produces. Xenon-133 emits high-speed electrons or beta rays and gamma rays, a penetrating form of electromagnetic radiation akin to x-rays. It also has a boiling point much higher than nitrogen and oxygen, which are the major constituents of air, and Alvarez knew that this disparity would facilitate the process of separating minute amounts of xenon from the enormous volume of air that would be sampled. But Alvarez realized that he needed more information about the family of chemically inert gases (the so-called noble gases) to which xenon belonged. He moved his office to a new location to be nearer the University library where the reference materials he required were readily accessible. In addition, he decided to seek the help of industrial experts.

At that time, the Cleveland plant of the General Electric Company possessed the country's largest liquid air processing facility for producing noble gases, such as argon, used in gas-discharge lamps. Arthur H. Compton, the eminent physicist and Nobel laureate who was head of the Metallurgical Laboratory, had many contacts at General Electric, and, through his intercession, arrangements were made for Alvarez to consult with the

General Electric scientists who had designed the liquid air plant. By that time, Alvarez had recruited a youthful physicist, Julius Tabin, to help in developing the necessary equipment. Together they visited General Electric's noble gas experts. It soon became apparent that General Electric could be of assistance not only in the conceptual stages of the work, but also in building part of the apparatus. General Electric's technical personnel were heavily engaged in wartime production, but Alvarez was armed with the Manhattan Engineer District's triple-A priority which elicited prompt attention to his project. Accordingly, as Tabin later wrote, General Electric scientists "assisted in the design and construction of the miniaturized liquid air plant which provided the coolant necessary to condense the air being monitored in our detection scheme."[15]

The xenon-detection system in its final form consisted of two components: an air-sampling apparatus designed to fit into the aircraft, and ground-based laboratory equipment to analyze the collected samples of air. The airborne apparatus used activated charcoal to extract xenon and other heavy gases such as radon (which are always present in minute amounts in the atmosphere) from the sampled air. After an overflight, the activated charcoal was removed and taken to the laboratory where it was heated to boil off these gases. Xenon was then extracted from this effluent using a technique that relied on the disparity of boiling points of the various gases present. The xenon thus obtained was subsequently tested for the presence of the characteristic xenon-133 radiation using standard detection equipment. This method of extracting xenon always produced some nonradioactive xenon as an end-product because this gas occurs naturally in the atmosphere (at one part per 20 million). But the radioisotope xenon-133 would only be found if there were an operating atomic reactor in the area where the air sample was obtained.

The sampling apparatus had to be airworthy, and it had to be physically configured to fit the aircraft. This meant that it was necessary to specify the aircraft in order to complete the design of the air-sampler. At that time, the Army Air Force had in its inventory a sleek, new lightweight attack bomber, the Douglas A-26 Invader. Its design had been begun in 1941, and by 1944 a substantial number had been built, although it had not been deployed overseas. It was then one of the fastest bombers in the world. It had been designed to fly swiftly to a target, descend to tree-top level to drop its bombs, and then climb rapidly to avoid antiaircraft fire. Thus, it was well suited to a mission profile requiring it to fly to a suspected atomic site at high speed to avoid German defenses, descend and execute a low-level pattern to sweep the air in the vicinity of the site, and return swiftly to base. It also possessed the range (1,400 miles) to reach suspected atomic sites deep within Germany. Because of its desirable features, the A-26 was selected as the platform for the airborne component of the xenon detection system. The air-sampling apparatus was, therefore, designed to fit into the bombardier's compartment of an A-26. Flight tests were scheduled for the summer of 1944 when construction of the system would be complete.

TESTING THE WATERS

In the spring of 1944, Groves, impatient to obtain positive evidence of German atomic developments, called for a method of radiological surveillance that would give more immediate results than the xenon-sampling scheme, which was not ready to be deployed. The urgency of Groves's need for such data reflected his concern over the potential atomic danger menacing the troops in England who were girding themselves for an attack on Europe. In later recalling his preinvasion fears, he wrote that the danger "grew out of the possibility that the Germans might use radioactive material to block the cross-channel attack of the Allied forces."[16] Once again the specter of German radioactive poisons had been raised.

In response to Groves's anxiety, his aide, Major Robert R. Furman, initiated correspondence with Alvarez early in May on an alternative method of radiological surveillance. "The general project," he wrote, "is to obtain samples of water from Lake Constance and the upper reaches of the Rhine River which are accessible from Switzerland."[17] The reasoning behind this scheme was that, if the Germans were operating an atomic reactor, they were probably using water from the Rhine as a coolant. This presented the possibility of finding traces of reactor-produced radioactivity in the water. The Office of Strategic Services had a station in Berne, Switzerland (headed by Allen Dulles, who was later to become a leading figure in American intelligence), which controlled agents who had access to the Rhine. At Furman's request, Alvarez drafted a telegram that was sent to Dulles in Switzerland outlining the procedures agents should use to obtain water samples.[18]

SURVEILLANCE OPERATIONS

The first of the two methods of radiological surveillance developed by the Manhattan Engineer District, the xenon-detection scheme, was used in Europe in the fall of 1944. In order to guide the A-26 overflights, the suspected sites of atomic activity had been pinpointed by the painstaking work of the staff in the London office of the Manhattan Engineer District's Foreign Intelligence Section. To accomplish this task, information gleaned from overt sources, such as German books, newspapers, and technical journals, had been collated with data from aerial photoreconnaissance and reports from U.S. and British intelligence agencies. The A-26s carrying Alvarez's air-sampling apparatus made flight after dangerous flight over these suspect locales, but no xenon-133 was detected. This, of course, tended to confirm the conclusion drawn from other intelligence sources that the Germans had no atomic reactors in operation.

The second method of radiological surveillance developed by the Manhattan Engineer District, water sampling, was also carried out in the fall of 1944. The record is unclear as to whether or not the waters of Lake Constance and the upper Rhine were sampled as called for in the original plan. But samples *were* obtained from the lower Rhine by the Alsos team that followed the Allied armies into France.

In October of 1944, the team leader, Colonel Boris T. Pash, received a radiogram from the Manhattan Engineer District's Washington headquarters directing him to sample the Rhine. The first Allied forces had just reached that waterway and captured the Nijmegen Bridge. The heavy fighting was over, but German troops were still active in the area of the bridge. In his memoirs, Colonel Pash has reconstructed the conversation that ensued when his subordinate, Captain Bob Blake, appeared at the Nijmegen Bridge and, with bucket in hand, approached the officer commanding the troops defending the crossing:

'Sir, with your permission I'd like to go out on the bridge and get some Rhine water.'

'I don't say you're crazy, Captain. But those who sent you definitely are!'

Undaunted, Bob Blake continued. 'If the Colonel will not object, I'd like to go on the bridge to get water from mid-stream as well as from both sides.' [Blake was religiously following the sampling procedure laid down by Alvarez.]

'Now I know that *you're* crazy too. But go ahead. You've lived this long so you must be lucky.'[19]

The water samples were duly flown back to Washington, together with a bottle of French wine, sent along as a joke, with the inscription "Test this for activity, too." What was intended, of course, was that the recipient would "test" the wine by drinking it. A radiogram soon appeared: "Water negative. Wine shows activity. Send more. Action."[20] And Pash was directed to obtain more wine samples. Thinking that this was Washington's way of continuing the joke with a ploy to get more wine, Pash attempted to resolve the matter with radiograms to the Manhattan Engineer District headquarters. But to no avail. Radioactivity had been found in the wine, and the Manhattan Engineer District scientists wanted to be sure it was not due to man-made radioisotopes. Thus, members of the Alsos team were diverted to (and by) the task of sampling the products of the French wine country.

As the Allied advance continued, the sites of German atomic research, with their voluminous records and even the atomic scientists themselves, became accessible to Alsos. The Alsos team was also able to receive reports directly from agents to augment the emerging picture of the German atomic program.[21] The enormous amount of evidence amassed by Alsos confirmed the accuracy of the picture of the German atomic program that had been built up by the Allies from intelligence sources long before the invasion.

Indeed, by the spring of 1945, with the ineffectualness of the German program confirmed, Alsos, under Groves's prodding, turned to activities such as the confiscation of German uranium and the sequestration of German atomic scientists. These activities were aimed primarily at thwarting the atomic ambitions of a country Groves perceived as a potential enemy, Russia. As Groves later wrote: "Our principal concern at this point was to keep information and atomic scientists from falling into the hands of the Russians."[22]

This was not a new concern for Groves. In 1942, when he assumed command of the Manhattan Engineer District, he had laid down the goals for Manhattan Engineer District security: "They were threefold," he later wrote, "first, to keep the Germans from knowing anything about our efforts . . . next, to do all we could to ensure complete surprise when the bomb was first used . . . and, finally . . . to keep the Russians from learning of our discoveries."[23]

Alsos's leader, Colonel Pash, was familiar with Groves's views on the threat posed by Russian espionage. Early in the war, Pash, a former high school teacher, had been assigned to Army Intelligence partly because he spoke fluent Russian, a legacy from his emigré father who was metropolitan of the Eastern Orthodox Church in North America. Pash had first met Groves in San Francisco, in early 1943, at a time when the Manhattan Engineer District's security was being handled by Army Intelligence. Pash was investigating the first attempt by Russian agents, aided by American Communists, to penetrate the Manhattan Engineer District's security shield. Their purpose, according to Pash, was "to steal our atomic secrets."[24] Groves later wrote that he was impressed by the "thorough competence and drive" Pash displayed in prosecuting this case.[25] Pash's sensitivity to the possibility of Russian espionage recommended him to Groves, and, when responsibility for the Manhattan Engineer District's security was transferred from Army Intelligence to the Manhattan Engineer District later in 1943, Groves requested that Pash serve as his security chief.

Throughout the war, therefore, Groves's security measures, including compartmentalization, had been designed to safeguard U.S. atomic secrets not only from Germany and Japan, but from Russia as well. Groves was not content with merely guarding against Russian espionage; he also attempted to penetrate Russian security in order to obtain information about their atomic program. In 1943, he orchestrated a somewhat Machiavellian scheme to obtain such information by shipping uranium to Russia!

In January of that year, the Lend Lease Administration had received an order from a Russian purchasing agent for over 450 pounds of uranium compounds. Several U.S. companies offered to supply the Russians with this commodity, but uranium had been placed on the War Production Board's critical list. Therefore, the Russian order was, at first, turned down. Groves heard of these negotiations and intervened to honor the Russian request. He reasoned that to refuse would provide the Russians with inferential knowledge of the status of the United States' atomic program. More important, Groves hoped that the uranium shipment could be tracked to its destination, thus identifying the location of the Russian atomic research center.[26]

Records indicative of the relative success of this plan are unavailable. Nevertheless, the incident (which was later the subject of congressional inquiry) illustrates the trend of Groves's thinking even at this early date. Thus, his orders to Pash in 1945, aimed at denying the Russians access to the results of German atomic research, came as no surprise to Pash since they were, in effect, an overseas extension of Groves's domestic security policies.

In summary, the following features characterized the U.S. stance on atomic intelligence at the end of World War II: centralization of responsibility, integration of atomic scientists as advisors into the intelligence function at the level of collection as well as analysis, coordination (albeit not always frictionless) with British atomic intelligence, concentration on Russia as the primary target, and possession of innovative technical means of gathering atomic intelligence, notably a radiological method of overhead surveillance.

All of these characteristics, except the first one, were largely due to initiatives undertaken by Groves. The notion of centralized responsibility for atomic intelligence did not survive the postwar dismantling and reassembly of the U.S. intelligence community. But the other features continued to characterize the United States' atomic intelligence effort in the postwar era.

NOTES

1. J. Tabin interview, January 5, 1991.

2. W. Lawren, *The General and the Bomb*, 1988, p. 138.

3. It is widely held that the American effort to develop an atomic bomb was a direct response to the fear that the Germans could and would build one. See, for example, R. D. Hewlett and O. E. Anderson, *The New World, 1939/1946*, 1962, pp. 69, 119–120; R. Rhodes, *The Making of the Atomic Bomb*, 1988 [1986], p. 325; and R. Jungk, *Brighter Than a Thousand Suns*, 1958, p. 343. See also S. A. Goudsmit, *Alsos*, 1983 [1947] and, more recently, T. Powers, *Heisenberg's War*, 1993. Goudsmit also notes that scientists in the United States, including many who were European refugees, worried that even if the Germans did not yet have an atomic bomb, they might launch a "radioactive attack" on the United States. See *Alsos*, pp. 7–8. See also Powers, *Heisenberg's War*, pp. 204, 353–355.

4. In 1986 a catastrophic reactor accident heavily contaminated much of the area around Chernobyl in the former Soviet Union.

5. B. J. Bernstein, "Oppenheimer and the Radioactivity Poison Plan," *Technology Review* 88, May–June 1985, pp. 14–17.

6. F. H. Hinsley, *British Intelligence in the Second World War*, Vol. 2, 1981, pp. 125–128.

7. L. R. Groves, *Now It Can Be Told*, 1962, p. 194.

8. Rhodes, *The Making of the Atomic Bomb*, p. 512. The quotation is from the Conant subcommittee report.

9. In recalling the way he felt about this subject, Groves later wrote that Hitler and his followers "would not hesitate to expose these same citizens to excessive radiation." See *Now It Can Be Told*, p. 200. Groves's belief that a totalitarian regime would not be concerned about a risk of excessive radiation to its workers was borne out by the Russian effort to build their atomic bomb. In this regard, see S. J. Zaloga, *Target America*, 1993, pp. 50–52, and p. 284, n.55.

10. At the time the Manhattan Engineer District was established in August 1942, intelligence on foreign atomic research was collected by the Office of Naval Intelligence, Army Intelligence (G-2), and the newly formed Office of Strategic Services. But atomic information was gathered independently by these agencies in the course of their general operations, not as part of a coordinated effort that targeted the German atomic program. The failure of the agencies to coordinate their efforts was largely the result of interorganizational battles over "turf." See Groves, *Now It Can Be Told*, p. 185.

11. Groves, *Now It Can Be Told*, p. 191.

12. For an account of the Alsos missions, see Groves, *Now It Can Be Told*, pp. 189–190. See also Goudsmit, *Alsos*, and Powers, *Heisenberg's War*.

13. L. W. Alvarez, *Alvarez*, 1987, p. 120.

14. Some complaints and problems stemming from compartmentalization are described by R. W. Clark, *The Greatest Power on Earth*, 1980, p. 139, and R. R. Hanson, *U.S. Nuclear Weapons*, 1988, p. 212 n.41.

15. J. Tabin, letter to L. C. Zumwalt, January 2, 1990, copy in the authors' collection. The noble gas detected by Alvarez's system is incorrectly identified as krypton-85 in two histories of the period: Lawren, *The General and the Bomb*, 1988, p. 135; and L. M. Libby, *The Uranium People*, 1979, p. 146.

16. Groves, *Now It Can Be Told*, p. 199.

17. R. R. Furman to L. Alvarez, May 12, 1944, Records Center/Archives, LANL. Powers states that the suggestion for testing river waters for signs of radioactivity came first from Oppenheimer in a letter to Furman, dated September 23, 1943. See Powers, *Heisenberg's War*, pp. 224, 528 n.29 and 362.

18. L. Alvarez to R. R. Furman, May 17, 1944, Records Center/Archives, LANL.

19. B. T. Pash, *The Alsos Mission*, 1969, p. 133.

20. See Goudsmit, *Alsos*, pp. 21–24 and Powers, *Heisenberg's War*, pp. 362 and 559 n.45.

21. Goudsmit, *Alsos*, p. 185.

22. Groves, *Now It Can Be Told*, p. 141.

23. Ibid., p. 173.

24. Pash, *The Alsos Mission*, p. 9.

25. Groves, *Now It Can Be Told*, p. 173.

26. Lawren, *The General and the Bomb*, p. 266. See also Herken, *The Winning Weapon*, p. 107.

CHAPTER 2

POSTWAR HINDRANCES TO RAPID DEVELOPMENT

In the wake of World War II, American military planners vowed never again to be caught by a surprise attack, especially an "atomic Pearl Harbor" that might result in the nation's defeat. Two months after the surrender of Japan, the Joint Chiefs of Staff issued a position paper stating that "to maintain the atomic bomb as the exclusive secret of this country over a considerable period of time will not be possible." It thus seemed provident to assume other nations would attempt to develop atomic weapons. And several months later, a report of the Joint Intelligence Committee emphasized the urgent need "to provide for intelligence concerning the capabilities and intentions of nations . . . regarding atomic warfare and related matters."[1]

A few military leaders, notably Generals Leslie Groves and Curtis E. LeMay, who had been made head of Army Air Force research, had concluded that it was unlikely that any country would attempt to produce atomic bombs without first testing the design. Therefore, they believed that technical means for secretly monitoring foreign atomic bomb tests should be developed as an adjunct to conventional methods of acquiring intelligence on atomic progress in other countries. A few months after the end of the war, using the scientific resources they controlled, both Groves and LeMay independently initiated technical work on methods for detecting a foreign bomb test at long range, methods on which a centrally coordinated covert monitoring system could be based. But in the immediate postwar years, several factors slowed the technical and organizational development of such a system. As will be described, three of these factors appear paramount: a limited stockpile of atomic bombs, organizational turmoil within the intelligence community, and the Murray Hill Area project.

A LIMITED STOCKPILE

The first factor, which affected the pace of the technical development of a monitoring system designed to detect a foreign atomic bomb test, needs little explanation. Although research on new and existing methods of long-range detection could be, and was, conducted in the absence of atomic explosions, such explosions were essential to assess the effectiveness of each technical advance. To a degree, this could be accomplished in the laboratory, but a definitive evaluation of performance could only be made using an atomic explosion. The American atomic bomb tests, however, were extremely expensive, potentially hazardous to personnel, and required extensive planning and coordination at many government levels from the president on down, attributes that, in the late 1940s, tended to make such testing infrequent. An overriding consideration that inhibited testing in the immediate postwar period, however, was the parlous state of the atomic stockpile. When David Lilienthal became chairman of the Atomic Energy Commission in January 1947, he was shocked to find that the nation had only one atomic bomb that, if assembled, "stood a good chance of being operable," a fact that he duly reported to the president.[2]

In short, the pace of the development of long-range detection methods was strongly modulated by the rate of testing atomic bombs. The limited number of such tests in the postwar years slowed the technical development of a monitoring capability.

ORGANIZATIONAL TURMOIL

The second factor, the postwar disarray of the American intelligence community, inhibited the organizational development of a centralized and coordinated monitoring system. This disarray was caused by the rapid dismantling of America's wartime intelligence services. In trying to avoid the putative evils of a powerful intelligence arm, the Office of Strategic Services, which might have served as a framework on which to build a nonmilitary centralized intelligence organization, was abolished. And the military intelligence services were crippled by the enormous loss of trained manpower that occurred when the war ended. A national debate over intelligence ensued in which the jurisdiction of the organizations within the intelligence community became ambiguous. The turf battles that followed did not create an organizational climate that was conducive to establishing a new intelligence organization for monitoring foreign atomic bomb tests or to assigning responsibility for this activity to one of the existing intelligence services. In order to understand this, it is necessary to review the tortuous process of reorganization that took place between 1945 and 1947.

The Joint Chiefs of Staff, the military intelligence agencies, the State Department, and the Bureau of the Budget, each had their own ideas about the nature of the United States' peacetime intelligence establishment. These governmental bodies put forth various proposals and counterproposals for restructuring intelligence operations. The result, according to the official history of the Central Intelligence Agency, was that "From the close of the war in August 1945 until January 1946, the debate over the organization of national intelligence raged."[3] President Harry S Truman ended the debate by implementing a modified version of the Joint Chiefs of Staff plan.

In January 1946, Truman established the National Intelligence Authority, an executive body comprised of the secretaries of War, Navy, State, and a presidential representative. Under the aegis of this body, a Central Intelligence Group was created whose primary task was to collate, evaluate, and disseminate intelligence obtained from other agencies, although it quickly acquired an intelligence-gathering capability of its own as well. As director of the Central Intelligence Group, Truman named Rear Admiral Sidney S. Souers, a former business executive whose wartime posts had included that of assistant director of naval intelligence.[4]

When Souers began his tenure as director of Central Intelligence in early 1946, many U.S. leaders perceived Russia as posing a potential atomic threat. This perception was based largely on the public stance Russian officials had taken on atomic matters. The Russians had failed to respond positively to a press leak in the *New York Times* in September 1945, which revealed that the United States, at the highest levels of government, was seriously considering sharing atomic information with Russia as part of a plan to control atomic weapons.[5] In October, at a London meeting of Allied foreign ministers, the Russian delegate had made a point of declaring that Russia would not be intimidated by the United States' possession of atomic weapons, thus implying that his country would soon have its own bomb.[6] This implication was reinforced in November when the Russian foreign minister gave a speech warning that technical secrets could not long remain the sole possession of one country.[7]

Russian emphasis on the ephemeral nature of the U.S. atomic monopoly made U.S. military leaders uneasily aware of the paucity of information on developments within Russia. In an early directive to the Central Intelligence Group, Souers ordered it to focus its intelligence-gathering capabilities on that country.[8] These capabilities resided largely in the Strategic Services Unit which Souers had acquired from the War Department and which was incorporated in the structure of the Central Intelligence Group as the Office of Special Operations. Described by one commentator as "pruned and rebuilt," the Strategic Services Unit included the former Secret Intelligence Branch of the defunct Office of Strategic Services.[9]

The Secret Intelligence Branch had worked closely with Groves's Foreign Intelligence Section during the war. By the spring of 1946, however, the collaboration between the Foreign Intelligence Section and the remnants of the Secret Intelligence Branch that existed in the Office of Special Operations had begun to falter because the notion that Groves had sole responsibility for foreign atomic intelligence appeared to conflict with the Central Intelligence Group's mandate to collate all intelligence related to national security. The collaboration of Army Intelligence and the Office of Naval Intelligence with the Foreign Intelligence Section also began to wear thin partly because these agencies had to use their dwindling resources to satisfy their military clients and partly because of resentment over the postwar continuation of Groves's policy of compartmentalization. These agencies increasingly saw this policy as an excuse to make the exchange of information with the Foreign Intelligence Section "a one-way street."[10]

Therefore, during the period between the war's end and the spring of 1946, the amount of information the Foreign Intelligence Section received from the clandestine networks of other intelligence agencies steadily declined. Moreover, during the war the Foreign Intelligence Section had not created networks of this type. Thus, after the demobilization of Alsos, the Foreign Intelligence Section essentially consisted of a small staff based at the Manhattan Engineer District's Washington headquarters. It also had a few people posted overseas whose "sole job," according to Secretary of War Robert Patterson, was "to watch what was going on in foreign countries in the development of atomic energy."[11] But because the Foreign Intelligence Section ended the war with no covert networks of its own, it appears that its overseas personnel collected much of their information from overt sources such as books, scientific journals, and the like.

The collection and evaluation of foreign atomic intelligence had not been the only or, perhaps, even the chief task of the Foreign Intelligence Section. A great deal of effort had been expended on an attempt to locate and, where possible, control the world's known supplies of uranium. Groves had first proposed this idea in June 1943, and at that time he had initiated a project, designated Murray Hill Area, to determine as quickly as possible the worldwide distribution of uranium. After the war ended, the evaluation of the results of this project was continued with even greater urgency.[12]

In the months that followed the war's end, the preoccupation of the Foreign Intelligence Section with the uranium project, coupled with the drying up of its customary sources of covert intelligence, reduced its intelligence "product" significantly. These deficiencies in production were exacerbated by an overly restrictive policy on dissemination. During the war, Groves had reported to the president through the secretary of War, and apparently he saw no reason to widen the circle of clients for atomic intelligence simply because the war was over. This "intelligence gap" did not escape the notice of the Joint Intelligence Committee of the Joint Chiefs of Staff. The Joint Intelligence Committee, comprised of representatives from the military services, State Department, and Foreign Economics Administration, was responsible for synthesizing intelligence from various sources for the Joint Chiefs of Staff use. In an April 1946 meeting of the Joint

Intelligence Committee, one of its members, Lieutenant General Hoyt T. Vandenberg, noted that there was a general dissatisfaction with the information on foreign atomic developments. Vandenberg went on to state that the "JIC [Joint Intelligence Committee] had an inferred obligation to furnish JCS [Joint Chiefs of Staff] with atomic intelligence."[13] This was the opening gun in an ultimately successful campaign to have the Foreign Intelligence Section transferred from the Manhattan Engineer District to the Central Intelligence Group.

A month later the Joint Intelligence Committee prepared a report on foreign atomic intelligence that identified the need to obtain such intelligence as a major problem. The report included an appendix which noted that, in order to ensure full utilization of the Manhattan Engineer District's information on foreign atomic developments, the Foreign Intelligence Section should coordinate its intelligence with the Joint Intelligence Committee. It went on to suggest that Groves be directed to "assist the intelligence agency of the Joint Chiefs of Staff to the fullest extent" in evaluating the atomic capabilities and intentions of other nations.[14]

This statement by the Joint Intelligence Committee was not an attack on Groves's policy of compartmentalization per se but, rather, a statement that the Joint Chiefs of Staff had a "need to know." Indeed, the rationale of atomic security that allowed the Foreign Intelligence Section to avoid coordinating its intelligence with the Joint Intelligence Committee during the war was clearly inappropriate in a postwar world characterized by Russian expansionism. By the spring of 1946, it had become obvious that Groves's method of handling atomic intelligence could seriously hamper the work of the Joint Chiefs of Staff, charged as it was with the responsibility for advising the president on matters affecting national security.

Quite apart from not receiving foreign atomic intelligence from the Manhattan Engineer District, the Joint Chiefs of Staff was not privy to some of the most important aspects of *domestic* atomic development. For example, the Joint Chiefs of Staff had not been told how many atomic bombs the United States possessed or the rate of bomb production, although these were crucial parameters in the strategic military plans prepared by that body. Ironically, it was to prove easier for the Joint Chiefs of Staff to obtain access to the Manhattan Engineer District's foreign intelligence data than to learn the size of the U.S. atomic arsenal. (The Foreign Intelligence Section files became accessible to the Joint Chiefs of Staff in February 1947, but the official channels needed to apprise the Joint Chiefs of Staff of the number of bombs the United States possessed were not established until April of that year.)[15]

It appears, however, that Groves cannot be blamed for failing to communicate data on the atomic arsenal to his immediate superiors. In the late fall of 1945, Groves tried to acquaint Robert P. Patterson and General Dwight D. Eisenhower, recently named secretary of War and Army Chief of Staff, respectively, with the atomic program. If Groves's memoirs are to be believed, both men requested that, to avoid an inadvertent revelation on their part, they not be given any secret information, "particularly about the production rates and the number of bombs on hand."[16] Groves claims that this policy remained unchanged until a member of the Senate threatened to complain publicly that the information should not be restricted to one man. "Personally," Groves later wrote, "I was relieved since I had always thought they should have it."[17]

Even if this explains why the Joint Chiefs of Staff lacked vital information on the United States' atomic capabilities, it remains true that Groves seemingly adopted a policy of reluctantly complying with changes in handling foreign atomic intelligence that were mandated by others, rather than initiating such obviously needed changes himself. This policy might be attributed to a desire to maintain control of information as a means of retaining power. Yet it seems odd that Groves, who was singularly apt at bureaucratic manipulation, would attempt to use a method of self-aggrandizement that was, on the face of it, logically insupportable and certain to raise the ire of his military superiors. From another perspective, it seems even odder that the Joint Chiefs of Staff would wait almost a year before complaining about Groves's policies. The picture of the Joint Chiefs and their staffs carefully formulating contingency war plans, with little or no foreign atomic intelligence and without knowing how many atom bombs the United States possessed, is almost surrealistic. Yet, this is the picture supported by the available documentation. In fact, the Joint Chiefs of Staff completed a theoretical plan for war with Russia in June 1946, code named Pincher, that was based on assumptions about U.S. atomic capabilities that bore little resemblance to reality.[18]

Perhaps Groves's failure to initiate changes in his security policy and the Joint Chiefs of Staff's belatedness in complaining about this failure can both be ascribed to the suddenness with which the atomic age superimposed its special exigencies and novel possibilities on the inevitably chaotic process of bureaucratic realignment from war to peace.

Another possible explanatory factor—and one that would leave no trace in the historical record—was the degree to which informal means, such as one-on-one conversations between Groves and other officials, were a substitute for the formal mechanisms that were eventually established. Groves found such informal communications congenial to his style of management, and, where intelligence matters were concerned, he apparently thought that it was usual practice to avoid written directions. For example, in referring to one such directive in his memoirs, he noted, "As was customary, nothing was put in writing."[19]

A third factor was the extent to which the urgency of the need for foreign intelligence was muted by the results of the Murray Hill Area project, from which it could be inferred that the United States' atomic monopoly might last a decade or more. This belief, promulgated by Groves, was accepted by some influential U.S. policy-makers.

Whatever the explanation, it appears that the Joint Intelligence Committee's first steps toward reorganizing the collection, evaluation, and dissemination of atomic intelligence were not taken until the spring of 1946. And in the summer of that year three events occurred which influenced the reorganization process: in June, Admiral Souers was replaced by General Vandenberg as director of Central Intelligence; early in July, the Joint Research and Development Board was set up; and, later that month, the McMahon Bill was passed creating the Atomic Energy Commission.

The first event caused little stir. Souers had accepted the post of director of Central Intelligence as a temporary appointment with the understanding that he would resign after the basic structure of the Central Intelligence Group was established. Vandenberg brought to the job a thorough understanding of Washington bureaucracy, a background in intelligence (he had been director of Military Intelligence and a member of the Joint Intelligence Committee), and certain political connections (he was nephew to Arthur Vandenberg, chairman of the powerful Senate Foreign Relations Committee). It is perhaps noteworthy that General Vandenberg had been wartime commander of the Ninth Air Force, whose A-26s flew the xenon-detection missions over Germany for Groves's Foreign Intelligence Section. Vandenberg was also an early critic of the way atomic intelligence was organized, and, as director of Central Intelligence, he made the transfer of the Manhattan Engineer District's intelligence unit to the Central Intelligence Group one of his first priorities.

The second event also caused little debate. The Joint Research and Development Board was the successor organization to the wartime Office of Scientific Research and Development, although, unlike its predecessor, it had no operational functions. Both the old Office of Scientific Research and Development and the new Joint Research and Development Board were headed by Vannevar Bush, president of the Carnegie Institution and formerly dean of Engineering at the Massachusetts Institute of Technology. The Office of Scientific Research and Development had been strongly linked to the intelligence community, and during the war it had collected intelligence on foreign scientific developments.[20] From the outset, the Joint Research and Development Board was also intimately involved with intelligence, although its primary purpose was to oversee scientific research and development for the military services in order to avoid duplication and improve coordination of effort. The Joint Research and Development Board established a number of committees in August 1946 to handle specific areas of science. One of these, the Committee on Atomic Energy, became a key advisory body in the field of atomic intelligence.[21]

The third event, the passage of the McMahon Bill, which transferred the Manhattan Engineer District from military to civilian control, established the Atomic Energy Commission as an executive body with five members each holding equal power. All were appointed by the president, who designated one as chairman to act as spokesman for the group. Groves had favored the idea of civilian control of atomic energy but had opposed the concept of group management embodied by the McMahon Bill. In commenting on the ineffectiveness of this concept, he later pointed out that "ever since the tribunes of Rome no executive group has ever functioned well."[22] But his objections were not heeded.

The distribution of authority among five commissioners, as mandated by the bill, was to have an effect on the reorganization of atomic intelligence because one of the newly appointed commissioners, Rear Admiral Lewis L. Strauss, was interested in—or according to some commentators, obsessed by—security matters. Strauss subsequently became a strong supporter of the effort to develop the United States' atomic intelligence capability. Other facets of the McMahon Bill that influenced the U.S. intelligence effort were its security provisions that, in effect, inhibited the close collaboration of U.S. and British intelligence services on matters dealing with atomic energy. Since this collaboration had proved fruitful in the past, a not insignificant effort was expended during the next few years to continue to coordinate U.S. and British atomic intelligence despite the bill's provisions.[23]

After their appointment, the Atomic Energy commissioners had requested that they not be required to assume their responsibilities until January of the coming year. Groves agreed to stay on in the interim as the Manhattan Engineer District's administrative head. The period from July 1946, when the McMahon Bill was passed, to January 1947, when the Atomic Energy Commission took over the reins of the Manhattan Engineer District's management, proved to be a difficult time for Groves.[24] The difficulties Groves encountered had been exacerbated because he was disliked by some of the commissioners, a fact that he noted in his memoirs.[25] On their part, according to the Atomic Energy Commission's official history, the commissioners found that Groves's "personality and attitude imposed additional obstacles" to the transfer of authority from the Army to the Atomic Energy Commission.[26]

In retrospect, Groves's testy attitude during the interim period of the transfer can be understood in the context of the basic problem he had been wrestling with since the end of the war—that is, the preservation of the Manhattan Engineer District as a viable organization in the face of major losses in personnel and the lack of any clear-cut policy directives from the president or the War Department. Despite the fact that this problem had occupied much of Groves's attention, two aspects of the research undertaken by the Manhattan Engineer District's intelligence unit had continued to engage his interest: the results of the Murray Hill Area project and the further development of methods of radiological surveillance to detect certain Soviet operations such as the test of an atomic bomb. Ironically, among top government officials, the seeming success of the former appears to have blunted their enthusiasm for the latter.

MURRAY HILL AREA

The third factor that tended to slow the development of an operational monitoring system was the work of the Murray Hill Area Office of the Manhattan Engineer District, which provided a persuasive rationale for projecting a relatively long duration for the American monopoly, a projection that was favored by Truman administration insiders. Thus, although administration leaders offered no resistance to the idea that a system of surveillance to detect the test of a Soviet atomic bomb would eventually be needed, they did not believe that its deployment was an urgent matter. As a result, there was no "top-down" impetus to develop such a surveillance capability. No directive was issued by the president or his top officials to create a monitoring agency or to assign this task to an existing agency. Such a directive would have sidestepped the organizational confusion that discouraged lower level administrators from establishing a centralized monitoring agency. Instead, it was necessary to wait for the organizational "dust" to settle before the lower echelons of the bureaucracy felt capable of finding an organizational "home" for the monitoring function.

The rationality of the argument for a relatively remote date for the advent of a Soviet atomic bomb, which was provided by the Murray Hill Area project, has not been emphasized in histories of the period, although all agree that President Truman and administration insiders accepted the argument. This has led some commentators to dismiss Truman and General Groves, who was the leading proponent of the argument based on the Murray Hill Area project, as xenophobes who were ideologically unable to admit that the Soviets had the ability to make an atomic bomb. These commentators usually contrast the estimate presented by Groves with that of many atomic scientists of the day, who had proposed a relatively early date for the advent of a Soviet bomb.

At the time, however, most atomic scientists were unaware of the Murray Hill Area project because of its supersecret status. Moreover, those scientists who knew about the Murray Hill Area project, such as Dr. J. Robert Oppenheimer, scientific leader of the wartime Manhattan Project, and most of the other members of the Atomic Energy Commission's prestigious General Advisory Committee, believed the American atomic monopoly would be relatively long-enduring. To explain the persuasive rationale that lay behind the estimate of a relatively remote date for the appearance of a Soviet bomb, it is pertinent to review the history of Murray Hill Area and related projects.[27]

Shortly after becoming head of the Manhattan Engineer District in 1942, Groves came to believe that his was a twofold task: first, to build an atomic bomb, and, second, to make sure that other nations could not. The means to accomplish the second part of this task—that is, U.S. control of the world's supply of high-grade uranium ore—suggested itself early in 1943, when Groves confronted the problem of securing uranium for the United States' atomic program.

One of Groves's first acts as head of the Manhattan Engineer District had been to purchase the 1,250 tons of uranium ore that had been stored in a Staten Island warehouse by a Belgian firm, Union Minière.[28] At the time, this represented over half of the 2,000 tons known to exist in North America.[29] Most of this ore had been produced by the Shinkolobwe mine in the Belgian Congo, and Groves had been made acutely aware that this mine was virtually the only known source of such high-grade ore. U.S. deposits were of very poor quality, and, although the ore produced by Canadian mines was better, it fell short of the extraordinary richness of the Congo ore. Moreover, it was clear that in the immediate future the United States would remain dependent on ore from the Congo because the technology to make rapid use of low-grade ore was not sufficiently developed.

U.S. dependence on high-grade ore suggested that control of this commodity could be the key to maintaining a monopoly on atomic weapons well into the postwar period. Groves put this idea before the Military Policy Committee, a four-man oversight body comprised of Vannevar Bush as chairman, Dr. James B. Conant, a distinguished chemist and president of Harvard, and representatives from the Army and Navy. The Military Policy Committee approved Groves's plan to attempt to "corner the market" in uranium, and the project was initiated in the spring of 1943. Consistent with the "Manhattan Engineer District," the project's designation, "Murray Hill Area," was a neighborhood in Manhattan.[30]

The concept of monopolizing uranium as a means of retaining control of atomic weapons was not original with Groves. The brilliant Hungarian emigré scientist, Leo Szilard, had espoused this idea earlier, although by the time the scheme was implemented he had come to doubt its practicality.[31] Nevertheless, the fact that two such eminent scientists as Bush and Conant were on the committee that approved the plan suggests that, at the time, this idea was considered feasible, as does the fact that money was provided on a lavish scale. The initial funding—paid into Groves's personal bank account to avoid the legality of obtaining congressional approval for disbursements—was in excess of $37 million, or equivalent to about $285 million in 1994 dollars.[32]

Groves selected Major Paul L. Guarin, an experienced geologist who had worked for Shell Oil Company before the war, to head the project. To carry out the enormous task of determining the worldwide distribution of uranium, a contract was let to Union Carbide and Chemical Company.[33]

By mid-1944, it had become apparent that thorium could also be used to make atomic bombs. Thus, Groves approached the British to ensure Anglo-American control of uranium and thorium located within British territories and dominions. This was accomplished through the agency of the Combined Development Trust, a jointly run Anglo-American body created for this purpose in June, 1944.[34]

Throughout 1944 and 1945, Murray Hill Area and the Combined Development Trust clandestinely acquired existing supplies of uranium and the future output of producing countries. Where necessary, Murray Hill Area representatives negotiated agreements with foreign governments, usually without the knowledge of the State Department.[35]

Areas of Europe and Asia that were under German or Russian domination could be surveyed only through bibliographic data, and, of course, these areas were beyond U.S. control. However, existing information suggested that Europe's only uranium was to be found in Czechoslovakian mines at Joachimstal. But the general grade of this ore, which was known, had led Groves to conclude that this "was not a particularly significant source."[36] It was not the Czechoslovakian ore that had made the wartime German atomic program a threat, but, rather, the 1,200 tons of very high-grade ore mined in the Congo that was seized by the Germans when they occupied Belgium in 1940. In fact, Groves later wrote that this ore from the Congo constituted "the bulk of the uranium supplies in Europe."[37]

Prompted, in part, by the monopolistic goal of the Murray Hill Area project, Groves gave the confiscation of the Congo ore that was still in German hands a very high priority. Early in 1945 he initiated Operation Harborage to seize this material. The importance given to obtaining the German uranium is evidenced by the fact that an airborne division and two armored divisions were attached to the Sixth Army for this operation. The planned assault was unnecessary, however. The town of Strassfurt, where the uranium was stored, was occupied without a major military action, and an Alsos team seized the complete supply on April 22, 1945.[38]

Three days later, Groves briefed President Truman on the accomplishments of the Manhattan Engineer District. In citing the work of Murray Hill Area, Groves indicated that the United States and Britain had achieved a virtual monopoly on high-grade uranium ore.[39]

It could be inferred from the results of the Murray Hill Area project that the American monopoly on atomic weapons, based on control of the raw material (high-grade uranium and thorium ore), was assured for some years to come. That Groves had already made this inference became apparent a few weeks later at a meeting of the Interim Committee. The Committee, whose members included Bush, Conant, and the soon-to-be appointed Secretary of State James F. Byrnes, had been set up in the spring of 1945 by Secretary of War Stimson to advise the president and Congress on postwar atomic energy matters. At the May 18 Committee meeting, to which General

Groves was invited, the question of when Russia would catch up to the United States in the field of atomic energy was introduced. Conant reiterated the estimate of three to four years which he and Bush had proposed in a memorandum to Stimson written in September 1944. "This premise," Conant later wrote, "was violently opposed by the General, who felt that twenty years was a better figure."[40]

Bush and Conant were key members of the Military Policy Committee that had approved the Murray Hill Area project in 1943. They had come to believe, however, that, while this project was useful as a means of securing raw material for the United States' atomic program, it could not be relied on to prolong the monopoly on atomic weapons for a period as long as a decade. They had stated this belief in their 1944 memorandum, and one of the reasons they adduced to support this opinion was that atomic bombs would be superseded by hydrogen bombs. Hydrogen, they pointed out, was freely available, and hydrogen bombs would be capable of producing enormous explosive power with only a small amount of uranium or thorium as a detonator.[41]

In view of the embryonic state of U.S. research on hydrogen bombs at that time, it seems surprising that Bush and Conant would cite the possible development of these weapons in support of their argument. It appears, however, that they were attempting to emphasize the point that scientific and technological advances would inevitably put an end to the American atomic monopoly. A month later, the fragility of the monopoly was emphasized in the Franck Report. In June 1945, six scientists, led by the Nobel laureate James Franck, submitted a report on the social and political consequences of atomic energy to the secretary of war. The report cited the fact that the principles on which the bomb was based were well known to scientists in other countries. This meant that "atomic bombs could not remain a 'secret weapon' for more than a few years." Leo Szilard, who had once believed that control of uranium ore might be the key to ensuring a monopoly on atomic weapons, was a member of the group that prepared the Franck Report. His disillusionment with this idea was reflected in a section of the report which warned that Czechoslovakian uranium ore was outside U.S. control and that Russia was so huge that the probability that no uranium reserves would be found in that country was small. The report concluded that an arms race with Russia could not be avoided "by cornering the raw materials required for such a race."[42]

Based on the projections of Murray Hill Area's geologists and experts in mining and ore processing, Groves disagreed with this assessment. He had told the Interim Committee that it would take ten to twenty years for the Russians to produce a bomb, and he steadfastly held to this estimate in the coming years, until the first Russian bomb test in 1949.[43] The estimate presented by Groves was considerably longer than the three to four years proposed by Bush and Conant and the estimates of other scientists, which ranged around five years.[44]

A puzzling question is why so many key people in the administration and in Congress accepted the estimate given by Groves in the face of criticisms by Bush, Conant, and other scientists, such as those who prepared the Franck Report. In his book *The Winning Weapon*, Gregg Herken suggests that the pertinent question is not "why" but "how" the scientists' opinions were ignored. He lists three factors that bear on this question: first, the secrecy that cloaked all atomic matters prevented scientists from knowing about, hence criticizing, Murray Hill Area; second, officials like Stimson, who might have allied themselves with the scientists in questioning the Murray Hill Area results, were gone by late 1945; and third, Groves's position, as the only person who knew all of the Manhattan Engineer District's secrets, lent a special weight to his pronouncements.[45]

These points are well taken, and additional factors can be adduced to explain the wide acceptance of Groves' estimate of the duration of the American atomic monopoly. One such factor was the confusion over the meaning of the term *atomic secret*. By the fall of 1945, it was apparent that only the United States was capable of building atomic bombs, a fact which suggested that the United States possessed an "atomic secret." But the notion of "atomic secret" was ill defined and confusing to the public and to most government officials as well.[46]

Groves, however, had no uncertainties about the basis of the American monopoly on atomic weapons. The estimate he presented for the duration of the monopoly was not based on the misguided idea that there was a "secret" in the sense of a formula that, if carefully guarded, would allow the United States to maintain control of atomic weapons in the decades to come. Even David Lilienthal, the first chairman of the Atomic Energy Commission, who cordially disliked Groves, exculpated him on this count. In a 1979 interview, Lilienthal stated, "Groves thought the idea of an 'atomic secret' was a joke."[47] In presenting the case in high-level briefings for a long duration of the atomic monopoly, Groves was able to share this joke with his listeners by pointing out that the *real* problem the Russians faced in attempting to make a bomb was a shortage of high-grade ore and the inability to make rapid use of the low-grade ore available to them. Thus, one reason why top officials found the argument he presented compelling was that it provided them with a clear idea of just what constituted the "true" secret of the United States' atomic supremacy.

A second factor that influenced the acceptance of Groves's estimate was its technologically authoritative nature. Groves could present it as the sophisticated result of detailed assessments by Foreign Intelligence Section analysts, based chiefly on the results of the studies supervised by the Murray Hill Area Office. Thus, another reason why his arguments were persuasive is that he could present them as being based not on his personal opinions, but on the results of an organizational effort. He emphasized that this was a painstaking evaluation carried out over a period of several years by a team of analysts whose expertise lay not in atomic science (like that of his critics), but in the branches of geology, mining, and chemical engineering that were necessary for this task.

This appeal to an expertise different from that possessed by his scientist-critics was apparent in the report Groves gave to U.S. officials in December 1945 on the results of the worldwide geologic surveys conducted under the supervision of Murray Hill Area. He indicated that his analysts had determined that 97 percent of the world's high-grade uranium ore from producing countries was under Anglo-American control. Most of the relatively lower grade ores capable of development with existing technology were under British or Swedish control. Of the widely distributed low-grade ore that would require a great deal of time and effort to develop, 35 percent was under Anglo-American control, with the remainder divided between South America and Russia. Only two industrialized countries, Sweden and Russia, had sufficient ore to challenge Anglo-American dominance, and only Russia could be considered a threat. But existing data indicated that Russian ores were of very low grade. To use such ore would require the costly and protracted development of radically new extraction techniques.[48]

These Murray Hill Area assessments and projections were directly related to any estimate of when the Russian bomb would appear. Three elements were required to make a bomb: knowledge of the relevant advances in atomic physics, raw materials (uranium or thorium), and the technological infrastructure (the processing facilities and plants needed to make the bomb). It was clear that a bomb could not be produced unless the Russians possessed all three elements. Hence, the one requiring the longest time to acquire would become the gating item. The time to acquire this item would thus become the *minimum* duration of the American atomic monopoly if the Russians developed the other two elements, not sequentially but in parallel. The parallel approach had been used in developing the American bomb, but, because of the wartime disruption of its educational system, Russia lacked trained technicians. According to Groves, this meant that the Russians "would not be able to work on the scientific and industrial problems simultaneously."[49] Consequently, the American atomic monopoly might endure for some time *after* the minimum projected period.

In a meeting a few weeks earlier with congressional and administration leaders, Groves had presented an analysis in which the gating item was Russian acquisition of raw materials. It was not expected that the Russians could develop the low-grade ores available to them for at least a decade. Some years later, in a discussion of Russian supplies of raw materials, Groves confirmed the fact that the estimate he had presented for the duration of the United States' monopoly on atomic weapons "was based primarily on the supplies of [fissionable] materials that would be available to them."[50] The projections of the time for the Russians to acquire the other two items were considerably shorter.

With regard to Russian acquisition of the knowledge of the physics of the bomb, Manhattan Engineer District scientists were virtually unanimous in asserting that this knowledge would be common currency within a few years throughout the world's scientific community. In 1945, the success of Russian espionage in obtaining this knowledge was not realized, but Russian physicists were known to have been both highly competent and active in atomic research before the war. Thus, they could be counted on to produce, within a few years, the theoretical and experimental knowledge needed to make a bomb.

The projection of the time needed to create the necessary technological infrastructure was only somewhat longer, assuming that the major facilities were built in parallel. In June 1945, the Interim Committee had met with the four men representing the companies that had been responsible for designing and building plants and equipment used to make the atomic bomb. The topic of discussion was the time it would take the Russians to duplicate these facilities. Walter S. Carpenter, president of DuPont, estimated five years to construct a plutonium plant like that at Hanford. James White, president of Tennessee Eastman, would not assay an estimate of the time the Russians would need to build an electromagnetic separation facility because he doubted they had the technicians needed to make the precision equipment. But George Bucher, president of Westinghouse, reckoned they could have an electromagnetic facility operating within three years, if they had the help of German technicians. James Rafferty, vice president of Union Carbide and Chemical, estimated that it would take the Russians ten years to build a gaseous-diffusion plant starting from scratch, but only three if they were able to obtain the needed technology through espionage.[51]

In the event, the Russians did obtain the help of German technicians, and there is evidence that suggests that Russian spies were able to ferret out the technology and the construction details of Manhattan Engineer District facilities such as the reactor at Hanford.[52] It is noteworthy that the projections of these experts, where they took account of German help or espionage, turned out to be fairly accurate in the light of what is now known about the Russian atomic program.

The estimated time for the Russians to obtain the raw materials for a bomb was not accurate, however. The overlong time period rested on two assumptions that were basic to the efforts of the Murray Hill Area project: first, that high-grade ore would not be acquired by Russia either from domestic or foreign deposits, and, second, that Russian improvements in the extraction technology for low-grade ores would not exceed those anticipated in the projected development of U.S. ore processing techniques. The first assumption could be made to sound plausible in the light of quantitative geological statistics. It was claimed that rich finds like that at Shink-olobwe were rare.[53] But the second assumption necessarily relied on a qualitative assessment of Russian capabilities in those branches of science and engineering involved in ore processing. This assessment of Russian capabilities was at the heart of the criticism leveled by Bush and Conant at the estimate offered by Groves. In a 1945 memorandum to Bush describing the argument for a long duration of the U.S. atomic monopoly, Conant wrote that its advocate, Groves, "was basing his long estimate on a very poor view of Russian ability, which I think is a highly unsafe assumption."[54]

Most of Groves's listeners, however, were prepared to accept the proposition that the Russians would be unable to devise revolutionary new extraction techniques that would enable them to make speedy use of their low-grade ores. Groves could point to the fact that most of the uranium for the United States' atomic bomb program had come from a single mine in Shinkolobwe because the best U.S. technology would not allow the rapid exploitation of the huge reserves of low-grade ore under Anglo-American control. Could the Russians do better? Strategic assessments of Russian industrial capacity, woefully shattered by the war, suggested that this was unlikely.

The possibility that Russian ore technology would leapfrog that of the United States was made to appear remote because of Russia's backwardness, as Groves emphasized in his discussions with government leaders. In a 1979 interview, Lilienthal recalled that Groves's attitude was "not a downgrading of the Russians generally," but, rather, "a downgrading emotionally of the Russian technical capacity—as an ignorant, clumsy, backward country. He overdid that."[55] Lilienthal's last comment may have been the result of hindsight because, at the time, Groves's remarks about Russian backwardness appear to have fallen on receptive ears.

An important factor that influenced the ready acceptance of Groves's views on Russian capabilities among the restricted group of top government officials that constituted his audience was that these individuals were drawn from segments of American society that had been enculturated to perceive non-Western nations as backward. Nor did this perception lack empirical confirmation. Throughout the late nineteenth century and the first third of the twentieth, the correlation between a nation's technological capabilities and the steady proliferation of the products of these capabilities, such as automobiles and indoor plumbing, were a matter of common experience. By this measure, Russia appeared technologically backward indeed.

But in the immediate postwar years, it was not generally realized that embedded in the primitive Russian economy was a cadre of world-class technologists whose efforts could be focused solely on armaments by a totalitarian government. This possibility fell outside the tradition of Western technological supremacy, a tradition distinctly on the wane by 1945, but, nevertheless, one with which many U.S. leaders had grown up. Therefore, another reason why the estimate presented by Groves became widely accepted in high government circles was that one of its key assumptions was consonant with the cultural values of those segments of society from which many of these individuals came.

Knowledge of Murray Hill Area and the rationale for the estimate based on it were restricted to those in the higher echelons of government. However, within this select group, the estimate was widely adopted. As one history of the period notes, it "permeated the Truman administration to such an extent that the president himself became convinced that the Russians would not have a bomb for some years, if ever."[56] Indeed, by 1946, even Bush and Conant had changed their minds and espoused this estimate. Conant became a convert because further consideration had led him to believe that the Russians really did lack the raw materials and the technological infrastructure. Bush came to believe that the Russians would not be willing to pay the huge price economically that a crash program to produce the bomb in a short time would entail.[57]

Some other scientists who were insiders in terms of their overall knowledge of the Manhattan Engineer District's activities, including Murray Hill Area, also shared the belief that the United States' monopoly on atomic bombs would not end soon. For example, in 1948, J. Robert Oppenheimer, then one of the nation's most influential scientists, declared that the Russians would not have the bomb "for a long time to come."[58] Most of the scientists on the Atomic Energy Commission's General Advisory Committee agreed with this estimate.

In sum, the factors that have been discussed—confusion over the "atomic secret," the technologically authoritative nature of the estimate, and the cultural predisposition to view non-Western nations as technologically primitive—contributed to making the argument for a long-enduring American atomic monopoly compelling. A fourth factor has been cited by other commentators: the tendency for forecasts to reflect interests. Thus, another reason why government officials widely adopted this prediction was that it made their jobs easier. If the Russian bomb was a decade away, some troublesome issues in the areas of defense and foreign policy could be postponed or ignored. And concerns over security were lessened because the "real" secret of the bomb could not be stolen. If the United States itself did not possess the technology that would allow the rapid utilization of low-grade ores, espionage could hardly provide the Russians with this information.

The estimate based on Murray Hill Area also had important implications for U.S. intelligence because the rationale for the estimate provided a kind of intelligence paradigm or model from which forecasts of Russian actions could be derived. Comparison of these predictions with real events would then either tend to confirm or contradict the paradigm. For example, the paradigm asserted that both the United States and Russia needed high-grade uranium ore and that the Russians lacked this commodity. Thus, it was to be expected that the Russians would attempt to disrupt or interdict U.S. supplies of high-grade ore and that they would make strenuous efforts to obtain such ore for themselves. They would attempt to acquire rich ore by prospecting in areas they controlled and by diplomatic means in ore-producing areas outside their control. Because they were unlikely to acquire rich ore by these methods, they would also exploit to the utmost the existing low-grade deposits available to them in Czechoslovakia and any new deposits of low-grade ore discovered through prospecting.

The actions of the Russians appeared to bear out these predictions. In October 1945, the Russian claim that they should be given trusteeship over some portion of former Italian colonies in Africa was perceived by Secretary of State Byrnes as the first step in an attempt to obtain African uranium ore. A month later, the Russian government officially requested the Czecho-slovakians to supply them with the output of the Joachimstal mines. Political upheavals in the Congo, fostered partly by Communist interventions, were seen as a Russian attempt to disrupt the supply of ore to the United States. In 1946, reports were received of energetic Russian uranium prospecting along the German-Czech border and the discovery of new deposits.[59]

Some historians have expressed surprise that such "warnings," portend-ing the erosion of the U.S. atomic monopoly, were not heeded in the United States.[60] Seen through the eyes of top government officials, however, these events merely fulfilled the predictions of the intelligence paradigm and tended to confirm the estimate given by Groves. Thus, these events were viewed not as warnings that the American monopoly would soon end, but as a confirmation that it would last for the predicted period.

The major concern was the possibility, albeit remote, that the grade of the new ore deposits found by the Russians would be richer than the low-grade ores already available to them. According to the paradigm, speedy produc-tion of the raw materials for a bomb was prevented by the lengthy time needed to process low-grade ore. Hence, the discovery of new deposits that merely added to Russian supplies of low-grade ore would not appreciably advance Russian progress toward the bomb. But the time for them to produce a bomb would be greatly reduced if high-grade ore were found. To cover this possibility, it appears that the methods of clandestine intelli-gence-gathering were employed. Documents relating to U.S. intelligence activities in Russia in the 1945–1949 period remain classified. But years later David Lilienthal recalled that, to find out if any such usable uranium deposits had been discovered, "we did send covert operations into Russia and they came back and said there isn't any."[61]

Therefore, both Russian actions and U.S. intelligence-gathering efforts tended to confirm the intelligence paradigm derived from Groves's argu-ment, which, in turn, was based on his knowledge of Murray Hill Area. In hindsight, it is clear that one or both of the assumptions on which the estimate was based were not valid. Several historians have claimed, without citing supporting data, that the idea that the United States had a monopoly on high-grade uranium ore was simply wrong.[62] In fact, no evidence has surfaced to suggest that the Russians discovered or acquired high-grade uranium similar in quality to that produced by the Congo mines. There is, however, evidence that they made advances in ore processing.[63] This technical advance, alone or in combination with the discovery of somewhat better grades of ore, allowed them to produce enough raw material for a bomb in about one-half the time that had been estimated by projections based on Murray Hill Area and related projects. This ore-processing technology could not have been acquired from the United States because large-scale efforts to improve techniques for extracting uranium from low-grade ores were not even begun by the Atomic Energy Commission until 1948.[64] It appears, therefore, that Conant was correct in 1945 when he said that basing the estimate of the duration of the United States' atomic monopoly on "a poor view" of Russian capabilities would be "highly unsafe." The fact that he later recanted and espoused this estimate was, at once, a blow to Conant's reputation for prescience and a testimonial to the persuasiveness of the argument on which the estimate was based.

It should be noted that, from time to time, the military intelligence services produced estimates that were more alarming than that presented by Groves, but these estimates were relatively unconvincing. Because of Soviet security measures, intelligence analysts were reduced to formulating their projections primarily on the basis of the American experience in making the bomb rather than on the basis of any hard facts about Soviet atomic progress. Analysts thus had wide latitude in assigning values to key parameters involved in bomb construction. Not surprisingly then, different organizations produced disparate estimates of the Soviet timetable. In keeping with the tendency of forecasts to reflect interests, the estimates appeared to be consonant with the goals of the organizations doing the analysis.

For example, in a report that appeared in July 1947, naval intelligence concluded that the Soviets would not have a bomb until 1952 or later, while the Army Air Force's intelligence arm dissented, claiming that the Soviets could be on the brink of success and might produce a bomb in 1949.[65] The fact that the Navy favored a later date than the Army Air Force supported the differing positions of these two organizations in interservice and congressional debates over the relative utility of surface ships versus strategic bombers, the desirability of creating an independent Air Force, and the budgetary goals of both services. Hence, it was easy for administration insiders, who alone were privy to the basis for the estimate presented by Groves, to ignore the projections of the military intelligence agencies on the grounds that they appeared self-serving rather than factually based.

In assessing the influence of the estimate derived from Murray Hill Area on the development of United States atomic intelligence capabilities, the important point is that it was believed by administration insiders. One result of the mind-set created by this belief was that no one in a top leadership position called for the establishment of technological surveillance that would warn of threatening atomic activities within Russia that were not easily hidden, such as the test of a bomb. If the advent of the Russian bomb was a decade away, these were issues that could safely be postponed.

It is pertinent to contrast the view of administration insiders on the probable date for a Soviet bomb not only with that of atomic scientists but also with that of the public at large. In the immediate post–World War II period, it was common currency in the United States that some other country had, or would soon have, the bomb. The source of this belief was seemingly impeccable: scientists who had worked on the bomb during the war had said over and over again in popularized writings that the American atomic monopoly was ephemeral. They emphasized that there was no "secret" to making atomic bombs, except, perhaps, insofar as it was necessary to master a host of technological details, including industrial processes, such as the technique for making superpure graphite, which had nothing to do with nuclear physics. Uranium ore, a necessary ingredient for making the American bombs, was thought to be available in many countries. Thus, other nations, they pointed out, had the resources and would soon possess the technology to produce atomic bombs.[66]

This message, amplified by the media, made a deep impression on the public, as attested to by surveys social scientists made in 1946 and 1947. For example, a 1947 study about the thinking of the American public on matters relating to atomic weapons revealed that 75 percent of the people who held an opinion on the subject believed that some other country already had the bomb or would have one within three years, that is, by 1950 or earlier. The great majority of the respondents believed the "other" country to be the Soviet Union.[67]

The "public" belief that a Soviet bomb was imminent was shared by many military officers and public officials who were not among the handful of Washington insiders who alone were privy to Murray Hill Area and the rationale it provided for a relatively long-enduring atomic monopoly. Some of the military officers and officials who espoused the public belief in the fragility of the atomic monopoly became supporters of the monitoring concept. As will be seen, a few who were in a position to do so initiated the development of long-range detection techniques to monitor the Soviet Union.

It might be thought that General Groves, as the arch-proponent of a long-enduring American atomic monopoly, would not be numbered among those who favored the immediate development of long-range detection techniques. But to think thus would be to misread Groves's character. Despite the complacency engendered in administration leaders by Murray Hill Area, Groves himself was not complacent. Complacency was foreign to his nature and to his military and engineering training, all of which had conspired to make him a man for whom preparedness was a way of life.

Examples of Groves's obsession with preparedness are easily found. In 1943, he had believed a Japanese bomb so improbable that it was unnecessary to seek further intelligence on their atomic program, and, by 1944, he had discounted the possibility of a German atomic bomb as well. Yet, according to Alvarez, Groves had secretly maintained a standby group of atomic scientists and a long-range aircraft fitted with radiation instruments ready to fly "anyplace where the Germans or Japanese might have dropped an atomic bomb."[68] Operation Peppermint, launched to detect German use of radioactivity against the Allied troops during the invasion of Europe, also exemplified Groves's penchant for preparing for contingencies that he considered to be remote.[69] Thus, although he considered it most unlikely that a Russian bomb would appear within the next ten years, he initiated and nurtured the development of technical ideas for monitoring from afar the test of a Russian bomb.

NOTES

1. Quotations are from "Report of the Joint Chiefs of Staff," October 19, 1945, p. 3, and "Intelligence Regarding Foreign Atomic Development," May 25, 1946, p. 2. RG 218, Joint Chiefs of Staff, Decimal File 1948–50, Box 165, Folder CCS 471.6, NA.

2. Lilienthal, quoted in Herken, *The Winning Weapon*, p. 197.

3. A. B. Darling, *The Central Intelligence Agency*, 1990, p. 10.

4. Ibid., pp. 67, 70.

5. *New York Times*, September 22, 1945, pp. 1, 3.

6. J. Prados, *The Soviet Estimate*, 1986 [1982], p. 15.

7. Ibid., pp. 17, 18.

8. Ibid., p. 6.

9. Darling, *The Central Intelligence Agency*, p. 76.

10. "Minutes of NIA Meeting," August 21, 1946. RG 218, JCS Records, Box 20, File 32, NA.

11. Ibid.

12. Herken, *The Winning Weapon*, p. 109.

13. "Minutes of JIC Meeting," April 29, 1946, p. 1. RG 218, JCS Records, Box 166, File 471.6, NA.

14. JIC Report, "Intelligence Regarding Foreign Atomic Development," May 25, 1946, Appendix. RG 218, JCS Records, Box 166, File 471.6, NA.

15. Darling, *The Central Intelligence Agency*, p. 165; Herken, *The Winning Weapon*, p. 196.

16. Groves, *Now It Can Be Told*, p. 380.

17. Ibid.

18. Herken, *The Winning Weapon*, pp. 219, 213.

19. Groves, *Now It Can Be Told*, pp. 186, 187.

20. Darling, *The Central Intelligence Agency*, p. 163.

21. Joint Research and Development Board Directive, "Formation of a Committee on Atomic Energy," August 15, 1946, RG 330, Secretary of Defense, RDB-1946–53, Box 7, Folder: Atomic Energy, NA.

22. Groves, *Now It Can Be Told*, p. 395.

23. W. B. Mann interview, April 11, 1990.

24. Groves later wrote, "I was a caretaker who could make no major decisions during a period in which major decisions were vital." See *Now It Can Be Told*, p. 397.

25. Groves, *Now It Can Be Told*, p. 397.

26. Hewlett and Anderson, *The New World*, p. 644.

27. For a detailed history of Murray Hill Area, see J. E. Helmreich, *Gathering Rare Ores*, 1986.

28. Rhodes, *The Making of the Atomic Bomb*, p. 427.

29. Hewlett and Anderson, *The New World*, p. 65.

30. Ibid., p. 285.

31. Herken, *The Winning Weapon*, p. 110.

32. Groves, *Now It Can Be Told*, p. 177.

33. Ibid., pp. 180, 181.

34. Hewlett and Anderson, *The New World*, p. 286. See also J. M. Sherwin, *A World Destroyed*, 1975, p. 104.

35. Herken, *The Winning Weapon*, p. 102.

36. Groves, *Now It Can Be Told*, p. 197.

37. Ibid., p. 239.

38. Ibid.

39. Herken, *The Winning Weapon*, p. 14.

40. J. B. Conant to V. Bush, May 18, 1945, RG 77, Office of Scientific Research and Development, Folder 12, Bush-Conant, NA. See also Rhodes, *The Making of the Atomic Bomb*, p. 633.

41. V. Bush and J. B. Conant to the Secretary of War, "Salient Points Concerning Future Handling of Subject of Atomic Bomb," September 30, 1944. RG 77, Office of Scientific Research and Development, Folder 69, Harrison–Bundy, NA. The text of this memorandum is in M. J. Sherwin, *A World Destroyed*, 1975, pp. 286–288. See also Herken, *The Winning Weapon*, p. 359 no.35.

42. J. Franck et al., "A Report to the Secretary of War, June 1945," June 11, 1945. The text of this report is in Jungk, *Brighter Than a Thousand Suns*, pp. 335–346.

43. Lawren, *The General and the Bomb*, p. 267.

44. The magnitude of this disparity and, in hindsight, the fact that the estimate Groves presented was wrong has been the subject of comment, much of which is based on misconceptions. For example, R. W. Clark (*The Greatest Power on Earth*, 1980, p. 189) has suggested that "since the whole [United States] atomic project had been fostered by fears that the Germans were mining the uranium deposits of Czechoslovakia," Groves's estimate was bound to be wrong once the Russians controlled Czechoslovakia. Actually, the fear of the wartime German atomic program was predicated on the 1,200 tons of high-grade Congo ore in German hands,

not on German access to Czechoslovakian mines, which were considered to be a relatively insignificant source.

45. Herken, *The Winning Weapon*, pp. 110, 111.

46. D. E. Lilienthal, *The Journals of David E. Lilienthal*, Vol. II, 1964, pp. 10, 11. See also Herken, *The Winning Weapon*, p. 98.

47. Lilienthal, quoted in Herken, *The Winning Weapon*, p. 112.

48. L. R. Groves to R. P. Patterson, December 3, 1945, RG 77, Office of Scientific Research and Development, Folder 40, Harrison–Bundy, NA. See also Herken, *The Winning Weapon*, p. 109.

49. Groves, quoted in Herken, *The Winning Weapon*, p. 112.

50. *Transcript of Hearing in the Matter of J. Robert Oppenheimer*, 1954, pp. 175–176. See also Herken, *The Winning Weapon*, p. 341.

51. Minutes of Interim Committee Meeting, June 1, 1945. RG 77, Office of Scientific Research and Development, Box 100, NA. See also Rhodes, *The Making of the Atomic Bomb*, pp. 649, 650.

52. A. Kramish, *Atomic Energy in the Soviet Union*, 1959, pp. 112, 113.

53. Not everyone agreed with this reading of the statistics. In addressing the Council on Foreign Relations in 1945, Manhattan Engineer District physicist Isidor Rabi claimed that deposits of high-grade uranium ores were "comparatively widespread." See D. Yergin, *Shattered Peace*, 1977, p. 136.

54. Conant to Bush, May 18, 1945. See also Rhodes, *The Making of the Atomic Bomb*, p. 633.

55. Lilienthal, quoted in Herken, *The Winning Weapon*, p. 112 n.

56. P. Pringle and J. Spigelman, *The Nuclear Barons*, 1981, p. 43.

57. Herken, *The Winning Weapon*, p. 400 n.4.

58. Oppenheimer, quoted in R. Gilpin, *American Scientists and Nuclear Weapons Policy*, 1962, p. 75. Gilpin indicates that other General Advisory Committee members agreed with Oppenheimer's statement on the United States' atomic monopoly.

59. Herken, *The Winning Weapon*, pp. 128, 380, 394.

60. Ibid., p. 128, and Jungk, *Brighter Than a Thousand Suns*, p. 255.

61. Lilienthal, quoted in Herken, *The Winning Weapon*, p. 358 n.29.

62. M. Bundy, *Danger and Survival*, 1988, p. 135; and Pringle and Spigelman, *The Nuclear Barons*, pp. 42, 43.

63. CIA report "Status of the Soviet Atomic Energy Program," July 4, 1950. RG 218, JCS Records, Box NSC, Folder: Atomic Weapons, NA. See also Herken, *The Winning Weapon*, p. 332 n. According to this CIA report, Russian refining methods using very low-grade uranium ore "were employed extensively and were unexpectedly successful. . . . Thus, the Soviets are now in a position where the supply of uranium is no longer critical."

64. Hewlett and Duncan, *Atomic Shield*, p. 149.

65. "The Capabilities of the USSR in Regard to Atomic Weapons," JIC 395/1, July 8, 1947. RG 218, JCS, Decimal File 1948–50, Box 166, Folder 471.6, Control of Atomic Weapons, Section 5, NA.

66. See, for example, R. E. Marshak, E. C. Nelson, and L. I. Schiff, *Our Atomic World*, 1946, p. 24.

67. S. Eberhart, "How the American People Feel about the Bomb," *Bulletin of the Atomic Scientists* 3, June 1947, p. 146.

68. Alvarez, *Alvarez*, p. 121.

69. A. V. Peterson, "Peppermint," in A. C. Brown and C. B. MacDonald, *The Secret History of the Atomic Bomb*, 1977, pp. 234–238.

CHAPTER 3

TECHNICAL PROGRESS:
1945–1946

Despite the existence of factors in the postwar environment that tended to slow progress toward the creation of an operational monitoring system, progress *was* made. Research on various methods of detecting foreign atomic operations at long range was conducted in the final months of the war, and this type of research was continued after Japan's surrender.

In the spring of 1945, an aircraft fitted with the xenon-detection apparatus developed by Alvarez had been flown over the Manhattan Engineer District facility in Hanford, Washington, to map the flow of gaseous emissions from the plutonium-producing reactor.[1] The range at which the emissions of xenon and other noble gases could be reliably detected fell short of the thousands of miles required to monitor similar reactors deep inside Russia. And the kind of low-level flights over suspected reactor sites that had been carried out in Germany during the war could not, for obvious reasons, be made over Russia in peacetime. Later, with the advent of sophisticated means of obtaining air samples inside Russia, the measurement of the noble gases was once again to become a major intelligence-gathering tool, but such means of sample collection did not exist in 1945. Hence, the approach based on measuring gas emissions from reactors was, for the time being, abandoned.

Long-range detection methods based on the effects produced by an atomic bomb test, rather than on the effluents of reactor operation, appeared to be a more promising approach to the problem of technical surveillance. The first information on potential remote-sensing techniques for detecting an atomic explosion was obtained from Trinity.

TRINITY

On July 16, 1945, the test of the first atomic bomb, code named Trinity, prompted research which suggested that atomic explosions, unlike atomic reactors, might be detected at very long range. Ironically, the experiment that revealed this possibility was begun somewhat belatedly as an after-thought, and it was undertaken primarily for the purpose of learning about the atmospheric disturbances produced by the bomb.

Manhattan Engineer District scientists had instrumented the test site at Alamogordo, New Mexico, and had employed aircraft to facilitate visual scientific observations. But information on the effects of the bomb at distances of more than a few miles from the testing ground was gathered for legal, not scientific, reasons. Measurements with barographs and radiation survey meters had been obtained in locales up to 200 miles from ground-zero by Army personnel. These data on blast and radiation were recorded as evidence to be used in settling claims against the Army for damage to structures and radiation injuries incurred by civilians. No provision had been made to extend such measurements to greater distances because it was considered unlikely that effects beyond about 200 miles could be the cause of litigation. To make sure of this, seismographs at various distances from Alamogordo (El Paso, 80 miles; Tucson, 300 miles; and Denver, 500 miles) were set up by the Manhattan Engineer District and monitored, but these instruments failed to detect the earthwaves produced by the blast.[2]

The lack of long-range measurements was remedied some days after the test when two scientists, Anthony Turkevitch and John Magee, at the Manhattan Engineer District's Metallurgical Laboratory in Chicago, suggested that the dust from the Trinity bomb might cause atmospheric disturbances akin to those produced by volcanic eruptions. Turkevitch later wrote, "Many of us had read the Royal Society report on the Krakatoa volcanic explosion [which occurred in 1883] as an indication of the type of long range effects we might expect." These phenomena appeared worthy of investigation and, as Turkevitch later recalled, "the Los Alamos Project hastily outfitted an airplane with filters."[3]

The report on this research, written in October 1945, indicates that one of the aims was to sample particulates in the air over the west coast of the United States to ascertain that the high-altitude dust from the Trinity explosion had girdled the earth. Trinity dust that had circled the world was expected at the longitude of Bakersfield, California, by August 8 or 9. The dust from the Hiroshima explosion on August 5 was due to arrive at the Bakersfield longitude after August 10, from its point of origin directly across the Pacific. The dust from the Nagasaki bomb was not expected to reach North America. The report states that Nagasaki dust "was dispersed because of local weather conditions."[4] Thus, any radioactive dust encountered before August 10 could only be from the Trinity bomb. But it proved impossible for logistic reasons to make any flights before that date. Therefore, the dust that showed up was attributed to Hiroshima rather than Trinity.

A B-29 bomber from the Second Air Force was outfitted with a tube connected to an air scoop atop the fuselage. The tube led to a perforated metal cylinder, 2 feet in diameter and 2 feet long, mounted in the bomb bay. The inside of the cylinder was lined with soft tissue paper of the type that had been used in air filters at the Trinity test site. The bomber was flown at various times between August 10 and 15, at altitudes between 15,000 and 30,000 feet. The flights were generally along the line between Bakersfield, Seattle, and Ketchikan, in southern Alaska. After each flight, the filter paper was removed and wrapped around a thin-walled "Geiger counter" (a tubelike device used to measure radioactivity) to determine whether or not radioactive dust had been picked up.

Traces of such dust were found, and the nature of the data suggested that it was "reasonable that the radioactivity was due to . . . Hiroshima." However, since Trinity had produced about thirty times as much dust as Hiroshima, this conclusion was questionable. Another obfuscating factor was that the Manhattan Engineer District's Hanford facility was located near the flight paths, so the possibility existed that part of the radioactivity might have come from operations at Hanford. Nevertheless, despite such shortcomings, it was concluded that this approach "would seem to be a practical means of detecting an atomic bomb explosion almost anywhere *with proper meteorological conditions* [emphasis added]."[5]

The meteorological caveat reflected ideas that were then held about the nature of atomic explosions. These ideas would later have an important bearing on subsequent assessments of the feasibility of radiological methods of long-range detection. At the time, however, the question of feasibility was not carefully evaluated. In this regard, it is pertinent to note that no thought was given to chemically analyzing the dust. Indeed, the small sample size would have precluded this in any case.

Turkevitch later recalled his reaction to the experiment: "The results . . . were ambiguous. Some non-natural radioactivity was found on filters . . . but it was not clear whether it came from Hanford. The levels of activity were very low." The importance of the experiment, he later wrote, was that it "represents the first attempt to capture nuclear explosion debris by airplane filters."[6]

The Manhattan Engineer District report on this experiment was classified as secret, but the Trinity bomb produced effects at long range that were measured by university scientists who were able to publish their results in the open literature. Radiological effects at great distances were made possible by the fact that the top of the dust cloud created by the explosion reached very high altitudes where it encountered the recently discovered jet stream.

The path taken by the high-altitude portion of the cloud became roughly known because of data obtained by a naval officer and scientist, Albert Coven, who, for purely scientific reasons, was making measurements of atmospheric radioactivity at his home near Annapolis, Maryland, some 1,400 miles from Alamogordo. About thirteen hours after the detonation, the high-altitude cloud was carried by jet stream air currents at a speed of about 110 mph over the east coast of the United States. When dust, settling to the ground from the cloud overhead, reached Coven's backyard, his instrument began to record an increase over the natural background radiation. The explanation for this increase was not obvious because the Trinity explosion was secret. Hence, it was only after the bombs were dropped on Japan and information about Trinity was released on August 7 that Coven could interpret his data. His results were published in the open literature in the fall of 1945. Coven's data confirmed one of the findings of the secret Los Alamos study—namely, that the radioactivity produced by a bomb could be carried long distances by high-altitude winds.[7]

The release of information about Trinity prompted a scientist at the California Institute of Technology in Pasadena, California, to reexamine the July 16 records of the Institute's ten seismic stations. Records of instruments for measuring pressure waves transmitted through the air (micro-barographs) and through the ground (seismographs) were available. The stations were located 262 to 682 miles from ground-zero. The explosion was detected by microbarographs at distances of up to 652 miles and (barely) registered on a seismograph at 682 miles. These results were communicated to the Manhattan Engineer District and later published in the open literature.[8] The fact that the seismograph stations set up by the Manhattan Engineer District failed to react to the blast at similar distances suggested that the geological path of the earthwaves and the sensitivity of the instruments were important factors in detectability.

Other reports of effects caused by Trinity reached the Manhattan Engineer District, although, at the time, they were not published. Scientists at the Coast and Geodetic Survey Observatory, at Tucson, Arizona, found that their routine measurements of the electrical conductivity of air rose dramatically five days after Trinity. (Air conductivity is directly related to atmospheric radioactivity.) And the Eastman Kodak Company reported to the Manhattan Engineer District that it had found man-made radioisotopes at a distance of 1,000 miles from Alamogordo in the straw-board used to package photographic film. The analysis of this finding was published in the open literature four years later, causing a serious rent in the curtain of secrecy that, by then, had been drawn over all aspects of long-range detection.[9]

Therefore, shortly after the surrender of Japan, Manhattan Engineer District headquarters was aware that Trinity had been detected at nearly 700 miles by measuring sonic effects (using microbarographs) and through earth tremors (using seismographs) and at 1,400 miles by measuring radiological effects (using Geiger tubes and air-conductivity instruments). The somewhat flawed measurements carried out by Manhattan Engineer District scientists indicated that the radioactive dust from Hiroshima might have been detected at a distance of 11,000 miles. The ambiguity of the latter finding, however, indicated that further experimental research on long-range detection was needed. And for such research, more atomic explosions were required.

More were soon forthcoming. By the late fall of 1945, General Groves was involved in the discussions of plans for a jointly run Army-Navy atomic bomb test in the Pacific. The notion of using these detonations to evaluate long-range detection techniques, as the basis of a system for monitoring foreign atomic explosions, was accepted as part of the program.

Also in the fall of 1945, a new player entered the game: the Army Air Force, independent of the Manhattan Engineer District, decided to conduct its own research on a unique sonic method of detecting foreign atomic bombs at long range. Paradoxically, this development effort, code named Mogul, was an offshoot of the Navy's wartime research on sound transmission in the ocean.

PROJECT MOGUL

The concept on which Mogul was based originated in World War II, during which the Navy greatly expanded its basic research program on antisubmarine warfare through contracts with university and industrial laboratories. As a participant in this research program, the Woods Hole Oceanographic Institution in Massachusetts carried out investigations of sound propagation in the ocean. In 1944, one of these investigations was spearheaded by Professor W. Maurice Ewing, a geophysicist whose research at Lehigh University and Columbia University in the 1930s had earned him a solid reputation in this field. At Woods Hole, under a contract with the Bureau of Ships, he had investigated the possibility of long-range transmission of underwater sound.

By 1944, Ewing had proved that at a depth of 4,000 feet there existed an oceanic "sound channel"—that is, a layer of water within which sound waves can travel unlimited distances without contacting the surface or the ocean bottom. Ewing's calculations had predicted that at the axis of the channel (whose boundaries are formed by the characteristics of the variation of temperature and pressure with depth) all acoustic waves leaving a source of sound within an angle of 12 degrees from the horizontal would be refracted back and forth across the axis and would thus travel great distances without diminution in intensity. Experiment confirmed Ewing's prediction. A four-pound bomb detonated at a depth of 4,000 feet was readily heard at a distance of 2,300 miles using a hydrophone positioned at the same depth, an enormous range compared to anything previously achieved.[10]

Moreover, the sound from an explosion within the channel exhibited a characteristic pattern of intensity over time that facilitated accurate triangulation with a network of three listening stations. Hence, the location of a source of sound at depths of a few thousand feet could be pinpointed with precision by stations thousands of miles from the disturbance. Ewing's discovery had applications in hydrographic surveying, but, given the depth limitations of the submarine technology of that day, there was no immediate military use for the oceanic sound channel.[11]

In September 1944, Ewing had speculated that a similar sound channel might be present in the earth's atmosphere. As in the case of the oceanic sound channel, there were no immediate wartime applications for an atmospheric channel since preliminary estimates suggested that the channel, if it existed at all, would be at altitudes above those generally used by the military airplanes of the time. However, a potential use for the hypothesized channel soon appeared when, in November 1944, the existence of the German V-2 weapon was made public.[12]

Mysterious explosions that had rocked London over the past weeks were revealed to have been caused by huge rockets weighing 13 tons at liftoff. In describing the vehicle to the House of Commons, Prime Minister Winston Churchill stated: "It rises to a height of sixty to seventy miles and its speed makes it impossible to sound air raid warnings."[13]

These rockets descended on their targets from enormous heights at over 3,500 miles per hour. Hence, they were impossible to track with the remote-sensing techniques, such as radar, then available. Since their trajectories could not be determined and the launchers were not readily identifiable by aerial reconnaissance, it was virtually impossible to locate the launching sites. This advantage was stressed by "informed sources in Berlin" who were quoted in American newspapers as saying, "It is impossible to guess where the launching sites of the V-2 are."[14]

Though sparse, the published technical information on the V-2 suggested that the rocket rose nearly vertically and traversed the sound channel hypothesized by Ewing within seconds after liftoff. It might thus be possible to detect the engine noise within the channel at great distances and to triangulate its position and, by inference, that of the launch site below.

Ewing discussed his idea about a stratospheric sound channel with the Woods Hole Oceanographic Institution's director, who regarded it as very important, but one that was outside his organization's ocean-oriented research program. The war in Europe ended a few months later, before Ewing had been able to carry the matter further.

After the surrender of Japan, Ewing prepared to return to his teaching post at Columbia University. In order to continue to pursue his research interests, he decided to seek government sponsorship to investigate the possibility that a stratospheric sound channel existed. In mid-October 1945, he met with Army Air Corps Commanding General Carl Spaatz in Washington, D.C. Spaatz had just returned from the Pacific theater where, as head of the United States Strategic Air Force, he had presided over the nuclear bombing of Japan. Spaatz thought Ewing's ideas about a sound channel merited further consideration, so he asked for a written proposal.

At the time, it was generally recognized that three new "secret" weapons that had emerged during the war were of seminal importance for future military development: jet aircraft, rockets, and, of course, atomic bombs. Sufficient information on jets and rockets was available to indicate that these vehicles would interact with the postulated sound channel, and Ewing stressed the implications of this interaction in his proposal to Spaatz. In commenting on the military applications of the sound channel, he wrote:

It is my belief that a large rocket or jet propulsion motor passing the axis of the channel would . . . be detectable by listening at several thousand miles, and subject to location if heard by three suitably chosen stations. In time of war this triangulation could locate launching sites of the enemy, and in peace time it is conceivable that suitably chosen stations could monitor the entire world.[15]

Ewing composed this paragraph a few weeks after Hiroshima, when the world first learned of the existence of the atomic bomb. Technical data on the only atomic explosion that had been scientifically observed, Trinity, was still a carefully guarded secret, available only to top-level bomb project personnel and a handful of military officers with a "need-to-know." Against this background, Ewing's failure to speculate on the possibility that the sound channel might play a role in locating foreign atomic explosions is understandable. At the Pentagon, however, Ewing's proposal was placed in the hands of Colonel Roscoe C. Wilson, one of the few officers privy to what was then known about atomic explosions.

Wilson had just been made deputy chief of staff for Research and Development. He was a career officer who had obtained an advanced degree in engineering after graduating from West Point in 1928. In the years that followed, he served in positions that required a technical background. In 1943, he was assigned to the Manhattan Engineer District, where his primary duty had been to make certain that when the atomic bomb became available there would be a suitable aircraft to carry it.[16]

In the context of his postwar involvement with U.S. efforts to covertly monitor atomic explosions in the Soviet Union, it is noteworthy that in 1944 Wilson participated in Groves's attempt to detect the possible existence of German nuclear reactors using bombers equipped to collect xenon.[17]

Wilson had also played a role in selecting the Trinity site for the first test of an atomic bomb, and, after the war, he was part of the team sent to Hiroshima and Nagasaki to survey the effects produced by these weapons. He was thus among the very few people at that time who were familiar with both the concept of using instrumental means to covertly monitor foreign atomic developments and the existing data on atomic explosions.

In November 1945, Wilson presented the results of his evaluation of Ewing's proposal to Spaatz. He wrote, "If Mr. Ewing is correct in his belief,

the existence of an atmospheric sound axis is of great military importance."[18] Wilson decided to initiate a program at the Army Air Force Watson
Laboratories, at Fort Monmouth, New Jersey, aimed at using the atmospheric sound channel primarily to detect atomic explosions at long range.

Early in 1946, Colonel Marcellus Duffy of Watson Laboratories' Operational Engineering Laboratory conferred with Ewing, who had returned to
Columbia University, to outline a program, code named Mogul, that would
explore and apply the atmospheric channel for this military purpose. A
contract was subsequently awarded to Columbia University to retain Ewing's services as researcher and advisor.

Mogul was necessarily a research-oriented project, since experiments to
prove the existence of the channel remained to be carried out. Ideally, to
accomplish this, measurements would be made of the intensity of sound
versus the distance between a source and receiver, both of which were
positioned at the axis of the hypothesized sound channel. One obstacle to
such an experiment was the fact that theoretical projections, based on
existing atmospheric data, placed the axis of the channel at 45,000 feet,
considerably above the operational ceiling of most airplanes of that day.
Wartime experiments on the depth variability of the oceanic sound channel,
however, gave grounds for optimism that meaningful results regarding
atmospheric sound channel properties could be obtained at lower altitudes.

The B-29 bomber was capable of flying at altitudes somewhat above
30,000 feet for fairly protracted periods of time. Thus, in the initial plans
made for Mogul, based on the use of a B-29, experiments were projected
involving the deployment of a bomb that would detonate at high altitude
to act as the sound source (bombs weighing from 100 to 500 pounds were
envisaged) and the use of balloon-borne sound receivers floating at a
similar altitude.

Unfortunately, balloons capable of floating at constant level within the
stratosphere at the appropriate altitude had yet to be developed. Plans to
initiate research on such aerostats were made, and in the interim it was
decided to push ahead with experimental work using a less-than-ideal
arrangement in which the sound source was located within the posited
atmospheric channel while the receivers were located on the ground.

By the spring of 1946, nearly a dozen people at Watson Laboratories were
employed on Mogul. They were led by Dr. James Peoples, formerly a Lehigh
University physicist, whose wartime experience had included work with
Ewing on the ocean sound channel experiments. In April 1946, Alpert P.
Crary, who had been one of Ewing's graduate students, joined Watson
Laboratories and became responsible for fieldwork on Mogul. In the pre-
vious ten years, Crary had been employed by several oil-prospecting firms
as a geophysicist, and he had been part of Ewing's team at Woods Hole.

Crary had been able to purchase and suitably modify ground-based
apparatus for measuring sound from a high-altitude source. In May, Peo-

ples had sent a team equipped with this apparatus to White Sands Proving Ground in New Mexico to attempt to detect sound waves produced at high altitude by the flights of captured V-2 rockets.

It should be noted that the military purposes of Mogul were classified Top Secret.[19] One purpose (as originally posited by Ewing) was to detect and locate rockets as they passed through the sound channel. A second and even more important goal was to use the channel to sonically monitor foreign atomic explosions. The initial plans for this project had thus included obtaining "large [explosive] charges to simulate 'A' test."[20] But at the time these plans were being formulated, preparations for the atomic tests in the Pacific, code named Crossroads, were initiated. Hence, instead of attempting to simulate an atomic test with conventional explosives, the possibility of using real atomic bombs was at hand. Therefore, plans were made to send a group to the Pacific to measure sonic effects produced at various distances from the atomic test site on Bikini.

A more novel experiment was also envisaged. This was inspired by a study, made sixty-three years earlier, of the eruption of Krakatoa, located in the Indian Ocean near the Sunda Strait, on August 27, 1883. This huge explosion, estimated to have released energy equivalent to about 500 megatons of TNT, produced atmospheric pressure waves that were observed as seven distinct pulses of decreasing intensity at various weather stations around the world.

To explain this pattern, it was assumed that the initial pressure pulse originating from the explosion was focused into the atmospheric region of minimum sound velocity where it traveled with little attenuation to the antipode (a point in northwestern Colombia, South America). At the antipode, focusing and reflection occurred, and the reflected wave returned to Krakatoa, where the process was repeated. This cycle continued with ever decreasing strength to produce seven measurable pulses, the first being roughly 100 times as intense as the seventh.

Crude calculations revealed that the much smaller blast from an atomic bomb might produce a sound pulse of detectable intensity at the antipode of the detonation site. The antipode of Bikini, the test site for Crossroads, was Ascension Island in the Atlantic. Hence, plans were made to send Crary and two assistants to Ascension with suitable equipment for detecting the sonic effects produced by the Crossroads bombs. It was expected that, if successful, this experiment would provide useful data on sound refraction and attenuation in the atmosphere that could be helpful in assessing the feasibility of Mogul.

OPERATION CROSSROADS

The evidence of the awesome power of the atomic bombs dropped on Japan had triggered a resurgence of the battleship versus bomber debate that had so exercised Army and Navy leaders in the 1930s. If a single atom bomb was powerful enough to wipe out a city, it could also be used to destroy a fleet, or so it seemed to the Army Air Force. The Navy was not so sure. To settle the question, the Army Air Force generals, in September 1945, had requested the Joint Chiefs of Staff to allow them to atom bomb some captured Japanese battleships. In October, Navy admirals countered with a proposal for a more comprehensive test jointly run by both services under the aegis of the Joint Chiefs of Staff. The Joint Chiefs agreed, and the plan for Operation Crossroads, approved by the president in January 1946, called for three bomb tests. The tests were designated as Able (an airburst), Baker (underwater but near the surface), and Charlie (deep underwater), and they were to take place on Bikini atoll in the Pacific in May 1946. The Army, Navy, and Manhattan Engineer District's civilian laboratories were to jointly plan and execute the tests.

The aim of Crossroads was to gain information of direct military value, that is, data that would aid in designing future ships, aircraft, and tactics for atomic warfare. Scientific experiments of general value were to be permitted only to the extent that they would not interfere with the military goal. But General Groves was interested in furthering the development of potential monitoring methods; therefore, research on long-range detection as a tool for monitoring foreign atomic explosions was included in the Crossroads program. To carry out this research, the Army and Navy set up a jointly manned Remote Measurements Section, headed by Commander George Vaux, who reported directly to William Parsons, newly promoted to rear admiral, who had been named technical director and deputy commander of Crossroads.[21]

Although it had been the first arm of government to initiate research on atomic energy (beginning in 1939), the Navy had been virtually excluded from the Manhattan Project.[22] Hence, few Navy scientists and even fewer scientifically oriented naval officers like Parsons had been able to acquire direct experience in the physics and chemistry of atomic weapons production. But the office of the chief of Naval Operations, which was involved in the planning of Crossroads, could draw on the services of Reserve officers who, although they were not atomic scientists, possessed technical expertise that was useful in assessing the effects produced by atomic bombs.

One such scientist-officer was Commander Roger Revelle, a geophysicist of note who was then serving in the Office of Research and Inventions (later, the Office of Naval Research). Early in 1946, the chief of Naval Operations directed Revelle to enlist the help of scientists at the Naval Research Laboratory in planning the Navy's contribution to the program of the Remote Measurements Section.[23]

Revelle visited the Naval Research Laboratory and met with key people in the physical science divisions. According to Dr. Herbert Friedman, who was at the meeting, Revelle minced no words but went directly to the point.

"He asked us," Friedman later recalled, "how we would go about setting up technical means of detecting the first Russian bomb, if and when it appeared."[24]

The Naval Research Laboratory harbored versatile and talented scientists with skills in areas related to long-range detection. Some Naval Research Laboratory physicists had developed apparatus capable of being carried aloft by balloons to measure cosmic rays at high altitude, while others had specialized in designing radiation-measuring equipment. Based on such past work, they were able to propose the development of specialized instruments that were incorporated into the experimental program of the Remote Measurements Section. These included various types of radiation survey meters and a "radiosonde"—that is, a balloon-borne monitoring device capable of radioing data on radioactivity at high altitudes to a recorder on the ground.[25]

Vaux was thus able to obtain personnel and equipment from naval facilities such as the Naval Research Laboratory and the Naval Ordnance Laboratory for his Remote Measurements Section. The wide range of expertise provided in this way allowed the Section to assess various approaches to the problem of long-range detection. These included sonic methods based on the use of microbarographs and various radiological techniques employing both ground-based and balloon-borne detectors.

Groves had assigned Lieutenant Colonel David Parker, from the Manhattan Engineer District's Military Operations Division, to work with Commander Vaux. In addition, a major part of the Army's contribution to the Remote Measurements program was conducted through Groves's Manhattan Engineer District headquarters. Groves had contracted with the Standard Oil Company for the services of Standard's Berkeley, California, laboratory in measuring the radioactivity of special filters that had been designed for collecting radioactivity at high altitude via aircraft. He had also assigned Major Paul Krueger, Corps of Engineers, the task of overseeing a remote air-sampling program that was to be carried out using Army Air Force bombers equipped with the filters.[26]

By May, the preparations for the Remote Measurements program were complete. But the bomb tests, originally scheduled to begin in mid-May, were delayed by President Truman primarily for political reasons. Thus, the Able test did not occur until July 1, 1946. A bomb was dropped from a B-29 at 30,000 feet over the Bikini lagoon where a fleet of obsolete U.S. and Japanese ships had been assembled. Some thirty-eight vessels of various types, ranging in size from submarines to battleships, were blasted by the equivalent of 23,000 tons of high explosive when the bomb was detonated 520 feet above them.[27]

Although it was classified as an airburst, the bomb was detonated only a few hundred feet higher than the bomb-tower used in Trinity. Since the explosion was relatively close to the surface of the lagoon, the cloud it created contained, in addition to the elements of the bomb itself, vaporized material from the ships and the water. The mushroom cloud rose to 40,000 feet, and as it ascended it was rapidly dispersed by shearing winds at different altitudes.

Drone aircraft were flown into the cloud to collect dust and air samples, and, as the cloud dispersed, it was tracked out to about 500 miles by aircraft fitted with equipment to measure airborne radioactivity. At greater distances, sonic data, using microbarographs, and data on radioactivity were obtained at stations located on various Pacific islands. Measurements of radioactivity at ground level were obtained with sensitive Geiger counters and at high altitudes by instrumented airplanes and by balloon-borne radiosondes.[28] These efforts provided useful results, but the instruments for measuring atmospheric pressure waves that were deployed by scientists from Watson Laboratories, including Crary's team on Ascension, did not yield data that would allow them to draw any useful conclusions about the feasibility of the Mogul concept.

The Manhattan Engineer District air-sampling program, supervised by Major Krueger, had commenced well before Crossroads to establish a baseline for samples collected during and after the tests. Filter-equipped Army Air Force B-29 bombers were deployed over locales at various distances from Bikini: Guam (1,600 miles), Okinawa (1,900), Hawaii (2,100), Spokane, Washington (4,500), Tucson, Arizona (4,700), Tampa, Florida (6,500), and Panama (8,000).

These planes carried special oil-bathed filters designed to pick up the minute dust particles created by the explosion. The planes made daily flights of five hours duration over each of the seven locales at an altitude of 30,000 feet. After each flight, the filter was removed and air-shipped to the office of the Corps of Engineers in Berkeley, California, where the area engineer would carefully remove all identifying labels (except for a coded number) before delivering the filters to the Standard Oil Company laboratory.

In accord with Groves's compartmentalization policy, the Standard Oil Company technicians had been told nothing of the nature or purpose of the samples they were given to measure. Knowing that Groves was preoccupied with secrecy, Krueger later reported that "from personal observation in Berkeley the security on the program was good. . . . Also, the air units concerned did not know what happened to their filters after shipment."[29] Thus, neither the air crews who collected the samples nor the laboratory technicians who measured them had been made aware of the nature of the program in which they were participating. This emphasis on secrecy and compartmentalization, begun by Groves, was to continue to characterize all aspects of long-range detection in the years to come.

The Able test sank four ships and seriously damaged seven. The remaining vessels suffered light damage and were used in the test that followed on July 25. The Baker bomb was suspended underwater from one of the ships at a depth not far below the surface. The lagoon was fairly shallow, nowhere deeper than 200 feet, so when the bomb was detonated a portion of the floor of the lagoon beneath the bomb was also vaporized together with huge quantities of water. Instead of the mushroom cloud that had characterized all previous atomic explosions, a pillar of water rose more than a mile in the air. Gases burst through the top of the column to form a cloud. As the enormous cylinder of water fell back into the lagoon, the cloud shed a radioactive rain that blanketed the fleet. A few ships were sunk, but virtually all the remaining vessels—even those that appeared undamaged—had become "radioactive stoves" that were completely dysfunctional.[30]

After several days, the radioactivity of the reef and lagoon had dropped to levels that, in 1946, were considered tolerable. But the tendency of radioactivity to build up on the evaporators and ventilation systems of task force ships made it necessary to abandon the Bikini site by mid-August. The originally scheduled third test, Charlie, had been canceled before Crossroads began owing to the parlous state of the stockpile of atomic bombs. But the contamination of the test site would have prevented a third test at Bikini in any case.[31]

Paradoxically, the fact that the Bikini area was so contaminated meant that stations remote from ground-zero reported virtually no radioactivity from Baker. Most of the radioactivity produced by the bomb was washed into the sea in the immediate area by the cloud of rain and mist created during the explosion. This finding was important in the subsequent evaluation of the limitations of radiological methods of long-range detection.

As in the case of the Trinity explosion, reports of the detection of airborne radioactivity from Crossroads by university scientists soon appeared in the press. Indeed, such news items surfaced several months before the long-range detection data obtained by Army and Navy experimenters were collated and analyzed.

These press reports reinforced the belief, fostered by similar findings on Trinity, that atomic explosions were easily detectable at great distances by measuring radioactivity in the atmosphere. Reporters had pointed out that atomic bombs differed from ordinary bombs not only because, weight-for-weight, they were enormously more powerful in terms of blast effect, but also because they produced airborne radioactivity or "fallout." Moreover, the popular press had drawn attention to unclassified reports in scientific journals, indicating that the blast effects and the fallout from the earlier Trinity test and from Crossroads had been detected thousands of miles from the test sites.

The media, however, failed to address some of the real problems faced by government scientists interested in developing reliable remote-sensing methods to detect an atomic test within the Soviet Union. One problem concerned atmospheric radioactivity. It was well known that natural occurrences such as volcanic activity could produce significant departures from the normal, naturally occurring atmospheric radioactivity. However, in and of itself, an increase in atmospheric radioactivity, even if known or assumed to be man-made, would not prove an atomic bomb had been detonated. Instead, it could be due to a reactor accident, a chemical explosion involving radioactive waste, or the purposeful test of a "radioactivity bomb," made by mixing reactor-produced radioactivity with an ordinary high explosive. All of these events could produce an increase in atmospheric radioactivity at long range, but would not necessarily indicate the capability to make an atomic bomb.

Moreover, reports of readings obtained with seismographs during the bomb tests that were cited by the media, far from showing that these instruments could reliably detect an atomic bomb detonated at long range, actually showed just the opposite. For example, the Baker bomb tested at Bikini registered on a seismograph in California, 4,500 miles away, but the Able bomb of this test series, detonated at the same location, was not detected seismically in California. The Baker bomb was detonated beneath the ocean surface within the lagoon; hence, the explosion was closely coupled to the floor of the lagoon. Able was an airburst, detonated 520 feet above the ocean's surface. Because of the air gap, the explosion was weakly coupled to the sea and earth below, producing earthwaves that were undetectable seismically at distances of more than a few hundred miles.

Moreover, seismographs were incapable of differentiating between a weakly coupled and/or low-yield atomic explosion and one caused by the detonation of a massive quantity of an ordinary explosive. The Soviets, for example, were known in the past to have used conventional explosives on a huge scale in civil engineering projects such as dam building.

Although the media reports suggested just the opposite, the only firm conclusion that could be derived from scientific journals on measuring the effects produced by atomic bombs was that *none* of the instruments and techniques described would have been capable of detecting unambiguously an atomic bomb test deep within the borders of the Soviet Union. Given the vast area encompassed by these borders, many sites suitable for testing bombs were known to exist. Some of these sites were located 2,000 miles and more from areas accessible to Western monitoring techniques.

Nevertheless, the belief that atomic explosions were easily detectable at great distances by measuring radioactivity in the atmosphere was shared by some government officials. This is exemplified by a memorandum written shortly after Crossroads in August 1946 by Major General Curtis LeMay, head of Army Air Force research. Addressed to Brigadier General N. B. Harbold of the air intelligence staff, the subject of the memorandum was "Intelligence on A-Bomb Explosions in Foreign Countries." LeMay began by noting the importance to national security of detecting an atomic bomb detonated in another country. In referring to Mogul, he indicated that, under his aegis, the Army Air Force was exploring a method of detecting such explosions. He pointed out that "It is not yet known whether this proposed method is at all practicable." He went on to say:

There is another method which has been shown to be practicable; namely, to monitor continuously the counting rate of an unshielded Geiger-Mueller Counter and to watch for anomalous increases in the counting rate. Following the explosion of the test bomb in New Mexico in July 1945 an observer on the eastern seaboard reported (Letter to Editor of Physical Review) a temporary increase of about 50 per cent in the rate of a counter set up at his home. A news release from the University of California (Washington Post, 8 August 1946) states that the radioactivity from the second explosion at Bikini had reached San Francisco, 4700 miles away, with measurable intensity.[32]

LeMay then discussed various devices, in addition to Geiger counters, that might be developed to measure anomalous increases in airborne radioactivity and proposed that "such monitor units be installed throughout the world in sites controlled by the United States." He also revealed his ignorance of the Manhattan Engineer District's activities by stating: "It is probable that a world-wide monitor system of the type proposed here has been considered by the Manhattan District, and it may even have been initiated."[33] LeMay's lack of knowledge about the Manhattan Engineer District's monitoring efforts is, of course, not surprising in view of the rigid policy of compartmentalization of information imposed by General Groves.

If the key points of this memorandum are analyzed, they reveal: first, LeMay's call for a worldwide monitoring system was based on the widely promulgated (but erroneous) belief that atomic explosions could readily be detected and identified as such at long range by measuring an anomalous increase in atmospheric radioactivity; second, although he held an important position in the military hierarchy, he was not privy to the Manhattan Engineer District's activities; and third, because it "has been shown to be practicable" to detect foreign atomic explosions radiologically, he thought that the Manhattan Engineer District had probably already established such a monitoring system. LeMay also echoed the belief, held by most Americans, about the imminence of a Soviet bomb. He wrote: "The need for a monitor system of some kind will, according to some estimates, become acute within the next year or so."[34]

LeMay's memorandum is noteworthy because it is structured around the same beliefs, assumptions, and areas of ignorance that, eight months later, would characterize the response of the newly appointed Atomic Energy commissioners when confronted by the issue of monitoring foreign atomic bomb explosions. The few military officers and scientists who had an in-depth understanding of the monitoring problem reacted to LeMay's memorandum (and, eight months later, to Atomic Energy Commissioner

Strauss's similar memorandum) by pointing out that more research would be needed before an effective monitoring system could be deployed.

The recipient of LeMay's memorandum, General Harbold, discussed LeMay's suggestions with Groves's deputy, with the Army Air Force liaison officer to the Manhattan Engineer District, and with a representative of the Central Intelligence Group. Harbold soon learned that adequate radiological methods for monitoring atomic explosions did not exist and remained to be developed. In forwarding LeMay's memorandum to Brigadier General John A. Samford of the air intelligence staff, Harbold wrote a covering letter in which he indicated that, contrary to the assertion in LeMay's memorandum, monitoring anomalous increases in airborne radioactivity to detect an atomic explosion "has not been shown to be practicable."

He further indicated that the first scientifically planned trial of the method suggested by LeMay was carried out at Crossroads, but the results had not been analyzed. Finally, in a statement that revealed the trend of official thought on the issue of organizational responsibility for monitoring, he wrote: "Consensus of opinion among those consulted is that it should be the function of CIG [Central Intelligence Group] to determine the feasibility of the [radiological] Geiger counter method, and—if found feasible—to inaugurate and monitor a project based on same."[35]

General Samford agreed with this assessment. On September 5, 1946, he sent a memorandum to General Vandenberg, head of the Central Intelligence Group. Samford expressed the concern of the Army Air Force's intelligence staff over the problem of monitoring foreign atomic explosions. He pointed out that, because atomic intelligence had recently been transferred to the Central Intelligence Group, it would be appropriate for the Central Intelligence Group to supervise and coordinate the further efforts on monitoring.[36] Plans for such efforts soon appeared.

In the fall of 1946, assessments of the various methods of long-range detection that had been studied during Crossroads, together with recommendations for future action, appeared in the form of three reports: on September 10, Parker reported to Groves and Colonel Kenneth Nichols, Groves's deputy, the results of his contribution to Vaux's Remote Measurement program; on September 18, Krueger reported to Groves the details of the remote air-sampling project; and on September 27, Vaux submitted his report on the activities of the Remote Measurements Section to Parsons, technical director of the Crossroads Joint Task Force.

Parker indicated that, although much information on long-range detection had been garnered, shortcomings in the instruments were apparent. He concluded that these problems appeared soluble and that remote measurements could be effected. Parker went on to point out a serious organizational difficulty; namely, that no arrangements had been made to continue research on the instruments tested at Crossroads. This comment highlighted what would turn out to be a major stumbling block in the further development of long-range detection, that is, the lack of an organizational "home" for this surveillance concept.[37]

Krueger's air-sampling project was, in the light of later events, the most important part of the Remote Measurements program. From June 24 to August 21 planes of the Army Air Force had made 357 flights over the seven locales that had been selected. (These long-duration flights had not been without danger—one set of filters was lost when the plane crashed.) No radioactivity had been detected after July 28, indicating that all the radioactivity that was observed had been produced by Able. Although traces of radioactivity were picked up at some of the more distant locales, unambiguous results were obtained only in areas relatively close to Bikini. Krueger concluded: "It is possible by monitoring air currents at various points around the world to determine if an atomic bomb has been detonated in the air. By detailed analysis of wind conditions, it may also be possible to determine the direction to the blast and, by additional judicious reasoning, approximately when and where it was detonated."[38] But Krueger added two qualifiers to this conclusion: first, that radiological methods would probably be reliable only at "ranges of 2000 miles or less," and, second, that "it does not appear feasible to monitor underwater explosions by these methods."[39] The latter conclusion stemmed from the fact that Baker had not been detected.

Krueger was not a trained nuclear scientist; hence, his assessment is somewhat naive. The measurements on which he based his conclusions merely noted anomalously high radioactivity in some atmospheric dust samples. Other measurements, not carried out, would have been necessary to prove that the radioactivity could be attributed with certainty to an atomic bomb explosion as opposed to, for example, a reactor accident. Such measurements, however, required sample processing techniques that, at the time, had not been developed.

Despite these shortcomings, which were not well understood at the time, the air-sampling program was valuable in that it conclusively demonstrated that radioactive dust from the explosion of an atomic bomb near the earth's surface could be carried substantial distances by stratospheric winds.

The measurements, however, *did not* demonstrate that an airburst at higher altitudes, where the fireball was well above the surface of the earth, would produce radioactive dust at long range that would be collectable in a quantity sufficient to allow the identification of specific radioisotopes through chemical analysis. Uncertainty over this point later became a major cause of concern in assessing the feasibility of radiological monitoring.

Krueger recommended that "Additional studies should be made as a preliminary step to establishing a program of monitoring potential atomic bomb explosions with the system used during this operation as the basis of the program." Like Parker, Krueger also foresaw organizational problems in carrying out this recommendation and added the suggestion that "all basic data be turned over to the FIS for study and exploitation."[40] The Foreign Intelligence Section (FIS), with its history of involvement in technological surveillance, was, indeed, a logical choice as the agency to continue the work on long-range detection.

Vaux's report echoed Parker and Krueger's assessments. Vaux indicated that all the methods used required further development, but such development promised to be successful. The results obtained by the Remote Measurements Section, he concluded, merited followup.[41]

The reports on the Remote Measurements program were, of course, secret. But the Crossroads tests had been publicly announced. Hence, as indicated earlier, the effects produced by the Bikini bombs were measured by nongovernment scientists, and their reports soon appeared in the open literature.

Published reports that dealt with radiological effects were consonant with Krueger's conclusion that such effects could not be *reliably* detected at distances beyond 2,000 miles. Meteorological conditions appeared to influence the findings reported in Texas.[42] The measurement of a group of scientists in Oklahoma led them to conclude that "No general correlation of total radioactive intensity with the explosion of the [Able] bomb was obtained."[43] And in New York, measurements of atmospheric ionization (a parameter related to radioactivity in the air) showed "no effects from the Bikini tests."[44]

Although seismic effects from Baker were observed on the west coast of the United States, it was apparent that the then existing seismographs could only detect bombs exploded on or quite near the earth's surface.[45] The Able bomb, which was an airburst, was not detected. Moreover, intelligence summaries indicated that highly sensitive instruments in seismic stations close to Bikini had failed to register earth tremors from Able.[46]

The information available after Crossroads, therefore, suggested that the most promising method of long-range detection was radiological, although the means for distinguishing a bomb test from, say, a reactor accident had not been developed. The feasibility of Mogul had yet to be demonstrated, and other sonic methods required the development of better microbarographs. Seismic techniques appeared to be unable to detect airbursts.

General Hoyt Vandenberg, recently made head of the Central Intelligence Group, was aware of the Remote Measurements program conducted during Crossroads. On September 27, 1946, he wrote to Groves requesting information on the feasibility of long-range detection and advice on how to proceed further with this concept.[47] By that time, Groves had already directed Colonel Lyle E. Seeman, the newly appointed associate director of the Manhattan Engineer District's Los Alamos laboratory, to draft a program for developing a long-range monitoring system.

On November 4, Seeman met with Vaux to review the Crossroads findings and to draft a program in accord with Groves's directive. Regarding the possibility of long-range detection, Seeman wrote that "No instruments exist at present to assure success." Nevertheless, he concluded that instruments could be designed that would provide such assurance. Both radiological and sonic detection methods had been tested at Crossroads, and, for both, better instruments were required. He recommended that the radiosonde, the radiation-measuring devices used in aircraft, and the microbarographs used in sonic detection be further developed. Following up Krueger's suggestion that backtracking the jet stream could roughly locate the explosion, Seeman recommended that a special study be made of the effect of weather on high-altitude air currents. Seeman also described his vision of the monitoring system that would result from this development work: "The future measurement net appears to be selected existing ground stations, with supplementary aircraft flights should ground measurements indicate desirability. For this a central controller appears necessary together with prompt communications."[48]

It seems that Seeman may have espoused the Navy's view of long-range detection that, apparently, Vaux ably presented to him. The "existing ground stations" cited by Seeman were almost all located at naval stations in the Pacific and on the United States west coast. After Crossroads, the Navy had begun to equip these stations with specially designed Geiger counters and ancillary equipment that had exceptionally high sensitivity to airborne radioactivity.[49] These instruments were not ready at the time of the Bikini test, but in the coming months they promised to give the Navy an "edge" over the Army and the Army Air Force in long-range detection.

Seeman apparently felt unable to unravel the knotty organizational problems associated with long-range detection. He recommended that the Manhattan Engineer District carry out the instrument development and initial procurement, but he stopped short of suggesting who or what agency should bear overall responsibility for developing and deploying the "future net" of monitoring stations. There was good reason for his reticence on this point.

Owing to the transfer of the Manhattan Engineer District from the Army to the Atomic Energy Commission in July, the interpretations of the results of the Crossroads Remote Measurements program took place in the five-month interim period during which Groves headed the Manhattan Engineer District as a "caretaker" manager. In the past, Groves had assigned responsibility for technological surveillance to his Foreign Intelligence Section. But the responsibilities and position of the Foreign Intelligence Section within the intelligence community had become ambiguous. Vandenberg had already begun a campaign to have the Foreign Intelligence Section transferred to the Central Intelligence Group. Moreover, the 1943 directive making Groves responsible for atomic intelligence seemingly conflicted with the authority of the newly created Central Intelligence Group. No individual or organization had a clear-cut mandate to conduct the needed research on remote measurement instruments and to establish an operational monitoring system based on such research. Thus, although technical work on long-range detection techniques continued, the organizational development of a centralized monitoring system had to await the further reorganization of the United States intelligence community.

NOTES

1. J. Tabin interview, January 5, 1991; and J. Tabin, letter to L. R. Zumwalt, January 2, 1990. Copy in the authors' collection. Our thanks to Dr. Tabin for allowing us to quote from this letter.

2. B. C. Hacker, *The Dragon's Tail*, 1987, p. 92; and E. C. Truslow and R. C. Smith, *Project Y*, 1983, pp. 243–246.

3. A. Turkevitch, letter to D. Jacobson, March 27, 1990, copy in the authors' collection. As previously noted, the general idea of monitoring the air to detect characteristic effluents produced by certain foreign atomic operations was conceived and developed by Alvarez in 1943–1944. His version of this idea was based on the detection of a certain gas that is produced by the fissioning process. This work was compartmentalized; hence, other Manhattan Project scientists were not privy to it. It thus appears that one of his colleagues, Anthony Turkevitch, independently thought of this idea. According to the memoirs of Stanislaw Ulam, "The idea . . . that [monitoring] could be done by examining air samples from the atmosphere for the presence of certain gases which came from uranium fission . . . came from Tony Turkevitch, a physical chemist from Chicago. I remember his mentioning such a plan in my presence in Los Alamos during the war" (S. M. Ulam, *Adventures of a Mathematician*, 1976, p. 210). It appears that the extension of this concept from the detection of gaseous products to the detection of airborne radioactive *particulates* resulted from the 1945 experiment suggested by Turkevitch and Magee, which showed that such particulates could travel long distances from the site of an atomic explosion.

4. J. M. Blair et al., "Detection of Nuclear-Explosion Dust in the Atmosphere," October 2, 1945, p. 3. Report No. LA-418, Report Library, LANL.

5. Ibid., p. 7.

6. A. Turkevitch, letter to D. Jacobson, March 27, 1990. Copy in the authors' collection.

7. A. W. Coven, "Evidence of Increased Radioactivity of the Atmosphere After the Atomic Bomb Test in New Mexico," *Physical Review* 68, 1945, p. 279.

8. B. Gutenberg, "Interpretations of Records Obtained from the New Mexico Atomic Bomb Test, July 16, 1945," *Bulletin of the Seismological Society of America* 36, 1946, p. 327.

9. "Long Range Detection of Atomic Explosions," *Intelligence Review*, No. 37, October 24, 1946, p. 58, Naval Aide Files, HHPL; and J. H. Webb, "Fogging of Photographic Film by Radioactive Contaminants in Cardboard Packaging Materials," *Physical Review* 76, 1949, p. 375.

10. M. Ewing, "Long Range Sound Transmission in the Atmosphere," October 14, 1945, p. 1. General LeMay Collection, microfilm roll No. 1760, frames 1978–1983, AFHRC.

11. Ewing, "Long Range Sound Transmission in the Atmosphere," October 14, 1945, p. 2.

12. D. H. DeVorkin, "War Heads into Peace Heads," *Journal of the British Interplanetary Society* 45, 1992, p. 439.

13. *New York Times*, November 11, 1944, p. 1.

14. Ibid., November 10, 1944, p. 5.

15. Ewing, "Long Range Sound Transmission in the Atmosphere," October 14, 1945, p. 3.

16. *Transcript of Hearing in the Matter of J. Robert Oppenheimer*, p. 681.

17. Ibid., p. 692.

18. R. C. Wilson to General Carl Spaatz, November 2, 1945. Microfilm roll No. 1760, frame 1976, AFHRC.

19. Colonel D. P. Graul to Commanding General, AMC, June 14, 1946; and Brigadier General E. O'Donnell to Commanding General, AMC, July 8, 1946. Record Group 342, decimal file 360.2, Upper Air Research, Box 3939, NA.

20. "Memorandum B," May 14, 1946, p. 2. Microfilm roll No. A1760, Frames 1970–71, AFHRC.

21. G. Vaux to Technical Director, September 27, 1946, Central Records/Archives, LANL; and Hacker, *The Dragon's Tail*, p. 118.

22. R. Gunn to Director, "Submarine Submerged Propulsion—Uranium Power Source," June 1, 1939; and R. Gunn, "Early History of Atomic Powered Submarine at NRL, 1939–1946," December 13, 1947, Strauss Papers, HHPL.

23. R. Revelle, interview, June 23, 1990.

24. H. Friedman, interview, June 27, 1989.

25. L. E. Seeman to K. D. Nichols, "Remote Measurements for Atomic Explosions," November 6, 1949, p. 2. Central Records/Archives, LANL.

26. P. Krueger to L. R. Groves, "Remote Air Sampling," September 18, 1946, p. 1. Central Records/Archives, LANL.

27. Hacker, *The Dragon's Tail*, pp. 124, 133.

28. Ibid., p. 132; and Seeman to Nichols, "Remote Measurements for Atomic Explosions," November 6, 1949, p. 2.

29. Krueger to Groves, "Remote Air Sampling," September 18, 1946, p. 4.

30. Hacker, *The Dragon's Tail*, pp. 137, 152.

31. Ibid., pp. 146, 147.

32. Lieutenant General C. E. LeMay to Brigadier General N. B. Harbold, August 12, 1946, p. 1. RG 18, Entry 1, Box 625, Decimal File 350.09, Intelligence/Counterintelligence, 1946–1947, Vol. I, NA.

33. Lieutenant General LeMay to Brigadier General Harbold, August 12, 1946, p. 2.

34. Ibid.

35. Brigadier General N. B. Harbold to Major General G. C. McDonald, August 30, 1946. RG 18, Entry 1, Box 625, Decimal File 350.09, Intelligence/Counterintelligence, 1946–1947, Vol. I, NA.

36. Brigadier General J. A. Samford to Director, CIG, September 5, 1946. RG 18, Entry 1, Box 625, Decimal File 350.09, Intelligence/Counterintelligence, 1947–1947, Vol. I, NA.

37. Seeman to Nichols, "Remote Measurements for Atomic Explosions," November 6, 1949, p. 1.

38. Krueger to Groves, "Remote Air Sampling," September 18, 1946, p. 1.

39. Ibid., p. 2.

40. Ibid., p. 4.

41. Seeman to Nichols, "Remote Measurements for Atomic Explosions," November 6, 1949, p. 1.

42. G. Herzog, "Gamma Ray Anomaly Following the Test of July 1, 1946," *Physical Review* 70, 1946, p. 227; and D. Weeks and D. F. Weeks, "Effort to Observe Anomalous Gamma Rays Connected with the Atomic Bomb Test of July 1, 1946," *Physical Review* 70, 1946, p. 565.

43. R. E. Fearson et al., "Results of Atmospheric Analysis Done at Tulsa, Oklahoma, During the Period Neighboring the Time of the Second Bikini Bomb Test," *Physical Review* 70, 1946, p. 564.

44. V. F. Hess and S. J. Lugar, "Ionization of the Atmosphere in the New York Area Before and After the Bikini Bomb Test," *Physical Review* 70, 1946, p. 564.

45. B. Gutenberg and C. F. Richter, "Seismic Waves from Bomb Tests," *Transactions of the American Geophysical Union* 27, December 1946, p. 776.

46. "Long Range Detection of Atomic Explosions," *Intelligence Review*, No. 37, October 24, 1946, p. 54.

47. Seeman to Nichols, "Remote Measurements for Atomic Explosions," November 6, 1949, p. 1.

48. Ibid.

49. H. Friedman, interview, June 27, 1989.

CHAPTER 4

THE ORGANIZATIONAL DUST SETTLES

Although scientific and technological progress towards a system of long-range monitoring was made during the Trinity and Crossroads Operations, the consolidation and utilization of the findings from these tests had to await the stabilization of the network of organizations, agencies, and offices that would further develop and eventually deploy the monitoring system. This stabilization was completed early in 1947, and it led to the resolution of organizational issues that had inhibited the creation of a centralized monitoring system. The relevant aspects of the process of stabilization can be best understood in the context of postwar changes in the evaluation and dissemination of atomic intelligence.

ATOMIC INTELLIGENCE REORGANIZED

The collection of scientific intelligence had been an important part of U.S. intelligence activities during World War II. Collated scientific information garnered by the various agencies had been constantly reviewed by the Reading Panel of the Joint Chiefs of Staff. The Panel, comprised of representatives of the Army, Navy, and Office of Scientific Research and Development, had kept the Joint Chiefs of Staff abreast of the latest scientific and technical advances and the new weaponry based on these advances. Information on foreign atomic programs had constituted an important category of scientific intelligence, but for security reasons it had not been handled like information in other categories such as radar or rocketry. Instead, it had been treated as the unique province of General Groves, who had limited its distribution to a few top leaders.

The revelation of U.S. atomic developments that followed the atomic bombing of Japan had eliminated the need for the kind of "internal" security that had justified insulating virtually all government bodies except the Manhattan Engineer District from atomic matters. This changed condition did not immediately lead to changes in handling atomic intelligence, however, chiefly because the key figures were preoccupied. The Joint Chiefs of Staff and the officials of other interested agencies were caught up in the debate over the creation of a permanent, nonmilitary intelligence organization, and Groves was grappling with the problem of preserving the rapidly disintegrating Manhattan Engineer District as a viable organization. By the spring of 1946, however, the Joint Chiefs of Staff and the intelligence chiefs had begun to feel that atomic intelligence, though of vital importance, should be treated like other categories of scientific intelligence in terms of collection, evaluation, and dissemination.

Groves had never really controlled the collection of atomic intelligence (apart from that produced by the technical surveillance methods developed by the Manhattan Engineer District). But his Foreign Intelligence Section did carry out the evaluation of raw intelligence and the dissemination of the finished intelligence product. But evaluation and dissemination of all categories of intelligence, except for some specialized areas retained by the military agencies, were the primary tasks assigned to the Central Intelligence Group that was created early in early 1946. In intent, if not in performance, the Central Intelligence Group made the Foreign Intelligence Section superfluous.

But as an embryonic organization, the Central Intelligence Group did not have the capability, in terms of scientifically trained personnel and the supporting infrastructure of procedures and background files, to evaluate atomic intelligence. The Central Intelligence Group would either have to create this capability slowly, starting from scratch, or acquire it quickly by assimilating the Foreign Intelligence Section. The Central Intelligence Group had earlier faced a similar problem in the area of intelligence collection, a problem that was largely solved by the transfer of the Strategic Services Unit from the War Department to the Central Intelligence Group. As director of Central Intelligence, Vandenberg opted to use the same approach to solve the problem of evaluating atomic intelligence. In August 1946, he sought the approval of the National Intelligence Authority to transfer the personnel and files of the Foreign Intelligence Section to the Central Intelligence Group. The members of the National Intelligence Authority, except for Acting Secretary of State Dean Acheson, agreed to sign a directive that would effect the transfer. Acheson refused to sign and called for a fuller discussion of the matter. This discussion took place at a meeting of the National Intelligence Authority on August 21, 1946, to which Vandenberg was invited.

The remarks of the participants during the meeting revealed the rationales on which their positions were based. Secretary of War Patterson opened the meeting by declaring that it was senseless to allow the Foreign Intelligence Section to continue to operate in isolation, and that he had discussed this problem with Groves. Patterson called for an immediate transfer of the Foreign Intelligence Section to the Central Intelligence Group in order to close a dangerous communication gap. He emphasized the point that under the present arrangement "if General Groves knew that the Russians were prepared to use atomic bombs, the members of the NIA [National Intelligence Authority] would not know about it." He went on to indicate that the Foreign Intelligence Section was an Army intelligence unit whose activities were unrelated to the Manhattan Engineer District's operational functions and, as such, need not be considered an integral part of the Manhattan Engineer District organization that was to be turned over to the Atomic Energy Commission in January 1947. Consequently, it was unnecessary to delay the transfer of the Foreign Intelligence Section pending review by the yet-to-be-confirmed commissioners. Both Admiral Leahy, the president's representative, and Secretary of the Navy Forrestal agreed with the substance of Patterson's argument, but Acheson demurred.[1]

Acheson declared that the Foreign Intelligence Section *did* perform some tasks, involving the supply of raw materials, that were directly related to the Manhattan Engineer District's operational functions. Hence, the fate of the Foreign Intelligence Section was of concern to the Atomic Energy Commission. In an allusion to Murray Hill Area, he pointed out that "it was one function of General Groves's intelligence group to find out where uranium ore is and how to get it to this country or to deny it to others." He went on to say that President Truman was aware of the Foreign Intelligence Section's role in preserving the United States' monopoly on uranium and that Truman had thought that the Foreign Intelligence Section should not be transferred until the commissioners had the opportunity to review the matter.[2]

Vandenberg, who was anxious to augment the capabilities of the Central Intelligence Group, supported Patterson's claim that the transfer was urgently needed. He stated that General Groves had asserted that the Foreign Intelligence Section's best source of raw intelligence had been the Strategic Services Unit, which was now incorporated into the Central Intelligence Group as the Office of Special Operations. The Office of Special Operations, however, lacked the ability to evaluate the atomic intelligence it collected. Hence, the Central Intelligence Group could not disseminate finished intelligence on foreign atomic developments as called for by its charter. Moreover, it would be an unnecessary security risk for the Central Intelligence Group to duplicate the Foreign Intelligence Section's evaluation capability. In this regard, Vandenberg indicated that Foreign Intelligence Section evaluators were part of Groves's personal staff who knew "all that Groves knows." If the Central Intelligence Group set up evaluators in parallel, they would have to be told "basic atomic secreta" which would jeopardize security by enlarging the circle of people having this knowledge.[3]

Acheson relented and agreed to the transfer if it was first cleared by the president. The directive was cabled to Truman, who was attending the peace conference in Paris. Truman replied that he wished to postpone any action until he returned to Washington.[4] Since other difficulties delayed the transfer after Truman's return, this decision had the effect of deferring the removal of the Foreign Intelligence Section until the Atomic Energy Commission could consider the matter.

At the time of the National Intelligence Authority meeting, Operation Crossroads, which had just ended, was receiving much media attention. Photos, eyewitness accounts by reporters, and government news releases that described the explosions and their aftereffects, had been widely promulgated to satisfy a curious public. No information about the experiments on long-range detection had been released, but leaders of the intelligence community were aware of the Remote Measurements program that had been carried out during Crossroads. A few weeks after the National Intelligence Authority meeting, Vandenberg sent a memorandum to Groves requesting information on the performance of the long-range detection methods that had been tried and advice on what action should be taken to further develop a technological surveillance system based on these methods. Groves had assigned the task of preparing a complete program for the further development of a long-range monitoring system to Colonel Seeman, and an extract copy of Seeman's memorandum outlining the proposed program had been sent to Vandenberg in reply to his query.[5]

The technical nature of Seeman's proposed monitoring network reinforced Vandenberg's awareness of the Central Intelligence Group's urgent need for expertise in a broad spectrum of scientific specialties. Therefore, in addition to attempting to acquire the Foreign Intelligence Section, Vandenberg also sought to establish a close working relationship with the Joint Research and Development Board chaired by Vannevar Bush. Vandenberg sent the Joint Research and Development Board a report prepared by the Central Intelligence Group's Office of Reports and Estimates, which outlined the capabilities of the Russians in developing atomic and conventional weapons. This report was read at a Joint Research and Development Board meeting in November, and Vandenberg sent his scientific consultant, Dr. H. P. Robertson, to the meeting to answer questions on the report. The Board was favorably impressed by the quality of the Office of Reports and Estimates report, and Robertson was able to use the meeting to discuss ways the Board and the Central Intelligence Group could help one another.[6]

One outcome of the ensuing discussion was an offer by the Board to find a head for a new branch to be created in the Central Intelligence Group for the evaluation of scientific intelligence. The Board also agreed to help recruit the experts necessary to staff the branch. Vandenberg and Bush subsequently formalized these ideas in January 1947 in a written agreement. It provided that a newly created Scientific Branch in the Office of

Reports and Estimates would develop a national program of scientific intelligence. The head of the Scientific Branch would serve as advisor to the director of Central Intelligence on scientific intelligence and also have direct access to the activities of the Joint Research and Development Board insofar as they pertained to the work of the Scientific Branch. On its part, the Central Intelligence Group undertook to provide the Joint Research and Development Board with the intelligence it required to perform its functions. In return, the Joint Research and Development Board agreed to supply the Central Intelligence Group with qualified personnel, special facilities, and counsel on scientific matters.[7]

The close relationship between the Central Intelligence Group and the Joint Research and Development Board (which was an advisory body to the military secretaries) appeared timely in view of a heightened interest in monitoring atomic explosions at the War Department. Evidence of this interest appeared in a periodical called the *Intelligence Review*, an influential publication produced by the Intelligence Division of the War Department General Staff. The *Review* was classified as secret. Each issue was sequentially numbered and sent only to selected members of the intelligence community and top government officials to keep these individuals abreast of the latest developments in intelligence.

The October 1946 issue of the *Review* contained an article entitled "Long Range Detections of Atomic Explosions." The anonymous author was aware that research on this topic had been carried out during Crossroads, but in August, when the article was prepared, the results of this research had not been analyzed. Despite the lack of Crossroads data, the article provided a sophisticated overview of various remote measurement techniques. It also provided, for the first time, a clear-cut statement of the rationale for creating a monitoring system based on methods of long-range detection:

With the advent of atomic weapons . . . the military and civilian leaders of a government must know with the greatest possible accuracy the capability of any potentially hostile nation to wage atomic war. . . . It is highly improbable that any nation, once it has constructed an atomic bomb, would risk war or push full-scale production of bombs until it had actually conducted a test explosion. . . . There may be many indications and more or less reliable reports [of such a test], but to confirm this by scientific measurement would be most desirable.[8]

The article considered four methods of detecting atomic explosions at long range: sonic, seismic, radiological, and radio transmissivity. These methods were based, respectively, on shock waves in the air, earth tremors, the radioactive cloud, and disruption of the E-layer (the lowest level of atmospheric ionization that affects radio transmission). After analyzing the sparse data on these various phenomena for the Trinity, Hiroshima, and Nagasaki bombs, the author concluded that measurement of the radiological effects was the most promising approach, but that even in the case of radiological measurements, "the phenomena that can definitely be attributed to the atom bomb occurred only within a few hundred miles or less from the explosion."[9] However, the author pointed out that the range might be considerably extended if radioactivity at high altitudes was measured using balloons or airplanes. By the time this article appeared in October, the Crossroads results were available, and they generally confirmed the conclusions and predictions of the author.

The *Review* article had carefully avoided specifying the country that would be targeted by any technological surveillance system based on long-range detection, identifying it only as "a potentially hostile nation." But to the readers of the *Review*, who had access to information from clandestine sources, "potentially hostile nation" was a euphemism for "Russia." This view of Russia was supported by the intelligence data that were being collected. For example, shortly after the *Review* article appeared, the War Department logged an espionage report from "a penetration source which in the past has been reliable and accurate."[10]

In the parlance of intelligence, a "penetration source" refers to a member of a foreign governmental organization (civilian or military) who reports on that organization's activities to an adversary's intelligence service.[11] In this case, the source may have been a Russian officer, since the report dealt with the text of an indoctrination lecture for high-ranking military officers. The lecture, given on October 26, 1946, by Lieutenant General Kujukov, was on Russian atomic bomb strategy. Kujukov had indicated that Russia's diplomatic efforts toward world disarmament were a ruse to buy the time needed for its atomic bomb program to produce a weapon. He had gone on to say that this approach would remain viable "as long as Russia stands firm in its refusal to agree to international inspection."[12]

Espionage reports such as this, coupled with the overt actions of the Russian government, made it inevitable that any evaluation of the usefulness of long-range detection methods would be carried out in the context of their ability to monitor events at distances that would be meaningful only if Russia were the target country, that is, at distances of 3,000 or 4,000 miles.

By the winter of 1946, analysis of the data from the five bombs that had been exploded indicated that, with existing instruments, reliable detection at such ranges was simply not possible. But the Crossroads experimenters believed that the instruments could be improved to achieve greater range. This opinion was echoed by Norris E. Bradbury, the newly appointed head of the Manhattan Project's Los Alamos laboratory and a respected scientist. His wartime work for the Manhattan Engineer District made him well qualified to critique the reports of Parker and Krueger, which he received in November. Bradbury cited the ambiguity in some of the ground-station radioactivity measurements and in the results of the filter-sampling method using B-29 bombers. Nevertheless, he concluded that the radiological approach "undoubtedly has promise."[13] The realization of this "promise" was the aim of the development program Seeman had prepared at Groves's direction. But the organizational question of what agencies would carry out this program had remained unresolved simply because larger organizational issues remained to be settled.

In December, the details of the Atomic Energy Commission's takeover of the Manhattan Engineer District were formulated. Regarding the intelligence functions, Groves had recommended that the Foreign Intelligence Section be transferred to the Central Intelligence Group, and he was supported in this position by Patterson. The commissioners, however, resisted the Foreign Intelligence Section's immediate transfer. The problem was that, through Murray Hill Area and the Combined Development Trust, the Foreign Intelligence Section in effect controlled the raw materials for atomic bombs. Groves wanted the Atomic Energy Commission to control raw materials only if they accepted membership in the Combined Development Trust. The commissioners were loath to do this without first reviewing the functions of the Combined Development Trust in the context of the provisions of the McMahon Act that governed the international exchange of atomic information. This review was completed late in December, and, on December 26, the Atomic Energy Commission agreed to join the Combined Development Trust and to allow the War Department to retain the Foreign Intelligence Section, if the Atomic Energy Commission could have access to the Foreign Intelligence Section files. On December 31, the Atomic Energy Commission assumed control of the Manhattan Engineer District.[14]

Groves was shunted to a new post, chief of the Armed Forces Special Weapons Project, that had been established in early January 1947 by the War and Navy departments. The Armed Forces Special Weapons Project had been created to carry out military functions related to atomic weapons. Chief among these functions was that of handling atomic bombs delivered to military units. The Armed Forces Special Weapons Project had been made responsible for developing the equipment and training the personnel needed for this task.[15]

On February 22, Groves was given an additional job; he was appointed to the Military Liaison Committee. The Military Liaison Committee had been set up in August 1946 by the Navy and War secretaries as the Atomic Energy Commission's contact with the armed forces on a policy-making level. Groves appears to have guided the initial appointments to the Military Liaison Committee that were made by the secretaries, and he may have had a hand in his own appointment in 1947. As a member of the Military Liaison Committee, he was able to influence Atomic Energy Commission coordination with the military in ways that facilitated his job as head of the Armed Forces Special Weapons Project.

On January 10, 1947, Bush and Vandenberg issued their plan for the cooperation of the Central Intelligence Group with the Joint Research and

Development Board in the area of scientific intelligence. On January 23, the Scientific Branch was formally established within the Central Intelligence Group. In February, the Atomic Energy Commission examined Foreign Intelligence Section files and arranged with the War Department to obtain the files of Murray Hill Area and the Combined Development Trust. The remaining Foreign Intelligence Section files and its personnel were transferred from the War Department to the Central Intelligence Group on February 18. The former Foreign Intelligence Section personnel became the core of the Nuclear Energy Group that was created within Scientific Branch on March 28. The charter of the Nuclear Energy Group called for them to "prepare estimates of the capabilities and intentions of foreign countries in the field of nuclear energy."[16]

The National Intelligence Authority's stance was that the director of Central Intelligence should coordinate all intelligence on foreign developments in atomic energy. This mandate did not provide Vandenberg with the kind of exclusive control over the evaluation and dissemination of atomic intelligence that Groves had exercised during the war, but it did ensure a continuing role for Groves's intelligence unit, which had been incorporated in the structure of the Central Intelligence Group. Therefore, in the spring of 1947, the organizational issues that had inhibited the development of long-range detection had been largely resolved.[17]

THE STATUS OF MONITORING

Before proceeding to trace the further progress toward creating an effective system for monitoring a potential Russian atomic bomb test, it is pertinent to summarize the situation as it existed in March 1947 in three areas: the system rationale; the scientific and technological basis for the system; and the organizational context.

The rationale for the system owed some of its elements to features that characterized Groves's use of radiological surveillance in World War II and, indeed, that characterized wartime technological surveillance methods generally. Namely, they were a way of checking conclusions derived from conventional intelligence activities, and the information they provided was free from the duplicity or incompetence that sometimes made information from human sources misleading.

Other elements of the system rationale reflected beliefs commonly held after the war. It was generally accepted that sooner or later other nations would acquire atomic bombs and that they would probably test a bomb before beginning full-scale production. Thus, timely detection of the first test would reveal that an important milestone had been reached in the process of developing a stockpile of atomic weapons. From the intelligence perspective, this kind of information was of great value because it could be used as a checkpoint to "calibrate" a model of the process of stockpile development built on information gathered from other sources. Such a checkpoint would considerably enhance the ability of the model to accurately predict future actions.

These various elements of the system rationale were implicit in the planning of the Remote Measurements program of Crossroads and were more explicitly stated in the *Intelligence Review* in October 1946. An unstated, but "understood" element of the rationale was that Russia would be the target of any monitoring system.

The scientific and technological basis for the system had been established before Crossroads by examining the effects produced by the Trinity, Hiroshima, and Nagasaki bombs. These effects suggested that seismic, sonic, and radiological methods could be used to detect atomic explosions. Each of these approaches was explored during Crossroads, and each was found to have important limitations. Airbursts could not be detected seismically, and underwater (and, by implication, underground) atomic explosions could not be detected radiologically at long range. Neither seismic nor sonic methods could distinguish atomic bomb explosions from those produced by massive amounts of ordinary explosives. Existing radiological methods could make this distinction but could not differentiate between an atomic bomb explosion, a reactor accident, or a chemical explosion involving radioisotopes. Moreover, such radiological methods appeared to be reliable only at ranges of less than 2,000 miles, considerably short of the 3,000 to 4,000 miles necessary to detect an explosion deep within Russia.

The latter figures were derived from the fact that the jet stream, which traveled from west to east across Russia, could not be sampled by aircraft until it had passed over the east coast of Russian territory. Atomic debris, carried by the jet stream from a site in the Russian heartland, would thus travel 3,00 to 4,000 miles before it could be sampled *without* flying over Russian territory. Existing aircraft, of course, were incapable of flying over Russia on a routine basis, as would be required of a monitoring system, without prohibitive losses owing to Russian countermeasures. Moreover, unpredictable meteorological phenomena had been known to affect the usefulness of radiological methods at great distances from the explosion. Upper-air wind currents had appeared to be an important factor in dispersing the debris from the high-altitude airburst at Nagasaki (which was not detected on the west coast of the United States), but not the airburst at Hiroshima (whose debris was detected in the United States).

The experiments at Crossroads had revealed that the instruments used to measure seismic, sonic, and radiological effects had serious shortcomings that limited the range at which these effects could be detected. Despite the fact that an adequate monitoring system based on long-range detection methods could not be deployed using existing instruments, the exploratory research had suggested that it was probably feasible to develop such a monitoring system. A program for such development had been prepared in October 1946, but it was not initiated, chiefly because the relevant organizations were in a state of flux.

These details of the rationale and scientific basis for a system capable of monitoring a potential Russian atomic bomb were not well known in the spring of 1947. Secrecy and compartmentalization had limited this information to a relatively small number of individuals. Members of the intelligence community had been given a clear idea of the rationale and a general overview of the scientific basis for such a monitoring system by the *Intelligence Review* article. Some individuals on the staffs of the War and Navy departments who had participated in the planning of Crossroads, such as Admiral Parsons and Commander Revelle, had a comprehensive understanding of the technical aspects of monitoring. The civilian scientists working for Army and Navy laboratories were focused on their special areas of expertise and, because of compartmentalization, they lacked an understanding of all aspects of the monitoring problem. The crews of Army Air Force planes who flew air-sampling missions during Crossroads and the personnel of Army and Navy units that supported the Remote Measurements program were not informed of the nature of the program in which they had participated. Intelligence chiefs, such as Vandenberg, had been briefed on the rationale, scientific basis, and plans for further development, but they had not participated directly in activities related to long-range detection.

Thus, the depth of knowledge that individuals in various organizations had about long-range detection varied, as did the level of their involvement with the work that had been carried out in this area. A few individuals possessed a knowledge of the rationale and scientific basis of the monitoring system, and some had also played important roles in the long-range detection activities conducted thus far. These individuals were affiliated with six governmental bodies that subsequently constituted the organizational context in which the initial development of a monitoring system based on long-range detection occurred.

The Nuclear Energy Group of the Central Intelligence Group. As former members of the Foreign Intelligence Section, the personnel of the Nuclear Energy Group were familiar with wartime radiological surveillance efforts, postwar research by Manhattan Engineer District scientists, the interagency research during Crossroads, and the future program for developing a monitoring system prepared at Groves's direction. The Nuclear Energy Group thus constituted a kind of "organizational memory" for long-range detection.

The Military Liaison Committee. This six-man body was evenly divided between Army and Navy Officers. Two of its members, Groves and Parsons, were knowledgeable in all areas of long-range detection.

The Research and Development Board's Committee on Atomic Energy. This nine-man committee consisted of the six officers who comprised the Military Liaison Committee plus three civilians: James B. Conant (chair), J. Robert Oppenheimer, and Crawford W. Greenwalt, vice president of DuPont Company. As Military Liaison Committee members, Groves and Parsons were also members of the Committee on Atomic Energy.

The Atomic Energy Commission's Los Alamos Laboratory. Key scientists had been involved in Crossroads, and the laboratory's associate director, Colonel Lyle Seeman, had prepared the plan for the future development of long-range detection methods.

The Naval Research Laboratory. Naval Research Laboratory scientists had supported the Navy's efforts on the Remote Measurements program before and during Crossroads. Some Naval Research Laboratory scientists had continued research on methods of long-range detection after Crossroads.

The Army Air Force's Watson Laboratories. Through Project Mogul, influential officers in the Army Air Force's research and development arm and in air intelligence became involved in the monitoring problem at an early date. General LeMay and some of his staff, such as Colonel Roscoe Wilson and Colonel Montgomery Canterbury, later became influential figures in the development of the monitoring system.

By March 1947, these groups had become linked chiefly through the Military Liaison Committee. Structurally, the Nuclear Energy Group was a subsection of the Scientific Branch of the Central Intelligence Group's Office of Reports and Estimates, and the Committee on Atomic Energy was a subgroup within the Joint Research and Development Board. The close linkage between the Central Intelligence Group and the Joint Research and Development Board that had been forged by Vandenberg had allowed the Nuclear Energy Group access to the advice of the Committee on Atomic Energy. The Military Liaison Committee, whose members were also on the Committee on Atomic Energy, included Groves and Parsons who were, of course, quintessential insiders on atomic energy matters.

Rear Admiral William S. Parsons was an Annapolis graduate with a strong scientific background. In 1943, at Bush's suggestion, Groves had made Parsons head of the Manhattan Project's Ordnance Division. He flew in the *Enola Gay* as weaponeer and performed the final arming sequence on Little Boy just before it was dropped on Hiroshima. He had been technical director at Crossroads and had directly supervised the head of the Remote Measurements Section.

Of the remaining four Military Liaison Committee members (Lieutenant General Lewis Brereton, Colonel John Hinds, Rear Admiral Ralph Ofstie, and Rear Admiral Thorwald Solberg), only Solberg, who was interested in atomic power for ships, had background experience in atomic energy.

The Committee on Atomic Energy, therefore, possessed three military members who were knowledgeable on atomic matters. Of the three civilian members, two, Conant and Oppenheimer, were insiders of long standing. The third civilian, Greenwalt, had been the DuPont executive in charge of that company's effort to design and build the Manhattan Engineer District's Hanford facility. As members of the Committee on Atomic Energy, Groves and Parsons constituted that body's repository of the history of long-range detection. This topic, however, had not been on the agendas of the few Committee on Atomic Energy meetings that had been held prior to April.[18]

Both Conant and Oppenheimer were also members of the influential General Advisory Committee that had been set up to counsel the Atomic Energy Commission. This nine-man committee was chaired by Oppenheimer, and, in addition to Conant, its membership included some of the nation's leading scientists such as Enrico Fermi. By April, the General Advisory Committee had held three meetings, but the topic of long-range detection had not been discussed. During the February meeting, however, General Advisory Committee members had acknowledged that the effects produced by the five atomic bombs that had been exploded could not be "accurately predicted in advance," although it was known that each explosion had produced effects at long range.[19] Thus, it appears that although Conant and Oppenheimer had not been directly involved in long-range detection experiments, they were aware in a general way of the measurements that had been made of effects produced by atomic explosions at great distances.

The Atomic Energy Commission's Los Alamos laboratory had been a key organization in the Crossroads test, and its director, Norris Bradbury, had reviewed the findings of the Remote Measurements program. Bradbury's office staff had set up a file on long-range detection that included the Navy, Army, and Manhattan Engineer District reports on this topic.[20] But the newly appointed Atomic Energy commissioners themselves had no knowledge of the work that had been carried out in this area.

Of the five commissioners only one, Dr. Robert F. Bacher, was a scientist. Bacher was an eminent physicist who had headed the Manhattan Project's wartime Experimental Physics Division. But because of compartmentalization, Bacher had not been involved in the wartime use of radiological surveillance methods. He had left the Manhattan Engineer District after the war to head the Laboratory of Nuclear Studies at Cornell University. Hence, Bacher had also not been involved in postwar work on such methods.

One of the commissioners, Lewis L. Strauss, was an investment banker who had resigned his partnership in the Wall Street firm of Kuhn, Loeb, and Company to join the Atomic Energy Commission. As will be described later, Strauss's actions regarding the development of the long-range detection system evince a keen interest in security and intelligence. Thus, it is pertinent to describe his background in these areas. In 1926, as a participant in a Navy program to recruit reserve officers from the professional classes, Strauss was commissioned as a lieutenant commander in the Naval Reserve. At that time, he was a rising young account executive and partner at Kuhn, Loeb. Another partner in this firm was Sir William Wiseman, who had headed the American office of the British Secret Intelligence Service in New York during World War I. Strauss and Wiseman became close friends and perhaps because of Wiseman's "war stories" and contacts, Strauss became interested in intelligence work and chose to serve as a reserve officer attached to the New York branch of the Office of Naval Intelligence.

Little is known about his intelligence activities during his fifteen years as a Reserve officer except for one curious incident that has recently come to light. During World War I, Strauss had been secretary to Herbert Hoover, who was then head of the Belgian Relief, and Strauss maintained close ties to Hoover after joining Kuhn, Loeb. In 1930, President Hoover used Strauss as a deniable link to Naval Intelligence in New York in a scenario reminiscent of the Watergate incident forty-two years later. Hoover "unofficially" asked Strauss to arrange for Naval Intelligence agents to break into the offices of an employee of the Democratic campaign manager to copy some documents allegedly harmful to Hoover's reputation. This bizarre plot was apparently carried out with Strauss's help.[21]

In 1941, when Strauss was called to active duty, he was assigned to the Bureau of Ordnance. His wartime service did not involve intelligence work, nor did he have contact with or knowledge of the Manhattan Project, but it included an eighteen-month stint as aide to Navy Secretary James Forrestal, during which he rose to the rank of rear admiral. Strauss became a close friend of Forrestal and in the last three months of 1945 Forrestal had made him the Navy's representative on the Interim Committee on Atomic Energy. In this post he acquired knowledge of some of the wartime activities of the Manhattan Engineer District, but by 1946 he had resigned his commission to return to civilian life. Thus, like Bacher, he had no involvement with the Manhattan Engineer District's wartime or postwar work on radiological surveillance methods as applied to the long range detection of foreign atomic activities.

The remaining commissioners were not scientists, had known little about the Manhattan Engineer District during or after the war, and were relatively unversed in atomic matters. Although the commissioners had held a number of formal meetings since the takeover of the Manhattan Engineer District in January, the subject of a monitoring system based on long range detection had not arisen.[22]

By March 1947, the reorganization of atomic intelligence and the resolution of other organizational issues that had inhibited further development of a monitoring system had largely been accomplished. A stable organizational context had evolved that was capable of facilitating such development. Earlier, the rationale for a monitoring system had been enunciated and accepted by the intelligence community; the scientific and technological basis for the system had been established; and a research program that would lead to the creation of a monitoring system had been drawn up. All the pieces were in place, and the only organizational problem remaining was the selection of an agency to carry out the program.

As head of the Central Intelligence Group, Lieutenant General Hoyt Vandenberg took the first step toward resolving the question of organizational responsibility for monitoring. On March 14, 1947, he sent letters to the Navy and War departments, the Atomic Energy Commission, and the Joint Research and Development Board suggesting that they furnish representatives to sit on a committee that would study the problem and recommend a solution. He emphasized the need for immediate action by pointing out that "there exists an urgent and high priority, in the interests of national intelligence, for the development and coordinated employment of equipment capable of locating and measuring foreign experiments of large scale nuclear explosions or other detectable activities in the nuclear field."[23]

STRAUSS INTERVENES

On April 11, Lewis Strauss addressed a memorandum to his fellow commissioners which repeated Vandenberg's call for action on monitoring. Strauss wrote:

It would be interesting to know whether the intelligence arrangements of the Manhattan District made any provision in the past for continuing monitoring of radioactivity in the upper atmosphere. This would be perhaps the only means that we would have for discovering that a test of an atomic weapon had been made by any other nation. It is to be presumed that any other country going into a large-scale manufacture of atomic weapons would be under the necessity of conducting at least one test to "prove" the weapon. If the Combined [sic] Intelligence Group has no such monitoring system in effect, it is perhaps incumbent upon us to bring to their attention the desirability of such an immediate step and, in default of their action, to initiate it ourselves at once.[24]

This memorandum echoes some of the same beliefs and areas of ignorance expressed in General LeMay's memorandum written eight months earlier on August 12, 1946.[25] Not surprisingly, it provoked a similar reaction from individuals who were aware of the problems of monitoring and the efforts that had been made to solve them.

On April 30, at the fourth joint meeting of the Atomic Energy commissioners and members of the Military Liaison Committee, Strauss's memorandum was discussed. In referring to the results of the remote-sensing or monitoring experiments already conducted, Admiral Parsons of the Military Liaison Committee stated that "although seismic records are of some value, they do not give a good signature and cannot be relied upon." He indicated that the best approach for monitoring airbursts would be to detect the atmospheric pressure wave that could be used to locate the site of the blast. Aircraft could then be appropriately directed to search for airborne fission products that would identify the atomic nature of the explosion. He went on to say that "The Armed Services will produce a plan for extensive monitoring based on Crossroads experience, and will inform the Atomic Energy Commission."[26]

Strauss was later to recall that he did not find this statement reassuring and that he made further inquiries in the course of which he met separately with General Vandenberg, Colonel Seeman, and Navy Secretary Forrestal. These meetings were not documented at the time by either party, so our knowledge of them is derived solely from a series of memoranda Strauss wrote eight years later in which he attempted to recall the salient points that were discussed.

In meetings with Vandenberg and, later, with Seeman, Strauss found that no monitoring system had been established, but apparently he learned little more.[27] A few months before these meetings, Seeman, then assistant director of the Manhattan Engineer District's Los Alamos Laboratory, had been asked by Groves to prepare a plan for developing a monitoring system. Seeman had left his post at Los Alamos when it was taken over by the Atomic Energy Commission in January to join the Central Intelligence Group. Strauss recalled that when he questioned Seeman about monitoring, Seeman had replied that "everything was in hand," but the details "could not be revealed to the Commissioners."[28]

Inasmuch as Seeman was privy to the considerable amount of information amassed on remote-sensing instruments and techniques during Crossroads and to the conclusions reached about the performance of various monitoring methods, it appears that his refusal to supply details to Strauss was due to the standard practice of compartmentalizing information for security reasons. Apparently, Strauss chose to interpret Seeman's reticence as a sign that monitoring efforts thus far had been dilatory and inconclusive.

This reaction is understandable if, as his April 11 memorandum seems to show, Strauss shared the widely promulgated beliefs then current about atomic bombs, including notions about the imminence of a Russian bomb test and the ease with which it could be detected by measuring atmospheric radioactivity. Within the framework of such beliefs, it appears logical to attribute the failure to establish a monitoring system to bureaucratic foot-dragging.

The meeting with Seeman left Strauss unsatisfied. "Following this," he recalled, "I began to have conferences with Secretary Forrestal with a view to seeing whether responsibility could not be assigned to the Navy to fly regular [monitoring] flights in the Arctic and off the Asiatic continental land mass in the Pacific."[29] The notion that instruments and technologies for monitoring already existed, but had simply not been utilized, seems to have been behind this inquiry.

Apart from the fact that adequate monitoring methods had yet to be developed, the Navy had few planes in its inventory that were suitable for such missions and that could match the range of the Army Air Force's B-29s. Moreover, weather-instrumented B-29s (designated WB-29s) of the Army Air Force's Air Weather Service had already begun to make weather reconnaissance flights in the very regions named by Strauss.[30] The Army Air

Force thus had both suitable aircraft and experience in flying the hazardous routes that would be needed to intercept eastward-flowing air currents from Asia.

Strauss's appeal to the Navy rather than the Army Air Force thus appears to have been dictated by his close relationship with Forrestal rather than by practicalities. Nothing came of Strauss's inquiry about the use of naval aircraft, but on his visits to Forrestal he was able to confirm what he had learned from Vandenberg—there was no operational monitoring system, nor had the responsibility for creating one been assigned to a specific government agency.

The latter deficiency was about to be remedied. In May, in response to Vandenberg's memorandum of March 14, the Army, the Navy, the Central Intelligence Group, the Atomic Energy Commission, and the Joint Research and Development Board named representatives to sit on a committee that would find an organizational "home" for the monitoring effort.

NOTES

1. "Minutes of the Sixth NIA Meeting," August 21, 1946, pp. 1, 2. RG 218, JCS Records, Box 20, Folder 132, NIA Papers, NA.

2. Ibid., p. 1.

3. Ibid., p. 4.

4. Ibid.

5. Seeman to Nichols, "Remote Measurements for Atomic Explosions," November 6, 1949, p. 2.

6. Darling, *The Central Intelligence Agency*, p. 164.

7. Ibid.

8. "Long Range Detection of Atomic Explosions," *Intelligence Review*, No. 37, October 24, 1946, pp. 54, 55.

9. Ibid., p. 59.

10. M. Clark to War Department for JCS, "Atomic Strategy," November 8, 1946, JCS Records, Record Group 218, Box 166, Folder 471.6, NA.

11. G.J.A. O'Toole, *Encyclopedia of American Intelligence and Espionage*, 1988, p. 367.

12. Clark to War Department for JCS, "Atomic Strategy," November 8, 1946.

13. N. E. Bradbury to H. C. Gee, "Remote Measuring," November 25, 1946, p. 1, Central Records/Archives, LANL.

14. Hewlett and Anderson, *The New World*, pp. 651, 652.

15. Hewlett and Duncan, *Atomic Shield*, p. 666.

16. Darling, *The Central Intelligence Agency*, pp. 164, 165, 227.

17. Ibid.

18. C. E. Rolander to K. D. Nichols, "Review of CAE-RDB Minutes," January 29, 1954. Strauss Papers, AEC Series, Box 76, Folder-Oppenheimer Correspondence, HHPL.

19. R. T. Syles, *The Nuclear Oracles*, 1947, p. 171.

20. Bradbury to Gee, "Remote Measuring," November 25, 1946, p. 1.

21. See R. Pfau, *No Sacrifice Too Great*, 1984, p. 41, and B. J. Bernstein, "Hoovergate," *American Heritage* 43, May–June 1992, pp. 106–110. Pfau argues that Strauss's concern for security was so great that it justified for him participation in clandestine and illegal activities and "shows that for Strauss, despite his integrity and high standard of honesty, *no sacrifice was too great*, even though it violated the law, if he could convince himself it was necessary for the nation's security" (p. 42, emphasis added). Apparently, Pfau was so impressed with this feature of Strauss's character that he found it fitting to use it in the title of his study.

22. Strauss commissioned Philip Farley, the Atomic Energy Commission's recording secretary, to examine the minutes of early Atomic Energy Commission meetings for any mention of long-range detection or monitoring. The first meeting at which this occurred was that held on July 23, 1947. See P. J. Farley to Strauss, "References to Long Range Detection," July 20, 1948. Strauss Papers, AEC Series, Box 113, Folder: Tests and Testing 1947–54, HHPL.

23. Lieutenant General H. S. Vandenberg, quoted in "Report of Operation Fitzwilliam," Vol. I, n.d. (classified on December 9, 1949), p. 1. Period 1948, microfilm reel No. A15941, Frames 4–495, AFHRC.

24. L. L. Strauss to Commissioners, April 11, 1947, Strauss Papers, AEC Series, Box 76, Folder-Monitoring Soviet Tests, HHPL.

25. See the discussion in Chapter 2 regarding common beliefs about detecting nuclear explosions radiologically. It may well be that Strauss, like LeMay, shared the popular view of the practicability of radiological monitoring. However, it is also possible that Strauss may have been influenced by intelligence reports on this matter, for example, the October 1946 *Intelligence Review* article previously cited. Indeed, that Strauss might have seen the *Intelligence Review* article, or been made aware of it, is consistent with the fact that, according to his assistant at the Atomic Energy Commission, William T. Golden, he was known to be "very security minded . . . knew his way around the Navy and the intelligence community . . . [and] was trusted and respected by the military" (W. T. Golden interview, January 4, 1990). It is also consistent with certain similarities in wording between the intelligence report and Strauss's memo.

26. "Notes on the Meeting of the AEC and the MLC" fourth Meeting, April 30, 1947, p. 4. RG 326, Atomic Energy Commission, Folder: Minutes of Meetings, NA.

27. L. L. Strauss, October 20, 1955, memorandum to the files, p. 1. Strauss Papers, AEC Series, Box 113, Folder: Tests and Testing 1955–1957, HHPL.

28. L. L. Strauss, February 5, 1955, memorandum to the files, p. 1. Strauss Papers, AEC Series, Box 113, Folder: Tests and Testing 1955–57. HHPL.

29. Ibid., p. 2.

30. C. C. Bates and J. F. Fuller, *America's Weather Warriors*, 1986, p. 137.

CHAPTER 5

BIRTH OF THE MONITORING ORGANIZATION

The search for a "home" for the long-range detection system was slowed by organizational factors. By its nature, the task of developing and operating such a system required technical skills and facilities that were, at that time, scattered among the various branches of government. It was thus necessary to mount an interagency effort to bring the full range of expertise within the government to bear on the problem of deciding how to organize the system.

This effort, however, took place during a period of dramatic structural changes within the government. Moreover, these changes had brought to a head rivalries that had festered for years. Although the conflicts at the departmental level between the Army, Navy, and the yearning-to-be-free Army Air Force were more obvious, the hidden struggles within the intelligence community were perhaps more bitterly fought. The monitoring project, involving as it did the military and civilian intelligence services as well as the various technical branches of the military, surfaced in the middle of both departmental and intelligence service battles.

The process of creating the monitoring organization, therefore, involved controversies and conflicts between different agencies over who was to have responsibility for further developing and operating the long-range detection system and the resolution of these conflicts through a series of compromises that permitted accommodation of disparate interests. In addition to jurisdictional disputes, questions about funding the system's operation had to be decided. As will be seen, factors external to the disputes, such as the restructuring of the military services mandated by the National Security Act, influenced both the outcome of the decision-making process and its timing.

ASSIGNING ORGANIZATIONAL RESPONSIBILITY

Under the auspices of the Central Intelligence Group, individuals designated by the agencies contacted by Vandenberg were organized as the Long Range Detection Committee. Chaired by Colonel Lyle Seeman, the twelve-man Committee sought to make recommendations on how to expeditiously develop and operate a monitoring system. Only organizations that had already worked on the monitoring problem were represented on the Committee, presumably because their delegates could draw on this experience and make appropriate recommendations with a minimum of deliberation.[1]

The use of an interagency committee to study the problem, however, was fraught with danger. With the exception of the Joint Research and Development Board, all of the organizations represented on the Committee—the Army, the Navy, the Atomic Energy Commission, and the Central Intelligence Group—had (or in the case of the Atomic Energy Commission, expected to have) units within their structure that specialized in collecting and evaluating scientific intelligence. And, at that time, the intelligence organizations of the departments of War, Navy, State, and Commerce and the Federal Bureau of Investigation were engaged in disputes over "turf" that had gridlocked some previous attempts at interagency cooperation.

As the "new boy on the block," the Central Intelligence Group was at the center of most of these controversies. Throughout his tenure as director of Central Intelligence, Vandenberg had waged an uphill battle to define the role of the Central Intelligence Group vis-à-vis the older, well-established intelligence services. This contest became especially visible in the conduct of interagency projects requiring frequent interactions between the Central Intelligence Group and the military intelligence services.

For example, one such interagency effort was known as the Defense Project. Originally proposed in March 1946 by Colonel J. R. Lovell of Army Intelligence, the project was an attempt to use the resources of all the intelligence services to produce "the highest quality of intelligence on the U.S.S.R. in the shortest possible time."[2] This endeavor was controversial from the outset, but, from an organizational perspective, it had the merit of revealing areas that each service regarded as its own, areas in which they expected to maintain control of information-gathering, evaluation, and dissemination.

The ill-defined boundaries of these areas sometimes produced interagency friction. To cite just one instance, the War Department's air intelligence staff was interested in the Defense Project only insofar as it provided a means to assess Soviet air capability, a process that included evaluating the ability of the Soviet air arm to deliver atomic bombs and, by extension, the Soviet capability to produce such bombs. Evaluating the latter capability, of course, also fell within the purview of the Central Intelligence Group's Nuclear Intelligence Group. Overlap areas such as this became friction points, and the intelligence effort in these areas was inhibited by uncertainties about agency responsibility.

By the spring of 1947, the outlines of another turf battle had begun to emerge in the field of scientific intelligence which, by then, included atomic intelligence. At issue was the distinction between intelligence on the new or improved weapons that were available to a foreign power either as development prototypes or production versions, and intelligence on the basic research being carried out that could lead to new or improved weapons. The military intelligence services tended to regard the first area, which would appear to include monitoring foreign atomic bomb tests, as their "territory," leaving only the area of basic research to the Central Intelligence Group.[3]

Moreover, reports that the Atomic Energy Commission was contemplating the formation of its own intelligence unit had reached the Central Intelligence Group, a move that would add yet another contender to the battle over turf. Vandenberg had expressed his concern over this possibility even before the Atomic Energy Commission superseded Manhattan Engineer District. In September 1946, at a meeting of the National Intelligence Authority, he had indicated that "it was rumored that the [Atomic Energy] Commission would conduct foreign [intelligence] operations that would involve a grave danger of conflict with similar CIG [Central Intelligence Group] operations."[4]

Although Vandenberg was well aware of the issue of organizational jurisdiction and the difficulties it sometimes introduced into interagency collaboration, he opted to use an interagency committee as the mechanism for deciding how to go about creating a monitoring system. He may have felt that he had little choice since the existing expertise on monitoring did not reside in a single organization, but, rather, it was distributed among several.

By May 1947, the various agencies had responded to Vandenberg's memorandum of March 14 by designating representatives to serve on the Long Range Detection Committee, but, by then, he had left the Central Intelligence Group. His successor was Rear Admiral Roscoe H. Hillenkoetter, who became director of Central Intelligence on May 1.[5] As an old hand in intelligence, Hillenkoetter was familiar with the service rivalry that characterized the intelligence community and the consequent difficulties that, at times, had plagued interagency collaboration. But there is no evidence to suggest that he had reservations about the approach to the monitoring problem that had been initiated by Vandenberg. Instead, perhaps because of the note of urgency sounded in Vandenberg's memorandum—a note that was later echoed by Atomic Energy Commissioner Strauss—Hillenkoetter elected to push ahead quickly and a meeting of the Long Range Detection Committee was scheduled for May 21.

Some of the committee members who assembled on May 21, such as Lieutenant Colonel David Parker, had participated in carrying out the monitoring experiments during Crossroads. Others, such as the Central Intelligence Group's Colonel Lyle Seeman and Colonel Benjamin G. Holzman, who headed the Army Air Force's geophysical research program, had been privy to the reports on Crossroads and had acquired a broad knowledge of the technical aspects of monitoring. One of the Atomic Energy Commission's delegates, Colonel Kiern, could not attend this first meeting. Not surprisingly, in view of Commissioner Strauss's keen interest in monitoring, Strauss's aide, William Golden, was assigned to sit in for Kiern. At the time, Golden knew little about the technical difficulties of monitoring, but his presence signalled Strauss's continuing concern about the lack of a monitoring system.

The Committee quickly reduced the monitoring problem to two components: first, developing suitable methods for reliably detecting and unambiguously identifying foreign atomic bomb explosions; and, second, fixing responsibility and establishing the organizational means for developing and operating the monitoring system. Three approaches were cited as having the potential to satisfy the first component of the problem: sonic (measuring the atmospheric pressure-waves produced by the blast), seismic (detecting the pressure-waves transmitted through the earth), and radiological (air-sampling followed by radioassay and radiochemical analyses of airborne particulates). Instruments existed for implementing each of these approaches, but improved performance was needed to meet the requirements of an operational monitoring system. It was estimated that two years would be required to develop instruments and techniques to the point where a reliable monitoring system could be based on them.[6]

Regarding the second component of the problem, assigning responsibility for monitoring, the Committee agreed that dividing responsibility among several agencies or using the mechanism of an operating committee would be less effective than placing the responsibility on a single agency. Thought was given to the potential organizational recipients of the information produced by the monitoring system. These "clients" included the Joint Chiefs of Staff, the Atomic Energy Commission, the Central Intelligence Group, the Joint Research and Development Board, the Armed Forces Special Weapons Project, Army, Navy, and Air intelligence, and the State Department. The Committee recognized that it would be necessary for each of these client organizations to assist the monitoring agency with its mission. For example, the Atomic Energy Commission would have to provide expertise in radiochemical analysis, the State Department would be called on to aid in establishing monitoring stations in foreign countries, and so on. Moreover, the assistance of nonclient agencies, such as the Weather Bureau and the Bureau of Standards, would be needed as well.[7]

The only organizations that were, in theory, capable of mobilizing the resources of these various agencies in the service of the monitoring mission were the Joint Chiefs of Staff and the Central Intelligence Group. But the Joint Chiefs of Staff was not an operating body, and it could be expected to command a monitoring effort through a balanced committee of Army and Navy officers, an approach that would tend to make responsibility overly diffuse. Thus, by elimination, the Central Intelligence Group appeared to be the appropriate organization to assume responsibility for monitoring. This, of course, was not a new idea—it had been expressed by some War Department and Atomic Energy Commission staff members in the fall of 1946.[8]

The first meeting of the Committee ended without producing any specific recommendations, but, before the next meeting on June 6, the Committee members apparently had second thoughts about the question of assigning responsibility for monitoring to the Central Intelligence Group. Neither Vandenberg nor his successor, Hillenkoetter, had expressed a desire for the Central Intelligence Group to become the monitoring agency. Indeed, at that time, the entire Washington-based staff of this organization, including professional and clerical personnel, numbered fewer than 1,200 persons, all of whom were hard-pressed to carry out the tasks already assigned.[9] And although in theory the Central Intelligence Group had a broad mandate to act as the coordinating agency in matters of intelligence, in practice its ability to influence the behavior of other agencies or to marshall their support in collaborative enterprises had proved to be quite limited.

Lurking in the background was the fact that the White House was still tinkering with the text of the National Security Act of 1947. In addition to creating an independent Air Force and unifying all three military arms under a single Defense Department, the Act would transform the Central Intelligence Group into the Central Intelligence Agency and formally list its functions. Its supporters hoped this legislation would clarify the status of the Central Intelligence Group relative to the other organizations that comprised the intelligence community. Attempts to define jurisdictional areas of the Central Intelligence Group vis-à-vis the intelligence activities of the military departments (War and Navy) had proved to be especially contentious. In the context of this organizational turf battle, it seems unlikely that the military departments wanted to see the Central Intelligence Group aggrandized by the assumption of control over monitoring.

Moreover, the War Department's Army Air Force already possessed what was considered to be one of the most important, and certainly the most costly, components of the projected monitoring system—that is, the airplanes and experienced air crews capable of carrying out the air-sampling program. Cost was not overtly discussed by the Committee as an inhibiting factor, but, according to the recollections of one of the Committee members, William Golden, cost was an important influence that tempered the desire or, indeed, the willingness of the various agencies represented on the Committee to assume responsibility for monitoring. Golden later recalled that "Money was a very important question, as always. The military had things to do with all their funds. And they were the ones that were going to have to fly the airplanes. That was an important issue. Beyond the [bureaucratic] inertia and beyond the interservice rivalry between the Navy, Army, and Air Force was the question of who was going to pay for it."[10]

The WB-29s of the Army Air Force's Air Weather Service were already patrolling the northern Pacific and Arctic regions using routes that were considered ideal for satisfying the air-sampling needs of the projected monitoring system. The Army Air Force could thus combine the air-sampling missions with its regular weather reconnaissance flights which were, of course, already within its budget. This meant that the marginal cost to the Army Air Force of assuming responsibility for monitoring would be considerably less than that which would be imposed on other agencies to which this responsibility might be assigned. Thoughts such as these may have mitigated the Navy's desire to become the monitoring agency, but, as will be seen, the Navy was not willing to concede that monitoring fell outside its purview.

The notion that the Army Air Force should be the monitoring agency was supported by General Vandenberg who, upon leaving the Central Intelligence Group, had returned to the War Department as Army Air Force deputy commander and chief of air staff—a career move that positioned him to become vice-chief of staff and, later, chief of staff of the Air Force after it became an independent military arm. Some objections had been raised within the Army Air Force to Vandenberg's position on monitoring on the grounds that aircraft used for air-sampling would be diverted from the strategic bombing mission. These objections became muted when it was pointed out that a military requirement for weather reconnaissance existed independently of the monitoring assignment. Thus, monitoring would be done by planes already discounted from the strategic force.[11]

Indeed, the Army Air Force's Air Weather Service was anxious to become involved in monitoring. After the war, the Air Weather Service had lost its status as an independent command and was submerged within the much larger Air Transport Command. In March 1947, partly as a dramatic gesture to reaffirm the identity of the Air Weather Service as an essential and forward-thinking organization, and partly to fill a real gap in the acquisition of weather data, its chief, Brigadier General Don Yates, had begun to make weather reconnaissance flights over the North Pole. As a support service, the Air Weather Service had often found itself at a disadvantage relative to the combat arms in competing for scarce Army Air Force resources. Thus, Yates welcomed the publicity afforded by the polar flights as a means of highlighting the importance of the Air Weather Service's mission.

In March 1947, Colonel Holzman had foreseen the value of the Air Weather Service in the projected monitoring system and had explored the problem of air-sampling with Dr. Ralph P. Johnson, deputy director of the Atomic Energy Commission's Division of Research. Holzman had apparently alerted Yates to the possibility that the Air Weather Service might be called upon to function as part of the monitoring system, and a report on the subject of monitoring by using an Air Weather Service air-sampling program was prepared. Yates evidently hoped that involvement in the monitoring program would raise the status of the Air Weather Service within the pecking order of the Army Air Force (an expectation that was later realized).[12]

The degree to which such jurisdictional and cost considerations played an undocumented, behind-the-scenes role in influencing Committee members during the two weeks that elapsed between their first and second meetings is unclear. Nevertheless, when the Committee next met, the notion of assigning responsibility for monitoring to the Central Intelligence Group appears to have been abandoned.

At a meeting on June 6, the Committee concluded that instruments and techniques for monitoring were available, according to a summary report of their findings, "actually or potentially," and that the problem of monitoring could be solved by the coordinated use of sonic, seismic, and radiological methods. (This assessment was later to be decried as overoptimistic by the scientists of the Research and Development Board's Committee on Atomic Energy.) The location of the blast would be determined by sonic and seismic means. Aircraft could then be positioned so that airborne products of the explosion could be obtained, according to the summary, "by an aerial sampling technique as near to the scene of the explosion as practical." The final step would involve "determining the nature of the products of the explosion by radioassays and radiochemical analyses."[13]

With regard to the issue of creating the organizational means to develop and operate the monitoring system, the Committee recommended that the "Army Air Force should be assigned the overall titular responsibility for long range detection provided that, in the analysis and evaluation, the Armed Forces Special Weapons Project, Atomic Energy Commission, and other appropriate agencies were included."[14] The Committee forwarded their findings and recommendations to Hillenkoetter. On June 30, Hillenkoetter, acting in his capacity as director of Central Intelligence, sent a memorandum to the chairman of the Atomic Energy Commission and to the War and Navy secretaries, in which he outlined the conclusions and recommendations of the Committee. Hillenkoetter urged that "The conclusions of the Committee be accepted, and implemented forthwith by appropriate directive to the Army Air Force for overall responsibility, supported by request to other interested agencies for necessary cooperation and assistance to carry out the program."[15] Hillenkoetter also presented the

Committee's timetable for deploying a monitoring system. Paragraph seven of his memorandum stated: "A rough estimate indicates that approximately two years will be required to locate, install and operate the complete network of stations and facilities, capable of feeding data into Control Central."[16]

On July 10, Atomic Energy Commission Chairman Lilienthal sent a sharp rejoinder to Hillenkoetter expressing dismay over the proposed timetable. He wrote,

We cannot regard a two year period as acceptable or realistic and believe it necessary that means be found, as should be done in time of national emergency, to devise a practical solution as a matter of utmost urgency. It is understood that the technical issues have already been essentially resolved and that the problem is now one of organization and coordination.[17]

The last sentence displayed the same misconception about the state of the art in monitoring that characterized LeMay's memorandum of August 12, 1946, and Strauss's memorandum of April 11, 1947. Indeed, there is evidence that Strauss may have influenced the wording of this part of the memorandum, although it was sent out under Lilienthal's signature.[18] Hillenkoetter was apparently able to assuage concern about the timetable, for it did not prove to be an issue when the Commission met formally to consider Hillenkoetter's memorandum of June 30.

Although Lilienthal's memorandum had expressed anxiety over the proposed timetable, the notion of designating the Army Air Force as the monitoring agency was consonant with views that he and others in the Atomic Energy Commission already held. At the eighth joint meeting of the Atomic Energy Commission–Military Liaison Committee on June 18, Lilienthal had expressed the belief that the primary responsibility for establishing a monitoring system rested with the military. Carroll Wilson, Atomic Energy Commission general manager, had concurred, commenting that the Commission expected one of the armed forces to assume responsibility for monitoring.[19] The Atomic Energy Commission was then in the process of forming its own intelligence unit, but the functions of this new unit did not include monitoring, which was regarded as a military problem. The commissioners, as a body, had expressed no preference as to which military service should become the monitoring agency, although in April Strauss had suggested to Forrestal that the Navy should assume this role.

On July 18, Dr. Johnson, a member of the Commission's Research Division, sent Strauss a copy of a report entitled "Long Range Reconnaissance, Weather." Johnson had worked with Colonel Holzman on this report, which was taken from Colonel Holzman's files. It described the Army Air Force's Air Weather Service as being uniquely capable, in terms of men and equipment, to carry out the air-sampling program that was an essential part of the projected monitoring system.[20] This report was apparently convincing because Strauss later recalled that the Army Air Force was selected as the monitoring agency "because of the assumed importance of aviation to nuclear monitoring techniques."[21]

Strauss may have been sent this report because he was suspected of harboring a bias in favor of the Navy. This was not an illogical assumption in view of his long service as a Naval Reserve officer. But apparently any Navy bias he may have had was submerged by his concern over the urgency of the monitoring problem. According to his assistant, William Golden, Strauss's attitude on monitoring was to take "Whatever the path of least resistance . . . to get the monitoring system going at the earliest possible date."[22]

Perhaps because Strauss had succeeded in imparting his sense of urgency about monitoring to his fellow commissioners, the Atomic Energy Commission was the first of the organizations contacted by Hillenkoetter to act on his memorandum. On July 23, at the eighty-third meeting of the Atomic Energy Commission, the Commissioners formally approved the conclusions and recommendations presented by Hillenkoetter. (Ironically, Strauss was absent from this meeting.)[23] The approval of the military departments, however, was to take longer because special interests had to be considered.

On July 24, General Groves, head of the Armed Forces Special Weapons Project, wrote to General Dwight D. Eisenhower, Army chief of staff, informing him that while he concurred with most of the conclusions put forth in Hillenkoetter's June 30 memorandum, he believed that the Army Air Force should be responsible only for the collection of monitoring data and that, initially at least, responsibility for analysis and evaluation should be given to the Armed Forces Special Weapons Project acting in concert with the Central Intelligence Group.[24]

Groves, who was a member of the Military Liaison Committee, also succeeded in gaining the support of this Committee for his position. Then, in a characteristic manner, he took his case directly to Secretary of War Kenneth C. Royall.

Royall did not dismiss lightly Groves's concerns about the Army Air Force's ability to carry out the analysis and evaluation aspects of monitoring. On August 13, he wrote to Eisenhower, noting that the Armed Forces Special Weapons Project and Military Liaison Committee objected to making the Army Air Force the monitoring agency unless the analysis function was placed within the purview of the Armed Forces Special Weapons Project. He suggested that the problem might be sidestepped by suitably defining the Army Air Force's responsibilities in developing and operating the long-range detection system. In essence, the responsibility of the Army Air Force could be tempered by a requirement to coordinate with other agencies in analyzing the data produced by the monitoring system.[25] This, of course, corresponded closely to the original wording of the recommendation made by the Long Range Detection Committee.

Two weeks later, Groves impressively documented his case in a second memorandum to Royall. He made it clear that the approach to the development of a monitoring system that was outlined in the report of the Long Range Detection Committee, and which the Army Air Force would be expected to implement, had been originally conceived under his aegis and discussed in reports prepared by his staff at the Manhattan Engineer District prior to the takeover by the Atomic Energy Commission. Groves then went on to describe in detail the organizational steps that would be needed to carry out the plan.[26]

Apart from any desire he may have had to aggrandize the Armed Forces Special Weapons Project which he headed, Groves's view that the Army Air Force was not ready to assume responsibility for all aspects of the monitoring system was not without foundation. On the positive side, the Army Air Force had acquired, through Project Mogul, some expertise in sonic methods of detecting atomic explosions at long range, and its Air Weather Service could carry out the air-sampling missions that were an essential part of the monitoring system. On the negative side, it lacked the radiological expertise to evaluate the collected samples.

The knowledge of monitoring gained thus far suggested that the most crucial component of the system would be radiological analysis and evaluation. Atomic Energy Commission scientists had been consulted by Colonels Seeman and Holzman about radiological methods of long-range detection. The scientists' input had raised the concept of "monitoring atmospheric radioactivity" to a highly sophisticated level, well beyond the methodology that, earlier, had been connoted by this phrase. Such earlier methodology had involved two techniques: (1) exposing a Geiger counter to the air and noting its response, and (2) measuring the radiation intensity produced by the radioactivity of an atmospheric dust sample collected on a filter. Using such simple methods, a sharp increase in atmospheric radioactivity would be suggestive, but its cause would remain ambiguous since events other than atomic bomb explosions can produce such increases.

It was clear that if this ambiguity could be eliminated, the importance of radiological monitoring would be dramatically increased. The method proposed for definitively tracing an increase in atmospheric radioactivity to an atomic bomb explosion was a logical extension of the radiochemical method that had been developed and used by Atomic Energy Commission scientists in determining the "yield" of the Trinity and Crossroads bombs. (Yield is a function of the bomb's efficiency, and it is usually expressed in terms of equivalent tons of TNT.)

As applied to the monitoring problem, the proposed approach based on "multiple fission product analysis" would add a second level of processing to any airborne dust sample that displayed an anomalously high level of radioactivity. This processing involved chemically isolating the radioisotopes present in the dust and then subjecting them to radioassay and an evaluation technique that, in theory, would determine unequivocally whether or not they were produced by an atomic explosion. Although the practicability of this approach as a means of long-range detection had yet to be tested, it gave an entirely new meaning to the phrase "monitoring atmospheric radioactivity."

This proposed sample processing technique, however, required the capability to chemically analyze the minute quantities of airborne dust picked up at long distances from the explosion and the mathematical capability to construct a quantitative chain of inferences, based on radiochemical assays, that would allow the true nature of the explosion to be determined. The Army Air Force lacked these capabilities. The Armed Forces Special Weapons Project, on the other hand, had established working relationships with groups within the Atomic Energy Commission that could carry out the radiochemistry, and with scientists within the Central Intelligence Group who could help in evaluating the radiological data and the data produced by other components of the projected monitoring system as well.

Although Groves's position was not without merit, his influence within the War Department had declined markedly in the postwar period. He had battled unsuccessfully to retain military control of atomic weapons and had made many enemies in the process. By 1947, according to one of his biographers, "Groves was considered by many to be a pest and an obstruction."[27] It was easy for his detractors to see Groves's suggestions on monitoring as a belated attempt to expand the role of the organization he headed. In addition, the notion that the Armed Forces Special Weapons Project should evaluate samples collected by the Army Air Force could be legitimately criticized on the grounds that it would weaken the monitoring effort by diffusing responsibility among several organizations.

Considerations such as these may have prompted the second thoughts Royall had about incorporating elements of Groves's ideas into his response to Hillenkoetter's June 30 memorandum. He had first drafted a reply in which the Army Air Force would be required to coordinate with the Armed Forces Special Weapons Project and other agencies in analyzing and evaluating the data produced by the monitoring system. But in the final version, which was sent to Hillenkoetter on September 5, this requirement had been dropped.[28] Groves had lost another battle.

Navy Secretary Forrestal was more dilatory than Royall in responding to Hillenkoetter, perhaps because his reply had to be framed in the context of special interests that had developed within his department, interests that would be affected by Army Air Force control of monitoring. After Crossroads, the Navy had established the Naval Radiological Defense Laboratory, which had recruited scientists skilled in atomic physics and radiochemistry. In addition, several naval stations had been equipped with ground-based monitors capable of measuring any anomalous increase in radioactivity in the atmosphere. Scientists at the Naval Radiological Defense Laboratory, the Naval Ordnance Laboratory, and the Naval Research Laboratory had displayed a keen interest in monitoring. Some were carrying out further research in this area, building on the knowledge gained by the Navy's Remote Measurements Section during Crossroads.[29] Nor was the Navy's air intelligence arm indifferent to any extension of the intelligence jurisdiction of the Army Air Force which, under the National Security Act, was about to become the independent Air Force.[30]

As mid-September approached, the lack of response from the Navy alarmed Strauss. According to his memoirs, he visited his old friend Forrestal to plead the urgent need for action on the monitoring problem.[31] Another, perhaps more keenly felt, spur to action was the approaching September 18 deadline for implementing the traumatic organizational changes mandated by the National Security Act. The military departments were attempting to complete as much old business as possible before the waters were muddied by the alteration in command structure that the Act would require.

On September 15, as one of his last decisions as Navy secretary, Forrestal told Hillenkoetter and Royall that he concurred in the designation of the Army Air Force as the monitoring agency. (Forrestal became secretary of the newly established Defense Department on September 18.) Rear Admiral John E. Gingrich, chief of the Atomic Energy Commission's recently formed intelligence unit, informed Strauss of Forrestal's concurrence, adding that it contained an important proviso—namely, that "in the analysis and evaluation phases [of monitoring] the facilities and resources of existing government agencies are utilized to the maximum possible extent."[32] The actions later taken by the Navy suggest that Forrestal intended this proviso to mean that the assignment of overall responsibility for monitoring to the Army Air Force did not confer "exclusivity" and that the Navy would be free to continue its own monitoring efforts if the results were made available to the Army Air Force.

Immediately upon hearing from Forrestal, Royall sent Hillenkoetter a memorandum informing him that the Army Air Force would be ordered to assume responsibility for monitoring.[33] The next day, September 16, Army Chief of Staff Eisenhower issued instructions to the commanding general, Army Air Force, to assume "overall responsibility for detecting atomic explosions anywhere in the world. This responsibility is to include the collection, analysis and evaluation of the required scientific data and the appropriate dissemination of the resulting intelligence."[34] In accord with Forrestal's wishes, Eisenhower added the proviso that, in carrying out this responsibility, the Army Air Force "will utilize to the maximum existing personnel and facilities both within and without the War Department, [and] will establish appropriate arrangements with other interested agencies." The Navy had made its point.

The search for an organizational "home" for the monitoring system, beginning with Vandenberg's memorandum of March 14, had taken six months (almost to the day) to complete. The time required to assign responsibility for monitoring must be evaluated against the background of organizational uncertainty and strife. This time would have been shortened had the president opted to create a monitoring organization by executive fiat, either as a division within an existing branch of government or as an independent agency. This kind of "top-down" initiative characterized, for example, the establishment of the Central Intelligence Group. On the issue of monitoring, however, the president took no such action.

On the other hand, there was no resistance at the highest levels of government to the notion that a monitoring system should be established. On July 14, 1947, the Joint Chiefs of Staff had issued a document in which it was stated that "the Soviets will continue atomic research and will produce atomic weapons as soon and as rapidly as possible. In consequence the United States should exhaust every practicable means of gaining factual information concerning the development of atomic and other weapons of mass destruction within the USSR."[35] A monitoring system obviously fell in the category of "every practicable means," but the real question was: How soon would it be needed? The Joint Chiefs of Staff document noted the conflicting estimates of when the Soviet bomb would appear which had been produced by the intelligence services of the military departments. (The Navy and War Departments favored a 1952 date, while the Army Air Force claimed that the Soviets could have a bomb as early as 1949.)[36]

Such estimates, however, could be criticized on the grounds that they appeared to be self-serving. More telling, they were based primarily on suppositions rather than on hard intelligence data about Soviet atomic developments. The president and administration insiders continued to give more credence to the longer estimates, originally presented by General Groves, that were based on the Murray Hill Area concept. In the context of this belief, monitoring was not an urgent problem.

Without direction from the top, the assignment of the monitoring responsibility became a "bottom-up" process. The evidence suggests that administration leaders did nothing to hinder this process but, rather, allowed it to occur subject to the usual bureaucratic delays that, at this time, were exacerbated by the reorganization of the command structure.

Some military staffers and officials, including Generals LeMay, Samford, and Vandenberg, Admiral Hillenkoetter, and Commissioner Strauss, either rejected the Murray Hill Area concept or, in 1946–1947, were not privy to it (and the evidence suggests the latter). They shared the belief that the Soviet bomb was imminent and that monitoring was thus an urgent need. LeMay and Samford had expressed this view as early as the summer of 1946, and

Vandenberg had stated it in the spring of 1947, followed by Strauss and Hillenkoetter. Their emphasis on the need for action on monitoring undoubtedly played a positive role in the bureaucratic process by which responsibility for monitoring was finally assigned. On the other hand, the speed with which the final events in this process were enacted on September 15 and 16 suggests that the September 18 deadline for the organizational restructuring mandated by the National Security Act may also have been an important factor in ending bureaucratic waffling on this matter.

AFMSW-I IS FORMED

When the Army Air Force was transformed into a separate military department on September 18, 1947, General Vandenberg became vice chief of staff of the Air Force. Despite the heavy workload imposed by the Army Air Force–Air Force changeover, he found time to consider the organizational issues involved in carrying out the responsibility for monitoring. Initially, it had been thought that the monitoring activity should be placed under General LeMay, who headed Air Force research and development, but another option had appeared.[37]

The Air Force was in the process of clarifying its relationships with other organizations within the government, and the Air Force's deputy chief of staff for materiel had established a Special Weapons Group that interfaced with the Armed Forces Special Weapons Project and the Atomic Energy Commission on matters related to atomic bombs. The research on sonic long-range detection methods (Project Mogul) was being conducted under the aegis of the Materiel Command, and subgroups within the Command were among the Air Force units scheduled to participate in the next atomic test series, code named Sandstone. Moreover, the initial plans for Sandstone already included experiments on the various methods of long-range detection. Thus, the possibility of assigning the monitoring activity to the deputy chief of staff for materiel appeared to be worth considering.[38]

Vandenberg asked LeMay to study the problem and recommend the most appropriate placement within the Air Force for the monitoring task. In any event, Vandenberg could not take action on this matter until the Joint Chiefs of Staff had officially allocated responsibility for monitoring to the Air Force, a step required under the new command structure. In this case, however, it appears to have been purely pro forma because of Eisenhower's directive of September 16 assigning this responsibility to the Air Force's predecessor, the Army Air Force.[39]

Curiously, the Atomic Energy Commission had not been notified of Eisenhower's order designating the Army Air Force as the monitoring agency, although, through Gingrich, Strauss had been informed that such an order was anticipated. On September 29, Lilienthal sent a memorandum to Hillenkoetter and the military secretaries in which he displayed this informational lack by reaffirming that, with regard to monitoring, "The Commission will offer full cooperation to such agency as may be designated by proper authority." Probably at Strauss's prompting, Lilienthal added, "we should like to express again the urgency with which . . . certain phases of the program should be initiated."[40]

As the weeks passed without any further response from the military, Strauss became anxious over the apparent lack of progress on the monitoring problem. His anxiety was not assuaged when the issue surfaced at the seventeenth joint meeting of the Military Liaison Committee–Atomic Energy Commission on October 22. The military officers on the Military Liaison Committee (which included General Groves and Admiral Parsons) reported that they were not aware of the exact status of the plans for monitoring within the armed services, but that they expected it would require about eighteen months to set up an operational monitoring system.[41]

A few weeks later, at the eighteenth Military Liaison Committee–Atomic Energy Commission meeting held on November 5, the commissioners were finally given some information. General Groves announced that on October 16 the Joint Chiefs of Staff had assigned responsibility for long-range detection to the chief of staff of the Air Force. He went on to say that the Air Force was preparing an organizational plan for carrying out this responsibility.

Strauss, unhappy with what he perceived as the dilatory handling of this issue, asked Groves to explain the military's apparent lack of urgency. Alluding to the National Security Act and the new command structure that had been created, Groves replied that the Air Force was faced with many pressing organizational problems in addition to those associated with the monitoring program.

The commissioners had been involved in the preliminary planning for Sandstone, which was scheduled for the spring of 1948, and they were aware that the military expected to use the tests to study long-range detection techniques. Thus, Lilienthal asked Groves if the Air Force's monitoring organization would be ready by the time of the Sandstone tests to carry out the necessary experiments. Groves assured him that the Air Force realized that it was essential that these experiments be performed. To allay further anxiety on this point, Groves offered to keep the commissioners posted on the Air Force's monitoring activities.[42]

While waiting for official word from the Joint Chiefs of Staff, the Air Force had continued with the placement study, commissioned by Vandenberg, on how best to position the monitoring effort within the Air Force. On October 16, the Joint Chiefs of Staff confirmed the Air Force as the monitoring agency, and by early November, General LeMay's staff officer, Colonel Holzman, had completed the placement study and his recommendations were on LeMay's desk. After further study, Vandenberg subsequently accepted them. On November 14 he issued an order to the deputy chief of staff for materiel designating its Special Weapons Group as "monitoring office . . . charged with the overall responsibility of initiating and carrying out a program for the Long Range Detection of atomic explosives [sic]."[43]

Major General William E. Kepner, who headed the Special Weapons Group, wasted no time in briefing all the interested agencies on his organizational plans. On November 17, he held a conference attended by representatives of various Army, Navy, and Air Force organizations, the Armed Forces Special Weapons Project, the Central Intelligence Agency, the Atomic Energy Commission, and the Research and Development Board. (The National Security Act had transformed the Joint Research and Development Board into the Research and Development Board, although its membership and most of its functions were unchanged.) Kepner told the conferees that he proposed to form a section within his headquarters that would be responsible for the long-range detection program. In carrying out this task, the section would be advised by a committee comprised of delegates from each of the agencies present.

Kepner also pointed out that the important question of funding remained to be clarified. He indicated his belief that the Air Force would have to allocate funds for the proposed monitoring organization, but the source of funds for other participating agencies remained to be determined. It was also clear that the Air Force would soon have to begin spending money on equipment for the long-range detection experiments to be carried out during Sandstone, although the source of funds for instrumentation had not been determined. The immediate problem, which had yet to be resolved, was the question of who would pay for the Air Force participation in Sandstone.[44]

Perhaps the most important result of this conference was that some of the agencies represented agreed to begin work immediately on preparing the long-range experiments to be conducted at Sandstone. The Air Weather Service offered to provide filter-equipped aircraft to collect airborne dust samples, and the Army Signal Corps and the Naval Research Laboratory offered to develop ground-based air-sampling units and apparatus for measuring radioactivity. The Signal Corps and the Naval Electronics Laboratory agreed to provide sonic instruments, and the Air Materiel Command's Watson Laboratories had already made plans to conduct some experiments based on the Mogul concept. The Naval Ordnance Laboratory assumed the task of providing seismic instrumentation. The fact that these organizations began work on radiological, sonic, and seismic monitoring techniques in November 1947, before the organizational and funding problems had been resolved, was an important factor in meeting the spring deadline for testing these methods at Sandstone.[45]

The organizational arrangements proposed by Kepner were not universally admired. William Golden objected on the grounds that there did not seem to be a "single operating head who will have the special qualifications and interest, [and] the long tenure . . . requisite for dependable operation in this post." In complaining about this to Strauss, he noted that "I have had a brief, in-the-corridor conversation with [Atomic Energy Commissioner] Bob Bacher and John Manley [executive secretary of the Atomic Energy Commission's General Advisory Committee] on the subject [of monitoring]. They also are troubled about its status, and the latter reports that this disquietude is shared by the GAC."[46]

Strauss agreed with Golden's remarks, and he made this plain at the twentieth Military Liaison Committee–Atomic Energy Commission meeting on November 26. He pointed out that in Kepner's proposed schema the head of the monitoring organization would have an overly short tenure owing to the usual military practice of rotating officers in and out of posts, and the establishment of an Advisory Committee would tend to distribute the blame for inaction. Strauss also expressed chagrin that the Atomic Energy Commission had not received timely notification of the action on monitoring taken by the military secretaries. He complained that official word of the designation of the Army Air Force as the responsible agency did not reach the Atomic Energy Commission until "over two months later." Finally, he declared:

I am greatly dissatisfied personally with the lack of a sense of urgency in this matter. I would like to call attention to the following timetable. It has been 2 1/2 years since the first bomb was exploded. I cannot find any record that anything was done on the subject of monitoring between Hiroshima and the Spring of 1947 when we began to agitate for action. . . . When I think back to the radar monitoring at Pearl Harbor, I have the feeling that "This is where we came in."[47]

Strauss's allusion to the failure of the intelligence service to heed the warnings provided by radar of the imminent Japanese attack at Pearl Harbor was an expression of his concern about the possibility that an "atomic Pearl Harbor" might occur if a monitoring system was not quickly deployed.

Groves was at this meeting in his capacity as a member of the Military Liaison Committee. He was, of course, well aware of the considerable technical effort on monitoring that had been expended by the Manhattan Engineer District, the Navy, and the Army Air Force during the two and a half years that had elapsed since Hiroshima. But, apparently adhering to the principles of compartmentalization and need-to-know, he failed to enlighten Strauss on this point. Instead, he chose to address the question of organizational responsibility by informing the group that Major General Albert H. Hegenburger, a technically oriented officer who had contributed to prewar advances in avionics, would be placed in charge of monitoring as Kepner's deputy. Dr. Ellis Johnson, a Carnegie Institution scientist, was being sought as research director.[48]

It is unclear whether Strauss's objections to the proposed organizational structure were passed on to Kepner, but, if they were, Kepner ignored them, for the organization he established a few weeks later was essentially consonant with the plan he announced on November 17. Within the time constraint imposed by Sandstone, the Air Force could not possibly carry out the necessary monitoring experiments alone. Kepner's plan, involving a relatively loose command structure guided by an Advisory Committee composed of delegates from participating agencies, actually made sense (and, in the event, worked quite well). This outcome was not unexpected by his superiors, for Kepner was regarded as a competent administrator and his remarkably varied career as a professional soldier had demonstrated his ability to work effectively within the constraints of the military. Indeed, as one of the very few officers to have served in the Navy, Army, and Air Force, he was well suited to oversee the monitoring effort, which involved organizations from all three military services.[49]

In the last weeks of November, Kepner was busy putting to good use his contacts with individuals and organizations involved in the long-range detection experiments conducted during Operation Crossroads. He was also organizing a conference to discuss the arrangements for the long-range detection experiments that would be carried out at Sandstone. The conference, held on December 5 at Kepner's Washington headquarters, brought together representatives from over twenty-five government agencies and contract research groups. Although the meeting furthered the plans being made to coordinate the work of the various agencies, some of the attendees were left with a keen sense of the difficulties they faced. For example, Professor Stafford L. Warren, a University of California scientist and consultant to the Atomic Energy Commission, wrote: "In summary of this meeting it may be said that the practical attainment of good coverage by proper personnel and instruments of the forthcoming tests in the Pacific will be extremely difficult because of the current lack of both personnel and instruments and the short time for development of the proper instruments."[50]

Dr. Vannevar Bush, chairman of the Research and Development Board, was also sufficiently concerned about the deadline imposed by Sandstone to contact Defense Secretary Forrestal. "I question," he wrote, "whether all those now involved recognize its importance."[51]

Technical advances in monitoring could only be definitively evaluated and calibrated using a real atomic explosion. As mentioned previously, several factors, including the small size of the atomic stockpile, tended to make testing infrequent in the late 1940s. Tests were conducted only when overriding considerations made this necessary—in other words, only when the information to be gained was considered worth the cost.

For example, the purpose of Trinity was to learn whether the plutonium implosion bomb would work. Crossroads was needed to assess the future of a surface Navy, and the purpose of Sandstone was to try out dramatic improvements in bomb design. Monitoring experiments had been piggy-backed on the atomic tests as an additional dividend, never as the tests' primary purpose. Planning and preparation for the tests was complex, generally requiring six months or more. Changes in scheduling were extremely costly, and, in the case of Sandstone, it was decided not to delay the schedule to give monitoring personnel more time to prepare their experiments. Therefore, quite apart from any considerations about the urgent need for an operational monitoring system, the agencies involved in the monitoring project faced a real deadline.

Bush drafted a memorandum for Forrestal to send to Air Force Secretary Stuart Symington, stating that "it is essential that [monitoring] methods be pursued with full vigor on the detection of atomic explosions, and tried out at the time of the next test in the Pacific."[52] Forrestal accepted the text of Bush's memorandum and on December 9 sent it out over his signature. This memorandum was later cited in reports on Sandstone produced by the various military organizations involved in the monitoring program. Hence, it appears to have served the purpose intended by Bush, which was to impress on the Air Force and the other military departments that the deadlines had to be met and that their efforts to conduct monitoring experiments during Sandstone had the full support of the secretary of defense.

Following the conference on December 5, Major General Albert F. Hegenberger had been assigned to the Special Weapons Group to head the new monitoring organization and Kepner had succeeded in recruiting Dr. Ellis Johnson to spearhead the technical development of long-range detection techniques. On December 14, Kepner established a new section within his staff to carry out the monitoring mission commanded by Hegenberger with Johnson as technical director. This section was designated AFMSW-1 (which is decoded as follows: AF = *Air Force*, M = Deputy Chief of Staff for *Materiel*, SW-1 = *Special Weapons* Group, Section *One*). The monitoring organization had finally been born.

NOTES

1. The Committee consisted of four Army delegates (Brigadier General K. S. Hoag, Colonel W. M. Creasy, Colonel B. G. Holzman, and Lieutenant Colonel D. Parker); three Navy (Captain H. Rivero, Dr. A. K. Brewer and Dr. E. S. Gilfillan); two Atomic Energy Commission (Colonel D. J. Kiern and Dr. A. V. Peterson); two Central Intelligence Group (Colonel L. Seeman and Lieutenant Colonel F. A. Valente); and one representative from the Joint Research and Development Board (Dr. C. S. Piggott). See D. L. Northrup, "Detection of the First Soviet Nuclear Test," February 6, 1962, p. 3. Archives, DOE.

2. Colonel J. R. Lovell, quoted in Darling, *The Central Intelligence Agency*, p. 84.

3. W. M. Leary, *The Central Intelligence Agency: History and Documents*, 1984, p. 35.

4. "Minutes, 7th meeting, National Intelligence Authority," September 25, 1946. RG 218, JCS, Chairman's File, Box 20, Folder 132, NIA Papers, NA.

5. Like Vandenberg, Hillenkoetter had prior experience in intelligence. As a young officer he had been assigned to the Office of Naval Intelligence, and from 1933 to 1941, except for two years of sea duty, he had been attaché and diplomatic courier at the American Embassy in France. He had observed and reported on the fall of France in 1940 and had subsequently worked closely with the then embryonic French Underground. Assigned to sea duty in November 1941, he had headed intelligence on the staff of Pacific commander Admiral Nimitz before being given command of a ship in 1943. In 1946 he had returned to intelligence work, resuming the post of naval attaché at the Paris Embassy.

6. W. T. Golden to L. L. Strauss, "Long Range Detection Committee (Monitoring)," May 22, 1947, p. 1. Strauss Papers, AEC Series, Monitoring Soviet Tests, HHPL.

7. Golden to Strauss, "Long Range Detection Committee (Monitoring)," May 22, 1947, p. 2.

8. See Brigadier General Harbold to Major General McDonald, August 30, 1946, and Brigadier General Samford to Director, CIG, September 5, 1946.

9. Leary, *The Central Intelligence Agency: History and Documents*, p. 26.

10. W. T. Golden interview, January 4, 1990.

11. Prados, *The Soviet Estimate*, p. 19.

12. According to Bates and Fuller, *America's Weather Warriors*, p. 137: "Eventually, the aerial nuclear sampling mission carried far more clout with Headquarters USAF—which translated into additional aircraft and manpower—than did pure weather reconnaissance."

13. The quotations in this paragraph are from a summary of the findings of the Long Range Detection Committee in "Report of Operation Fitzwilliam," Vol. I, n.d. (classified on December 9, 1949), Introduction, pp. 1, 2. Period 1948, microfilm reel No. A15941, Frames 4–495, AFHRC.

14. "Report of Operation Fitzwilliam," Vol. I, n.d. (classified on December 9, 1949), Introduction, pp. 1, 2.

15. Ibid.

16. Hillenkoetter, quoted in D. Lilienthal to Director, CIG, July 10, 1947. Strauss Papers, AEC Series, Monitoring Soviet Tests, HHPL.

17. Ibid.

18. In 1955, Strauss recorded his recollections of his participation in the formation of the monitoring organization. In this memoir, he claims to have prepared a memorandum to the director, Central Intelligence. The portion of the text of this memorandum that he quotes is identical to a major part of the text of Lilienthal's memorandum to the director, Central Intelligence, written on July 10, 1947. This seems to suggest that he may have drafted the July 10, 1947, memorandum to be sent out under Lilienthal's signature. See L. L. Strauss, memorandum to the files, February 5, 1955, p. 4. Strauss Papers, AEC Series, Box 113, Folder: Tests and Testing, 1955–1957, HHPL.

19. L. L. Strauss, "History of the Long Range Detection System." July 21, 1948, p. 2. Strauss Papers, AEC Series, Monitoring Soviet Tests, HHPL.

20. L. L. Strauss to J. E. Gingrich, July 10, 1947, and L. L. Strauss to R. P. Johnson, July 10, 1947. Strauss Papers, AEC Series, Monitoring Soviet Tests, HHPL.

21. Strauss, *Men and Decisions*, p. 204.

22. W. T. Golden interview, January 4, 1990.

23. "Atomic Energy Commission Minutes of Meeting No. 83," July 23, 1947, p. 172. AEC Archives, DOE.

24. L. R. Groves to Army Chief of Staff, July 24, 1947. RG 165, Army Chief of Staff, Decimal File 147, Box 5, Folder 350.05, NA.

25. Secretary of War to Chief of Staff. August 13, 1947. RG 165, Army Chief of Staff, Decimal File 147, Box 5, Folder 350.05, NA.

26. L. R. Groves to Secretary of War, August 29, 1947. RG 165, Army Chief of Staff, Decimal File 1947, Box 5, Folder 350.05, NA.

27. Lawren, *The General and the Bomb*, p. 265.

28. K. C. Royall to Hillenkoetter, September 7, 1947. RG 165 Army Chief of Staff, Decimal File 1947, Box 5, Folder 350.05, NA.

29. H. Friedman interview, July 27, 1989.

30. Darling, *The Central Intelligence Agency*, p. 198.

31. Strauss, *Men and Decisions*, p. 203.

32. J. E. Gingrich to L. L. Strauss, September 15, 1947. Strauss Papers, AEC Series, HHPL.

33. K. C. Royall to Director, CIG, September 15, 1947. RG 165 Army Chief of Staff, Decimal File 1947, Box 5, Folder 350.05, NA.

34. General D. D. Eisenhower to Commanding General, AAF, September 16, 1947. RG 165; Army Chief of Staff, Decimal File 1947, Box 5, Folder 350.05, NA.

35. "JCS document 1764/1" July 14, 1947. RG 218, JCS, Decimal File 1948–1950. Box 166, Folder CCS 471.6, NA.

36. "The Capabilities of the USSR in Regard to Atomic Weapons," JIC 395/1, July 8, 1947, p. 4. RG 218, JCS, Decimal File 1948–1950, Box 166, Folder 471.6, Control of Atomic Weapons, Section 5, NA.

37. Gingrich to Strauss, September 15, 1947.

38. Secretary, Military Liaison Committee, to D. Lilienthal, October 10, 1947, and J. V. Forrestal to Air Force Chief of Staff, October 21, 1947. RG 218, JCS, Decimal File 1948–1950, Box 166; Folder CCS 471.6, NA.

39. D. L. Northrup, "Detection of the First Soviet Nuclear Test," February 6, 1962, pp. 3, 4. Archives, DOE.

40. D. Lilienthal to Director, CIG, September 29, 1947. RG 330; Secretary of Defense RDB, 1946–1953, Folder 100, Atomic Energy, Long Range Detection, NA.

41. Strauss, "History of the Long Range Detection System," July 21, 1948, p. 3.

42. Ibid.

43. H. S. Vandenberg to DCS, Materiel, November 14, 1947. Archives, DOD.

44. See Strauss, "History of the Long Range Detection System," July 21, 1948, p. 3; and W. T. Golden to L. L. Strauss, November 25, 1947. Strauss Papers, AEC Series, HHPL.

45. "Report of Operation Fitzwilliam," Vol. I, n.d. (classified on December 9, 1949), Introduction, p. 7.

46. Golden to Strauss, November 25, 1947.

47. L. L. Strauss, "Minutes of the MLC-AEC meeting of 26 November 1947," December 1, 1947. Strauss Papers, AEC Series, Monitoring Soviet Tests, HHPL.

48. Strauss, "History of the Long Range Detection System," July 21, 1948, pp. 3, 4.

49. Kepner's career started in 1909 when, at age 16, he began a four-year stint as a private in the Marine Corps. In 1916, after a brief period of college study, he was commissioned a second lieutenant in a cavalry unit of the Indiana National Guard and participated in the Mexican Border Campaign. Transferred in 1917 to the Regular Army, he served as an infantry officer in World War I. In 1920 he joined the Air Service, and during World War II he commanded the Eighth Fighter Command of the Army Air Force in the European theater. In 1946 he was commander of aviation in the joint Army-Navy task force that had carried out Operation Crossroads. In this post, he made valuable contacts with individuals and organizations involved in the long-range detection experiments conducted during this operation.

50. S. L. Warren to C. Tyler, AEC, December 18, 1947. RG 330, Secretary of Defense, RDB 1946–53, CAE, Box 7, Folder 100, Atomic Energy Long Range Detection, NA.

51. V. Bush to J. V. Forrestal, December 9, 1947. RG 330, Office of Secretary of Defense, Entry 199, Box 77, Folder CD 16–1–2, NA.

52. J. V. Forrestal to S. W. Symington, December 9, 1947. RG 330, Office of Secretary of Defense, Entry 199, Box 77 Folder CD 16–1–2, NA.

CHAPTER 6

TECHNICAL PROGRESS:
1946–1947

Although the bureaucratic paper shuffling required to launch the monitoring organization had preoccupied the high-level administrators of the involved agencies during most of 1947, scientists at some of these agencies had continued to work throughout this period on the technical aspect of monitoring. Research on Mogul had produced advances in sonic instrumentation and in the balloon technology needed to position this instrumentation within the hypothesized atmospheric sound channel. Work on the methodology of the proposed multiple fission product analysis promised to dramatically enhance the value of radiological monitoring. Yet, some doubts about the feasibility of establishing an effective monitoring system remained to be overcome.

TECHNICAL FEASIBILITY QUESTIONED

Despite the continuing scientific and technological efforts and the results obtained from earlier research on monitoring conducted before and during Crossroads, knowledgeable scientists were not impressed by the technical progress that had been made. For example, when General Kepner addressed a meeting of the Atomic Energy commissioners on December 18, 1947, to brief them on the organization and technical approaches to be used in the long-range detection program, Dr. Bacher, the only scientist on the Commission, had replied to Kepner's remarks by pointing out that the instrumental means of monitoring were still relatively undeveloped and presented problems of great difficulty. He opined that "intensive research and development was necessary to improve present methods and perhaps devise new methods of detection."[1]

An even gloomier prognostication was offered by the Research and Development Board's Committee on Atomic Energy, which included two scientists of unquestioned repute, Dr. James B. Conant and Dr. J. Robert Oppenheimer. At its tenth meeting, held on December 22, 1947, the Committee discussed the subject of long-range detection in executive session with General Kepner and Army and Navy representatives. In an allusion to the view expressed in Hillenkoetter's memorandum of June 30, Conant's report of this meeting noted that "The Committee finds that the view has been expressed that the techniques and instruments for detection and evaluation of the remote detonation of an atomic bomb are available either potentially or actually. The Committee feels that there are grave doubts as to whether this optimistic view is justified."[2]

The Committee also warned that radiological monitoring methods were of limited utility. In this regard, they listed five conditions under which an atomic bomb could be detonated: (1) surface detonation, as at Alamogordo; (2) surface detonation in the rain; (3) high-altitude burst, as in Japan; (4) an underground explosion; and (5) an underwater burst, as at Bikini (Test Baker).[3]

Of these five possible conditions, only the first offered some hope of being detectable radiologically, "at not too great distances." The Committee noted that seismic and sonic methods were "the only presently known methods thought to be hopeful for detecting an explosion under conditions where the detection of radioactive particulates would fail." But the Committee felt that the available information was "totally inadequate" to assess the feasibility of these two methods.[4]

Ironically, in December 1947, as America's top scientists agonized over the likely impracticability of detecting atomic explosions at long range using radiological methods, media pundits were busy reinforcing the notion that such detection was readily accomplished. For example, William Laurence of the *New York Times*, in a radio interview aired in December, answered a question about the possibility of detecting a Russian bomb test radiologically. He pontificated that this was so easy to do that "the possibility of anyone testing atomic bombs in secret from now on is practically nil as far as we know."[5]

The true state of affairs regarding the monitoring of foreign atomic bomb tests was so different from the media perception of this activity as to seem almost surreal. Indeed, the negative opinions of the feasibility of long-range detection methods offered by the Committee on Atomic Energy caused consternation within the Air Force, since it implied that the Air Force may have been charged with an impossible task. The Committee's opinions and recommendations carried considerable weight because, as part of the Research and Development Board, the Committee could influence the research agendas and budgets of the military departments. Regarding the Air Force's plans for detecting the Sandstone detonations at long range, the Committee recommended that the experiments should not be focused on long-range detection per se but, rather, on the more modest goal of maximizing the kind of information that would be useful as a guide for further research on monitoring techniques.

The Committee emphasized the value of the Sandstone experiments, but it was not in favor of delaying Sandstone to allow Kepner more time to prepare. Conant's report noted that "While [the Committee] considers the long range detection problems of the greatest importance, and feels the urgency of a research and development program which will obtain useful data from the Sandstone operations, it is unanimous in its agreement that this is not important enough to warrant any postponement of the Sandstone tests."[6]

The Committee's feelings about the urgency of the need for a monitoring system, as opposed to the need for research on monitoring methods, can be gauged by its response to a question from General Kepner on the duration of the American atomic monopoly. The Committee expressed the unanimous opinion that it would be "highly improbable" that a foreign country would detonate an atomic bomb until 1952 or later. Most of the Committee members, such as Conant, Oppenheimer, and Groves, were quintessential insiders on atomic matters. Hence, it is not surprising that the Committee's estimate of a relatively long duration for the atomic monopoly reflected the insider view based on the Murray Hill Area concept. Indeed, the Committee had pointed out that "its estimate . . . of the minimum time before Russia will have an atomic bomb is conservative."[7]

The less-than-optimistic view of scientists on the Committee on Atomic Energy about the technical aspects of monitoring differed sharply from that which had been expressed by some administrators, beginning with LeMay in 1946 and later by Strauss, Golden, and Lilienthal. Only a few months earlier, Lilienthal had stated his belief that "the technical issues have already been solved and that the problem is now one of organization and coordination."[8]

The disparate beliefs of these scientists and administrators about the technical practicability of monitoring methods appears to have arisen largely because the administrators had not been given (presumably because of the compartmentalization policy) detailed information on what had been learned about long-range detection over the past two years. For example, at the November Military Liaison Committee–Atomic Energy Commission meeting, Strauss had implied that there was no record "that anything was done on the subject of monitoring between Hiroshima and the spring of 1947."[9] Neither Groves nor Parsons, who were both present, had disabused Strauss of this notion.

To the degree that the technical difficulties of monitoring had been explained to them in briefings, these administrators appear to have lacked the scientific training necessary to correctly interpret this information. Instead, they tended to perceive the "difficulties" as the scientists' excessive concern that they had to iron out every last technical detail before getting on with the monitoring task. The belief that the scientists' position was merely an expression of technical preciosity appeared reasonable against the background of media reports that had repeatedly emphasized how easy it was to detect atomic explosions at great distances.

One administrator, William Golden, who was present at most of these briefings, later recalled that among the scientists

there was a sense that you have got to prove these things will work. The Strauss approach and mine was: Look, you know we just can't wait for that and we have just got to go ahead. . . . Lewis Strauss and I felt we could not delay on this. It was very important to know as early as we could when the Russians had something, and we did not believe, as General Groves and many others believed, that it was going to take the Russians ten years or more [to produce an atomic bomb]. . . . Some of this was temperament and, though it would not trouble Bob Bacher and Robert Oppenheimer, the financing was important. The military had a budget. They had all their money allocated and it would be preferable to prove something at a low level of expenditure before getting airplanes up and getting into real expense. That was not the view of Lewis Strauss, fortunately.[10]

According to Golden, he and Strauss opposed the "serial approach" of first perfecting monitoring techniques through research and only then deploying them in an operational system. Instead, they advocated the idea that research to improve monitoring methods should go on in parallel with setting up an operational monitoring system using existing techniques. This approach was apparently based on the notion that even an imperfect system was better than none.

If the views of administrators like Golden and Strauss were characterized by a lack of information about the technical status of monitoring, the scientists' perspective was characterized by a surfeit of such information. To the scientists of the Committee on Atomic Energy, this information indicated that a monitoring system based on existing techniques would not merely be less than perfect but, rather, that it would probably not work at all. To understand the scientists' beliefs about monitoring, it is necessary to review the technical work carried out after Crossroads and to examine the technical status of the various monitoring methods as of December 1947.

In November 1946, when the results of the long-range detection experiments at Crossroads had been analyzed, Colonel Seeman had concluded that "no instruments exist at present to insure success."[11] Regarding the methods that had been tested (seismic, sonic, and radiological), it was felt that "seismic records," as Admiral Parsons had pointed out, "cannot be relied upon."[12] Sonic techniques using ground-based microbarographs had displayed a disappointingly short range. The sonic method based on the Mogul concept, however, promised a dramatic increase in the range at which the blast would be detectable. Thus, although the Mogul experiments at Crossroads had been inconclusive, work on Mogul was continued. Radiological methods, as they were used at Crossroads, appeared to offer the possibility of relatively long-range detection but could not unambiguously identify the cause of an anomalous increase in atmospheric radioactivity as an atomic explosion. However, in the fall of 1946, Manhattan Engineer District scientists had suggested a methodology involving fission product analysis that could remedy this deficiency. Hence, this approach had been further explored in 1947 by Colonel Holzman in collaboration with scientists at the Atomic Energy Commission's Argonne Laboratory.[13]

The results of the work on Mogul and on fission product analysis conducted during 1947 had, therefore, provided part of the mosaic of information about monitoring that had been reviewed by the Committee on Atomic Energy in December of that year. The Committee's pessimistic conclusions stemmed from the fact that, although real progress had been made on both projects, both were plagued by serious uncertainties.

MOGUL AFTER CROSSROADS

Since Project Mogul had barely begun at the time of Crossroads, the Mogul team's experiments during this atomic test series were limited to using ground-based microphones in an attempt to see what use could be made of sound refraction in the atmosphere in detecting the explosions at long range. The most imaginative of the approaches used was the antipodal refraction experiment conducted by Albert Crary on Ascension Island. The failure of these experiments to produce positive results made it obvious that more definitive tests were needed in which the sound receivers would be positioned within the hypothesized atmospheric sound channel.

In his initial proposal to Spaatz, Professor Ewing had suggested a way to accomplish this: "Small stratospheric balloons," he wrote, "provided with radio means for transmission of sound impulses to a receiving station either fixed or mobile, probably provide the most readily available listening arrangement."[14]

This notion was duly adopted in the plans for Mogul, which called for the development of balloons capable of operating at "constant altitude 40–60,000 feet."[15] This terminology meant that the balloons had to loiter at some preset altitude between 40,000 and 60,000 feet for a "long" period of time. If they were to be useful as part of an operational monitoring system, as opposed to being merely test vehicles, this time period would have to be measured in days rather than hours. Moreover, the altitudes of interest were considerably higher than the operating range of conventional "sounding balloons," small instrumented balloons used in probing the atmosphere to collect geophysical data. Ewing's comment that aerostats would provide "the most readily available listening arrangement" would thus prove to be a highly relativistic statement. No suitable balloons then existed.

Of course, researchers had used small sounding balloons capable of lofting data-gathering instruments into the atmosphere since the 1890s. Mass-produced "radiosondes," balloon-borne instrument packages that radioed weather data to a ground receiver, had been widely used since the mid-1930s. The rubber balloons used in such applications, however, generally burst well below 40,000 feet, because of the deterioration of the rubber envelope caused by the high concentration of ozone found in this altitude region. This altitude limitation did not pose a problem for weather balloons since meteorologically useful data are obtained at much lower altitudes.

Nor was ozone a problem for the huge, costly manned balloons that made much-publicized journeys into the stratosphere in the 1930s, because the envelopes of these aerostats were much too thick to be weakened by ozone. Nevertheless, despite their ability to reach great heights, such stratospheric balloons were far too expensive to be used as unmanned, instrument-carrying vehicles. Thus, beginning in the late 1920s, researchers interested in probing the stratosphere had sought a suitably cheap, lightweight, ozone-resistant envelope material for their balloons. Trials with some of the plastics available prior to Word War II, such as cellophane, had been disappointing. In 1946, this problem had yet to be solved.[16]

Moreover, sounding balloons generally acquire data as they travel upward through the atmosphere, and they are allowed to ascend without limitation until they burst or lose their lift gas and descend. The balloons needed for Mogul, on the other hand, had to float at a preset level in the atmosphere. Such "constant altitude" balloons were not unknown. The United States had designed aerostats of this type during World War I for use as "balloon bombers"—small, unmanned balloons armed with incendiary bombs. These vehicles were designed to float at an altitude where the prevailing wind would carry them over an enemy country where the bombs could be automatically jettisoned to burn croplands and forests.

In World War II, the Japanese used this concept to bomb the United States. The Japanese balloon bombers had been designed to float from Japan to North America by incorporating a control mechanism that limited fluctuation in altitude to about 25 percent of the mean altitude. This deviation was larger than desirable for Mogul experiments. In addition, neither the small World War I balloon bombers nor the much larger World War II versions were capable of reaching the altitudes desired by the Mogul researchers. Thus, to make a balloon that would meet the Mogul requirements, it would be necessary to find a satisfactory material for the envelope and to develop a suitable altitude-control mechanism.

In the fall of 1946, Watson Laboratories had been reorganized, and Colonel Marcellus Duffy, who headed a newly formed Applied Propagation Sub-Division, continued as project officer in overall charge of Mogul. Duffy, a West Point graduate, had studied at Massachusetts Institute of Technology in the 1930s, where he had specialized in meteorology. Following these studies he had set up the Air Corps' first experimental radiosonde station at Wright-Patterson Air Base. Instrumentation of this type was an essential feature of the experiments planned for Mogul. Duffy therefore brought to the project a relevant technical background and managerial skills honed during the war when he had functioned as liaison between the chief signal officer and the head of the Air Corps on matters dealing with meteorological instruments.

Duffy's assistant during the war had been Captain Athelstan Spilhaus, a New York University geophysicist. After the war Spilhaus had returned to New York University where he was made director of research. Duffy had been impressed by Spilhaus's knowledge of meteorological instrumentation and the fact that, during the war, Spilhaus, together with his assistant, Lieutenant Charles B. Moore, had developed a cure for the excessive failure rate of weather balloons made from a wartime rubber substitute. Duffy decided to place the Mogul balloon development in Spilhaus's hands, a move that would allow Duffy's in-house staff to focus on other aspects of the Mogul project, which included theoretical studies of sound refraction in the atmosphere and the development of sonic instruments. In November 1946, New York University was awarded a research contract "to develop and fly constant-level balloons to carry instruments to altitudes from 10 to 20 km."[17]

At New York University, Spilhaus lost no time in assembling a Balloon Group, which included his former wartime assistant Charles Moore, to work on the development of the Mogul balloon and the associated telemetering devices. Spilhaus began by studying captured Japanese balloon bombers.

It was evident to Spilhaus that the envelope materials used by the Japanese, varnished paper or silk, had a weight per unit area that precluded their use in balloons designed to operate at 60,000 feet. The search for a more suitable envelope material ended when Moore joined the Group, since as a result of his previous work on this problem, he had already concluded that polyethylene, recently available in large, thin sheets, was the most promising material. Moore's initial tests on polyethylene, carried out at New York University, revealed that this material would be satisfactory.[18]

Although of technical interest, the altitude control mechanism used on the Japanese balloons had not been designed with the aim of achieving true constant-level flight. It consisted of a barometrically operated ballast reduction system that periodically jettisoned small sandbags. The Balloon Group designed a system that was similar in principle but had a fluid ballast that would allow the altitude to be continuously controlled within narrow limits. The Group worked feverishly on these ideas to have balloons ready by the summer of 1947.

Meanwhile, work on other aspects of Mogul had proceeded apace. In the summer and fall of 1946, Crary had conducted sonic tests off the coast of

New Jersey, using the B-29s that had been assigned to the project, to deploy bombs that were detonated at high altitude. Stratospherically refracted sound waves were detected at distances up to 180 miles. These "sky waves," as Crary dubbed them, were delayed relative to the calculated time of travel of direct waves at the same distance. Similar phenomena had been observed during the artillery barrages of World War I, but the more controlled experiments conducted by Crary provided data that tended to support the existence of an atmospheric sound channel.[19]

In the spring of 1947, a unique opportunity to test the sound channel concept appeared. On April 18, engineers of the British Army were scheduled to blow up German military installations on the island of Helgoland located in the North Sea about 20 miles off the Dutch-German coast. The tiny island fortress had protected Germany's northwest flank in two world wars and was reckoned to be one of the most heavily fortified spots in the world. Only one mile long and one-third of a mile wide, the island was crammed with radar stations, gun emplacements, and submarine pens. The amount of TNT used to destroy these facilities, as well as the German ammunition that was still stored in the island's caves, was expected to produce an explosion that was equivalent to 3 to 6 kilotons.[20] This, of course, was considerably less than the 23 kilotons produced by the Crossroads bombs, but crude calculations suggested that it would be worthwhile to attempt to detect the blast in the United States using the Mogul approach.

Although constant-level Mogul balloons were unavailable because they were still under development, Moore was able to cobble together twenty neoprene weather balloons to form a balloon train capable of reaching the stratosphere. Previous investigators had shown that, by automatically releasing some of the balloons at a preset altitude, constant-level flight could be achieved for a short time. Hence, Moore decided to adopt this technique.[21] In the past, neoprene balloons had displayed a resistance to ozone which generally allowed them to reach higher altitudes than natural rubber balloons, but ultraviolet light soon disintegrated them at high altitudes. For this experiment, however, it would be sufficient if they remained aloft for only a few hours.

The Mogul team from Watson Laboratories provided the balloon-borne microphone and radio transmitter. The radio receiver was mounted in an airplane that was to be flown below the balloons keeping within range of the transmitted signal.

When the balloons could not be launched in the New York area because of air traffic control regulations, arrangements were made to release them from the football field at Lehigh University in Pennsylvania where a trial flight was made on April 3. On April 18, a similar balloon train carrying a low-frequency microphone and transmitter was prepared for launching from the Watson Laboratories facility at Eatontown, New Jersey. A C-54 aircraft carrying the radio receiver was to orbit below the balloon train to pick up the transmitted data. But due to high winds, the launch was cancelled. Later experiments revealed that no positive results would have been obtained from this experiment, since the range (3,450 miles) and the explosive energy (about one-eighth as powerful as the Crossroads bombs) resulted in signals that were below the threshold of detectability of the apparatus then available.[22] Useful data *were* obtained by the Signal Corps, however, using ground-based acoustic sensors sited in Europe to monitor the Helgoland explosion.

By July, the first polyethylene balloons were ready, and on July 3 a balloon cluster was launched by the New York University field engineers at the Army Air Force Base in Alamogordo, New Mexico.[23] The flight lasted a little over three hours and reached a height of 18,500 feet. Although a Mogul payload was attached, the primary aim was to test the balloons. Indeed, in the annals of ballooning, this flight was of historic significance since it signalled a new beginning in balloon technology. The cluster of aerostats launched that day were the first in a long line of polyethylene balloons that were to be designed and built over the next decade by a number of agencies for research and military purposes. But, in the summer of 1947, much development work lay ahead for the New York University team before they would produce an adequate balloon-payload package for the Mogul project. Crary later recalled that "The last four months of 1947 at Alamogordo were hectic. We tested various types of balloons, even multiple balloons and several new types by commercial suppliers, to find combinations of characteristics that would be satisfactory."[24]

By December 1947, balloons were available that met the Mogul specifications. A few sonic tests using explosives detonated on the ground were made using these balloons, but the range officer at Alamogordo vetoed experiments in which bombs would be released from a plane and detonated at high altitude. Crary had thus made plans to conduct such tests in Alaska, but this lay in the future.

In sum, by the end of 1947, the Mogul experiments had provided new and valuable data on atmosphere acoustics, and the project had produced important advances in balloon technology. After two years of work, however, the feasibility of the Mogul concept for monitoring atomic explosions had remained undetermined.

Near the end of 1947, before AFMSW-1 became functional, a curious Mogul-related incident occurred involving duplication of research. This incident illustrated the degree of compartmentalization that then prevailed. It also highlighted the need for a single agency such as AFMSW-1 to oversee all aspects of monitoring to prevent the duplication of effort that can sometimes occur when the free flow of scientific information is restricted for security reasons.

In December 1947, a report was coauthored by two of America's leading physicists, Frederick Reines and Edward Teller, who were members of the

Theoretical Division of the Atomic Energy Commission's Los Alamos Laboratory. Its purpose, according to the authors, was to evaluate the feasibility of a suggestion that had been made by another scientist of renown, George Gamow, over eighteen months earlier. Gamow had proposed "that an atomic bomb explosion might be detected at the opposite side of the earth from the detonation point by a sensitive pressure measurement."[25]

The report asserted that data obtained during the eruption of Krakatoa in 1883 could be used to assess whether this approach might be used to detect an atomic explosion. To make a comparison between the pressure pulse produced at the antipode by Krakatoa and that produced by an atomic bomb, it was first necessary to calculate the fraction of the energy released at Krakatoa that went into blast waves. John von Neumann, an Atomic Energy Commission consultant and one of the world's premier scientists (later described in the official history of the Atomic Energy Commission as a "mathematical genius"), had provided the authors with an estimate of this parameter.

Using the Krakatoa data, the authors themselves calculated other important factors, including the attenuation of the sound pulse and the focusing in the antipodal region. Combined with von Neumann's estimate of the energy fraction, the authors' calculations suggested that atmospheric refraction of the pressure pulse produced by an atomic bomb, similar to those that would be detonated at Sandstone, "might be detected" by an antipodal pressure measurement. They also pointed out that the antipode of Eniwetok, where the Sandstone tests were to be conducted, was Ascension Island in the Atlantic. They had concluded that "although the possibility of such a simple scheme for detecting the occurrence of atomic bomb explosions is not definitely established by the rough considerations in this note, it is suggested that the situation is hopeful enough to warrant further study."[26]

The report was classified Top Secret, and only five copies were made. Presumably, one copy found its way through bureaucratic channels to the monitoring agency, AFMSW-1, where Ellis Johnson was planning and coordinating the long-range detection experiments for Sandstone. The experiment proposed in this report, however, was not carried out at Sandstone. It had, of course, been tried by the Mogul team's Albert Crary at Ascension eighteen months earlier during the Crossroads test at Bikini. (Ascension was in the antipodal region of both Bikini and Eniwetok.) Evidently, the negative results obtained by Crary were sufficiently convincing to preclude further trials. Crary later recalled that, at Ascension, "The several types of equipment which we used were quite adequate and covered all anticipated frequency ranges, but no recognizable signals were received. The Bikini bomb was no Krakatoa."[27]

Reines and Teller were apparently unaware of the Mogul team's much earlier work on this approach (which was classified Top Secret). Thus, Reines, Teller, and von Neumann, three of America's scientific luminaries, had been diverted from other tasks to prepare a useless report. At the time, Reines was involved with "dragon" theory.[28] Teller, later known as the father of the hydrogen bomb, was working on fusion, and von Neumann was laying the mathematical foundations for the Atomic Energy Commission's first electronic computers. The policy of compartmentalization was not without its costs.

MULTIPLE FISSION PRODUCT ANALYSIS

Theoretical work on radiological methods of long-range detection had been carried out in the winter of 1946–1947. In what had promised to be the single most important advance in such techniques thus far achieved, the concept of monitoring airborne radioactivity had been expanded to include a sample processing method based on multiple fission product analysis. This methodology eliminated the ambiguity that had plagued earlier radiological monitoring attempts that had relied solely on a gross measurement of atmospheric radioactivity.

The suggested technique required knowledge of the characteristic mix of fission products produced by an atomic explosion, information that was then highly classified. Indeed, at that time, the Atomic Energy Commission's Argonne and Los Alamos Laboratories were the repository of most of the world's knowledge about fission products. Although the methodology proposed by Atomic Energy Commission scientists required special knowledge and consummate skill in radiochemistry, the underlying rationale can be described in relatively simple terms.

Both atomic bombs and reactors produce fission products consisting of some 200 radioisotopes of thirty-four elements. If created in a bomb that is exploded at or above the earth's surface, these fission products become airborne; if created in a reactor, they may be spewed into the air by accident (á la Chernobyl) or purposefully (to simulate an atomic bomb explosion). In either case, airborne dust emanating from the locale of the event will contain fission products.

It is only necessary to analyze about a dozen of the 200 radioisotopes present in the dust to tell whether they were created by a bomb or a reactor. If created instantaneously by the explosion of an atomic bomb, the age, hence the "birthday," of each fission product analyzed will be the same. But the birthdays will differ if the fission products have been created slowly in a reactor by controlled fission.

To determine age, it is convenient for observational purposes if the radioisotopes selected for analysis emit easily measured beta rays and include some radioisotopes with very long "half-lives" (the time it takes for half the radioisotope present in a given sample to decay radioactively) and some whose half-lives are relatively "short," that is, with half lives of a few to tens of days. Ideally, the half-life of a short-lived radioisotope should not be so short that the radioisotope will virtually disappear before it can be measured, and not so long that it will not diminish appreciably in a few days. (Otherwise the measurement period becomes overly extended.)

After selecting the dozen or so radioisotopes to be analyzed, they must be chemically separated out from the dust sample and physically isolated so that the beta ray intensity can be measured. Chemical separation is very difficult and may be impossible if the sample size is small. It was chiefly the uncertainty about this step of the methodology that placed this approach squarely in the same category as the rest of the monitoring methods about whose practicability the Committee on Atomic Energy had "grave doubts."

At the instant of creation in an atomic explosion the radioisotopes bear a fixed relation to each other. Thus, the next step, after chemical separation of the radioisotopes at some arbitrary later time, is to calculate the amounts of each of the analyzed radioisotopes that would be required at an earlier time to produce the quantities actually observed. It is possible to do this because the half-life of each is known. The procedure requires comparing the observed beta ray intensities of short-lived radioisotopes to those that are long-lived. The ratio of intensities is then extrapolated backward to various earlier times until a time is found at which the projected percentage mix of the analyzed radioisotopes closely matches the known creation percentage of those radioisotopes produced by a uranium or plutonium bomb.

This procedure will give the same birthday for each fission product only if they were produced by an atomic bomb. If, on the other hand, no single birthday is found at which the correct mix of radioisotopes obtains, then it can be concluded that the radioisotopes are reactor-produced. If the fission products were created by an atomic explosion, the time of the explosion, the nature of the fissionable material, and some features of the bomb's design could be deduced from fission product analysis.

This method obviously had enormous potential. But the big question was: Would dust samples collected at a distance of several thousand miles from an atomic explosion provide fission products in the quantities needed for chemical analysis? The available evidence was not encouraging. Large-volume samples of fission products had been scooped from the soil near the detonation point of the Trinity bomb and subjected to chemical analysis and radioassay. But attempts to collect and measure airborne fission products from Trinity using filter-equipped B-29s had been unsuccessful.[29] Some radioactivity attributed to fission products from the Hiroshima bomb had been picked up over the west coast of the United States by filter-equipped B-29s, but only the radiation produced by the samples had been measured. These measurements suggested that fission products were present in the samples in very minute quantities and chemical separation had not even been attempted. No traces of the fission products from the Nagasaki bomb were found outside Japan, a fact that had been attributed to meteorological effects.[30]

During the Crossroads tests, drone airplanes were flown into the mushroom cloud within minutes of detonation, and large-volume samples of fission products were obtained that were suitable for chemical analysis. As in the Trinity test, radiochemical fission product analysis was used in determining the yield of the bombs.[31] On the other hand, the samples of airborne dust collected at points thousands of miles from the explosions by filter-equipped aircraft, as part of the long-range detection experiments, were not subjected to chemical analysis.[32] The intensity of the radiation the samples produced was measured, and, as in the case of the Hiroshima dust samples, the measurements suggested that the quantity of fission products in the samples collected at these distances was minute.

Moreover, the available evidence on airborne fission products was consonant with the model of an atomic explosion that was then extant. To understand this model, it is necessary to review what was then known about the radioactivity produced by an atomic bomb. One of the first references to this subject had appeared in a British report of 1941, which said that an atomic bomb explosion would "release large quantities of radioactive substances."[33] General Groves later claimed that he was "unaware of the existence of this report" and that concern about the bomb-produced airborne radioactivity first appeared at Manhattan Engineer District when the planning for Trinity began early in 1945. At that time, a Manhattan Engineer District scientist, Joseph Hirschfelder, had raised the possibility that airborne radioactivity might pose a problem.[34]

To explore the possible problems related to such airborne radioactivity, a simulated atomic explosion was detonated at Alamogordo on May 7, 1945, using 100 tons of TNT spiked with reactor-produced fission products (a kind of radioactivity bomb). Hirschfelder's subsequent analysis of this explosion revealed that the radioisotopes had been deposited on the surface of the sand particles kicked up by the blast. This finding led Hirschfelder to warn that, during the projected Trinity test, there might be a danger posed by "dust containing active material and fission products."[35] This kind of dust was later to be called "fallout."

Calculations of the potential hazard had indicated that the danger from airborne radioactivity produced by Trinity would be minimal, unless a heavy rainfall occurred during or immediately after the explosion. In his capacity as head of the Los Alamos Laboratory, Oppenheimer had then informed the Army that the Trinity explosion "should not normally lead to a deposition of a large fraction of either the initial active material or the radioactive products in the immediate vicinity."[36]

Hirschfelder's work on this problem had produced the first outlines of a model for an atomic explosion at or near the earth's surface where large amounts of blast-pulverized dirt are present in the lower portion of the fireball. In this model, according to Hirschfelder, "The upper and central portions of the ball of fire contain so little solid material that the rate of condensation is very slow." He went on to point out that the large amount of ionization present during the explosion "will tend to prevent agglomeration and thus help to disperse the active material."[37]

Considerations such as these suggested that if the bomb was detonated high in the air, where the fireball would be well away from the earth's surface, the slow condensation and ionization-inhibited agglomeration predicted by the model would tend to keep the fission products produced by the explosion in a finely divided state. This gaslike material created in such an "airburst" would be dispersed in the atmosphere, and there would be virtually no fallout.

A few weeks after Trinity, when the preliminary findings had appeared to confirm Hirschfelder's projections, Oppenheimer had predicted that the fallout from bombs dropped on Japan would not be a problem because of the height (1,850 feet) selected for detonation. He wrote: "With such high-firing heights it is not expected that radioactive contamination will reach the ground."[38] Later measurements in 1946 at Hiroshima and Nagasaki did not contradict this prediction, which was consonant with the model of an airburst that had evolved.

This model was based on the fact that the interior of the fireball was known to be hot enough to vaporize all the substances present, fission products, unchanged uranium or plutonium, and the bomb casing. As the fireball rose in the air and cooled, its constituents condensed, forming water droplets and a "gas" of submicron-sized particles that dispersed in the atmosphere.[39] Calculations indicated that both radioactive decay and the dilution factor introduced by dispersal in the air would cause the concentration of airborne radioactivity, at distances of a few hundred miles from the explosion, to fall to a level similar to that of the natural radioactivity that was always present in the air. This model of an airburst was reflected in the first official Atomic Energy Commission publication on the effects of atomic weapons, which stated that "if the height of the bomb exceeds a certain value, there will be no detectable fallout since no extraneous particles will be sucked into the cloud."[40]

The model clearly distinguished between the vaporized radioactive material created during the explosion and fallout (a distinction that had been blurred in media descriptions of atomic explosions). On the basis of the model, the term *fallout* referred to a phenomenon that occurred only when bombs were exploded near the earth's surface so that much of the fireball touched the ground, something that occurred at heights below about 500 feet. In this case, particles of dirt, much larger in size than the vaporized fission products, are sucked up into the fireball. Fission products rapidly condense on dirt particles, causing them to become radioactive in proportion to their surface area. Dirt particles are carried up to the base of the stratosphere (about 40,000 feet) and fall to the ground at rates dependent on their size.

At Trinity the particle size distribution was such that about half the radioactivity produced by the bomb fell to the ground within eight hours. The underwater explosion (Baker) at Crossroads produced little fallout in distant areas because most of the fission products, aside from those that remained in the water, were washed down into the sea near the detonation point by the huge column of water that had been thrust upward by the blast. The other Crossroads bomb (Able) and the bombs that had been dropped on Japan were airbursts that did not produce data contrary to the existing views about fallout.

Therefore, information on the five atomic explosions that had occurred through 1947 was compatible with the belief that fallout could be avoided by exploding a bomb at heights of 1,800 feet or so, and that at distances of more than a few hundred miles, the vaporized fission products could not be collected with existing techniques in amounts that would allow them to be analyzed chemically. Such analysis, of course, was essential, if the radiological approach was to become a reliable and unambiguous means of monitoring.

The Soviets might opt to test their first atomic bomb using an airburst, for example, by elevating the bomb on a tethered balloon, in which case it appeared that the resulting explosion would foil radiological methods of identification based on fission product analysis. Even if the Soviets exploded a bomb on or near the ground, there was no certainty that the fallout would provide fission products in chemically analyzable quantities at distances of several thousand miles.

Thus, when exposed to careful scientific scrutiny, the radiological method of monitoring shared one characteristic with seismic and sonic techniques: Its technical capability for detecting atomic explosions unambiguously at long range had not been established. It was apparent that much development work would be needed before a reliable monitoring system, based on these methods, could be deployed. The uncertainties surrounding long-range detection methods further highlighted the importance of Sandstone as a means of testing new monitoring methods, of accurately assessing the technical status of existing methods, and of identifying more precisely the areas in which further development was needed.

On the last day of 1947, the first operational step in mounting the program of long-range detection experiments during Sandstone was taken; that is, the program was assigned a code name. Effective December 31, 1947, the code name "Fitzwilliam" was given to "long range detection researches as related to Sandstone."[41] Experiments that would prove crucial for the timely detection of the first Soviet bomb test were about to begin.

NOTES

1. "Minutes of the 133rd Meeting of the AEC," December 18, 1947. RG 326, Atomic Energy Commission, Folder: Minutes of Meetings, NA, and Strauss, "History of the Long Range Detection System," July 21, 1948, p. 2.

2. J. B. Conant to V. Bush, January 2, 1948, pp. 1, 2. RDB 113/1, AE 21/1, Log No. 2448, Archives, DOD.

3. Conant to Bush, January 2, 1948, pp. 1, 2.

4. Ibid.

5. "Transcript of Remarks of W. Laurence on American Town Meeting of the Air," December 15, 1947. Strauss Papers, AEC-Monitoring of Soviet Tests, HHPL.

6. Conant to Bush, January 2, 1948, pp. 1, 2.

7. Ibid.

8. D. Lilienthal to Director, CIG, July 10, 1947, p. 2. Strauss Papers, AEC Series, Monitoring Soviet Tests, HHPL.

9. L. L. Strauss, "Minutes of the MLC-AEC meeting of 26 November 1947," December 1, 1947. Strauss Papers, AEC Series, Monitoring Soviet Tests, HHPL.

10. W. T. Golden interview, January 4, 1990.

11. Seeman to Nichols, "Remote Measurements for Atomic Explosions," November 6, 1949, p. 2.

12. "Notes on the Meeting of the AEC and the MLC," 4th Meeting, April 30, 1947, p. 4. RG 326, Atomic Energy Commission, Folder: Minutes of Meetings, NA.

13. S. L. Warren to C. Tyler, AEC, December 18, 1947. RG 330, Secretary of Defense, RDB 1946–1953, CAE, Box 7, Folder 100, Atomic Energy Long Range Detection, NA.

14. Ewing, "Long Range Sound Transmission in the Atmosphere," October 14, 1945, p. 3.

15. "Memorandum B," May 14, 1946, p. 2.

16. Some "lucky" flights by sounding balloons made in the early decades of this century reached remarkably high altitudes, despite the fact that most balloons burst at lower altitudes owing to the effects of ozone. See C. A. Ziegler, "Technology and the Process of Scientific Discovery: The Case of Cosmic Rays," *Technology and Culture* 30, 1989, pp. 524–963.

17. J. R. Smith and M. D. Murray, "Constant Level Balloons," November 15, 1949, p. 5. Technical Report 93.02, College of Engineering, New York University. Copy in the authors' collection. Our thanks to Professor C. B. Moore for providing this report.

18. C. B. Moore interview, July 29, 1992.

19. A. P. Crary, "Atmosphere Acoustics," n.d. Unpublished manuscript. Copy in the authors' collection. Our thanks to Mrs. Mildred Rodgers Crary for providing this document, which is part of a book by A. P. Crary, *On the Ice*, to be published by University of Ohio Press.

20. *New York Times*, April 18, 1947, p. 14.

21. This multiple balloon technique had been developed by cosmic ray investigators in 1941. See E. T. Clarke and S. A. Korff, "The Radiosonde: The Stratosphere Laboratory," *Journal of the Franklin Institute* 232, 1941, pp. 217–355.

22. C. B. Moore interview, July 29, 1992.

23. C. B. Moore et al., "Constant Level Balloon," April 1, 1948, p. 36. Technical Report No. 1, College of Engineering, New York University. Copy in authors' collection. Our thanks to Professor C. B. Moore for providing this report.

24. Crary, "Atmosphere Acoustics," n.d., p. 15.

25. F. Reines and E. Teller, "Possible Detection of Atomic Bomb Explosion by Pressure Measurements at the Antipodes," LA 667, Series A, December 29, 1947, p. 1. Copy in the authors' collection. Our thanks to Dr. Gregg Herken for providing this report.

26. Reines and Teller, "Possible Detection of Atomic Bomb Explosion by Pressure Measurements at the Antipodes," p. 4.

27. Crary, "Atmosphere Acoustics," n.d., p. 5.

28. Among the most spectacular experiments conducted during and after the war were those that involved working with an almost critical assembly of uranium hydride. Because of the danger of an atomic explosion if anything went amiss, this kind of experiment was called "tickling the dragon's tail" or, more simply, "dragon." The dragon was a crucial research tool. See Truslow and Smith, *Project Y*, p. 198.

29. Truslow and Smith, *Project Y*, pp. 244, 246.

30. J. M. Blair et al., "Detection of Nuclear-Explosion Dust in the Atmosphere," October 2, 1945, p. 3. Report No. LA-418, Report Library, LANL.

31. Truslow and Smith, *Project Y*, p. 275; and J. P. Delgado, "Operation Crossroads," *American History* 28, May–June 1993, p. 57.

32. Krueger to Groves, "Remote Air Sampling," September 18, 1946, p. 1.

33. The MAUD report is reproduced in R. Williams and P. Cantelon, eds., *The American Atom*, 1984, p. 19.

34. L. R. Groves, "Some Recollections of July 16, 1945," *Bulletin of the Atomic Scientists* 26, No. 6, June 1970, p. 22.

35. J. O. Hirschfelder and J. Magee to K. T. Bainbridge, June 16, 1945. Archives, LANL.

36. J. R. Oppenheimer to T. Farrell, May 11, 1945. RG 77, Med Records, Modern Military Section, NA.

37. J. O. Hirschfelder and J. Magee to K. T. Bainbridge, July 6, 1945, Archives, LANL.

38. J. R. Oppenheimer to T. Farrell and W. Parsons, July 25, 1945. Archives, LANL.

39. Atomic Energy Commission scientist Edward Teller used the word "gas" in popular writings to emphasize the minute size of particulate fission products. See, for example, E. Teller, "How Dangerous Are Atomic Weapons?," *Bulletin of the Atomic Scientists* 3, No. 1, January 1947, p. 35.

40. *Effects of Atomic Weapons*, 1950, p. 3.

41. R. H. Rathbun, to Distribution List, December 31, 1947. RG 330, Secretary of Defense RDB, 1946–1953, CAE, Box 7, Folder:100, Atomic Energy Long Range Detection, NA.

CHAPTER 7

SANDSTONE AND FITZWILLIAM

In the spring of 1948, the establishment of a Communist dictatorship in Czechoslovakia caused consternation in the West and exacerbated the already strained relations between the United States and Russia. This crisis led the Joint Chiefs of Staff to consider curtailing Sandstone. In a memorandum to the Joint Chiefs of Staff, former Secretary of State Cordell Hull opposed this action by pointing out that "The importance of completing the tests now planned is greater under a deteriorating international situation than it was under the situation existing at the time the tests were first approved."[1] This argument was found compelling, and the operation was allowed to continue as planned.

Operation Sandstone thus proceeded in an atmosphere of tension that grimly heightened its significance. The primary purpose of Sandstone was to test novel ideas for constructing atomic bombs, ideas that were expected to produce dramatic improvements in performance. According to the official history of the Atomic Energy Commission, the tests were to verify "the new design principles developed by Los Alamos scientists."[2] It is now known that these new principles involved the use of fissile cores made of a composite of plutonium and uranium and levitated cores designed to greatly increase bomb efficiency.[3] Three bombs (X-Ray, Yoke and Zebra) were to be detonated on Eniwetok between mid-April and mid-May. To carry out this program, an interservice organization called Joint Task Force Seven had been formed, commanded by Lieutenant General John E. Hull.

The Sandstone tests were also used by AFMSW-1 to assess the various monitoring techniques that were being considered for the long-range detection system (Operation Fitzwilliam). The way in which the leaders of this organization planned and executed Fitzwilliam, and the results of the Fitzwilliam experiments were primary determinants in shaping the operational monitoring system that was later deployed. Indeed, the network of government agencies and industrial firms that was established to carry out the crucial Fitzwilliam experiments became the organizational nucleus of the system. In this regard, the way in which industrial firms, particularly Tracerlab, became involved in this Top Secret operation provides a useful illustration of the sometimes contingent nature of organizational development.

AFMSW-1 MOBILIZES

Space had been found in the Pentagon for the newly created monitoring organization, AFMSW-1. Its chief, Major General Albert F. Hegenberger, and his staff were ensconced in Room 5b 318, adjacent to the headquarters of Task Force Seven. This arrangement encouraged close liaison between the two organizations and facilitated Hegenberger's task in carrying out Fitzwilliam. Despite Hegenberger's technical background, his role at AFMSW-1 was chiefly administrative. The task of directly supervising the scientific and engineering work of the organization fell to its technical director, Dr. Ellis A. Johnson.

It seems unlikely that Johnson was daunted by the somber prognostications of the Committee on Atomic Energy concerning the practicability of the various monitoring methods. He had a history of tackling difficult problems and solving them. A Massachusetts Institute of Technology graduate, he had worked at the Institute as an instructor for a few years before joining the Department of Terrestrial Magnetism at Carnegie Institution in 1935. On loan to the Naval Ordnance Laboratory to work on magnetic mines, he was at Pearl Harbor when the Japanese attacked in December 1941. While there, he played a key role in solving an alarming problem encountered by American submariners in the weeks following the attack—their torpedoes were failing to detonate after striking Japanese ships.

Johnson's work during the remainder of the war gained him friends in high places, both in the Navy and the Army Air Force, a fact that may have played a role in his selection as technical director of AFMSW-1. In 1942 he was commissioned a commander in the Naval Reserve and assigned to Admiral Nimitz's Pacific command as mining officer. Based on an extensive study of Japanese shipping that he had made, he proposed a strategic plan to employ mines on a huge scale using massed aircraft to deploy them, a concept that higher authority at first resisted. The Navy, for which mine laying was a low-priority task, lacked suitable aircraft. The Army Air Force had the planes but considered mine laying the Navy's job. To overcome this gridlock, Johnson had set about persuading General Curtis LeMay, who headed the 21st Bomber Command, to implement the mine-laying scheme. According to a colleague of Johnson's, Captain Ralph D. Bennett, "The factor that won LeMay over was when Ellis studied LeMay's bomber losses from flak over Japan. Ellis suggested different bombing levels that reduced plane losses considerably. It convinced the General that the tenacious physicist might have something of real value in his mining proposal."[4]

Both the study of LeMay's bomber losses and the mining proposal that had been made by Johnson were examples of a new mathematical approach to problem solving and performance maximization called operations research, a field that Johnson had pioneered at the Naval Ordnance Laboratory.[5] The mines were deployed by LeMay's bombers with spectacular results: by March of 1945, all major Japanese ports had been closed by the mines, which sank about a million tons of Japanese shipping in the final months of the war. Captain Bennett later recalled: "Bringing army and navy together as Ellis did was a great service to his country . . . without his persistence, I doubt that mines would have been used so successfully or would have played such a major part in victory."[6]

After the war, Johnson returned to Carnegie Institution's Department of Terrestrial Magnetism. Plucked from this post by Kepner because of his background in geophysics and a demonstrated ability to organize complex, interservice projects, Johnson was well suited to head the technical effort on monitoring. His leadership would prove to be an important factor in achieving the objectives of Fitzwilliam, which had become a crash program, and in recruiting able scientists for AFMSW-1.

Johnson gathered around him chiefly scientists he had worked with in the past and whose capabilities he knew. They included Dr. George Shortley, an Ohio State University physicist who had been one of the group working on magnetic mines at the Naval Ordnance Laboratory; Doyle Northrup, another Ordnance Laboratory researcher who had been with Johnson at Pearl Harbor; and Dr. William D. Urry, a radiochemist who had been Johnson's colleague at both the Massachusetts Institute of Technology and Carnegie Institution. By the beginning of 1948, the organizational structure planned by Kepner had been fleshed out by the recruitment of military and scientific personnel.

Administratively, AFMSW-1 was a staff organization. Therefore, although Johnson and his scientists were responsible for planning, they relied on the Special Projects Office of the Air Weather Service to carry out the operational aspects of Fitzwilliam. The latter included field coordination and support of participating agencies, the operation of specific detection systems as directed by AFMSW-1, and personnel training.

In prosecuting Fitzwilliam, one of the first problems to be overcome was the lack of money for instrumentation. In November 1947, Kepner had pointed out that the source of funds for this aspect of the program remained to be determined, and by the year's end this problem had become acute. Kepner, therefore, decided to seek the help of the Atomic Energy Commission. One obvious channel through which this could be accomplished was the Military Liaison Committee, which met regularly with the commissioners and which was represented on the Advisory Group that Kepner had established to guide policy decisions on long-range detection.

Unfortunately, General Groves, who was a prominent member of the Military Liaison Committee, had become extremely controversial, a state of affairs that was detrimental to the Committee's usefulness as a broker between the Air Force and the Commission. Lilienthal claimed, with some justification, that Groves was attempting to influence members of Congress to transfer custody of atomic weapons to the military. So bitter were the feelings on both sides that Lilienthal urged that Groves be ousted from all posts of influence. Toward the end of January, Groves, who had become aware that he was fighting a losing battle, announced his retirement from the Army.[7]

Early in January, however, this controversy was coming to a head, and Kepner decided to bypass the Military Liaison Committee in favor of a more direct approach. Two Air Force staff officers were detailed to sound out Commissioner Strauss, who was recognized as a staunch supporter of the monitoring program, on the possibility of obtaining funds.[8]

Strauss agreed to put their request before the Commission at the next meeting scheduled for January 6, 1948. At this meeting, Strauss reported that representatives of the Air Force monitoring program had informally advised him that they lacked a million dollars to purchase instrumentation. Strauss's proposal that the Commission make up this shortfall was unanimously approved, and, by the end of the month, the needed funds (which in the interim had increased to $1.5 million) were officially transferred to the Air Force.[9] With money in hand, the staff of AFMSW-1 began immediately to oversee the placement of equipment orders by the agencies participating in the monitoring effort.

Some of the planned Fitzwilliam experiments made use of apparatus that the participating agencies already possessed, such as the Army Signal Corps' acoustic network and the Coast and Geodetic Survey's seismographs. Many experiments, however, required the purchase of new, off-the-shelf instruments or, in some cases, the development of special apparatus. The task of procuring both standard and one-of-a-kind instrumentation was divided about equally among the Air Material Command, the Office of Naval Research, and the Army Signal Corps.[10]

Except for a few measurement systems built in-house, these agencies obtained all the standard and special equipment needed for the Fitzwilliam experiments from instrument manufacturing firms. Because of Fitzwilliam's Top Secret classification these firms were unaware of the existence of the project, and, when solicited by the government to bid on a procurement, they were given no information beyond that needed to specify the item sought. There were, however, two exceptions to this rule which provide some insight into the nature of the security that cloaked the activities of AFMSW-1 and the monitoring effort.

In November 1947, an enterprising entrepreneur, Frank Rieber, decided that his small New York City firm had something the government needed, an instrument capable of sonically detecting atomic explosions at long range. Rieber was already involved in government work on flight instrumentation for guided missiles, and through this channel he was placed in contact with William Golden, Strauss's aide at the Atomic Energy Commission. Golden suggested he write directly to Strauss.

In his letter to Strauss, Rieber asserted that the detection of a foreign atomic bomb test "must be of vital interest to your commission or to the defense authorities, or both." He had been driven to this conclusion "Because this possibility has had a certain amount of public discussion." He went on to point out that, "At this very moment, news headlines, and interviews with various scientists and seismologists, make reference to an alleged possibility of a bomb test in Russia." He ended with a plea to be allowed to place his firm's capabilities in the area of long-range detection at the disposal of the appropriate agency.[11]

As a result of this letter, Rieber was routed to the Naval Electronics Laboratory, one of the agencies participating in Fitzwilliam, where Navy scientists expressed a keen interest in the instrument he proposed. This was a microphone based on the "vibratron," a proprietary device developed by Rieber Research Laboratories to permit the recording of minute air pressure changes by measuring the frequency of a thin tungsten wire vibrating in a magnetic field. The wire was mounted in such a way that its tension (hence its vibration frequency) varied in proportion to the pressure change. This approach promised a great enhancement in sensitivity relative to other microphone types.[12]

Rieber worked with Navy scientists to develop a highly sensitive sonic pickup, and a number of these devices were scheduled to be used by Naval Electronics laboratory personnel in Operation Fitzwilliam. Unlike most of the other instrument suppliers, Rieber knew the ultimate purpose of the instruments he provided. He had, in effect, penetrated the security screen to become an insider.

Representatives of another small entrepreneurial firm, Tracerlab, also accomplished this feat. Indeed, they became quintessential insiders, for they succeeded in participating not only in the instrument development phase of Fitzwilliam but in the operational phase as well.

ENTER TRACERLAB

Late in January 1948, an engineer employed by Tracerlab Inc., a radiological equipment manufacturer located in Boston, Massachusetts, visited the Signal Corps' Fort Monmouth, New Jersey, laboratory complex to repair some previously purchased Tracerlab apparatus. While there, he learned that the Signal Corps was about to announce a very large "open bid" procurement for radiological instrumentation for some classified purpose. He hastened to inform Tracerlab's sales manager, Dana Atchley, of this opportunity. Atchley, in turn, informed the firm's chief scientist, Dr. Frederick Henriques, of this anomalously large procurement for a secret application.[13] As a result of this rather pedestrian train of events, Tracerlab was to become a participant in Fitzwilliam and, later, to play a crucial role in the detection of Joe-1.

Tracerlab was founded in March 1946, by a group of four entrepreneurs led by William Barbour, a Massachusetts Institute of Technology graduate and an ex-major in the Army Air Force with wartime experience in electronics, who became its president. An influential member of the group was Dr. Wendell C. Peacock, a young Massachusetts Institute of Technology-trained physicist who had worked briefly for the firm as consultant and provided some of the initial product concepts on which the firm was based. After it became apparent that these concepts would lead to viable products, Peacock, who was interested primarily in basic research, left Tracerlab to join the staff of the Atomic Energy Commission's Oak Ridge Laboratory, a move that was later instrumental in facilitating Tracerlab's participation in Fitzwilliam.

Tracerlab began by manufacturing equipment for measuring radioactivity and also offered a unique service in which radiochemicals obtained from the Atomic Energy Commission were reprocessed to meet the special needs of medical, industrial and university laboratories. In December 1947, the influential magazine *Fortune* had called Tracerlab "The first real business to be built out of by-products of the atomic bomb." By February 1948, when the firm first became involved with AFMSW-1, Tracerlab was not quite two years old, employed forty-five people, and its annualized sales were about $180,000.[14] Despite its small size, it had, according to *Fortune*, "established itself strongly in a field where, whether from sluggishness or disinterest, giants like GE and RCA have so far failed to move."[15]

One reason for its early success was that Barbour had succeeded in hiring some extremely able people: notably, Dr. Frederick Henriques, an innovative and energetic chemist who had headed research groups at Harvard University and the University of California at Berkeley, and Dana Atchley, Harvard-trained in science with wartime service as a lieutenant commander at the Naval Research Laboratory. In terms of its personnel as well as its birthday, Tracerlab was a young organization; with the exception of Barbour, all of its staff were under 30.

When the news of the impending Signal Corps order arrived at Tracerlab, Henriques and Atchley decided to visit Fort Monmouth to get further information. As a result, Atchley later recalled, "We got ourselves into one hell of a big thing."[16] At Monmouth they found that no one knew, or would reveal, the reason for the procurement, but they learned the name of the person who originated the order: Ellis Johnson. Through former Navy associates in the Pentagon, Atchley was able to track down Johnson. During the war, both Atchley and Henriques had been cleared by security agencies, so they were able to arrange to visit Johnson and his colleague, Dr. William D. Urry, chief of AFMSW-1's Nuclear Research Section, in their Pentagon office. Despite their clearances, Henriques and Atchley lacked the "need-to-know"; hence, they found that Johnson and Urry were unable to state directly the nature of the work in which they were engaged. Consequently, according to Henriques,

We had to winkle the information out of them. We go down there and these people are desperate. We started talking to them. I'm a scientist, they were scientists . . . and just in chatting it became obvious that what these people were trying to do was to use our own test procedures to mask setting up a means of detecting Russian bombs. The Pacific [Sandstone] tests would also be a calibration of these procedures. Dana [as Tracerlab sales manager] was busy counting up all the scalers, cutie pies [a type of radiation detection equipment], survey meters, and so forth, that they would need. Who was going to man this stuff? They said they had nobody to run it. So I said I'd collect the staff to run the stuff. It was our equipment . . . and who is more competent to keep it running than our people? This was the sales pitch.[17]

Early in February when this meeting took place, Johnson and Urry, if not "desperate," were working feverishly on the preparations for Fitzwilliam. In his first month at AFMSW-1, Johnson, although his technical strengths lay outside the field of nuclear physics, had become aware that the radiological technique based on multiple fission product analysis would be crucial to the success of the monitoring system. Because he lacked expertise in this area, he had recruited Urry, whose chemistry background included work with radioisotopes, to oversee the test of this technique during Sandstone.

Urry was apparently unimpressed by the Atomic Energy Commission's assurances that their scientists would provide the Air Force with assistance in this area. The Radiochemistry Group at Los Alamos, headed by Dr. Roderick Spence, had never performed bomb diagnostic work. Such radiochemistry had been conducted at Los Alamos during Crossroads, but for this purpose a temporary division called the B-14 Group had been formed to carry out chemical determinations of yield. This Group, however, had been disbanded after the tests. Many of the Atomic Energy Commission's most knowledgeable scientists in fission product analysis were totally immersed in preparing for the radiochemical work that would be required to "prove" the new principles of bomb design, which was, of course, the primary purpose of Sandstone.[18]

By the time of Henriques's unexpected visit, Johnson had had second thoughts about depending solely on the Commission's radiochemists to carry out the vital fission product analyses. The confident demeanor of Henriques and Atchley, who seemed eager to take on the difficult task of supplying a large number of radiological instruments, together with the field maintenance crews that would be needed to keep them operating during Fitzwilliam, sufficiently impressed Johnson to prompt him to inquire whether Tracerlab could carry out fission product analysis as well. Henriques indicated that this would be possible if Tracerlab could hire someone knowledgeable in this area. Henriques left the meeting with the impression that Tracerlab might be given a contract to provide the needed radiochemical service, if they could demonstrate the capability to perform it.[19]

On the following day, Henriques called Peacock at Oak Ridge to solicit his aid in winning a contract for instruments and services from AFMSW-1. Henriques had known Peacock during the war when they both worked on projects sponsored by the government's Office of Scientific Research and Development. He knew that Peacock had remained interested in the fortunes of Tracerlab. Peacock agreed to take time off to help prepare a proposal for AFMSW-1, and he also offered to recruit a chemist capable of performing fission product analysis.

While at Oak Ridge, Peacock had become friends with a highly competent and inventive scientist, Dr. Lloyd R. Zumwalt, who had been employed as a chemist at the Oak Ridge Laboratory since 1946. Before that, he had been a scientist–Army officer and had worked at the Manhattan Engineer District's Berkeley Laboratory and at the Clinton Laboratory (later the Oak Ridge National Laboratory) on the production of uranium for the first atomic bombs. By 1948, like many bomb-project scientists, he was planning to leave the Laboratory for employment in industry. So he readily agreed when Peacock asked for his aid in exploring the possibility of obtaining a government contract to establish a laboratory for fission product analysis at Tracerlab.

In mid-February, Henriques, Peacock, and Zumwalt were invited to meet with AFMSW-1 staff members at their Washington headquarters. The representatives of two other radiological instrument firms, considered by the Air Force as potential bidders on the equipment contract, were also at this meeting. Peacock, who had been briefed by Henriques on the assumed classified objectives of the procurement, was able to give an impressive exposition of the type of equipment, field maintenance services, and personnel training that would be needed to accomplish the aims of the program. He later recalled that "Eventually, during the mid-afternoon, they pulled these other two bidders out of the room and it became sole source."[20]

Zumwalt, whose presence was evidence that Tracerlab could provide scientists able to carry out fission product analysis, later recalled that "At this point it was disclosed what they were up to and I suppose that [prior to this meeting] I might have been given some sort of a quick security check."[21] As Atomic Energy Commission scientists, Peacock and Zumwalt already had security clearances. In the eyes of the Air Force this was an important qualification, for had they lacked such clearance the protracted period that would have been needed for a security investigation would have precluded their participation in the preparations for Fitzwilliam. Ellis Johnson apparently felt that their existing clearance allowed him to describe the radiochemical work that would be needed. According to Zumwalt:

In broad terms it was clearly outlined that I had to use the coming A-bomb test in the Pacific as the means of trying to see if long range detection was possible . . . they knew that Atomic Energy Commission scientists at Los Alamos had developed a technique of sampling clouds very close in and subjecting the samples to radiochemical analysis to get information on the efficiency of the given atomic explosion. It was hoped that perhaps sufficient radioactivity would be collected at some considerable distance in order to derive similar information.[22]

The subsequent discussion confirmed the assumption Henriques had made about the ultimate purpose of the program. Ellis Johnson indicated that the information gained at Sandstone would be used to provide the United States with an operational monitoring system. Moreover, according to Zumwalt, "It was clearly stated that we were looking for Russian explosions."[23] At that time, Zumwalt had no intimations that eighteen months later he would be the first American to learn that the Russians had tested an atomic bomb. But this lay in the future. The present was exciting enough, for Johnson decided to allot much of the radiological work required by Fitzwilliam to Tracerlab.

The firm was asked to provide a substantial portion of the radiological instrumentation together with technicians to maintain it. Some of this equipment would be used in the field by Tracerlab scientists, while the remainder would be operated by the personnel of participating government agencies. To train the latter, Tracerlab was given a contract to run a "school" at which trainees would be given a crash course on methods of measuring radioactivity. In addition, Tracerlab received a contract to carry out the radiochemical work needed to analyze the Fitzwilliam samples. Tracerlab thus became an integral part of the network of organizations and individuals that would conduct long-range detection experiments during Sandstone.

OPERATION FITZWILLIAM

The conferences so ably organized by Kepner in November and December of 1947 had resulted in a number of ideas for obtaining quantitative data on the performance of potential monitoring methods based on the sonic, seismic, and radiological effects produced by detonating an atomic bomb. In one form or another, of course, methods based on such effects had already been tried. In addition, the conferences generated ideas for novel techniques that made use of the optical and electromagnetic effects produced by atomic explosions, effects never before used in long-range detection. Trials of three of these new approaches, categorized as "exploratory," were to be conducted as part of Fitzwilliam: observation of the light flash reflected from the dark side of the moon following the detonation; measurement of the magnetic effects of the dynamo action in the ionosphere caused by blast-produced pressure waves; and detection of the "dimple" in the ionosphere caused by the explosion.

When the trials of these exploratory techniques were added to the experiments that had been suggested for testing sonic, seismic, and radiological methods of long-range detection, the resulting plan for Fitzwilliam became quite complicated. It called for AFMSW-1 to coordinate nineteen projects conducted by eight agencies at various locations within an area bounded in the east by the longitude of Tokyo, Japan, and in the west by that of Frankfurt, Germany, and the latitudes of Point Barrow, Alaska, and the Panama Canal. It was planned as a *big* operation in every sense of the word.[24]

And so it was. Fitzwilliam Forward, the headquarters staff whose job it was to supervise and coordinate the work of participating agencies, was established on Eniwetok on March 17, 1948. Teams from the various agencies began arriving soon after. General Kepner, chief of Fitzwilliam Forward, maintained an office aboard the USS *Mount McKinley*, anchored near Kwajalein. From this headquarters, he oversaw the redeployment of newly arrived teams from Kwajalein to their operational locations on surrounding islands. Almost all of the groups experienced difficulties in preparing their experiments. Most of the Fitzwilliam personnel had never participated in an atomic test and did not know exactly what to expect. Virtually all the projects had been hurriedly planned and equipped. One team from the Naval Electronics Laboratory, which was scheduled to establish sonic arrays at Guam and Hawaii, found it necessary to cancel their project.

The deployment to the Pacific was especially difficult for Tracerlab, the only nongovernment organization to play an important operational role in Fitzwilliam. This small firm had had only a few weeks in which to nearly double its size, in terms of personnel and facilities, and to substantially increase its capital financing. These steps were mandated by its participation in the program and were ably handled by Barbour, its president, and Henriques, its technical director.[25]

After receiving contracts for fission product analysis and for radiological instrumentation and field services, urgency became the keynote at Tracerlab. According to Henriques, the contracts were awarded so hastily that there was virtually "no negotiation, nothing! We didn't even estimate an overhead rate."[26] Henriques hired Zumwalt and Peacock on the basis of a highly paid, short-term labor contract. They, in turn, were able to recruit fifteen scientists from the Atomic Energy Commission's Oak Ridge Laboratory on the same basis. Atchley has described their activities in the weeks that followed:

We had to just jump in . . . literally in less than a month. And I guess our ignorance allowed us the confidence to do it. We had to come up with a scheme of detection; we had to hire a scientific staff; we also had to hire what I call a field engineering staff. We had to train it and train some military people and then get . . . out all over the Pacific . . . In essence, my role was to deal with the military; since I had been a lieutenant commander in the Navy, I could do that. I also hired all the field engineers—they're a type that like money and will go anywhere. So I probably hired about twenty. Henriques was responsible for getting the big scientific hitters. We all went to Fairfield-Suisun [now Travis Air Force Base in the Napa Valley region of California] and we trained on the type of equipment we were going to use.[27]

By the end of March, Tracerlab had established field stations in Hawaii (supervised by Peacock), Kwajalein (supervised by Zumwalt), and Guam (supervised by E. H. Turk) to perform preliminary radiochemical analyses on filter samples collected by aircraft and by ground stations. The firm's laboratory in Boston, Massachusetts, was expanded to perform additional analyses on samples forwarded from the field stations, and a special facility was created to analyze atmospheric gas samples collected by WB-29s of the Air Weather Service.

While Army engineers, Navy construction teams, and civilian contractors on Eniwetok were busy building the 200-foot tower for the X-Ray shot, the various Fitzwilliam groups set up their equipment and rehearsed their roles. In order to describe the seemingly bewildering variety of activities that constituted Operation Fitzwilliam after the detonations began, it is useful to focus on some specific questions which the operation was designed to answer.

Regarding radiological methods of long-range detection, the crucial question was: Would airborne fission products be collectable at great distances from the explosion in quantities sufficient for chemical analysis? As previously described, an affirmative answer would dramatically enhance the utility of this monitoring approach. To increase the probability of an affirmative answer, it was necessary to track the dispersal of the radioactive cloud created by the blast. Airborne particulates could then be collected from the remnants of the cloud, at sites thousands of miles from ground-zero, thus increasing the likelihood that filter samples would contain fission products. Tracking the cloud was accomplished by outfitting aircraft—that is, cloud-chasing bombers and high-altitude balloons—with instrumentation capable of automatically recording the radiation intensity produced by atmospheric radioactivity. Similar instrumentation was installed in shipboard and ground stations located in the expected path of the bomb-produced radioactivity.

A subsidiary question was: Would filter samples of airborne particulates collected by ground-based equipment provide the same results as filter-equipped aircraft? An affirmative answer to this question would greatly reduce the cost of an operational monitoring system since the expense of routine picket flights could be eliminated. Collection of such airborne dust samples was accomplished by equipping selected ground stations with an apparatus that incorporated a high-speed blower to force large volumes of air through a filter of the same type used in the aircraft sampling units.

The radiological projects carried out by various groups during Fitzwilliam can be understood in the context of the two questions just discussed. Within the operational zone of Joint Task Force Seven, which included much of the Pacific, the Naval Research Laboratory operated three bombers, equipped to measure radiation intensity, as cloud-chasers. The Laboratory also established balloon launching stations on two atolls, Majura and Rongerik, and on the Navy destroyers USS *Davison* and USS *Quick*, which were positioned to intercept the expected path of the radioactive clouds. The payloads of the balloons launched from these stations transmitted radiation intensity data obtained at altitudes of from 35,000 to 50,000 feet. The bombers provided the same data but at lower altitudes. Each balloon ground station was also equipped to record surface-level radiation intensity and to collect surface-level dust samples.[28]

The Air Weather Service's Major Harry C. Crim operated a similarly equipped ground station at Kwajalein, but with the addition of a precipitation sampling pan to collect rainwater for analysis. The notion of rainfall scavenging to concentrate airborne particulate debris from the explosion lay behind the use of the precipitation pan, which was installed not only at Kwajalein, but at other selected ground stations as well.

In addition to regular reconnaissance flights of filter-equipped aircraft from bases in Guam and elsewhere, the Air Weather Service also flew eight WB-29s from Kwajalein on special sampling missions (operational radius about 1,800 miles). These aircraft were equipped with radiation intensity recorders supplied by the Naval Research Laboratory. The planes were also provided with an apparatus to obtain atmospheric gas samples and filters to collect airborne particulates.

A new method of detecting the radioactive cloud was tested by a team from the Carnegie Institution led by Dr. O. H. Gish, working under contract from the Watson Laboratories. The aim was to measure simultaneously the conductivity and the voltage gradient of the atmosphere. By combining these two parameters, a value for the earth-to-air electrical current is obtained which provides a measure of air ionization, a variable related to atmospheric radioactivity. The apparatus was mounted in one of the WB-29 cloud-chasers. It was hoped that it would provide a more sensitive means of detecting and following the radioactive cloud produced by each explosion—more sensitive, that is, than standard radiation detectors, such as Geiger counters, which were mounted in other cloud-chasing aircraft. The special sampling flights, the cloud-chasing missions, and, indeed, the positioning of most of the ground stations that collected radiological data during Fitzwilliam were carried out in accord with the advice of Colonel Benjamin Holzman. Air Force meteorological officers Majors Stephen W.

Pounaras and Gerard M. Leies supervised various aspects of the Air Weather Service's participation in Fitzwilliam, which included responsibility for the coordination and support of all Fitzwilliam field experiments.

Outside the operational zone of the Task Force, an extensive array of ground stations had been established to detect the presence of the radioactive cloud from each detonation as it slowly dispersed toward the east. Within the continental United States, a network of ground-based automatic radiation intensity measuring devices had been established at fourteen cooperating universities and twelve naval stations selected on the basis of location to provide good geographic coverage. Outside the United States, such automatic devices were installed at forty-one government installations located at strategic points from the Panama Canal in the south, to the Aleutians in the north, to Bermuda in the west. In addition to these ground stations, the Air Weather Service operated filter-equipped WB-29s to collect airborne particulates by flying from bases on the west and east coasts of the United States, Bermuda, the Azores, and North Africa.

The eight aircraft detailed for special sampling missions were part of the Air Weather Service's 514th Reconnaissance Squadron. Flights from the 374th and 375th Reconnaissance Squadrons carried out cloud-chasing missions coordinated by the Joint Task Force meteorological officer, and filter-equipped planes of the 308th and 373rd Reconnaissance Squadrons conducted air-sampling flights. As part of Operation Fitzwilliam, these five squadrons flew 466 missions totaling 4,944 hours of flying time. These flights provided coverage of the Northern Hemisphere from the polar regions to the equator within an area extending eastward from Manila in the Philippines to Tripoli in North Africa.[29]

The far-eastern segment of the Air Weather Service's air-sampling program was code named Blueboy. Some idea of the scope of the missions in this area can be gained from the personal logbook of Arnold Ross, a chief radio operator for Flight C of the 373rd Squadron based at Lagens Air Force Base, in the Azores, who wrote "we flew high altitude (35,000 feet) missions through Egyptian airspace, up to the Turkish border, through the Mediterranean area, and on one occasion a 15 hour mission from Wheelus [Air Force Base, Tripoli] to the Cape Verde Islands."[30]

Regarding sonic and seismic methods of long-range detection, the crucial question was range: Would either technique detect an atomic explosion at distances of several thousand miles? An affirmative answer would mean that monitoring stations around the periphery of the Soviet Union could be depended on to alert other components of the monitoring system and to locate, by triangulation, the site of the explosion. Quantitative determination of the range of the best sonic pickups and the most sensitive long-range seismographs was accomplished by operating them at various distances, out to thousands of miles, from the test detonations.

A subsidiary question regarding seismic methods was: Might it eventually be possible to differentiate, at long range, atomic explosions from natural seismic events? An affirmative answer would mean that further seismic research was warranted since it was recognized that an important feature of an operational monitoring system would be the ability to detect and unambiguously identify underground atomic tests. It was not expected that seismic experiments made during Fitzwilliam would definitely answer this question. However, a step in this direction was accomplished by using short-range seismographs to learn more about the "signature" of seismic waves resulting from the explosions.

Within the zone of the Joint Task Force, William K. Cloud, of the Coast and Geodetic Survey, headed a team that operated short-range diagnostic seismographs on three atolls near Eniwetok (Runit, Parry, and Aniyaaii). A group from the Naval Ordnance Laboratory, led by Aaron Heller, placed both sonic pickups and seismographs on Kwajalein, Majuro, Bikini, Ailinglapalap, Mili, and Eniwetok and seismographs on Jaliut and Ebon.

The Army Signal Corps operated a sonic network that extended far beyond the zone of the Joint Task Force, consisting of stations at Kyoto, Japan; Hickam Field, Hawaii; Fort Lewis, California; Red Bank, New Jersey; and Frankfurt, Germany. To obtain the sensitivity needed to "hear" an atomic explosion at very long range, Signal Corps scientists had equipped each station with a detection array of twenty or more appropriately spaced acoustical sensors. The array provided directionality and a dramatic increase in signal-to-noise ratio relative to a single sensor.

In the zone of Joint Task Force Seven, three Mogul balloons operated by a mobile team from Watson Laboratories, headed by Albert Crary, were launched sequentially from Kwajalein to detect X-Ray at 450 miles; at Guam to detect Yoke at 1,200 miles; and at Hawaii to detect Zebra at 2,750 miles. In addition, Mogul balloons were launched from sites outside the zone of the Task Force at Holloman Air Force Base, New Mexico, and at Maxwell Field, Alabama, under the supervision of Dr. James Peoples of the Watson Laboratories.

The Mogul balloons then available had a polyethylene envelope with an inflated sea-level diameter of 20 feet. They carried a "dribble can" containing a kerosene-based mixture that was allowed to leak out at a controlled rate. This allowed the balloon to ascend at a rate of 500 feet per minute to 50,000 feet where it was held at a constant altitude by an automatic control that jettisoned liquid ballast at a rate that compensated for temperature effects and the helium lift gas lost by diffusion through the envelope.

The payloads of these balloons consisted of a microphone coupled to a transmitter which radioed sonic data to a standard Air Weather Service-type radiosonde receiver on the ground. In addition to the sonic balloon experiments, the team in the Pacific also operated a ground-level sonic detector, a modified World War II artillery sound-ranging apparatus, to pick up acoustic waves from the explosion that traveled along the earth's surface.[31]

Of the three exploratory techniques tried out during Fitzwilliam, the optical experiment can perhaps be considered "far out" in both the literal and metaphorical sense of this term, for there was little to recommend this method as an element in any future operational monitoring system. The general idea of the optical experiment conducted during Fitzwilliam was to find out if sufficient light from the explosions would be reflected from the dark side of the moon to be detectable by optical means at suitably located ground stations. In the context of a monitoring system, the problem with this approach is that a partial moon must be visible from both the site of the explosion and that of the observational apparatus. These conditions do not apply in daylight or when both sites are not sufficiently free of clouds or when the moon is full. Nevertheless, despite its impracticality as a monitoring technique, two teams from the Army Signal Corps, led by D. M. Crenshaw and D. J. Southard, set up equipment for this experiment on Eniwetok and Guam, consisting of telescopes coupled to photoelectric detectors and cameras.[32]

The remaining two exploratory experiments had the virtue that, if successful, they would guide future research in a direction that could result in a practically useful monitoring technique. Scientists at the Naval Ordnance Laboratory had constructed three high-sensitivity magnetometers designed to provide data on the electromagnetic effects produced in the vicinity of the explosions. A team led by Naval Ordnance technician C. J. Aronson installed two of these devices on Eniwetok and one on Kwajalein. The third exploratory experiment, based on measuring the "dimple" produced in the ionosphere by the explosion, was set up on Eniwetok by a group from Watson Laboratories headed by L. Elterman. The apparatus consisted of an "ionospherograph," a pulsed radiotransmitter that periodically swept the frequency band from 1 to 25 megahertz (million cycles per second). This was a standard type of apparatus for measuring the heights of various ionospheric layers in the atmosphere.[33]

The activities of all Fitzwilliam groups were orchestrated by a complex communications system that provided each team with precise timing information that allowed them to coordinate their experiments with each detonation. The timing information was highly classified. In this regard, it is noteworthy that both military and scientific leaders opposed any public announcement of the exact time of the explosions. They had adopted this position solely to foil Russian attempts to learn about American advances in atomic weapons by using the very monitoring methods being tested in Operation Fitzwilliam. Information about the new design features incorporated in the Sandstone bombs was thought to be especially vulnerable to being compromised by radiological monitoring carried out by Russian naval vessels and filter-equipped Russian aircraft.

An announcement of the approximate dates for the detonations had been released. Members of the press, some of whom were allowed on Navy ships to cover Operation Sandstone, wanted more precise notification in advance of each detonation. But, according to David Lilienthal, "the military and Van Bush [head of the Research and Development Board] were dead against it, as helping the Russians acquire information by filters." Lilienthal was told that airborne dust samples collected within a day or two after an explosion would be especially revealing because of the information that could be deduced from the analysis of short-lived radioisotopes. To Lilienthal, these facts suggested a compromise that would satisfy both the press and the military: "I came up with the idea of providing 72 hours leeway (which it is agreed will minimize the usefulness of filter results), but cutting down that time by giving the word to the press some 24 hours in advance, with a press conference to emphasize the importance of not breaking the release date."[34]

This compromise solution may have limited the information the Russians derived from air-sampling. Moreover, the controversy about announcing the time of the Sandstone explosions served to highlight the fact that monitoring was a two-way street, a lesson that was not forgotten by American scientists and administrators a decade later when the question of the desirability of a test ban treaty that would prohibit bomb testing in the atmosphere arose.

RESULTS: EXPECTED AND UNEXPECTED

In terms of its primary purpose, Operation Sandstone was considered a spectacular success. The new design principles had, indeed, resulted in more efficient bombs. The yield of Yoke, for example, was nearly 400 percent greater than that of the Hiroshima bomb. The results of Operation Fitzwilliam provided no such dramatic success. Nevertheless, despite the more modest achievements of Fitzwilliam, the experiments conducted as part of this operation laid the foundation for the "interim" monitoring system that detected Joe-1.

Before discussing these results, it should be noted that all of the Sandstone bombs were detonated on towers. Thus, during each explosion much of the fireball touched the ground, throwing huge quantities of dust coated with radioactive fission products into the air which later descended as fallout. Moreover, the fallout, which affected the performance of radiological monitoring methods, and the blast, which affected sonic and seismic methods, was not the same for each bomb because the energy released by the explosions differed markedly. The yield of X-Ray, detonated on April 14, was 37 kilotons; Yoke, detonated on April 30, was 49 kilotons; and Zebra, detonated on May 14, was 18 kilotons.[35] These differences in yield, as well as the fact that, in the case of radiological detection methods, the effects produced by earlier explosions influenced measurements associated with later detonations, were factors that the experimenters had to take into account in analyzing their data.

The records produced by seismographs indicated that these instruments were essentially useless for detecting bombs detonated on towers (and by extension, for detecting airbursts). Even the most powerful bomb, Yoke, could not be detected seismically beyond 500 miles. The distance at which ground-based sonic detection was effective was considerably greater, but fell short of the range needed to monitor explosions in the Soviet heartland. Yoke was detected sonically at 1,700 miles and Zebra at 1,000 miles.[36]

The sonic data obtained by Mogul balloons were especially disappointing. A report on the performance of Mogul equipment during Fitzwilliam stated that the data from balloon-borne microphones showed "no significant improvement over ground pick ups."[37] This finding sounded the death knell for Mogul. It had long been recognized that the day-to-day deployment of balloon-borne sonic detectors as part of a routine monitoring system would pose serious operational and security problems, but these difficulties had been accepted in view of the expected increase in range relative to ground-based detectors. The failure to realize this increase prompted the project officer to recommend that Mogul be terminated.

The results of the exploratory experiments were uniformly bad. The report on the optical method indicated that conditions necessary for observing the moon obtained only during Yoke. Moreover, during this explosion no reflected light was detected. The report concluded that the method was "unreliable for use in a long-range detection system."[38] Regarding the magnetic experiment, Aronson, the team leader, reported that "there was no indication of magnetic phenomena recorded."[39] The experiment to detect the "dimple" in the ionosphere was never carried out because it was found that the ionospherograph interfered with the telemetry of other experiments and with the radio control of drone aircraft.[40]

The bright spot in the results was the relative success of some of the radiological methods used to detect the explosions. The methods tested were of two types: those in which airborne radioactive debris was concentrated from large volumes of air by air-filtration or by rainfall scavenging; and those that depended on direct measurement of the radiation intensity produced by radioactivity in the atmosphere. The first of these methods proved capable of detecting about a thousand-fold lower concentration of airborne radioactivity than the second method.

In practice, this meant that ground-based equipment that measured only the radiation intensity produced by atmospheric radioactivity was relatively insensitive at distances beyond about 600 miles from the explosion. Nor could this range be much improved. According to the analysts' report, "the limits of detection were not due to the instruments per se, but to the small concentrations of debris which reaches ground level at distances greater than 600 miles from an explosion."[41] This assessment, based on explosions that produced massive fallout, finally laid to rest the notion that the radioactivity produced by atomic explosions in the atmosphere was so pervasive that it could be reliably detected near the ground at great distances using simple means, such as exposing a Geiger counter to the air.

Much longer ranges, however, were achieved using such instrumentation in aircraft. At altitudes between 25,000 and 35,000 feet it was found that standard radiation detectors, such as Geiger counters and even the more novel air conductivity device that was tried, were capable of detecting and tracking radioactive clouds out to distances of about 2,000 miles from the explosions at Eniwetok. Balloon-borne versions of such instrumentation proved capable of detecting the clouds at higher altitudes—to 80,000 feet—at distances up to 6,000 miles from Eniwetok.

Although airborne instrumentation of this kind was useful in tracking the radioactive cloud, it was recognized that this method would not in itself be a satisfactory means of monitoring foreign atomic explosions in the atmosphere because the data it produced could not unambiguously identify the nature of the event that caused the radioactive cloud. For this purpose the other radiological approach tested during Fitzwilliam, involving concentrating the airborne dust produced by the explosion, would be necessary. As stated in the report on Fitzwilliam, "it is only this procedure which results in samples of fission products which can be subject to chemical and physical studies, and it is only through these studies that unquestioned proof of an atomic explosion can result."[42]

As previously noted, the crucial question, which the Fitzwilliam radiological experiments were designed to answer, was: Would fission products be collectable at long range in sufficient quantity to allow such chemical and physical studies to be made? The data provided by Fitzwilliam provided an affirmative answer to this question, but the range over which various sampling methods were effective differed. Samples had been obtained from the many ground stations that had been equipped with high-speed blowers and filters to extract particulates from the air. According to the report on this sampling method, it was found that "as a means of collecting fission products it becomes marginal at 2,000 miles."[43] The approach based on rainfall-scavenged radioactive debris from the atmosphere proved to have a greater range: stations up to 9,000 miles from Eniwetok that had been equipped with precipitation collectors provided rain-scavenged fission products in quantities sufficient for chemical analysis. Filter-equipped aircraft, however, provided a truly long-range method of sampling. For example, according to the final report on Fitzwilliam, "samples of sufficient strength to allow radiochemical analysis were collected over Tripoli, approximately 12,000 miles from the site of the explosion."[44]

Unfortunately, these results provided a negative answer to the subsidiary question of whether ground-based sample collection could be substituted for the more costly aerial sampling in an operational monitoring system. The sensitivity of ground-based air filters was disappointing, and, while ground-based precipitation collectors were effective at longer range, this method of sampling was too unreliable. "It was," according to AFMSW-1 analysts, "dependent on two simultaneous phenomena; namely, that the trajectory [of the radioactive cloud] will pass over the rainwater collection station, and that precipitation will occur during this period."[45] On the other hand, the results of aerial sampling with filter-equipped aircraft indicated that both the range and reliability of this method would be adequate for detecting tower bursts when used as part of an operational monitoring system.

Conant's report of December 1947 on the findings of the Committee on Atomic Energy had indicated that, of the various ways to test an atomic bomb in the atmosphere, only detonation near the ground, such as a tower burst, would be unambiguously detectable using radiological means. The Committee was not, however, optimistic about the range of this approach. Indeed, in commenting on existing sonic, seismic, and radiological instrumentation, the Committee had predicted: "If the detonations were carried out precisely as at Alamogordo [i.e., the Trinity test, which was a tower burst], then the present instrumentation may be considered adequate at not too great distances," distances, in other words, that were not considered adequate for monitoring the Soviet Union.[46] The unexpectedly long range of the aerial sampling method was a welcome exception to the predictions made by the Committee, which had otherwise turned out to be depressingly accurate.

The fact that radiological monitoring from afar could be depended on to detect unambiguously a tower burst was encouraging, but the Committee had cited two additional ways of conducting a bomb test in the atmosphere that they believed could not be monitored radiologically at long range. Soviet use of the first of these, exploding the bomb during a rainstorm, was considered most improbable. Soviet scientists would be unlikely to purposely detonate a bomb in this way for the same reason American scientists went to great lengths to avoid testing in the rain; namely, to prevent the heavy contamination of many hundreds of square miles of the area surrounding the test site with rain-scavenged radioactivity. But the second radiologically undetectable method of testing in the atmosphere using a high-altitude airburst, which would produce little or no fallout, might well be used by the Soviets.

It was likely that the Soviets would opt to test in the atmosphere, using either a tower burst or an airburst, rather than underground. This seemed probable, even though an underground test would be undetectable. Radiological methods of detection would be useless in identifying an underground test and, although such a test would probably register on long range seismographs there was, at that time, no way to distinguish such a disturbance from natural seismic events. Thus, an underground atomic test could not be identified as such with the then existing monitoring methods.

Because the technology for underground testing had yet to be developed in the United States, it seemed reasonable to suppose the Soviets would not choose to complicate and possibly delay their bomb program by developing such a technology. Their scientists, as did American scientists until Fitzwilliam, probably believed that testing in the atmosphere would be undetectable if conducted at a site where airborne radioactivity would travel several thousand miles before entering areas subject to Western monitoring efforts. Thus, despite the security advantage of underground testing, if the pace of the Soviet bomb program were as frenetic as the American intelligence services had claimed, their first bomb would probably be tested in the atmosphere. This reasoning later became the basis for deploying the so-called interim long-range detection system.

AFMSW-1 analysts were acutely aware that Fitzwilliam had proved only that tower bursts, one of the two likely modes of atmospheric testing, could be reliably detected. The report on the findings of the Committee on Atomic Energy had cited the other mode, high-altitude airbursts, among the methods that could not be detected radiologically at appreciable distances from the explosion.

Just prior to Sandstone, one of the Committee members, J. Robert Oppenheimer, had reiterated this view about airbursts at a Fitzwilliam planning session held in March 1948. This meeting was organized by AFMSW-1 and included representatives from the Atomic Energy Commission, the Committee on Atomic Energy, and the Office of Naval Research, as well as scientists from the various organizations who were participating directly in Fitzwilliam.

Henriques was present at this meeting as Tracerlab's representative. He recalled that Oppenheimer was very negative when the topic of high-altitude airbursts arose: "Oppenheimer had come to the conclusion that there was no way you could detect radioactive isotopes at long distances from the explosion . . . since they were going to be atomized." Oppenheimer had stated, according to Henriques, that "the possibility of collecting fission fragments from a thousand miles away is very remote."[47]

This view, of course, was consonant with the extant model of an atomic explosion described in previous chapters, which indicated that, in a high-altitude airburst, bomb debris would be vaporized and subsequent agglomeration would be retarded to the extent that quantities of fission products sufficient for chemical analysis would not be collectable far from the site of the explosion. Thus, Oppenheimer's belief about airbursts was hardly idiosyncratic. Zumwalt later recalled that those of his colleagues at the Oak Ridge Laboratory who held an opinion on the subject expressed a similar view: "It was a sort of common belief that atomic debris would be, more or less, of atomic size . . . this was a kind of general belief [about airbursts] that people had."[48]

Following this meeting, Henriques, whose professional background included research on agglomeration in industrial processes, decided to make some theoretical calculations of the degree of agglomeration that might occur in an airburst. His figures suggested that the agglomeration of the condensate might produce larger sized (hence collectable) particles in substantial quantities. He showed these estimates to Johnson and Urry, who agreed that they looked promising. According to Henriques, "My calculations were not sufficient to demonstrate that fission products *would* be detected, but they did demonstrate that it might be possible."[49]

Prior to Sandstone, the Fitzwilliam results were not expected either to prove or disprove Henriques's hypothesis, since all the detonations were to be tower bursts that would create copious amounts of fallout. It appeared that final resolution of the question of the radiological detectability of airbursts would have to await atomic tests involving high-altitude detonations to be conducted at some later date, possibly a year or more in the future.

During Fitzwilliam the filters used to collect airborne particulates were sent to Tracerlab facilities where they were subjected to a novel autoradiographic procedure developed by Henriques and Peacock. The filters were placed in contact with a sheet of x-ray film. The radioactivity of minute particles trapped in the filter produced dark spots on the developed film, which then became a kind of map showing particle location. Tiny portions of the filter containing radioactive particles were located in this way and patiently removed by technicians working with microscopes and tweezers. Filter material was dissolved away and the remaining particles were then chemically analyzed, using methods developed by Zumwalt, to separate out selected fission products.

This approach had proved that a tower burst could be reliably detected and identified radiologically. In itself this result was not unexpected, although the very long range at which chemically analyzable quantities of fission products from a tower burst could be collected using filter-equipped aircraft was a pleasant surprise. As previously pointed out, however, it was not expected that the "dirty" Sandstone tests could validate Henriques's hypothesis that agglomeration might produce particles in an airburst that would be similarly collectable at long range. The reason was that it could not be anticipated that particles representing the agglomeration of fission products would be distinguishable from fallout particles.

But the unanticipated occurred. Through routine microscopic processing Peacock discovered that a few minute, but perfect, spheres were present among the jagged fallout particles. According to Zumwalt, "It was considered quite a revelation to find not only particulate matter, but matter in terms of shiny, metallic-looking spheres that were just beautiful."[50] Henriques's predictions on condensation and agglomeration had not anticipated the formation of perfect spheres. Nevertheless, such spheres validated his prediction, since they could only be explained on the basis that they were created by the coalescing of vaporized material of the bomb itself, free from any association with the dirt sucked up by the explosion. This meant that such spheres would also be present in a high-altitude airburst. Moreover, they were obviously collectable with existing methods at long range, and, most important, they were chemically analyzable.

This startling discovery allowed Dr. Ellis Johnson, in a meeting with members of the Atomic Energy Commission on July 8, to report good news. According to Strauss, Johnson indicated that "as a result of experiments conducted during Operation Sandstone, the Air Force was confident of being able to detect by radiological means an airburst."[51] Quite unexpectedly, the possibility of reliably monitoring a Soviet atomic bomb detonated in the atmosphere had become an immediate reality.

NOTES

1. C. Hull to Joint Chiefs of Staff, April 3, 1948. RG 330, Secretary of Defense, Administrative Secretary, Numerical File 1947–1950, Box 77, Folder CD-16–1–2, NA.

2. Hewlett and Duncan, *Atomic Shield*, p. 164.

3. R. Norris et al., "History of the Nuclear Stockpile," *Bulletin of the Atomic Scientists* 41, August 1985, pp. 106–109, p. 107.

4. R. D. Bennett, "An Unsung Hero of World War II," August 1, 1986, p. 7, unpublished manuscript. Copy in the authors' collection. Our thanks to Captain Ralph Bennett for providing this document.

5. Operations research, which is sometimes defined as the application of science to the solution of managerial and administrative problems, originated as a separate discipline in England during the 1930s as a result of efforts to maximize the performance of the radar net protecting Britain. Through contacts with British researchers, it was spread to the United States where it was employed by Ellis Johnson's mine warfare group at the Naval Ordnance Laboratory in 1942. Their scheme to mine the Inland Sea of Japan was the first large-scale use of operations research by American strategic planners.

6. Bennett, "An Unsung Hero of World War II," August 1, 1986, p. 8; R. D. Bennett interview, August 2, 1990.

7. Hewlett and Duncan, *Atomic Shield*, pp. 151, 152.

8. In describing the million dollar visit of the Air Force officers in his memoirs, Strauss states that he "volunteered to obligate myself for the amount." There is no reason to doubt his recollection of making this offer, but it appears most unlikely that it achieved the objective claimed for it by Strauss, namely, "so that contracts could be made firm immediately." There was no readily available accounting mechanism that would allow an Air Force Procurement Officer to finalize contracts on the basis of a pledge from an individual to donate or loan his private funds to the government. See Strauss, *Men and Decisions*, p. 204.

9. "Minutes, Atomic Energy Commission Meeting No. 137," January 6, 1948, pp. 1, 2. Archives, DOE. The Commission, at first, transferred $1.5 million to the Air Force from monies it had allotted to the Navy for Sandstone. But when General Hull, the military commander of the Sandstone Joint Task Force, objected, the Commission restored the allotment and transferred $1.5 million to the Air Force from its general funds. See "Minutes, Atomic Energy Commission Meeting No. 141," January 13, 1948, p. 18. Archives, DOE; and Strauss, "History of the Long Range Detection System," July 21, 1948, p. 4.

10. "Operations Plan—Fitzwilliam," March 20, 1948, p. A-3. In "Report of Operation Fitzwilliam," Vol. I, Tab A. Period 1948, Microfilm Roll A15941, Frames 4–495, AFHRC.

11. F. Rieber to Strauss, November 12, 1947; and V. H. Walker to F. Rieber, November 14, 1947. Strauss Papers, AEC Series, Box 15, Folder CIA, HHPL.

12. "Operations Plan—Fitzwilliam," March 20, 1948, p. A-6.

13. D. Atchley interview, July 21, 1981; F. C. Henriques interview, August 18, 1981.

14. C. A. Ziegler, "Looking-Glass Houses: A Study of the Process of Fissioning in an Innovative, Science-Based Firm," Ph.D. diss., Brandeis University, 1982, p. 214.

15. *Fortune*, December, 1947, pp. 121, 124.

16. D. Atchley interview, July 21, 1981.

17. F. C. Henriques interview, August 18, 1981.

18. R. W. Spence letter to C. A. Ziegler, November 30, 1988, copy in the authors' collection, and R. W. Spence interview, November 1, 1990.

19. F. C. Henriques interview, August 18, 1981.

20. W. C. Peacock interview, October 14, 1980.

21. L. R. Zumwalt interview, July 29, 1985.

22. Ibid.

23. Ibid.

24. "Operations Plan—Fitzwilliam," March 20, 1948, pp. A-1 to A-14.

25. W. E. Barbour interviews, October 25, 1980 and June 26, 1981.

26. F. C. Henriques interview, August 18, 1981.

27. D. Atchley interview, July 21, 1981.

28. "Fitzwilliam Forward Report," May 17, 1948, pp. 5–14. In "Report of Operation Fitzwilliam," Vol. I, Tab C.

29. "Project Firstrate, AWS Participation in Fitzwilliam," n.d., Appendix VI. In "Report of Operation Fitzwilliam," Vol. I, Tab D.

30. A. Ross, quoted in Richelson, *American Espionage and the Soviet Target*, p. 117.

31. "Sonic Balloon Test," Inclosure G to "Fitzwilliam Forward Report." In "Report of Operation Fitzwilliam," Vol. I, Tab C.

32. "Optical Detector Tests." Inclosure J to "Fitzwilliam Forward Report." In "Report of Operation Fitzwilliam," Vol. I, Tab C.

33. "Magnetic Detection Tests" and "Ionospherograph Tests." Inclosures M and O, respectively, "Fitzwilliam Forward Report." In "Report of Operation Fitzwilliam," Vol I, Tab C.

34. Lilienthal, *The Journals of David E. Lilienthal*, Vol. II, p. 315.

35. Hewlett and Duncan, *Atomic Shield*, p. 672.

36. "Design of Operation and Summary of Results," n.d., p. 13. In "Report of Operation Fitzwilliam," Vol. I.

37. A. C. Trakowski to M. Duffy, December 2, 1948, p. 2. Geophysics Directorate Archives, Program for Long Range Detection Files, AFPL.

38. "Optical Detection Tests," Inclosure J to "Fitzwilliam Forward Report," p. 2. In "Report of Operation Fitzwilliam," Vol. I, Tab C.

39. "Magnetic Detection Tests," Inclosure M, "Fitzwilliam Forward Report," p. 4. In "Report of Operation Fitzwilliam," Vol. I, Tab C.

40. "Ionospherograph Tests," Inclosure O, "Fitzwilliam Forward Report," p. 2. In "Report of Operation Fitzwilliam," Vol. I, Tab C.

41. "Design of Operation and Summary of Results," n.d., p. 14. In "Report of Operation Fitzwilliam," Vol. I.

42. Ibid.

43. Ibid., p. 15.

44. Ibid.

45. J. B. Conant to V. Bush, January 2, 1948, p. 1. RDB 113/1, AE 21/1, Log No. 2448. Archives, DOD.

46. Ibid.

47. F. C. Henriques interview, August 18, 1981.

48. L. R. Zumwalt interview, July 29, 1985.

49. F. C. Henriques interview, August 18, 1981.

50. L. R. Zumwalt interview, July 29, 1985.

51. Strauss, "History of the Long Range Detection Program," July 21, 1948, p. 5.

CHAPTER 8

PRESSURES TO BECOME OPERATIONAL

International events such as the Berlin blockade (1948) disturbed government leaders and created an atmosphere conducive to activating the atomic surveillance network. However, several other factors, including technical developments and vested interests, were crucial determinants in the process that led to the early deployment of a system based almost entirely on radiological monitoring. The unexpected success of radiological analysis led to a shift in the long-range detection program from a focus on research to one on operations, and other developments led to an expansion of its mission from detecting atomic bomb explosions to the gathering of intelligence on a range of atomic activities. These changes, in turn, led to a restructuring of AFMSW-1 to accommodate the task of operating the surveillance system.

THE WIDER CONTEXT OF DEPLOYMENT

While the atomic tests were being conducted in the Pacific in April and May of 1948, the political situation in Europe steadily worsened. Operation Fitzwilliam was officially terminated on June 6. A few weeks later, the Soviets blocked the overland access of the Western powers to Berlin. On June 25, the Berlin airlift began, marking a new and ominous escalation in the Cold War. The heightened tension between the United States and the Soviet Union gave added weight to the argument for establishing a so-called interim net—that is, an incomplete, less-than-ideal monitoring system, in parallel with a program to improve methods of long-range detection. Thus, following Johnson's announcement that it was technically feasible to detect an above-ground test of a Soviet atomic bomb, AFMSW-1 was pressured to deploy a system for monitoring the Soviet Union based solely on the radiological method that Fitzwilliam had shown to be adequate for this purpose.

Predictably, Strauss was a proponent of this course of action. At a meeting with the commissioners on July 8, Johnson responded to Strauss's query about radiological monitoring by indicating that "a program of long range detection on a routine unalerted basis would probably be initiated in September of this year and surely by January 1949."[1] Strauss was apparently unsatisfied with this reply and continued to entertain a belief in the fragility of the Air Force commitment to begin monitoring. Thus, a few weeks later, he reacted vigorously to a letter Hegenberger sent to Lilienthal on the topic of underground testing. Hegenberger wanted the opinion of the Commission's scientists on the correctness of two assumptions contained in the plans prepared by AFMSW-1—namely, that it would be technically feasible for the Soviets to eliminate all traces of an atomic explosion by siting it underground, and that such a test would provide them with adequate data on yield and other parameters of interest. A reply had been drafted for Lilienthal to sign which affirmed without comment the validity of these two assumptions.

When Strauss read the proposed reply, he immediately sent a memorandum to the other commissioners to protest its wording. "I am concerned," he wrote, "that the original [letter] and the reply seem to forecast a downgrading of the project of [radiological] monitoring, with our concurrence . . . this will, by inference, increase the importance of seismic and sonic methods."[2] He went on to raise the bugaboo that the United States could be fooled by a fake bomb test if the Air Force failed to deploy the radiological means at hand to prevent this outcome: "Any delay in the establishment of thoroughgoing air monitoring will incur the hazard that a test of a spurious weapon may be made by another power (accompanied by effects for observers) which, unless its true character be ascertained, would have unpredictable but important repercussions on our international relations."[3] Because of Strauss's protest, the Commission's reply to Hegenberger's letter was rewritten to include the points Strauss had raised.[4] Strauss, however, had misread the situation, and his exhortations were unnecessary. By July 19, the date of Hegenberger's inquiry, the interim radiological net concept was already included in the plan prepared by Johnson and his staff.

A contributing factor in Strauss's continuing concern about monitoring was that the Air Force did not keep the commissioners abreast of the rapidly changing plans and organizational restructuring associated with the evolving monitoring program. Nor, apparently, had Strauss been made aware that this evolution had been taking place over the past three years. Instead, he believed that the Commission, at his instigation, had initiated the program on long-range detection in the spring of 1947 and had played the leading role in keeping it alive despite bureaucratic inertia.

This belief is reflected in the four-page memorandum to the file entitled "History of the Long Range Detection Program," which he wrote on July 21, 1948. He begins this memorandum by characterizing it as "a chronological outline of matters related to the long range detection of atomic explosions to date." But his outline essentially begins with his own memorandum of April 11, 1947, which brought the subject to the attention of the newly-appointed Commissioners. Only a few lines are devoted to "some work" which preceded his April memorandum, such as the Crossroads experiments, whose "results were not conclusive." And he goes on to say that, "the subject was dormant until the Atomic Energy Commission pressed for its activation in the Spring of 1947."[5]

On July 27, Lilienthal forwarded a copy of Strauss's "History" to Bureau of the Budget Director James Webb, who had requested background information on this topic. In addition, Strauss distributed copies to other administration officials.[6] There is no evidence that this document had any significant influence on the course of events at the time, but it appears to have bolstered Strauss's reputation in government circles and biased statements about the origins of the monitoring system that have appeared in subsequent histories published over the past few decades.[7]

Hegenberger and Johnson's decision to initiate radiological monitoring antedated Strauss's protestations on this topic. The decision appears to have been dictated partly by the fact that the Joint Chiefs of Staff had advanced the target date for having a *complete* monitoring system in operation from 1951 to 1950, and partly by pressure from within the Air Force itself. Brigadier General Don Yates, chief of the Air Weather Service, correctly foresaw that the monitoring mission would greatly enhance the Service's status in the Air Force hierarchy, and the sooner the mission began, the sooner this change in status would become a reality. Yates had retained the air-sampling equipment in some of the WB-29s that had been used in Fitzwilliam in the hope that the airborne radiological experiments would prove successful. Thus, when these experiments produced favorable results, it was possible to state, in the Air Weather Service's report on Fitzwilliam, that the Service "was ready for transition into the interim net operation under Whitesmith."[8] Whitesmith was the code name for the long-range detection of foreign atomic explosions "on a routine surveillance basis."[9]

The extent to which Yates influenced the Air Force decision on monitoring is only partially documented because many of his interactions with the Air Force chief of staff on this topic were purposely unrecorded. With regard to these interactions, the final Air Weather Service report on Fitzwilliam states: "Much of the initial discussion of the detection program was kept in verbal form for security reasons."[10] Nevertheless, it seems reasonable to suppose that the deteriorating international situation, coupled with the unexpected success of airborne radiological detection, strengthened Yates's advocacy of the interim net concept.

Hegenberger and Johnson were not averse to the idea of deploying the interim radiological net immediately, but practical difficulties had to be overcome before the net could become a reality. The first of these was technical: although the Air Force had filter-equipped planes to collect the samples of airborne dust, there was no central laboratory to analyze them. After Fitzwilliam, the short-term contract with Tracerlab had ended, and the existing Air Force laboratories did not have the facilities and personnel to perform this task.

Another difficulty was organizational. Initially, AFMSW-1 had been structured to plan and implement a program of research rather than to oversee an operational monitoring system. The original program prepared by AFMSW-1 scientists called for two years of research on various methods of long-range detection to be followed by deployment of these methods as an atomic surveillance system focused on the Soviet Union. It had been anticipated that AFMSW-1 would be restructured to operate the surveillance network when monitoring began in 1951, the original target date assigned to Whitesmith by the Joint Chiefs of Staff.

Before Fitzwilliam, all the methods of long-range detection were expected to require considerable development to meet operational requirements. Little thought had been given to the possibility that one method might reach operational status long before the others. Indeed, little thought had been given to the way the various methods would be employed as part of an integrated system of surveillance. This shortcoming had been noted in the final report on Fitzwilliam:

In the rush of preparations for FITZWILLIAM . . . little attention could be paid to the analysis of the ultimate surveillance problem. Each of the pertinent scientific fields (i.e., nuclear, acoustic, seismographic, magnetographic, ionospheric, etc.) was regarded at that time as a distinct method of detection, complete in itself, rather than as one of many possible components of a complete surveillance system designed to provide the best possible determination of whether an atomic explosion has occurred.[11]

The unanticipated evidence provided by Fitzwilliam that radiological monitoring could detect a Soviet bomb test conducted in the atmosphere forced Johnson and his staff to rethink their plans. It now appeared that the most prudent course would be to establish an interim radiological network to which sonic and seismic components could be subsequently added as they became sufficiently developed to be used to monitor the Soviet Union. Their plans also included the establishment of experimental seismic and sonic networks, chiefly within the United States, using existing instrumentation known to be inadequate for monitoring the Soviet Union. Such networks, they argued, were needed to conduct trials of modifications intended to improve the apparatus and to train personnel to operate the "complete" radiological, seismic, and sonic surveillance network that would target the Soviet Union.

This new focus on using the various monitoring techniques as complementary elements of an integrated system prompted Hegenberger and

Johnson to emphasize in their reformulated program that while, in the short term, an interim radiological network would be worthwhile, in the long term the radiological method could not stand alone and the urgent need to develop adequate sonic and seismic methods remained unchanged.

Early in August, Hegenberger presented a new, revised plan to the Research and Development Board for review. This was not the first such plan submitted to the Board by the monitoring agency. To understand the response this plan elicited, it is necessary to review the brief history of the planning activities of AFMSW-1.

THE BEST LAID PLANS

In November 1947, shortly before the formation of AFMSW-1, an effort to fund the anticipated monitoring program was initiated. According to a later (1948) summary of this program by Dr. Karl T. Compton, an eminent physicist and president of Massachusetts Institute of Technology, who had been named the new chairman of the Research and Development Board, "Dr. Bush [the Board's original chairman] took up with the Director of the Budget the importance of following every possible lead to discover if and when the Russians explode an atomic bomb. Thirty million dollars was made available to carry on the research and prepare for a reconnaissance network, with the rough guess that an additional 13 million might be required to complete the network."[12] Bush's estimate for the cost of developing and deploying the monitoring network was apparently based on the plan that Colonel Seeman had prepared at Groves's order. The proposed amount was not insignificant. It represented about 9 percent of the total spent by all three military services on research during 1947, placing this program among the more costly military research projects to be undertaken in the coming year.[13]

In December 1947, after AFMSW-1 was created, Hegenberger assigned a high priority to having the funds set aside for the monitoring network allocated to his newly formed organization. The first step in accomplishing this goal was the preparation of an updated and more detailed research plan. Thus, one of Ellis Johnson's first tasks after joining AFMSW-1 as technical director was to produce a new version of the plan for creating a monitoring system, one that would incorporate information garnered at the interagency conference held by AFMSW-1 earlier that month. This task was deemed so important that Johnson opted to carry it out in parallel with the planning and implementation of Operation Fitzwilliam. This choice imposed a heavy workload on his already overburdened staff.

As a result of this effort, by the end of February 1948, the under secretary of the Air Force was able to report to Defense Secretary Forrestal that "a two-year research and development program, based on our present knowledge of the [monitoring] problem, has been prepared." He also emphasized the rudimentary nature of the "present knowledge" by stating: "We feel that the detection problem is still in the research and development phase. A tremendous effort on both aspects is still necessary."[14]

The second step that had to be completed before the funds could be allocated was getting the Research and Development Board to approve the plan. Under the National Security Act of 1947, the Board had become one of the staff organizations of the secretary of defense. Its primary role as coordinator of military research, however, remained unchanged, and military agencies were obliged to submit plans for proposed research to the Board. Depending on the nature of the plan, it was assigned to one or more of the Board's eighteen committees, each representing a specific scientific area, for review. The results of these deliberations were reported to the Board which resolved any differences among committees and produced a final evaluation. As the final reviewing body within the Defense Department, the Board's findings were considered as emanating from the secretary of defense.

This procedure was not only supposed to eliminate duplication of research by the military agencies, but also to result in modifying research plans in ways that increased their cost-effectiveness. Hence, the review process sometimes placed committee members in the position of carrying out a kind of cost-benefit analysis based in part on their technical knowledge and in part on information about the tactical and strategic needs of the military. Committee members were expected to possess the requisite technical knowledge, but they could obtain authoritative information about military needs only by soliciting the Joint Chiefs of Staff for guidance on a case-by-case basis. In reviewing projects where this situation arose, and the creation of the monitoring system was such a project, the time needed by the Joint Chiefs to formulate a response to a request for guidance was additive to that required by the ruminations of one or more of the Board's committees and the deliberations of the Board itself. The approval process could thus become a protracted "paper chase."

On April 14, Hegenberger sent the revised plan to the Board for an "informal" review by Bush and the Board's scientific staff. To ensure that the input of some of the best brains in the country had been included in the plan, its various elements had been further modified on the basis of information produced by conferences arranged in February and March by AFMSW-1. These meetings had gathered together over fifty scientists from academic, industrial, and government laboratories, who were selected because of their expertise in acoustics, seismology, meteorology, and radiological physics and chemistry. The total cost estimate that accompanied the plan, $43 million, was the same as the amount Bush had asked the director of the Budget to set aside before the plan existed.

Bush and members of his staff who reviewed the plan were not pleased with it. Possibly they felt that the plan had been fashioned to fit the available funding rather than to meet strictly technical requirements. They may also have found it difficult to accept a plan that had been formulated without taking into account the results of Fitzwilliam, especially since the Air Force had stressed the importance of this operation for uncovering areas where research was needed. Moreover, the preliminary Fitzwilliam data would be available in a matter of weeks. In any case, a few days after receiving the plan, Bush returned it to Hegenberger. On April 19, the Board's executive secretary sent Hegenberger a request to rewrite the plan to include "a more extended technical description" of the proposed research and a more detailed breakdown of cost for each element of the work.[15]

Defense Secretary Forrestal had requested the Board to treat the evaluation of the plan for the monitoring project as a matter of urgency. Thus, on April 19, the Board's executive secretary arranged for Board scientists to meet with Johnson to outline the scope of the problem to be reviewed. June 1 was set as a terminal date for the review process.[16]

This date was quite unrealistic, however, for it assumed that a rewritten plan could be submitted to the Board before the end of May. In the event, Johnson and his small staff were fully occupied through the end of June in making a preliminary assessment of the data produced by Fitzwilliam. By mid-July a completely rewritten plan had been prepared that incorporated new information derived from Fitzwilliam, but Johnson realized that resubmission of the plan would only be the beginning of an evaluation process that he had come to regard as overly complex and time consuming.

Frustrated in his attempt to initiate what he believed were urgently needed research projects, Johnson decided to resign. According to Lloyd Berkner, a former colleague of Johnson at the Carnegie Institution's Department of Terrestrial Magnetism, Johnson's resignation "came about when he found it impossible to effectuate his program. He used his resignation as a means to register his protest against the . . . method of handling research programs."[17] Johnson was replaced as technical director by Dr. George Shortley, who had headed the Research and Development Branch of AFMSW-1.

On August 10, Hegenberger's deputy, Brigadier General M. R. Nelson, held a briefing for members of the Advisory Group on the elements of the new plan. In this plan, radiological surveillance of the Soviet Union on a "research" basis would be initiated in parallel with a research program that emphasized sonic and seismic methods and the further development of radiological techniques. The characterization of the surveillance aspect of the plan as a "radiological research net" was apparently a pro forma acknowledgment that the Joint Chiefs of Staff had yet to authorize operational monitoring. Indeed, some eight months of paper-shuffling were to ensue before the Joint Chiefs would formally approve the establishment of the interim long-range detection system.

Nelson and Shortley presided over the briefing. Although he had resigned, Ellis Johnson was also present. The plan they proposed called for
$14.4 million to be spent on research, of which roughly equal amounts, in
terms of manpower and overhead costs, would be spent on developing
radiological, sonic, and seismic methods. However, the total cost of sonic
and seismic research was greatly inflated by the need to spend an additional
$5.6 million on conventional explosives to simulate, albeit on a very small
scale, the blast effects of atomic bombs. Such simulation was regarded as
the only solution to the problem of developing sonic and seismic monitoring methods in the absence of tests of real atomic bombs. In addition, $6.4
million was needed for contingencies and to cover the more than $2 million
already spent in fiscal 1948, which had to be repaid out of fiscal 1949
funding. Apart from research costs, $3.6 million had to be included to pay
for the operation of the surveillance network, even though the expense of
the air-sampling flights was borne by the Air Weather Service. These
amounts, totaling $30 million, were to be expended in fiscal 1949, and $13
million ($8 million for research and $5 million for operation of the surveillance network) would be needed in fiscal 1950.[18]

Technically, the plan placed considerable emphasis on seismic research,
since there was no possibility that an interim radiological net would detect
an underground test. Seismic methods appeared to offer the only hope of
detecting such a test, but much research would be needed to achieve an
adequate level of reliability. According to William Golden, who was present
at the briefing,

In connection with the seismic program, Dr. Johnson commented that, in the present state
of the art, there is only about a one to one chance of success in distinguishing between an
earthquake and a major explosion. Furthermore, he did not think there was much chance
by seismic methods of distinguishing a deep underground atomic explosion from a similar
explosion of an appropriate quantity of chemical explosives.[19]

Since an above-ground detonation could be detected radiologically, it
appeared clear that the addition of seismic means that could be relied on to
identify an underground atomic explosion at long range would result in a
"complete" monitoring system. Seemingly, the further addition of a sonic
network to such a monitoring system would be, to a degree, redundant.
Anticipating this chain of reasoning, Johnson and his staff had prepared a
rationale to justify the proposed program of sonic research, which was the
most costly element in the plan when the necessary explosives were included in the total.

According to this rationale, ground stations incorporating improved
versions of the Signal Corps' sonic detector arrays would alert the radiological surveillance net whenever a large above-ground explosion took
place in the Soviet Union. Since sonic signals could be used to locate the
burst, special flights could be vectored to intercept the radioactive cloud

when it crossed the border of the Soviet Union. The sonic network would thus enhance the reliability of the radiological method and reduce the need for continuous and costly flights patrolling the entire periphery of the Soviet Union. Moreover, relative to other methods, the sonic network could provide much more accurate information on the time and place of the explosion. For example, it was expected that sonic signals could time the burst to within ten minutes versus one to three days for the radiological method, and they could locate the burst to within 100 miles versus about 1,000 to 2,000 miles for the meteorological method of backtracking the radioactive cloud.

As a final point, it was stated that "An acoustic signal would assist in the clear distinction between two atomic bursts, in case a foreign power should attempt to disguise a burst by setting off the bomb shortly after we had set off one ourselves."[20] As will be seen, however, despite efforts to justify it, the need for sonic research was later to become a point of contention between the Air Force and the Research and Development Board.

Perhaps the most intriguing aspect of the plan was the inclusion of radiological research that, if successful, would mandate an expansion in the mission of the monitoring organization to include detection not only of a Soviet bomb test but also of activities that would necessarily precede such a test. In his report on this portion of the briefing, William Golden stated

The Air Force is proposing a restatement of the mission of this project from "the detection of an atomic explosion anywhere in the world" to a broader one which is to cover the detection not only of atomic explosions but of uranium mining, U-235 separation, reactor operation and plutonium production, and stockpiling of fissionable materials. It is understood that a letter on this subject is in the mill.[21]

As part of the proposed research to carry out this expanded mission, the notion of detecting the radioisotopes of some of the noble gases emitted during certain atomic operations, originally invented by Alvarez and used in World War II, had been revived.

Following this briefing, the plan was submitted to the Research and Development Board for review. However, in anticipation that the plan would be approved, Hegenberger and Johnson had already initiated steps to overcome the technical and organizational difficulties associated with the operation of the interim radiological net. The chief technical difficulty—the lack of analytic facilities—was removed by negotiating a new contract with Tracerlab for radiochemical analysis of airborne dust samples collected by the planes of the Air Weather Service.

After Fitzwilliam, the short-term contract with Tracerlab had ended, and the temporary personnel hired by the firm had departed. Some key scientists, however, such as Dr. Lloyd Zumwalt, had been retained in the expectation that another Air Force contract might be forthcoming. Thus, Tracerlab had positioned itself to react swiftly to the Air Force's needs. Under the new contract, Tracerlab was to provide personnel to operate the laboratory for analyzing Air Force samples, located at Hickam Air Force Base, Hawaii. Dr. Frederick Henriques, the firm's technical director, became the principal investigator on the contract and supervised a second sample-analysis laboratory established at Tracerlab's headquarters in Boston.[22]

The organizational difficulty associated with implementing the interim net concept was addressed by activating the planned-for but hitherto neglected Operations Branch within AFMSW-1. By the end of July, another important organizational change had taken place. AFMSW-1 was transferred within the Air Force from the deputy chief of staff for materiel to the deputy chief of staff for operations, where it was placed under the aegis of a newly established Atomic Energy Office. This, of course, necessitated a change in name: AFMSW-1 became AFOAT-1 (decoded as *Air Force*, deputy chief of staff for *Operations*, *Atomic* Energy Office, Section *One*).

This transfer signified that, in the eyes of the Air Force, the scope of activity of the monitoring organization had broadened. Instead of focusing exclusively on research and development of monitoring methods, AFOAT-1 was expected to oversee operations involving the use of at least one of these methods as part of an interim long-range detection system. Symbolic of the renamed monitoring agency's intent to begin the operational phase of its mission was the assignment, on July 28, of the code word Workbag, to denote "the participation of the Air Weather Service in a radiological research net operated under the auspices of AFOAT-1."[23] At this point in time (July 1948), the activities of Hegenberger and his scientific staff became bifurcated. Part of their efforts over the next year were devoted to the creation of an interim surveillance net that targeted the Soviet Union and to the research needed to improve this system. But much of their time was absorbed by another activity: obtaining Research and Development Board approval for the proposed AFOAT-1 program. Because the process of winning Board approval reshaped the AFOAT-1 program, it is this process that will be analyzed and described in the next chapter. The tale of how the interim net was implemented will be deferred to Chapter 10, when it can be understood in the context of the reorientation of AFOAT-1's mission that resulted from the interactions of AFOAT-1 scientists with members of the Board's reviewing bodies.

NOTES

1. Strauss, "History of the Long Range Detection Program," July 21, 1948, p. 5.
2. L. L. Strauss to Commissioners, August 9, 1948. Strauss Papers, Atomic Energy Box, HHPL.
3. Ibid.
4. L. L. Strauss, memorandum to the file, August 10, 1948. Strauss Papers, Atomic Energy Box HHPL.
5. Strauss, "History of the Long Range Detection Program," July 21, 1948, p. 1.

6. D. Lilienthal to J. Webb, July 27, 1948. Strauss Papers, AEC series, Box 113, Folder on Tests and Testing, HHPL.

7. Although Strauss's 1948 "History" was classified Top Secret, it was sent to influential members of the administration and Congress who were cleared to receive such material. This document apparently left no doubt in the minds of some of its readers about Strauss's role in the development of the monitoring system. For example, in 1953, Carl Durham, chairman of the Joint Congressional Committee on Atomic Energy, indicated in a press release that Strauss was the "One individual" who was largely responsible for the creation of the monitoring system. Strauss also made this claim himself. In a letter to President Dwight D. Eisenhower, he referred to "a monitoring system which I initiated in the fall of 1947." See "Statement of the Joint Committee on Atomic Energy," January 17, 1953, issued by Chairman Carl Durham; and L. L. Strauss to President Dwight D. Eisenhower, August 7, 1953. Both documents are in the Strauss Papers, AEC Series, Folder on Tests and Testing, HHPL. Regarding its historiographic effect, Strauss's 1948 "History" is cited in many histories of the period, including the official history of the Atomic Energy Commission. See Hewlett and Duncan, *Atomic Shield*, pp. 130, 131.

8. "Project Firstrate, AWS Participation in Fitzwilliam," p. 46. In "Report of Operation Fitzwilliam," Vol. I, Tab D.

9. R. H. Rathbun, "Annex A and B," December 31, 1947. RG 330, Secretary of Defense, Box 7, Folder 100 Atomic Energy/Long Range Detection, NA.

10. "Project Firstrate, AWS Participation in Fitzwilliam," p. 4. In "Report of Operation Fitzwilliam," Vol. I, Tab D.

11. "Report of Operation Fitzwilliam," p. 8. In "Report of Operation Fitzwilliam," Vol. I, Tab D.

12. K. T. Compton, memorandum to the files, November 29, 1948. RG 330, Secretary of Defense, Research and Development Board, Box 7, Folder 13W4, NA.

13. The total spent on military research in 1947 was about $500 million. See Yergin, *Shattered Peace*, p. 267.

14. A. S. Burrows to J. V. Forrestal, "Long Range Detection of Atomic Explosions," February 20, 1948, p. 1. RG 330, Secretary of Defense, Box 77, Folder CD-16–1–2, NA.

15. L. R. Hafstad to Major General A. Hegenberger, "Air Force Whitesmith Project," April 19, 1948. RG 330, Secretary of Defense, Research and Development Board, Box 7, Folder 13W4, NA.

16. F. H. Richardson to Director, Planning Division, Research and Development Board, April 19, 1948. RG 330, Secretary of Defense, Research and Development Board, Box 7, Folder 13W4, NA.

17. L. B. Berkner to M. Tuve, August 3, 1948. Berkner Papers, Manuscript-Division, Box 3, Personal Correspondence, 1948, LC. Another factor that probably influenced Johnson's decision to resign was, according to Berkner, the fact that "General Tony McAuliffe offered Johnson the opportunity to demonstrate what could be done in accordance with his ideas by making him head of the Army's operational [*sic*] research set-up." Johnson accepted this job and became eminent in the field of operations research.

18. "United States Air Force Fiscal Year 1949 Program for Long Range Detection," n.d., pp. 3–5. RG 330, Box 7, Folder 100 Atomic Energy/Workbag, NA. Also

see W. T. Golden to L. L. Strauss, August 12, 1948, pp. 1, 2. Strauss Papers, AEC Series, Box 113, Folder on Tests and Testing, HHPL.

19. Golden to Strauss, August 12, 1948, p. 2.

20. "United States Air Force Fiscal Year 1949 Program for Long Range Detection," n.d., p. 28.

21. Golden to Strauss, August 12, 1948, p. 2.

22. F. C. Henriques interview, August 18, 1981.

23. W. J. Morgan, to Distribution List, "Annex A and B," July 28, 1948. RG 330, Secretary of Defense, Box 77, Folder CD-16–1–2, NA.

CHAPTER 9

VESTED INTERESTS AND COMMITTEE POLITICS

In setting up the Joint Research and Development Board in 1946, its chairman, Vannevar Bush, had originally envisaged it not as an evaluating body but as a court of arbitration for adjudicating the overlapping research interests of the military services to prevent duplication of effort. It soon became apparent that, although the Board accomplished this purpose, another problem remained: namely, the failure of the services to critically evaluate their own programs.

This problem was later addressed by the 1947 Security Act, which not only removed the "Joint" from the Board's designation, but also expanded its mandate to include evaluating the technical merits of research proposed by the services. Since military research involved money, manpower, and prestige, the services were strongly motivated to defend and aggrandize their proposed research programs. In the process of judging the "quality" of these programs, the Board and its committees became an arena for interservice rivalries. One of the more bizarre outcomes sometimes produced by such rivalries, and by the ponderous and overly decentralized Board review process, was that a panel composed of a few "disinterested" civilian scientists could wind up arbitrating the fate of a program that affected the nation's military posture and strategy.

The Board's shortcomings were later recognized, but this occurred after the period in which the research program of AFOAT-1 was under review. Hence, this program was subjected to the full array of organizational mechanisms involved in the Board review process. This process will be described in detail because the resulting modifications of AFOAT-1's research program were an important factor in shaping the kind of organization it eventually became.

THE PAPER CHASE

On August 18, 1948, the Research and Development Board girded itself for the task of reviewing the newly written plan submitted by Hegenberger. The members of the Board's Committee on Atomic Energy had the appropriate security clearances and had been periodically commenting on the development of long-range detection methods since December of 1947. Hence, the Board had assumed that the Committee would be a reviewing body for the plan. However, the plan's emphasis on seismic and sonic research made it necessary to enlist the help of reviewers with expertise in these fields.

According to the Board's Planning Division director, the Board thus decided on "a division of this job" which called for the plan to be reviewed by the Committee on Geophysical Sciences as well as the Committee on Atomic Energy. The Planning Division would then be responsible for coordinating and controlling the work of both these committees on the monitoring project. Although this administrative scheme seemed reasonable, the Division's director felt impelled to record his misgivings about this approach by pointing out that the review process "is difficult to follow when the actions are divided between a number of agencies of the Board."[1] His misgivings were to be amply borne out in the months that followed.

A practical difficulty appeared immediately: The members of the Committee on Geophysical Sciences were not cleared to handle material at the level of classification of the plan. The Board's executive secretary thus decided to proceed sequentially by allowing the Committee on Atomic Energy to review the plan while security agencies were conducting the necessary investigations to clear the members of the Committee on Geophysical Sciences.

The following day, August 19, the Committee on Atomic Energy met to consider the AFOAT-1 plan.[2] On its own, the Committee on Atomic Energy felt competent to evaluate and approve, with no dissenting members, the radiological portion of the plan "in principle and amount." They also approved unanimously the research aspects of the seismic portion of the plan, but not the part of the program that called for deploying an experimental seismic network. With regard to the proposed sonic work, however, a serious breach opened up between the Navy and Army members, who saw little value in the sonic method, and the Air Force members, who strongly advocated further sonic research.

The possibility that further disagreements would have overtones of service rivalry may have been a factor in the Committee's later decision to form an ad hoc panel of university scientists to make a separate review of the AFOAT-1 plan. In any case, on September 16, the Board accepted the Committee's findings with the reservation that the Air Force should use the Committee's conclusions and recommendations as interim guidelines until a joint meeting with the Committee on Geophysical Sciences could be held.

The Board also decided to solicit guidance concerning the strategic requirements associated with monitoring from the Joint Chiefs of Staff. Specifically,
the Board wanted to know how soon routine monitoring would be needed
and what priority the monitoring program should enjoy vis-à-vis other
projects under review.[3]

At the next meeting of the Committee on Atomic Energy, on September
23, 1948, General Schlatter was able to report that AFOAT-1 would soon
submit a revised plan for the seismic and sonic research that would conform
to the Committee's conclusions and recommendations. Apparently to ensure that this revised plan would receive expert scrutiny, free from any taint
of interservice rivalry, the Committee voted unanimously to establish an ad
hoc panel that would be capable of technically evaluating the radiological,
seismic, and sonic portions of the plan as an integrated whole.[4]

In revising the seismic and sonic portions of the plan, AFOAT-1 had
rearranged the scheduling of the research projects and the deployment of
experimental networks so that they were carried out sequentially rather
than in parallel, with deployment occurring only if the research was successful.[5]

On October 15, AFOAT-1 duly submitted the revised plan to the Board.
This plan was given to both the Committee on Atomic Energy and the (now
cleared) Committee on Geophysical Sciences for review. The following
week, on October 22, the Committee on Atomic Energy formed the so-called
Loomis Panel to review the AFOAT-1 plan. This Panel was made up of
university scientists and included Alfred L. Loomis as chairman, Charles P.
Boner, Joseph C. Boyce, and L. Don Leet.

At a meeting on October 29 (to which members of the Committee on
Atomic Energy had been invited), the plan was reviewed by the Committee
on Geophysical Sciences. The Committee concluded that the revised seismic and sonic programs were "an adequate scientific approach to the basic
problem." Since the Joint Chiefs had not responded to the Board's request
for guidance, the Committee decided, "on the assumption that the ultimate
objectives are of the highest priority," that the proposed sonic and seismic
programs were justified. It further recommended that the Air Force begin
work on the program immediately.[6] These recommendations, however,
were later to be contradicted by the findings of the Loomis Panel, an
outcome that was to pique the members of the Committee on Geophysical
Sciences.

Needless to say, the problem of producing a final evaluation of the
AFOAT-1 plan was becoming quite complicated. This evaluation, of course,
was only one of a myriad of projects being reviewed by the Board, and its
chairman, Vannevar Bush, began to find himself physically unequal to the
demands of his position. In October, his failing health forced him to resign.
Indeed, his physical condition had deteriorated markedly, apparently from
stress-related causes. In a letter to his successor, Karl T. Compton, he wrote:

"I evidently walked pretty close to the edge and I guess it is fortunate that I did not walk over the brink."[7]

After Compton took over as the Board's chairman, he made a strenuous effort to understand the underlying issues of each project under review. After conferring with AFOAT-1 staff members on November 28, he prepared an accurate and succinct summary of the status of the monitoring program in the form of a two-page memorandum to the file. In this document, he noted that, "on the basis of intelligence reports," the Joint Chiefs of Staff had changed the target date for deploying a complete monitoring system from 1951 to 1950.[8] He further recorded that, because of this change, the plan submitted by AFOAT-1 was based on the belief that it would be "impossible to meet this dateline unless both research and the establishment of the [experimental seismic/sonic] network, with training for personnel, are carried out in parallel."

In commenting negatively on this approach, Compton wrote: "We felt, however, that the state of the seismic and acoustic art and the probability of success were not sufficient at this time to justify a simultaneous establishment of the surveyance network." Essentially, the Board had concurred with the recommendation of the Committee on Atomic Energy that the approach of the seismic and sonic portions of the program be changed to one based on establishing "the feasibility by research first, and to follow this, if successful, by the network." The reasons given by Compton for this conclusion were "expense, uncertainty of success, and doubt regarding the 1950 date line."[9]

The doubt about the new target date set by the Joint Chiefs of Staff is especially significant because it illustrates the tenacity with which insiders clung to a belief in a relatively remote date for the advent of a Russian bomb, a belief fostered by the Murray Hill Area concept. Compton wrote: "We in RDB [the Research and Development Board] deem it very unlikely that Russia will be in a position to test an atomic bomb as early as 1950, or within several years after that date."[10] This belief, of course, echoed "the unanimous opinion" of the Committee on Atomic Energy on this topic expressed a year earlier.[11] One implication of this belief was that there was ample time to conduct research on seismic and sonic monitoring methods before deploying them.

On December 10, 1948, the Loomis Panel submitted its report to the Committee on Atomic Energy. The report began by pointing out that "Nowhere has the Panel found an evaluation or, in fact, a plain statement of the importance which should attach to the detection of a foreign detonation." The Board was unable to provide this information because the Joint Chiefs had yet to respond to Bush's request for guidance in this area. Lacking such guidance, the Panel elected to divide its recommendations into two parts: Plan A to be undertaken if "medium priority" is assigned to the program, and Plan B if the priority is "very high."

A second problem faced by the Panel was the lack of knowledge about whether the Soviets would test their first bomb in the atmosphere or underground. The Panel opined that "an underground test is highly unlikely."[12] They came to this conclusion because they believed that it did not appear probable that such a detonation could provide Soviet scientists with sufficient information on the characteristics of the bomb. (In the previous August, Atomic Energy Commission scientists had given Hegenberger a contrary opinion on this point.) Therefore, the Panel proceeded to evaluate the AFOAT-1 plan on the basis that the first Soviet bomb test would be an above-ground detonation.

With regard to the radiological research incorporated in the plan, the Panel focused its attention on the methods proposed for the detection of Soviet atomic activities other than the test of a bomb, such as reactor operation or plutonium production. They stressed the importance of this research, although they urged that the development of such methods should proceed strictly on "a bread-board scale" until results warranted their use in the field.

The Panel approved the establishment of the radiological surveillance net and, with few reservations, approved the research in the radiological portion of the plan. However, it recommended cuts in much of the seismic research and suggested that most of the sonic program could be canceled. In accord with the earlier conclusion of the Committee on Atomic Energy, the report stated that "the Panel is not inclined to view the prosecution of an extensive research and development effort in the acoustic field as being very important."

Oppenheimer evidently shared the Panel's dim view of seismic and sonic research, and, apparently, he made this abundantly clear to the Air Force members of the Committee—to his cost. One such member, General Wilson, later recalled that, to accomplish the monitoring mission, "the Air Force felt that it required quite an elaborate system of devices. . . . Dr. Oppenheimer was not enthusiastic about 2 out of 3 of these devices." He went on to note that, because of Oppenheimer's influence on the opinions of other Committee members, "the overall effect was to deny to the Air Force the mechanism which we felt was essential to determine when this bomb went off."[13]

The members of the Committee on Atomic Energy spent the following week analyzing the Loomis Report. At a meeting of the Committee on December 17, a motion by Oppenheimer, seconded by Conant, to accept and approve the Loomis Report was carried. The Committee also authorized its chairman to review the execution of the Report's recommendations by the Air Force.[14]

The practical effect of this action was to confirm the prior approval given to AFOAT-1 by both committees for most of the proposed fiscal 1949 program, including the seismic and sonic research. However, it curtailed the seismic and sonic program for fiscal 1950 and 1951, in accord with the Loomis Report's recommendations. The Air Force disagreed sharply with these cutbacks and appealed to the Board for a reevaluation.[15]

Understandably, the unilateral action of the Committee on Atomic Energy in approving the Loomis Report provoked the ire of the members of the Committee on Geophysical Sciences (which had been renamed the Committee on Geophysics and Geography). The executive director of the Committee wrote a stiff note to the Board suggesting that in the future "due considerations be given to the contributions that the Committee on Geophysics and Geography can make to the programs on long range detection and that it be appropriately represented in matters that are of joint interest to both the Committee on Atomic Energy and the Committee on Geophysics and Geography."[16] The Board noted this complaint, and its executive secretary duly informed the chairman of the Committee on Atomic Energy that his Committee would be expected to coordinate their future actions with the Committee on Geophysics and Geography.

By the end of January 1949, AFOAT-1 was able to produce a rewritten sonic and seismic plan that conformed to most of the recommendations in the Loomis Report. The new plan was prepared by Doyle Northrup, who conducted most of the negotiations with Panel members. Dr. Shortley had resigned as technical director of AFOAT-1, and Hegenberger had assigned the functions of this position to Northrup, another of Ellis Johnson's proteges, who was AFOAT-1's deputy chief for science.

In writing this plan, Northrup took great pains to justify the Air Force's belief that the proposed sonic program was urgently needed. He repeated the rationale given in the earlier plan and added new information derived from recent analyses of the Fitzwilliam experiments. This information illustrated, on the one hand, that it would be dangerous to rely solely on radiological methods and, on the other, that adequate sonic instrumentation would soon be available. He also rejected the notion that there was ample time to conduct research on the monitoring system simply because it was assumed the Soviets would not have the bomb until after 1950. "The only safe date for the completion of such a system," he wrote, "is as soon as intelligent and energetic use of manpower can accomplish it."[17]

The plan also contained a statement suggesting that Northrup and his staff were becoming wearied by the task of rewriting the AFOAT-1 program to conform to the findings of repeated reviews. He wrote: "It is believed that the point of diminishing returns has been reached in benefit to the program from further review and modification."[18] His plea for "final" approval, however, was ignored.

Northrup's plan called for expenditures of $23 million—$9.5 million in fiscal 1949 and $13.5 million in 1950. At a meeting on February 2, the Committee on Atomic Energy was told by General Schlatter, one of the Air Force members, that the currently proposed Air Force program was consonant with the Panel's recommendations except for an area involving about $1 million that was still under review.

During this meeting, the Committee once again split along service lines. In attempting to define the priority status of the AFOAT-1 program, the Board had asked the Committee to place the program in one of two categories: "1.a," which referred to programs "of the greatest importance and necessary to national security" or "1.b," which referred to programs "not as urgent as those in 1.a above, but nevertheless important to the national security." Oppenheimer and the Army and Navy members voted against the Air Force members to place the AFOAT-1 program in the 1.b category.[19]

On February 7, 1949, the Board's assistant executive secretary, who had been charged with the tasks of coordinating and tracking the review process for the AFOAT-1 plan, figuratively threw up his hands. In a memorandum to the Board's executive secretary, he noted that two of the Board's committees "have considered the Air Force program and made conflicting recommendations." He went on to point out that two additional committees (on Electronics and on Basic Physical Science) might, "on their own initiative," be expected to examine the AFOAT-1 plan. "It appears to be an administrative impossibility," he wrote, "to expect cooperative action from four interested Committees." He thus concluded that "only one Committee should be made responsible . . . the present policy of placing responsibility on several autonomous Committees or even jointly is not a workable solution."[20]

The response of the Board's executive secretary was to limit the review process to the two committees already involved and to again inform the respective chairmen that recommendations should only be formulated after joint consultation.[21]

On February 23, at its nineteenth meeting, the Board considered the findings and recommendations submitted by the two committees. Despite the declaration by the Board's assistant executive secretary that the two committees had produced "conflicting recommendations," the committees chose to issue a joint recommendation to the effect that the seismic and sonic research outlined in Plan A (medium priority) of the Loomis Report should be approved by the Board.

This bit of legerdemain in the service of "consistency" ignored the fact that the Air Force plan approved initially by the Committee on Geophysics and Geography differed markedly from Plan A of the Loomis Report, which was approved by the Committee on Atomic Energy and which provided a relatively lower level of funding. Hence, by accepting the joint recommendations of the two committees, the Board would be, in effect, cutting AFOAT-1's budget.[22]

The Air Force reacted vigorously to the joint recommendations. The Air Force member of the Board, General McNarney, invited Hegenberger to appear before the Board to review the AFOAT-1 program, including the latest information produced by analyses of the Fitzwilliam experiments. In his presentation, Hegenberger pointed out that the Loomis Panel did not have access to the most recent information derived from Fitzwilliam and that, therefore, some of its recommendations were questionable. He particularly emphasized that the Panel had failed to give appropriate weight to the complementarity of radiological and sonic methods in their evaluation.[23]

The Board was shaken by this counterattack, and it decided to instruct the two committees to further review the plan "in the light of recent information reported by the Air Force." Regarding its own "final" evaluation, the Board voted to defer further action pending the completion of the review by the committees.[24]

Northrup's latest version of the AFOAT-1 plan, "Technical Memo No. 31," was sent to both committees, and a joint meeting of the committees was scheduled for April 7, at which they would formulate their combined recommendations for the Board.

Upon receiving the plan, Dr. W. E. Wrather, chairman of the Committee on Geophysics and Geography, decided that his committee could not proceed further with the review process without appropriate guidance. He immediately contacted Compton requesting the answers to two vital questions: first, whether or not the monitoring program was considered "of top level priority" and, second, "with what degree of precision the location of a large explosion is to be known and within what time from such an occurrence."[25]

It seems bizarre that some seven months of deliberations on the appropriateness of the AFOAT-1 plans had taken place in the absence of authoritative answers to these questions. Knowledge of priority was an essential contextual element in deciding the funding level of the entire program. And it was necessary to know the required accuracy for the time and place of a foreign bomb test in order to evaluate the need for sonic methods of detection, since one of the chief arguments for developing sonic methods was their greater precision in determining these parameters, relative to other methods.

It appears unlikely that eminent scientists such as Compton, Oppenheimer, and Conant, and military leaders of the caliber of Hegenberger, Wilson, and Parsons failed to realize that the answers to these questions went to the heart of the matter. It is difficult to escape the conclusion that one reason the involved parties had not pressed for answers was that their own answers to these questions had been smuggled into their respective positions on the monitoring program in the form of assumptions. And each feared that "authoritative," but not necessarily "correct," answers by the Joint Chiefs might undermine their position. Indeed, when the Joint Chiefs finally spoke, their pronouncements did have this effect.

There seems little doubt that some of the participants in the reviewing process regarded themselves at least as competent as the Joint Chiefs to formulate their own answer to the question of priority. Members of the Board and of the Committee on Atomic Energy felt confident in dismissing the view of the Joint Chiefs that a Soviet bomb might be tested as early as 1950. The Committee had not hesitated to pass a resolution on the matter of priority that, in effect, evaluated the importance of the monitoring program to national security.

In the face of Wrather's reluctance to continue the reviewing process without appropriate guidance, Compton decided to pressure the Joint Chiefs to respond to Bush's September 1948 inquiry about priority and Wrather's question on the accuracy needed in determining the time and place of a Soviet bomb test. In a memorandum to the Joint Chiefs on this matter, he stressed that it was "extremely important" to have their response by March 30.[26]

Meanwhile, William Webster, chairman of the Committee on Atomic Energy, made a proposal to Compton that appears to suggest that some Committee members were beginning to find the task of evaluating the monitoring program somewhat wearisome. Webster pointed out that the only remaining point in contention between the AFOAT-1 plan and the recommendations of the Loomis Panel (involving about $1 million for seismic research) had been settled by negotiations between Northrup and Panel scientist Don Leet. But this compromise had yet to be approved by the Committee. Webster indicated that, as chairman, he had been authorized to approve the compromise that would finalize the evaluation for the Board. However, in return for his action in this matter, Webster requested that the Board call off the joint meeting of the committees scheduled for April 7. Instead, he proposed that the Loomis Panel be reconstituted and designated as a joint group for both committees to perform a final evaluation. The Panel's recommendations, together with any comments the committees might feel inclined to make, could then be submitted to the Board.[27]

This proposal, phrased in terms of a *quid pro quo*, was not well received. In summarizing this incident in a memorandum to the file, Compton wrote that Webster's suggestion "has apparently aroused quite a bit of feeling, being interpreted as blackmail." Compton decided to adopt Webster's suggestion for having a joint panel review the plan, but he opted to allow the meeting of the two committees to take place as scheduled because he wanted the question of the reliability of the radiological method examined further by both committees.[28]

The Air Force, however, did not want the Board to further defer approval of the fiscal 1949 portion of the program. Air Force finance officers were becoming concerned by the fact that the actual expenditures of AFOAT-1 during the past year had yet to be authorized by the Board. At a conference between Northrup, staff members of the Board, and some members of both reviewing committees, it was agreed that this authorization could be effected by including an agenda item for the next meeting of the Board to "Note and approve AFOAT-1 implementation of the FY1949 program as described in AFOAT-1 Technical Memo No. 31."[29]

To emphasize the Air Force's anxiety on this matter, the assistant deputy chief of staff, Major General E. M. Powers, wrote to Compton to inform him that the Air Force had assumed that it was permissible to execute the portion of the plan that had been approved by the Committee on Atomic Energy on December 17, 1948. He urged that the Board formally endorse the expenditures that had been made under this assumption. In reply, Compton informed Powers that a resolution to this effect would be considered at the next meeting of the Board, which was to take place on March 29.[30]

On March 28, 1949, the Joint Chiefs of Staff responded to Compton's request for guidance. On the question of priority they stated:

From the military point of view the receipt of positive and timely information indicating that an atomic explosion had in fact been accomplished by a potentially unfriendly nation is of great importance to our national security. Conventional intelligence methods cannot be relied upon to provide conclusive and timely information of the atomic potential of a foreign power. It is therefore necessary to supplement conventional intelligence by other methods.[31]

However, they went on to say that they did not feel that they were "in a position to comment on the technical factors which must determine the research and development effort . . . to make the surveillance system effective."

This statement, in effect, allowed the Research and Development Board to determine the priority of the monitoring program in accord with a Master Plan that had been created by the Board. This Master Plan had various categories, one of which was "Intelligence Planning and Operations," designated as IO. Within this category was a subcategory for "Long Range Detection of Foreign Atomic Explosions," designated IO-7.[32]

The Board had rated IO-7 (the monitoring program) at a lower level than desired by the Air Force, and this had remained a point of contention. (The Board's action was consonant with that of the Committee on Atomic Energy, which assigned, on February 22, 1949, a "1.b" priority level.) The Board, however, interpreted the pronouncement of the Joint Chiefs as a validation of their right, in this case, to assign priority. Therefore, the Board's executive secretary promptly informed Compton that the Board's priority level for the monitoring program was "not in agreement with the priority assumed by the Air Force." He also lost no time in notifying AFOAT-1 and the chairmen of both reviewing Committees of this fact. He went on to request that "future planning and implementation of the AFOAT-1 program should be made on a realistic basis consistent with the recent action of the Joint Chiefs of Staff."[33]

In their response to Compton's memorandum, the Joint Chiefs also sanctioned the operational status of the interim radiological net. It appears, however, that this action was necessitated by AFOAT-1's insistence that the new target date of 1950 for a complete operational monitoring system could not be met due to the cutbacks that had been imposed on its proposed program. In this regard, the Joint Chiefs stated that

The Joint Nuclear Energy Intelligence Committee composed of representatives of the Central Intelligence Agency, the Atomic Energy Commission, and the armed forces have estimated that the earliest date by which the Soviets may have exploded their first test bomb is mid-1950. From an operational point of view, the target date for readiness of a reliable system of surveillance reasonably capable of detecting any atomic explosion occurring within USSR controlled territory should be not later than the estimated earliest possible date. However, inasmuch as the Joint Chiefs of Staff are informed that this date as currently estimated, may not be technically feasible of attainment, the system of surveillance should be placed in operation as soon as practicable. Thereafter the system should be perfected as rapidly as possible, keeping abreast of new developments.[34]

Despite their apparent expectations of "new developments," the answer given by the Joint Chiefs to the second question posed in Compton's memorandum had the effect of weakening the rationale for one such development—that is, the sonic method. In replying to the request for guidance on the question of locating and timing a foreign bomb test, they wrote: "From strategic considerations alone an accuracy in the determination of the time of an atomic explosion such that an explosion could be determined within one or two months after the event is acceptable. Knowledge as to the geographic location of such an explosion is desirable but not essential, provided the explosion can be determined to be within the USSR or Soviet sphere."[35] This statement tended to vitiate one of the principal arguments for developing sonic methods of long-range detection, since the required accuracy of time and place, as indicated by the Joint Chiefs, was well within the capability of existing radiological and meteorological techniques.

The remarks of the Joint Chiefs provided a new framework within which AFOAT-1, the Board, and its various reviewing bodies could formulate a mutually acceptable plan. Indeed, the Joint Chiefs, the last of the major players in the paper chase, had produced a document that would catalyze a significant change of emphasis in the AFOAT-1 program.

THE BONER PANEL

On March 29, 1949, the Research and Development Board held its twentieth meeting. In a fuzzily worded resolution, its members approved the portion of the AFOAT-1 plan for fiscal 1949 that had received prior approval by the Committee on Atomic Energy. This action authorized expenditures that had already been made by the Air Force, but the status of further expenditures against the unspent funds budgeted for approved projects in 1949 remained unclear. Indeed, the secretary of the Air Force later indicated that he was "in need of more spelling out as to just what had been agreed to by the Board."[36]

The Board's subsequent actions, however, made it clear that their approval of fiscal 1949 expenditures, even if obligated but not paid out by the Air Force, was contingent on the findings of a further evaluation of the AFOAT-1 program. In this interpretation, reductions in the fiscal 1949 budgets of approved projects were a possibility (and, in the event, a reality).

The Board also passed a resolution directing the chairmen of the Committees on Atomic Energy and Geophysics and Geography, in consultation, to form a joint panel to evaluate the AFOAT-1 program for 1949, 1950, and 1951. In the words of the Board's executive secretary, the Committee on Atomic Energy was to be "administratively responsible" for the panel, but the panel's report on the AFOAT-1 program would be subject to the scrutiny and approval of both committees, whose chairmen would then forward the panel's findings to the Board as the joint recommendations of their respective Committees.[37]

It is pertinent to point out that in the more usual process of review, the Board depended on its committees to perform the active role of making a detailed analysis and evaluation of submitted research programs. The Board itself generally acted as a kind of filter through which Committee recommendations passed or were not passed or were passed in modified form to emerge as the Board's final word. In the case of the monitoring program, the Board's decision to form a joint panel had the effect of delegating the usual functions of the committees to the panel. The committees themselves were thus relegated to performing the more passive role of accepting, rejecting, or modifying the recommendations of the panel. The committees, therefore, became merely an additional layer of filtration between the panel and the Board.

In tracing the events that led to this situation, one factor that seems to have been instrumental was that both the Committee on Atomic Energy and the Committee on Geophysics and Geography were comprised—as was the Board itself—chiefly of representatives of the three military services. The Board's dependence on its committees to take the active role in the evaluation process minimized the problem of interservice rivalry on the Board itself only if the civilian members of the committees could successfully arbitrate any split of the military members along service lines. Such arbitration did not occur in the two committees reviewing the monitoring program. It is not surprising, therefore, that the Committee on Atomic Energy, on its own initiative, formed the Loomis Panel as a way of obtaining a supposedly dispassionate appraisal of the AFOAT-1 program. (The possibility that the Panel members may have had their own interests and agendas was apparently ignored.)

In later adopting the approach of creating a joint panel to evaluate the AFOAT-1 plan, the Board seems to have been addressing both the problem of interservice rivalry and the problem of evaluating a program whose scientific complexity required a range of reviewer expertise that extended beyond the purview of the members of a single committee. Perhaps the most interesting aspect of the implementation of this scheme was that a small group of five relatively unknown university scientists were to emerge as the most active players in a decision-making process that determined the future course of a government program of considerable strategic importance.

Webster and Wrather collaborated in forming a new joint panel in accord with the Board's directive. The Loomis Panel had been allowed to lapse, but two of the members of the new joint panel, Dr. Charles P. Boner, University of Texas, and Dr. Joseph D. Boyce, Hudson Institute, had served on the Loomis Panel. Another member, Dr. Athelstan Spilhaus, University of Minnesota, had supervised the balloon development work for Project Mogul, one of the first attempts to develop sonic methods of long-range detection. The remaining two members were Dr. James B. MacElwane, University of St. Louis, and Dr. J. B. Fisk, Harvard University. Boner chaired the Panel, which, not surprisingly, was later referred to as the Boner Panel or the Joint Panel, although its official designation, in accord with the Board's Master Plan, was the IO-7 Panel.

Several significant events had occurred in the larger national context in which the Panel's deliberations were to take place. Tensions had lessened somewhat following the end of the Berlin Blockade in May. In March, Forrestal had resigned as secretary of defense. His replacement was Louis A. Johnson, who had been assistant secretary of war prior to World War II and the president's personal envoy early in the war. Perhaps his chief qualification for the post of defense secretary was that he had long occupied a high position in the Democratic party. Johnson was a strong supporter of the "austerity budget" that had limited postwar military spending, a policy that he was later to admit was based erroneously on an overlong estimate for the American atomic monopoly's duration.[38]

Johnson's appointment to this post resulted in a greater "top-down" emphasis on budget cuts within the Defense Department. His policies affecting organizations within the Defense Department, such as the Research and Development Board, thus mandated careful scrutiny of all projected spending. This "belt-tightening" milieu in which the Boner Panel conducted its evaluation of the AFOAT-1 program was, therefore, conducive to further cuts in the proposed research.

The Panel was also influenced by the documentation it received. The members were provided with all the relevant memoranda, minutes, and reports containing previous evaluations and comments on the monitoring program by the Board, the two committees, and the Joint Chiefs. A common theme expressed in these documents by Army, Navy, and civilian reviewers was that the Air Force's perception of the monitoring mission displayed a kind of "tunnel vision" that failed to take into account the larger picture. This perception, in the opinion of reviewers, caused the Air Force to overestimate the importance of knowing that the Soviets had tested a bomb. Other information, they argued, was even more important, such as determining the rate of bomb production. Such information could be obtained through conventional intelligence-gathering activities aided by instrumental means to detect atomic operations. To be sure, the development of methods to detect atomic operations had been included in the projects outlined by AFOAT-1, but only as a relatively small part of the overall program. Reviewers felt that this allocation of effort was disproportionate to the value of the information such methods could provide.

For example, in commenting on the points emphasized in the Loomis Report, Compton had noted that "the panel has recommended that considerable effort be expended along conventional approaches to this intelligence problem, in view of the high cost of guaranteed detection by instrumental means. In its opinion, the detection and identification of a foreign bomb detonation is only one part of a larger and more difficult intelligence problem."[39] This view was also consonant with that of some members of the intelligence community. For instance, in a memorandum entitled "Atomic Energy Program of the USSR," Admiral Hillenkoetter, director of Central Intelligence, wrote:

In order to estimate the capability of the USSR to wage atomic warfare, it is necessary to know, not only the events that preceded the date when the first bomb is detonated, but also the capability for bomb production thereafter. In consequence, our intelligence should furnish a comprehensive picture of the Soviet program, including the means and methods by which weapons are produced, the capacity of the production installations and the rate of supply of uranium to the USSR.[40]

A second theme that informed the various evaluations that had been made of the Air Force's plans was that research on seismic and sonic methods was an overly expensive "luxury." Existing methods of radiological detection appeared adequate for detecting a Soviet bomb test since it seemed highly unlikely such a test would be conducted underground or underwater.

The negative view of seismic and sonic methods expressed by reviewers seems to have been shared by some members of the Joint Chiefs of Staff. According to historian Kenneth Condit, the chief reason for the delay in responding to the Board's request for guidance in the matter of the long-range detection program was "[d]isagreement within the Joint Chiefs of Staff." Condit also indicates the nature of this disagreement. "At one point," he argues, Army General Omar Bradley "contended that the Air Force was relying too heavily on untried seismic and acoustic methods at the expense of conventional intelligence."[41]

Bradley's position on this matter suggests that the disagreement among the Joint Chiefs mirrored the split between military members of the Committee on Atomic Energy, where the Air Force members sometimes found that their vision of the monitoring program was at variance with that of the Army and Navy members.

While these two themes, in various guises, had appeared in evaluations made prior to that of the Loomis Panel, they were highlighted in Loomis's report on the Panel's findings. In assessing the importance of knowing the time of the first Soviet bomb test relative to learning about other kinds of atomic operations, Loomis had stated: "The Panel does not feel that the exact date when the first bomb is tested is of overriding importance compared with knowledge of other significant developments and progress in the atomic field."[42]

Regarding seismic research, which was considered expensive and by no means certain to succeed, the Panel suggested that the need for such research was based solely on a "worst case scenario" in which the Soviets would explode their first bomb underground. The Panel felt that it was more "realistic" to proceed on the assumption that the first test would be conducted above the ground as an airburst detectable by radiological means. "The Eniwetok [Sandstone] air bursts," the report noted, "have been reliably detected [radiologically] halfway around the world." The Panel was even less enthusiastic about sonic research, which they did not regard as "very important."[43]

Such considerations led the Loomis Panel to recommend a restructuring of the Air Force program to make the effort to develop methods of obtaining information about atomic operations more effective and to decrease the level of research on sonic and seismic methods of detecting a bomb test. The Air Force had accepted some of the Panel's recommendations "as is" and others only after they had been modified by negotiation, but stopped short of reorienting the basic plan.

In resisting the kind of restructuring of its program that had been suggested by the Loomis Panel, the Air Force had been able to cite the high priority originally assigned to the task of detecting a Soviet bomb test, and to point out that the mission assigned to the Air Force to carry out this task remained unchanged. The belated guidance of the Joint Chiefs, however, appeared consonant with the two themes that had characterized the various evaluations that had been made of the Air Force program. These themes supported the reviewers' vision of what the Air Force program should be.

The radiological monitoring net to detect a Soviet bomb test was a crucial element of this vision, as was more effective research on methods of detecting other atomic operations. There was little space in this picture, however, for seismic research and even less for the development of sonic methods. Regarding the latter, one of the Air Force's arguments for sonic detection—its accuracy in determining the time and place of an explosion—had been vitiated by the comments of the Joint Chiefs. Also unpersuasive was the argument that sonic detection would alert the system, thus reducing the need for expensive monitoring flights patrolling around the border of the Soviet Union. Fitzwilliam experiments suggested that patrolling the entire periphery of the Soviet Union was unnecessary. Instead, positioning aircraft below the high-altitude jet stream flowing out of the Soviet Union appeared to be a reliable and effective monitoring approach that could be inexpensively implemented with relatively few aircraft.

Needless to say, parts of this vision were not shared by the Air Force. But the nature of the recommendations made later by the Boner Panel suggests that their perusal of the documentation in which this vision is spelled out strongly influenced their findings. Indeed, shortly after the Boner Panel began its deliberations, the Committee on Atomic Energy passed a resolution that appears to have been prompted by the vision of the monitoring program just described.

The Committee's resolution resulted from their review of yet another new version of the AFOAT-1 plan prepared on June 3 by Doyle Northrup. This plan, which outlined AFOAT-1 projects for fiscal years 1949, 1950, and 1951, was heavily weighted in favor of seismic and sonic research in 1950 and 1951.[44] In reviewing the budget estimates for the last two years of the plan, at a meeting on June 9, 1949, the Committee passed the following resolution by a majority vote (Conant, Oppenheimer, and the Army and Navy members for, the two Air Force members, Generals Schlatter and Wilson, against): "The CAE [Committee on Atomic Energy] believes that the 20 odd million dollars, the total approximate cost of the research and development in the program of LRD [long-range detection] might be spent more wisely in other fields of research and development."[45]

If implemented, the effect of this resolution would be to eliminate the possibility of deploying effective seismic and sonic networks within a time scale that would be meaningful in tracking the initial buildup of a Soviet atomic arsenal. The Air Force members were incensed by this action of the Committee. General Wilson later recalled his feelings about what he described as the "bitter wrangle" to develop and deploy seismic and sonic methods to supplement radiological methods of surveillance: "The Air Force was frantic because it was charged with the job of detecting this first explosion and it felt that all three methods had to be developed and put in place or it would fall down on the job."[46] The Air Force members of the Committee submitted a written dissent, but the significance of this resolution was not lost on the members of the Boner Panel who were preparing to embark on an ambitious program of evaluation.

The Air Force was not inclined to wait for the Boner Panel's evaluation. Under Air Force pressure, the Board had provisionally approved the Air Force fiscal 1949 program totaling $8.5 million, although the unspent portion of this amount was subject to further review. However, the funds needed for operations in fiscal 1950 had yet to be authorized by the Board, and the Air Force demanded action on this matter.

At its twenty-third meeting on July 12, 1949, the Board, in the words of its executive secretary, "gave special consideration to the planned obligations for the support of Special Project IO-7 [the AFOAT-1 program]."[47] This resulted in Board approval of a $9.4 million budget for fiscal 1950, but this approval was conditional. The Air Force could begin projects that would be paid with fiscal 1950 funds, but the unexpended portion of these funds would be subject to further scrutiny by the Board in the light of the Boner Panel's findings.

Dr. Boner, on first hearing of this action by the Board, was not pleased. He expressed some reservations as to whether the Panel's efforts to evaluate the Air Force program would be worthwhile, since the Board had directed the Air Force to proceed with all phases of its fiscal 1950 program. However, he was assured that the Board could take steps to modify the program pending a report from the Panel.[48]

The efforts of the Boner Panel were nothing if not thorough. They read all the relevant documentation and were briefed at length by AFOAT-1 scientists on the technical progress that had been made during the past year. They spent a number of days at the headquarters of AFOAT-1 in Washington, D.C., where they interviewed both civilian contractors and the representatives of government agencies who were working for AFOAT-1, and where they were exposed to briefings by members of various intelligence agencies who were cooperating with AFOAT-1.[49]

On September 1, 1949, the Boner Panel submitted its report to the Board. To a marked degree, the Panel's recommendations reflected the vision of the monitoring program that had informed the evaluations by previous reviewers. Indeed, the first recommendation in their report implied that government planners had given insufficient thought to tracking Soviet atomic progress by a combination of conventional intelligence activities and technical methods of covert surveillance that would detect atomic operations such as uranium mining and the production of fissile material. Such an approach might obviate the need for an expensive monitoring system to detect a Soviet bomb test. The Panel's report thus began by suggesting that "The Joint Chiefs of Staff be requested to reevaluate the necessity for detecting a foreign atomic explosion by instrumental means, with a view to canceling the research and development program on LRD [long-range detection]."[50]

In the event that military considerations dictated that the program to detect a Soviet bomb test be continued, the Panel provided guidelines for the fiscal 1949 and 1950 portions of the program. The first guideline addressed the problem of defining the Air Force's mission. AFOAT-1 representatives had pointed out that as long as the statement of their assigned mission remained unchanged, they had little choice but to insist that they be allowed to develop all possible methods to carry out the mission, even the development of projects that appeared dubious of success. To remedy this situation, the Panel recommended that "The Eisenhower directive charging the USAF [United States Air Force] with overall responsibility for LRD [long-range detection] be modified to recognize the limited capability of detection by instrumental means."[51]

The remaining guidelines lowered the ceiling on approved spending from $8.5 to $6.8 million in fiscal 1949, and from $9.4 to $4.84 million in fiscal 1950. The Panel's report provided a detailed project-by-project list of where these cuts should be made. Seismic and sonic work was generally cut back while radiological research and the effort to develop methods to detect atomic operations were left unchanged or, in the case of some projects in these categories, increased. These changes represented a significant reorientation of the AFOAT-1 program.

The Board reacted swiftly in sending the Panel's report to the two reviewing committees. In a memorandum to the chairmen of the committees, the Board's executive secretary had noted that, upon Board approval of the Panel's recommendations, "The Air Force will revise the program downward to provide a lesser capability of detection." Dr. Boner immediately took exception to this phrasing. He insisted that, far from producing lesser capability, "By reducing activity along unpromising lines, and by increasing effort in favorable areas, the Panel feels that its recommendations will materially improve the USAF capabilities for LRD."[52]

Anticipating that the Board would approve the Boner Panel's recommendations, the secretary of defense's Management Committee decided that the Air Force should not be allowed to make expenditures beyond the ceilings recommended by the Panel. At a meeting on September 14, at which the matter of immediate economies in military research was discussed, the Management Committee directed that a stop-order be issued to limit Air Force commitments on the monitoring program to the levels suggested by the Panel.[53]

By then, however, all had changed: although it was known only to a few top military leaders and administrative officials, during the previous week the Interim Surveillance Research Net had detected an atomic explosion within the Soviet Union.

NOTES

1. H. H. O'Bryan to Deputy Executive Secretary, August 18, 1948, p. 3. RG 330, Secretary of Defense, Research and Development Board, Box 7, Folder 100, Atomic Energy/Long Range Detection, NA.

2. The suggestion to have the Committee on Geophysical Science review the seismic and sonic portions of the monitoring program had been made to the Board by the Committee on Atomic Energy as early as December 1947. See J. B. Conant to V. Bush, January 2, 1948. Archives, DOD. In August 1948, the Committee on Atomic Energy was composed of three civilian members, J. B. Conant, J. R. Oppenheimer, C. E. Greenblatt, and six military members (who also comprised the Military Liaison Committee, a completely separate organization then chaired by D. F. Carpenter). The military members were Colonel J. H. Hinds, Major General K. D. Nichols (Army); Major General R. C. Wilson, Major General D. M. Schlatter (Air Force); Rear Admiral W. S. Parsons, Rear Admiral R. A. Ofstie (Navy). For the relationships between the Military Liaison Committee and the Committee on Atomic Energy, see D. F. Carpenter to V. Bush, April 27, 1948. RG 330, Box 3, Folder 12/3 Agenda, NA.

3. V. Bush to Joint Chiefs of Staff, September 16, 1948, Log No. 2970/RDB 113/2.3. Archives, DOD.

4. Rolander, "Review of CAE, R&DB Minutes," to Nichols, January 29, 1954, p. 4.

5. "FY1949 Program for Long Range Detection, Modified Seismic and Acoustic," October 14, 1948. RG 330, Entry 341, Box 7, Folder 100, Atomic Energy/Long Range Detection, NA.

6. C. S. Piggot, "Action of the Geophysical Sciences Committee Relative to Project Workbag, at the Meeting of 18 October 1948," November 2, 1948. RG 330 Secretary of Defense, Entry 341 Research and Development Board, Box 7, Folder 100, Atomic Energy/Workbag, NA.

7. V. Bush to K. T. Compton, October 28, 1948. Bush Papers Box 26, Folder 170, LC. Also quoted in A. A. Needell, *Science and Defense*, unpublished manuscript, January 9, 1990, Chapter 3, p. 42. Copy in the authors' collection. Our thanks to Dr. Needell for providing this manuscript.

8. Compton, memorandum to the files, November 29, 1948, p. 1.

9. Ibid.

10. Ibid., p. 2.

11. J. B. Conant to V. Bush, January 2, 1948, p. 2. RDB 113/1, AE 21/1, Log No. 2448, Archives, DOD. According to Conant's report on the meeting of the Committee on Atomic Energy held on December 22, 1947, he believed that the Committee's "estimate of three years [i.e., 1950] as the minimum time before Russia will have an atomic bomb is conservative."

12. A. L. Loomis, "The Report of the Ad Hoc Panel on Long Range Detection," December 10, 1948, p. 1. Archives, DOD.

13. R. C. Wilson, testimony at Oppenheimer Hearing. In *Transcript of Hearing in the Matter of J. Robert Oppenheimer*, p. 684.

14. Rolander, "Review of CAE, R&DB Minutes," to Nichols, January 29, 1954, p. 4; H. N. O'Bryan, "Status of the LRD Project of the Air Force," to R. F. Rinehart, February 8, 1949, p. 3. RG 330 Secretary of Defense, Research and Development Board, Box 7, Folder 100, Atomic Energy/Long Range Detection, NA.

15. W. Webster to Secretary of Defense, January 3, 1949. RG 330, Entry 341, Box 7, Folder 100, Atomic Energy/Long Range Detection, NA. In this brief note, Webster states: "The subject [of monitoring] is hot now. Our Committee [on Atomic Energy] of RDB has recommended cutting work back considerably below the level

the Air Force would like—they are appealing vigorously." Webster, an eminent engineer and utilities company executive, chaired the Committee and was also assistant for atomic energy to the secretary of defense.

16. H. E. Landsberg to Executive Secretary, RDB, February 1, 1949. RG 330, Entry 341, Box 7, Folder 100, Atomic Energy/Long Range Detection, NA.

17. D. L. Northrup, "Technical Memo No. 31," to A. F. Hegenberger, February 23, 1949, p. 3. RG 330, Entry 341, Box 7, Folder 100, Atomic Energy/Long Range Detection, NA.

18. Northrup, "Technical Memo No. 31," to A. F. Hegenberger, February 23, 1949, p. 2.

19. Rolander, "Review of CAE, R&DB Minutes," to Nichols, January 29, 1954, p. 5; H. M. Bryan, "Status of the LRD Project of the Air Force," to R. E. Rinehart, February 8, 1949. RG 330, Entry 341, Box 7, Folder 100, Atomic Energy/Long Range Detection, NA.

20. Bryan, "Status of the LRD Project of the Air Force," to Rinehart, February 8, 1949, pp. 1, 2.

21. J. B. Knapp to Executive Secretary, RDB, February 15, 1949. RG 330, Entry 341, Box 7, Folder 100, Atomic Energy/Long Range Detection, NA.

22. "Agenda, 19th Meeting, RDB, February 23, 1949, pp. 3, 4. RG 330, Entry 341, Box 7, Folder 100, Atomic Energy/Long Range Detection, NA.

23. R. F. Rinehart, "Board Action with Regard to the Report of the Loomis Panel," to D. L. Northrup, March 1, 1949, p. 2. RG 330, Entry 341, Box 7, Folder 100, Atomic Energy/Long Range Detection, NA.

24. Ibid., p. 2.

25. W. E. Wrather to K. T. Compton, March 8, 1949. RG 330 Entry 341, Box 7, Folder 100, Atomic Energy/Long Range Detection, NA.

26. K. T. Compton to the Joint Chiefs of Staff, March 12, 1949. RG 330, Entry 341, Box 7, Folder 100, Atomic Energy/Long Range Detection, NA.

27. H. M. O'Bryan to the Executive Secretary, RDB, March 14, 1949. RG 330, Entry 341, Box 7, Folder 100, Atomic Energy/Long Range Detection, NA.

28. K. T. Compton, memorandum to the file, March 14, 1949. RG 330, Entry 341, Box 7, Folder 100, Atomic Energy/Long Range Detection, NA.

29. R. M. Emberson to R. H. Rinehart, March 17, 1949. RG 330, Entry 341, Box 7, Folder 100, Atomic Energy/Long Range Detection, NA.

30. K. T. Compton to Major General E. M. Powers, March 17, 1949. RG 330, Entry 341, Box 7, Folder 100, Atomic Energy/Long Range Detection, NA.

31. JCS, "Long Range Detection Program," to Chairman, RDB, March 28, 1949, p. 1. RG 330, Entry 341, Box 7, Folder 100, Atomic Energy/Long Range Detection, NA.

32. "RDB Master Plan Classification System," n.d. RG 330, Entry 431, Box 46, Folder 102, Master Plan Classification System, NA.

33. E. F. Rinehart to K. T. Compton, March 28, 1949. RG 330 Entry 341, Box 7, Folder 100, Atomic Energy/Long Range Detection, NA.

34. JCS, "Long Range Detection Program," to Chairman, RDB, March 28, 1949, p. 2.

35. Ibid., p. 2.

36. J. F. Phillips to Director, Research and Development Office, April 15, 1949. RG 330, Entry 341, Box 7, Folder 100, Atomic Energy/Long Range Detection, NA.

37. R. F. Rinehart to Executive Director, Committee on Atomic Energy and Geophysics and Geography, March 16, 1949. RG 330, Entry 341, Box 7, Folder 100, Atomic Energy/Long Range Detection, NA; Rolander to Nichols, "Review of CAE, R&DB Minutes," January 29, 1954, p. 5.

38. Herken, *The Winning Weapon*, p. 326.

39. K. T. Compton to R. H. Hillenkoetter, March 23, 1949. RG 330, Entry 341, Box 7, Folder 100, Atomic Energy/Long Range Detection, NA.

40. R. H. Hillenkoetter, "Atomic Energy Program of the USSR," to Executive Secretary, National Security Council, April 20, 1949. RG 218, Box 130, Folder CCS 471.6 (8–15–45) Section 14, NA.

41. K. W. Condit, *History of the Joint Chiefs of Staff*, Vol. II, 1979, p. 526.

42. Loomis, "The Report of the Ad Hoc Panel on Long Range Detection," December 10, 1948, p. 1.

43. Ibid., pp. 3, 4.

44. D. L. Northrup, "Technical memo No. 34, " June 3, 1949. RG 330, Entry 341, Box 7, Folder 100, Atomic Energy, Long Range Detection, NA.

45. Rolander, "Review of CAE, R&DB Minutes," to Nichols, January 29, 1954, p. 6.

46. R. C. Wilson, testimony at the Oppenheimer Hearings. In *Transcript of Hearing in the Matter of J. Robert Oppenheimer*, p. 695. Wilson singled out Oppenheimer as chiefly responsible for the opposition to the Air Force's monitoring program, apparently because Wilson felt that Oppenheimer's negative opinion of the Air Force's sonic and seismic development program swayed his colleagues. Oppenheimer's actions as a member of the Committee on Atomic Energy and as a member of other advisory groups appeared sinister to Wilson—sufficiently so that he felt it necessary to report to General Cabell, chief of Air Force Intelligence, that Oppenheimer had exhibited a pattern of behavior that redounded negatively on national security. See C. A. Rolander to L. L. Strauss, February 25, 1954. Strauss Papers, AEC Series, Box 26, Folder on Oppenheimer correspondence, 1954, HHPL.

47. R. F. Rinehart to Chairmen, Committees on Atomic Energy and Geophysics and Geography, July 21, 1949. RG 330, Entry 341, Box 7, Folder 10, Atomic Energy/Long Range Detection, NA.

48. D. Z. Beckler to W. E. Wrather, July 22, 1949. RG 330, Entry 341, Box 7, Folder 100, Atomic Energy/Long Range Detection, NA.

49. Beckler to Wrather, July 22, 1949.

50. R. F. Rinehart to General J. T. McNarney, September 15, 1949. RG 330, Entry 341, Box 7, Folder 100, Atomic Energy/Long Range Detection, NA.

51. Ibid.

52. C. P. Boner to R. F. Rinehart, September 12, 1949. RG 330, Entry 341, Box 7, Folder 100, Atomic Energy/Long Range Detection, NA.

53. R. F. Rinehart to General D. L. Putt, September 15, 1949. RG 330, Entry 341, Box 7, Folder 100, Atomic Energy/Long Range Detection, NA.

CHAPTER 10

BUILDING ALLIANCES AND THE INTERIM NET

The so-called Interim Surveillance Research Net that detected Joe-1 was operated under the aegis of AFOAT-1 by the Air Weather Service, Tracerlab, and the Army Signal Corps. This interim net was only a component, albeit the major one, of a larger, loosely coupled organizational system coordinated by AFOAT-1 that included two "mini-nets," one operated by the United States Navy and one by Britain's Royal Air Force, and an array of consultative agencies, such as Los Alamos Laboratory's Radiochemistry Group and the Weather Bureau, that could be mobilized at need to provide special services.

In order to describe the formation of the interim net and the system of organizational alliances in which it was embedded, it is necessary to return to July 1948 and to examine the operational activities of AFOAT-1 over the ensuing twelve months. These activities took place in parallel with, and were modified by, the effort of Hegenberger and his staff to obtain Research and Development Board approval for AFOAT-1's program. The effort to obtain Board approval, as related in the previous chapter, displays the actors maneuvering within the web of constraints imposed by government rules and policies, but, as will be seen, in the process of establishing the interim net and its organizational context, some of the most sacrosanct rules were bent or broken in the service of the larger aim—the timely detection of the first Soviet bomb test.

IMPLEMENTING THE INTERIM NET CONCEPT

The formation of a major part of the interim net can be said to have begun with the planning of Fitzwilliam, for it was realized that the prevailing west winds traversing the Soviet Union would carry airborne fission products from a Soviet bomb test into the Pacific area where the Sandstone bombs were to be detonated. Meteorological conditions thus made it possible to position some of the radiological ground stations for Fitzwilliam in geographic locations that would not only be appropriate for monitoring the American bomb test in the Pacific, but a future Soviet test as well. According to the Fitzwilliam report: "The network for [radiological] ground stations for Project Fitzwilliam was determined by climatological studies of air mass trajectories prepared by the Director of Military Climatology, Air Weather Service. These stations were selected specifically for the Sandstone explosions so that the network would be readily adaptable to the Project Whitesmith net during the post-Sandstone period."[1]

When Fitzwilliam was terminated on June 6, 1948, most of the sonic and all of the seismic stations used in this operation were dismantled because their geographic location and the limited effectiveness of their apparatus made them useless for monitoring the Soviet Union. On the other hand, most of the radiological stations were left intact to be incorporated into Whitesmith, the program for routine nuclear surveillance of the Soviet Union. Indeed, the success of radiological monitoring experiments carried out during Fitzwilliam ensured that Whitesmith could begin much sooner than had been scheduled.

Anticipating an early start for Whitesmith, General Hegenberger had given Tracerlab a $20,000 follow-on contract to continue to service and maintain the apparatus in these radiological ground stations after the Fitzwilliam "roll-up"—that is, the terminal phase of the operation involving the redeployment of men and equipment to home bases. As part of this contract, the equipment in the laboratory at Hickam Air Force Base, Hawaii, that had been established during Fitzwilliam to analyze the fission products collected on airborne filters by Air Weather Service WB-29s was also maintained in operating condition. However, the initial contract given to Tracerlab for operating the Hickam laboratory and a similar laboratory at Tracerlab's Boston, Massachusetts, headquarters expired on July 1. In an unofficial aside to Dr. Frederick Henriques, Tracerlab's technical director, Ellis Johnson, had indicated that Tracerlab would soon be awarded another contract to operate these two facilities. On the basis of this assurance, Henriques retained some of the key scientists hired for Fitzwilliam on the Tracerlab payroll in anticipation of further analytical work from the Air Force.[2]

Meanwhile, during the Fitzwilliam roll-up in June, Johnson transformed the network of radiological ground stations into the configuration that would be used for Whitesmith. This task required only the closing of two Pacific stations (at Wake Island and at Henderson Field, Guadalcanal) whose equipment was moved to a new station established in the Atlantic at Lagens Air Force Base, Azores. The remaining twenty-three radiological ground stations used in Fitzwilliam plus the four Air Weather Service squadrons, some of whose WB-29s had been fitted with filter units, were considered ready for incorporation into Whitesmith. However, the key element in the net—the central laboratories needed to analyze the airborne material collected on filters mounted in aircraft and in ground-based units—was in a standby status and lacked the scientific staff needed to make them operational.

The Fitzwilliam radiological experiments had revealed that multiple fission product analysis, when applied to the material collected by filter-equipped aircraft, could provide a reliable method of unambiguously detecting an above-ground atomic bomb test at long range. However, these same experiments had exposed shortcomings in the methods of radio-chemical analysis that were used and gaps in the knowledge needed to interpret the results of analyses. Ellis Johnson and his staff, together with Tracerlab scientists, had outlined a research project to be conducted by Tracerlab that would remedy these deficiencies at a cost of about $1 million. This project, which included the operation of the two analytical laboratories, was incorporated into the overall plan for AFMSW-1 research that Johnson prepared in early July. But it was apparent that the funding for the program outlined in this plan would be sequestered pending review by the Research and Development Board.

AFMSW-1 had a small budget to cover its overhead and operational needs. Indeed, this was the source used to fund the equipment-maintenance contract that had been given to Tracerlab. As a stopgap measure, General Hegenberger authorized Johnson to dip once more into the overhead and operations budget to award Tracerlab a $15,200 contract to begin to operate the laboratories at Hickam Field and at Boston.[3] The reactivation of these facilities allowed the Air Weather Service network to become operational, and, by the end of July, its WB-29s had flown a few radiological surveillance missions for training purposes.[4]

The Air Weather Service's budget enabled it to carry out such flights and to operate the ground stations as part of its weather reconnaissance mission, but the maintenance of ground station equipment and the operation of the central analytical facilities had to be funded by AFMSW-1. Most of the $43 million that had been set aside for the monitoring program was unavailable to AFMSW-1 owing to Department of Defense program review procedures. Hence, the interim net was launched on the basis of $35,200 squeezed out of AFMSW-1's small overhead budget and given to Tracerlab to supply the services that would allow the net to begin operation.

By August, as has been previously described, the operational orientation of AFMSW-1 had been reinforced by placing it under the aegis of the deputy chief of staff for operations, a transfer that changed its designation to AFOAT-1. Partly in protest over cumbersome program review procedures that threatened to stymie the creation of an adequate monitoring system, Ellis Johnson had resigned. Indeed, Hegenberger and the new technical

director of AFOAT-1, Dr. George Shortley, could do little to advance
AFOAT's program until additional funds were made available.

On August 19, the Research and Development Board's Committee on
Atomic Energy approved the radiological portion of the AFOAT-1 plan, but
evinced reservations about the parts of the plan dealing with sonic and
seismic work. On September 16, the Board tentatively accepted the Committee's findings and suggested that the Air Force might use them as
guidelines pending further reviews by the Board and its subsidiary committees and panels.

Hegenberger and his scientific staff were quick to take advantage of this
advice by segregating projects from the AFOAT-1 plan that appeared to
have been specifically approved by the Committee or had, at least, not
evoked controversy. They then attempted to persuade Air Force financial
officers that the Board's remarks about "guidelines" constituted an authorization to allot funds for these projects from the money that had been
allocated for monitoring by the director of the Budget. The government
organizations and industrial firms that would carry out these projects at an
agreed-upon cost had already been selected as part of the process of
preparing the AFOAT-1 program plan, so lack of funds was the only bar to
implementation. (In the case of industrial firms, security considerations
allowed AFOAT-1 to award sole-source contracts.)

The Air Force approved $390,200 for overhead and operations costs,
including $35,200 to cover contracts already awarded to Tracerlab. It also
authorized a follow-on contract for $16,000 to Air Reduction Sales Com-
pany and the transfer of $25,000 to the Air Materiel Command's Watson
Laboratories to complete work that had been started during Fitzwilliam.
On December 17, the Committee on Atomic Energy endorsed the recommendations of the Loomis Panel. In effect, this reconfirmed the approval of
certain projects that had been voted by the Committee on August 19. The
Committee members' reaffirmation of their position on these projects convinced Air Force fiscal authorities that it was permissible to execute them,
and they authorized Hegenberger to spend an additional $2,383,415 for this
purpose.

Hegenberger lost no time in transferring funds to other government
organizations and in issuing contracts that would expend this money.
Therefore, by the year's end not only was the interim net operating on a
routine basis, but also some of the research needed to improve the system
was underway. Indeed, even Atomic Energy Commissioner Lewis Strauss,
who a few months earlier had been apprehensive that "monitoring is to be
downgraded" by the Air Force, was satisfied with the progress that had
been made.[5] On January 3, 1949, he had been briefed on AFOAT-1's activities and, that same day, he wrote to his old friend, Defense Secretary
Forrestal, to say that the Air Force's interim monitoring procedures "struck

me as impressive." He went on to point out that, "Although much remains to be done in this area, at least the door is no longer being left unguarded."[6]

Before describing the surveillance activities that so impressed Strauss, it is pertinent to examine first the research activities of AFOAT-1 because some developments stemming from this work were later incorporated in the interim net that detected Joe-1.

RESEARCH ACTIVITIES

From the outset, there was virtually no disagreement between AFOAT-1 and the Research and Development Board about the need to develop radiological monitoring methods. Thus, Air Force financial officers felt no qualms about allocating funds for this portion of the AFOAT-1 research plan. But General Hegenberger and Doyle Northrup, who in the fall of 1948 had superseded Shortley as technical director, were convinced that an adequate monitoring system had to employ sonic and seismic, as well as radiological, means of detection. Therefore, despite the fact that scientists serving on the Research and Development Board's committees and panels opposed much of the sonic and seismic research in the AFOAT-1 plan, some work in these areas was carried out.

Two methods for accomplishing this work appear to have been employed: first, certain activities that were labeled "operational" seem to have involved research; and second, money was channeled to other government agencies for "overhead" or "operational" activities that appear to have been developmental in nature. In addition, of course, the Board's reviewers had agreed with AFOAT-1 on the need for a few modest sonic and seismic studies. There was another area of agreement as well: both Hegenberger and Northrup espoused a concept that had received the enthusiastic endorsement of reviewers, namely, expanding the mission of AFOAT-1 to include monitoring foreign atomic activities such as the production of fissionable material. Therefore, by the beginning of 1949, AFOAT-1 had initiated research and development in four areas.

Radiological

Perhaps the most important research sponsored by AFOAT-1 was Tracerlab's multifaceted project to (1) investigate and develop techniques for the collection, radiochemical separation, and radioassay of airborne fission products, (2) devise high-sensitivity radiation detectors for use in conjunction with these techniques, (3) train Air Force personnel in the use of these methods, (4) operate laboratories for analyzing airborne radioactivity collected by the Air Weather Service, and (5) service radiological equipment in Air Force ground stations. Contracts totaling $1,107,000 had been awarded to Tracerlab to carry out this work. Next in importance was Project

Rainbarrel, for which the Naval Research laboratory received $150,000. The aim of this project was to develop techniques for the collection and measurement of fission products scavenged from the atmosphere by rain, snow, or ice. Several smaller projects totaling $70,000 to develop better filters and improved methods of detecting airborne radioactivity by measuring atmospheric conductivity were undertaken by the Geophysics Directorate of the Air Materiel Command.[7]

Sonic

As a result of the controversy between the Board's reviewers and AFOAT-1 over the value of sonic research, most of the projects specifically labeled as sonic had yet to be approved. But the Air Materiel Command's Watson laboratories received $110,000 from AFOAT-1's overhead and operations budget for "geophysical" work, including some sonic research. The Signal Corps' Evans Signal Laboratory was given $204,415 to "operate" part of the sonic net that was used in Fitzwilliam. This net originally consisted of six stations (Nome, Alaska; Hickam Field, Hawaii; San Francisco, California; Redbank, New Jersey; Grafenwöhr, Germany; Manila, Philippine Islands). The role played later by sonic stations in determining the time and location of Joe-1 suggests, however, that part of this money was used to relocate stations, establish new stations, and develop sonic detector arrays with improved range. As 1949 began, because of the controversial nature of sonic research, AFOAT-1 was still struggling to obtain authorization to give the Signal Corps an additional $250,000 to improve sonic instrumentation. Approval was also sought to fund three sonic projects: Naval Ordnance Laboratory, $400,000 to investigate the nature and source of infrasonic background noise to permit the design of highly sensitive sonic detectors ("infrasonic" refers to sound waves with a frequency less than 15 cycles per second); Naval Electronics Laboratory, $25,000 to improve the Rieber microbaragraph; Beers and Heroy (an industrial firm specializing in geophysical research) and the National Bureau of Standards, $100,000 each, to provide consulting services on sonic projects.[8]

Seismic

Although the urgency of seismic research was also controversial, the Research and Development Board's reviewers recognized that seismic monitoring methods would eventually be useful. Hence, funds had been authorized for three projects: Coast and Geodetic Survey, $50,000 to survey the geology of potential foreign and domestic sites for seismographic monitoring stations and $50,000 to develop low-cost explosives to be used for the detonations needed in developing improved seismic detection methods and equipment; Naval Ordnance Laboratory, $525,000 to design, develop, and furnish instruments for seismic background studies. AFOAT-1 was seeking authorization for four additional seismic projects: Coast and Geodetic Survey, $56,000 to collect, analyze, and interpret records from all seismic observatories under government control; Office of Naval Research, $100,000 to utilize the competence of university seismographers and seismic observatories in Whitesmith; Geotechnical Corporation, $475,000 to calibrate and modify velocity and displacement-type seismographs for installation in government and university observatories and to arrange for their operation on a continuous basis; and Beers and Heroy, $500,000 to conduct theoretical studies of seismic wave generation, propagation, and detection.[9]

Detection of Atomic Operations

The $16,000 contract with Air Reduction Sales Company to continue the development, begun during Fitzwilliam, of a small, portable apparatus for extracting noble gases, including xenon-133 and krypton-85, from the atmosphere was successful. AFOAT-1 sought authorization to award this company $225,000 to further develop, construct, and test a number of similar devices, some apparently designed to be used surreptitiously to sample the air near suspected atomic facilities within the Soviet Union. As part of this research, funds totaling $59,000 were given to the Weather Bureau to determine the relationship between meteorological conditions and the distribution of radioactive material in the atmosphere over the atomic facilities at Oak Ridge, Tennessee, and Hanford, Washington. The aim was to establish the feasibility of using air-sampling techniques to locate similar facilities in the Soviet Union and to determine the rate at which the Soviets were producing fissionable material. AFOAT-1 also sought authorization to award Armour Research Foundation $575,000 to "make a compilation of all possible methods of detecting atomic operations by scientific intelligence" and to evaluate by experimentation the requirements of the equipment needed for this purpose.[10]

During the first three months of 1949, AFOAT-1 added a few more projects in each of these four areas to its "wish list." At the end of March, the Research and Development Board conditionally approved most of AFOAT-1's past expenditures and the proposed commitments for fiscal 1949. This action allowed work to begin on the projects just described for which authorization was sought at the beginning of the year, although in some cases the amounts were considerably reduced by the Board's reviewers. In June, Northrup wrote an elaborate summary of current and projected projects in which he revealed that, despite the seemingly interminable program review process, and the fact that formal Board approval had not been received, AFOAT-1 had succeeded in initiating thirty-five research projects that were being carried out by fifteen organizations, including government agencies, industrial firms, universities, and foundations.[11]

By September 1949, some of the initial developments resulting from radiological, sonic, and seismic research had been incorporated into the interim net, although the actual detection of Joe-1 in early September was accomplished by radiological means. The research to devise methods of detecting foreign atomic operations was aimed at expanding AFOAT-1's mission, so it had no part to play in improving the interim net or in detecting Joe-1. Nevertheless, this work warrants further discussion because it created a capability that became an important element in the *raison d'etre* of AFOAT-1 *after* the discovery of the first Soviet bomb test.

The creation of this capability began with Fitzwilliam. The radiological experiments planned for this operation included the collection and analysis of the fission products of certain noble gases, such as xenon and krypton, that are produced when an atomic bomb is exploded. Such gaseous fission products are not trapped by the filters used to collect fission products in the form of airborne particulates or dust. Instead, it is necessary to extract the noble gases from large-volume air samples and to subject them to radioassay to determine the concentration of the radioisotopes of each gas.

Accordingly, during Fitzwilliam air samples were collected using a WB-29 fitted with a compressor which, at altitude, drew air from outside the aircraft and pumped it into storage cylinders. These cylinders had been rushed to Tracerlab's Boston headquarters where a laboratory, dubbed "the gas works," had been established to analyze the air they contained. The inadequacy of this gas collection scheme was soon revealed by the failure to find detectable amounts of gaseous fission products in the cylinders, and the use of this apparatus was discontinued after the first two Sandstone detonations. To solve this problem, a project, code named Beans, was hurriedly launched. Project Beans was an attempt to cryogenically extract the noble gases in liquid form from the atmosphere; the Cleveland Wire Works Company had been given the task of developing a flyable cryogenic sampling unit.[12]

This apparatus apparently proved successful in collecting gaseous fission products produced by the third Sandstone bomb, but after the results of Fitzwilliam were analyzed an anomaly appeared. Since the percent fission yield of xenon-133 was known to be more than an order of magnitude higher than that of krypton-85, it was expected that the air samples would contain a higher concentration of the former than the latter gas.[13] But, according to a report on the Fitzwilliam findings, "The expected high concentration of xenon was actually very low and krypton was high."[14] It was speculated that xenon-133 was being scavenged from the atmosphere by some unknown adsorption process that affected this gas but not krypton-85. Although this phenomenon was not well understood at the time, it was decided to concentrate future gas sampling on krypton-85. Air Reduction Sales Company was thus given a contract to develop a more compact version of the cryogenic collection unit having a peak efficiency for krypton-85. This work resulted in an instrument that was useful not only as part of the monitoring system for detecting atomic explosions, but also as part of the program to detect other types of atomic operations.

The collection and analysis of both xenon-133 and krypton-85, as byproducts of an atomic explosion, was retained as an element in the interim monitoring system whose sole purpose, of course, was to detect the test of a Soviet atomic bomb. But the fact that these gases are also produced during other atomic activity, such as reactor operation, was well known. Nor had the use that had been made of this fact during the war to locate the sites of German reactors via xenon-133 detection been forgotten. Therefore, in response to the need to obtain information about Soviet atomic operations, which had been expressed by the director of Central Intelligence, Hegenberger and Northrup decided to launch a program based on the detection of xenon-133 and other effluents of atomic facilities and processing plants, such as iodine-131, to locate the sites of Soviet production facilities and a separate program based on the measurement of krypton-85 to determine the rate of Soviet production of plutonium.

For various technical reasons, krypton-85 appeared to be more useful than xenon-133 for the latter purpose. Most important, it was highly unlikely that the Soviets would attempt to suppress the emission of any of the noble gases, including krypton-85, during the operation of their atomic facilities. Since the noble gases do not participate in chemical reactions, it is difficult and expensive to remove them from the effluent produced by large-scale atomic operations. Moreover, their chemical inertness also eliminates the need to suppress the emission of such gases for safety reasons since, unlike most other forms of radioactivity, they are not metabolized by living organisms. Krypton-85, for example, is for this reason among the less hazardous of radioisotopes. Indeed, no attempt had been made to suppress krypton-85 emissions at American atomic facilities, a counterintelligence oversight that was later remedied.

In order to locate the sites of Soviet atomic facilities by their characteristic emissions, it was necessary to obtain air samples surreptitiously within the Soviet Union. The development of compact, portable devices for this purpose by Air Reduction Sales Company allowed several schemes for accomplishing this to be considered. For example, a project to build the air-sampling equipment into an ordinary household refrigerator was apparently initiated.[15] Presumably, such refrigerators would be shipped into the Soviet Union as part of the furniture used by Western diplomatic personnel. However, the program to use krypton-85 measurements to determine the rate of plutonium production did not necessarily require obtaining air samples within the borders of the Soviet Union, although it did require air sampling on a worldwide scale. The importance of knowing the rate of Soviet plutonium production was, of course, the fact that this information could, in turn, be used to deduce the number of atomic bombs in the Soviet stockpile at any given time. The general idea, according to one commentator, was that "a reactor used to produce the plutonium for nuclear weapons also gave off the isotope krypton-85. As the U.S. knew how much krypton-85 was given off by Western reactors, any excess was from Russian reactors. This would allow an estimate to be made of Soviet plutonium production."[16]

If this is a fair statement of the concept behind AFOAT-1's effort to monitor the growth of the Soviet atomic stockpile, it fails to indicate the many difficulties involved in implementing this scheme. Krypton-85 is evolved not only during reactor operations, but also during other atomic activities, including the detonation of atomic bombs. Keeping track of such foreign and domestic activities was a daunting task. One of the problems was that by 1949 the atomic programs of both the British and French were in full swing. The provisions of the MacMahon Act had seriously disrupted the exchange of atomic information with the British, and the xenophobic tendencies of the French were exacerbated by the secrecy that cloaked their atomic program. In addition, because some of the leading French scientists in this program were avowed Communists, the United States had been unwilling to exchange much atomic information with France. These considerations made it difficult to assess the Western contribution to the atmospheric burden of krypton-85 by direct inquiry.[17]

Moreover, it would be necessary not only to continuously assess the Western contribution to atmospheric krypton-85, but also to simultaneously determine the total worldwide atmospheric burden of this gas. At any given time, the difference of these two parameters would then represent the Soviet contribution. Technically, however, the continuous determination of the total worldwide burden of krypton-85 was a very difficult task. It required a better understanding of the global mixing of atmospheric gases than that which existed in 1948 and an atmospheric air-sampling program that was global in scope. One scheme for obtaining such samples seems to have involved installing air-sampling devices in aircraft operated by the Military Air Transport Service and, perhaps, by commercial airlines. Hence, according to a report on the aims of the AFOAT-1 research contract with Air Reduction Sales Company, it was necessary to incorporate the cryogenic krypton-85 sampling device into "a package that can be used for continuous sampling in scheduled air traffic."[18]

Serious efforts to solve these and other problems associated with the krypton-85 method of determining the rate of Soviet plutonium production were begun in 1948 and 1949, but the payoff was several years away. By mid-1951, the report of the IO-7 (or Boner) Panel, which periodically reviewed the activities of AFOAT-1, stated:

Determination of the krypton-85 distribution in the atmosphere and the krypton-85 fission yield have not yet been accomplished with sufficient accuracy to assign definite limits to the Russian radiokrypton [or krypton-85] contribution. The Panel believes that by mid-1952 quantitative measurement of Russian krypton-85 generation with a precision equal to ten percent of the U.S. generation may be possible assuming no suppression of radiokrypton by the USSR.[19]

The Panel's report also went on to note "with satisfaction" that the new American reactors planned for plutonium production were designed to "include provision for krypton suppression."

AFOAT-1 research on the detection of Soviet atomic operations not only reoriented and expanded the mission of this organization, but it also played an important counterintelligence role in thwarting the possibility of a Soviet attempt to learn about American atomic progress via the krypton-85 method (assuming they invented this method on their own). The other side of this coin was that the elaborate and costly measures taken by AFOAT-1 to implement the krypton-85 method would be nullified if American security on this matter was compromised—an event that would surely cause the Soviets to suppress krypton-85 emissions. In this regard, the Boner Panel's prediction that this method would yield results by 1952 appears to have been correct. For according to one historian of intelligence methods, the fact that "the krypton-85 level was the key to U.S. ability to estimate Soviet nuclear weapons production was *the* deepest secret of AFOAT-1 as of late 1954."[20] Indeed, the "payoff" for much of the research initiated in 1948 on methods of detecting foreign atomic operations was only realized in the 1950s.

By mid-1951, the research initiated in 1948–1949 that was needed to establish the net in "complete" form for monitoring Soviet atomic bomb tests had also paid off. In addition to an enhanced radiological monitoring system, the net included sonic and seismic arrays capable of detecting atomic bomb tests deep within the Soviet Union. But, of course, this was not the case when Joe-1 was detected. Hence, to understand how this feat was accomplished, it is necessary to describe the interim net "on the ground" as it existed in 1948–1949.

THE INTERIM NET

The Air Weather Service's radiological monitoring system, which was by far the major component of the interim net, consisted of twenty-four ground stations, four WB-29 squadrons, and two central laboratories. The Fitzwilliam radiological experiments had revealed that, except for the precipitation tank (or rain-scavenging) method of collecting fission products, the sensitivity of ground-based methods of detecting and collecting airborne radioactivity was considerably less than that of aerial collection. Thus, the ground stations were operated primarily as backup for the more effective monitoring carried out by aircraft.

The ground stations contained equipment to detect radioactivity in the air and to collect airborne particulates. The simplest item of equipment, a shallow tank to collect precipitation, relied on the scavenging action of rain and snow on airborne material. High-speed blowers that drew air through filters were also employed to collect airborne dust particles. Fifteen of the stations were provided with apparatus, designed by Naval Research Laboratory scientists, to automatically record in continuous fashion the level of radioactivity near the station.

The ground stations at the four airfields at which the WB-29 squadrons were based contained an additional piece of equipment, a so-called wraparound counter. This consisted of a long cylindrical Geiger counter mounted within a massive lead shield that reduced unwanted background radiation. The sheetlike filters exposed during WB-29 flights were rolled into tubular form and slipped over the cylindrical Geiger counter, which then detected beta rays emitted by radioactive material that had been trapped by the filter. The output of the Geiger counter was recorded electronically for a preset time. A value of 100 counts per minute or less was not considered anomalous. If the count rate exceeded this value, the filter was sent to a central laboratory where, through chemical analysis and radioassay, the nature of the radioactive material was ascertained and its probable source identified.[21]

Such wraparound counters were located not only at the home base of each of the four WB-29 squadrons, but also at ten of the ground stations located at airfields that were the terminus of flights initiated from the home field. This arrangement allowed filters to be scanned immediately after the plane landed regardless of whether the filter was exposed on the outgoing or return leg of the mission.

The ground stations of the Air Weather Service were located in a huge arc extending along the western and eastern rims of the Pacific from Guam northward to the Philippines, Japan, Korea, and the Aleutians, with the northernmost point at Fort Barter, Alaska, then extending southward along the west coast of Alaska and the continental United States to the Canal Zone near the equator.[22] For most of the year, winds at high altitudes are westerly in the Northern Hemisphere; hence, these stations were well positioned to intercept airborne fission products carried eastward from the Soviet Union. However, for a month or two each year, these winds reverse direction. Therefore, to intercept winds blowing from the Soviet Union toward the west, stations were established in Bermuda, Washington, D.C., and the Azores, and monitoring flights were made eastward to Libya. (As will be seen, the northern regions of the Atlantic were monitored by the British.)

Aerial collection of fission products was a key element of the system. Routine flights of filter-equipped WB-29s were made by four Weather Reconnaissance Squadrons, VLR (Very Long Range): the 518th stationed at North Air Force Base, Guam; the 375th at Eielson Air Force Base, Alaska;

the 374th at Fairfield-Suisun (now Travis) Air Force Base, California; and the 373rd at Kindley Air Force Base, Bermuda. The long-range monitoring missions flown by these squadrons were able to sample airborne dust in the Northern Hemisphere from the pole to the equator in a huge area extending from Korea in the east to Libya in the west (although the North Atlantic region was not covered).

At the heart of the system were the two central laboratories located on the east and west coasts of the United States. In the fall of 1948, the laboratory at Hickam Field, Oahu, Hawaii, had been moved to a newly formed Western Division of Tracerlab that had been established in Berkeley, California. The east coast laboratory remained located at Tracerlab's Boston headquarters. In the year that followed the Sandstone tests, Tracerlab had made great strides in improving its methodology for the radiochemical analysis and radioassy of fission products. This progress, in large part, was due to the close relationship that had been formed with the Radiochemistry Group, headed by Dr. Roderick W. Spence, at the Atomic Energy Commission's Los Alamos Laboratory.

This crucial relationship appears to have been established in an almost offhand manner. In the summer of 1948, a Los Alamos Laboratory scientist, Alvin C. Graves, had attended a top secret seminar that was given at the nearby Kirtland Air Force Base by Dr. William D. Urry, head of AFOAT-1's Nuclear Research Section. Urry had described the effort being made by AFOAT-1 to derive information about atomic bombs from exceedingly small samples of bomb debris. Since this "bomb diagnostic" work was related to that of Spence's Group, Graves later suggested to Spence that he contact Urry. A meeting was arranged at Tracerlab's Boston headquarters where Spence met Urry and the Tracerlab scientists involved in fission product analysis, including Dr. Lloyd Zumwalt, who had been made head of Tracerlab's fission product analysis laboratory at Berkeley. Spence was surprised to learn that Tracerlab scientists did not have access to Atomic Energy Commission reports outlining the latest radiochemical procedures for fission product analysis. The policy of compartmentalization had apparently worked only too well. At Urry's urging, Spence agreed to work as closely as security regulations would permit with Tracerlab scientists in order to bring their radiochemical procedure up to date.[23]

Prior to Sandstone, Spence's Group had not been involved in bomb diagnostics, but late in 1947 Spence had been instructed to form a new section within his Group to carry out radiochemical diagnostic work for the Sandstone tests. Both Spence and members of his Group and Zumwalt and other Tracerlab scientists had, therefore, been in the Pacific during Sandstone performing similar radiochemical analyses, albeit for different purposes. Security regulations and compartmentalization had prevented each group from knowing about the other's work during Sandstone, but they worked closely on bomb diagnostics after the Boston meeting.

At first, Spence was worried about the security considerations involved in working with AFOAT-1 and Tracerlab. Regarding this collaboration, he later recalled that "Bill Urry had impressed on me that I was not to tell anybody, and he meant my boss, Al Graves, and Norris Bradbury, Director of the [Los Alamos] Laboratory. . . . This kind of thing went on for a while until I felt that this was no way to operate. So I told my bosses what I was doing even though Urry didn't want me to. This is an example of how closely things were kept under wraps."[24]

Another security problem that bothered Spence was the degree to which the compartmentalization policy could be compromised. According to Spence, "I was very concerned about what I should tell the AFOAT-1 people." To fully interpret information from fission product analysis of the Sandstone detonations, AFOAT-1 and Tracerlab needed information on the design and construction of the American atomic bombs. Such information was needed to "calibrate" the multiple fission product analysis technique if Tracerlab and AFOAT-1 scientists were later to correctly interpret the data obtained from analyzing the debris of a Soviet bomb. But Spence was aware that this was "some of the most closely guarded information in the Atomic Energy Commission, outside of stockpile information."[25]

To resolve this conundrum, Spence had recourse to a colonel on the Air Force staff in Washington who was a frequent visitor to the Los Alamos Laboratory's J Division, of which Spence's Group was a part. Spence had gotten to know him well, and he later recalled that one day, shortly after the Boston meeting with Urry and Tracerlab scientists,

I met [the colonel] in the hall just outside of the J Division's offices and I said, "Paul, what can I tell AFOAT-1 and Tracerlab involving classified and restricted data?" And he said, "Rod, you tell them anything they need to know to do their job." And [based] on that verbal conversation I essentially withheld nothing about the materials in our bombs . . . and they needed this information, no question. I gave them all this with nothing in writing. It was, in fact, probably the only way to do it. In those days if you tried to work it out through the system you might never have gotten there.[26]

In the summer of 1948, Spence had found Tracerlab radiochemical methods "rudimentary," and he had provided Zumwalt with informal write-ups that outlined techniques that had been developed at Atomic Energy Commission laboratories. One area to which Tracerlab scientists had given little thought was "chemical fractionation." This effect stems from the fact that the particle size distribution in bomb debris as sampled at long range can be affected by fallout of the heavier particles, inefficient collection of the smaller particles, and selective sampling procedures. The observed particle size distribution at great distances from the explosion thus becomes unrepresentative of the actual distribution produced at detonation. This distortion in the particle size distribution results in distortions in the relative amounts of refractory and volatile fission products in the sampled airborne particulates. This phenomenon is called chemical fractionation, and it can seriously interfere with the interpretation of the data produced by multiple fission product analysis. That is, parameters such as the time of the explosion, the nature of the fissionable material, and so on, become difficult to determine if fractionation is not mitigated or properly taken into account.[27]

By the time Joe-1 was detected, the difficulties presented by fractionation were by no means solved. But Tracerlab was aware of the problem, and because of two somewhat fortuitous factors it was possible to correctly analyze the Soviet bomb debris. First, the filters used in the WB-29s proved to be nearly 100 percent efficient over a very wide range of particle sizes, including those in the submicron region. Second, the Soviet bomb debris was picked up within a few days of the explosion and was thus "fresh" (relatively unchanged by time and distance). Another factor that contributed to the effectiveness of radiological monitoring was that, by the time Joe-1 appeared, Tracerlab's capability for analyzing fission products equaled that of the Atomic Energy Commission laboratories and for analyzing extremely small samples it was, perhaps, superior.

While the radiological component of the interim net was completely operational at the time of Joe-1, the same cannot be said of the seismic and sonic components. No seismic stations specifically designed for monitoring existed. Some information on seismic events occurring within the Soviet Union was available to AFOAT-1, but such data came only from cooperating government and university seismic observatories. In the event, Joe-1 was a tower burst that was undetected by existing seismographs.

The 1,800-mile range of existing sonic instrumentation was not adequate for monitoring the Soviet Union, but the AFOAT-1 program plan had called for sonic stations to be operated for training and experimentation. This appears to have been the purpose of most of the half dozen or so sonic stations operated by the Signal Corps, since they were originally located several thousand miles from the Soviet borders and were thus not well positioned to monitor that country. Indeed, AFOAT-1 was not alerted sonically by the Joe-1 detonation. But, as will be related in the next chapter, important information about Joe-1 was later derived from sonic records. This suggests that by mid-1949 the range of the sonic detector arrays had been increased, or at least two sonic stations had been established at locations that were much closer to the Soviet Union than were the stations of the original Signal Corps network used in Fitzwilliam (although documentation that would verify this remains to be declassified).

By the spring of 1949, the Data Analysis Center at AFOAT-1's Washington headquarters was routinely receiving data from a radiological monitoring system operated by the Air Weather Service and Tracerlab and from a couple of marginally effective sonic stations operated by the Signal Corps. In addition, AFOAT-1 scientists had access to the data of certain seismic observatories. But by August 1949 AFOAT-1 was also receiving information produced by two small-scale radiological monitoring systems that remain to be discussed. One of these "mini-nets" was operated by the United States Navy and the other by the British.

PROJECT RAINBARREL

As related in a previous chapter, the Navy had begun work on radiological methods of long-range detection or "remote sensing" of foreign atomic bomb tests in early 1946 as part of Operation Crossroads. The Navy had remained interested in the monitoring problem, and, in 1947, the Navy secretary endorsed the choice of the Air Force as the monitoring agency only on the condition that this decision did not confer "exclusivity" in this area. Therefore, following the extensive participation by naval laboratories in Fitzwilliam, the Navy continued to operate a monitoring system consisting of a few radiological ground stations that had been established at certain naval bases. In addition, some scientists at naval laboratories proceeded to pursue research interests involving the development of monitoring techniques. One such development, Project Rainbarrel, played a role in the detection of Joe-1.

In describing the origin of this project, a memorandum to the chief of Naval Operations noted that "the detection of an atomic bomb air burst by the atmospheric scavenging effect of rainfall was first visualized by Naval Research Laboratory personnel in the spring of 1948."[28] A physicist at the Laboratory, Dr. Herbert Friedman, had become familiar with the effect of rain scavenging in the course of his work on the radiation detectors that were used in Navy monitoring stations. These detectors had indicated anomalously high radiation levels whenever rain fell. Friedman had suggested to a colleague in the Laboratory's Chemistry Division, Dr. Peter King, that the extraction and analysis of radioactivity in rainwater would be a valuable addition to the Navy's monitoring effort.[29]

King and another chemist, Dr. Luther B. Lockhart, responded by developing a method of treating large volumes of collected rainwater to precipitate material present in the water. The fission products of selected elements in the precipitate were then chemically separated. They worked with Friedman to evolve a radioassay technique for determining the ratios of certain fission products by establishing the half-lives and by radiation absorption measurements. Ratio comparisons then allowed the date of fission (and detonation) to be calculated.[30]

The rainwater for their first experiments had been collected on June 6, 1948, on St. Thomas Island, located about 9,000 miles from Eniwetok, the site of the Sandstone explosion. Rain on St. Thomas was regularly collected in cisterns for drinking water, so it was readily possible to obtain a very large volume of water for analysis. Although the last Sandstone explosion had occurred two months prior to the date the samples were analyzed, the water yielded fission products in substantial amounts. Additional samples of rainwater were later obtained from Shemy, Alaska, and Moen Island in the Truk group.[31]

The Naval Research Laboratory scientists concluded from the results obtained by analyzing these samples that the methods they had developed were capable of providing information on the nature of the fission material used in the bomb and the time of detonation. Rain scavenging thus appeared to offer a relatively low-cost alternative to aerial sampling as a means of collecting analyzable quantities of fission products at great distances from the explosion. Accordingly, the director of the Naval Research Laboratory, H. A. Schade, proposed to the chief of Naval Operations that the Laboratory's rain-scavenging approach be incorporated into the Navy's monitoring activities.[32]

The existing Navy ground stations were already equipped with apparatus that automatically recorded the local level of radioactivity and with air blowers fitted with filter paper collectors to sample airborne particulates. Schade recommended that these devices continue to be used for "maintaining surveillance during periods without rainfall." The addition of rainwater collectors at these stations would provide the capability of collecting chemically analyzable amounts of fission products. Schade went on to point out that the ground stations, when so equipped, would provide the Navy with a radiological monitoring system for which the operating cost "would be very small."[33]

AFOAT-1's William Urry had visited Friedman and King in the fall of 1948 to review their work on water sample analysis. Friedman later recalled that Urry seemed to welcome the fact that the Naval Research Laboratory had developed a capability to analyze fission products, but in later meetings representatives of AFOAT-1 appeared reluctant to engage in an open exchange of information.[34] Indeed, neither Urry nor the other representatives of AFOAT-1 had discussed with the Naval Research Laboratory scientists the results of the Air Weather Service's experiments, carried out during Fitzwilliam, to collect fission products via rain scavenging.

During Fitzwilliam, the Air Weather Service had included small, shallow pans for rainwater collection in the equipment used at thirteen of the radiological ground stations operated by the Service for this operation. Collected rainwater had been evaporated on paper filters that were processed in the same manner as the filters used in ground-based air blowers and in aircraft. Compared to the sensitivity and dependability of aerial sample collection, AFOAT-1 scientists had found the results obtained from the Air Weather Service's rainwater collection pans unimpressive. The Navy's utilization of rain scavenging, however, was based on removing material from the water chemically rather than by evaporation and on the use of far larger water samples than were collected by the Air Weather Service. For example, the volume of the sample from St. Thomas that was used to establish the feasibility of the Navy's methods was 2,500 gallons. Because much larger water volumes were used, the sensitivity of the Navy's rain-scavenging method was greater than that of the Air Weather Service.

When evaluating rain scavenging as a monitoring technique in the final report on Fitzwilliam, AFOAT-1 scientists had, in fact, quoted the result of the Navy's rain-scavenging experiment on St. Thomas rather than the results obtained from the Air Weather Service's rainwater collection pans. The Fitzwilliam report noted that the sensitivity and range of the Navy's rain-scavenging approach was comparable to that achieved by aerial collection of bomb debris. For meteorological reasons, however, AFOAT-1 scientists had concluded that rain scavenging was too problematic to be an adequate substitute for aerial sampling.[35] Nevertheless, to supplement the more reliable aerial sampling program, rainwater collectors had been retained as part of the equipment in the Whitesmith ground stations operated by the Air Weather Service.[36]

The reluctance of AFOAT-1 scientists to exchange information on rain scavenging and on fission product analysis with Naval Research Laboratory scientists may have been dictated by inter-service rivalry or by the compartmentalization policy or, perhaps, by the desire to allow the Navy to develop separately an expertise in collecting and analyzing fission products. AFOAT-1 could then use Navy results as an independent check on the findings of Tracerlab. In fact, the radiochemical methods of fission product analysis developed by Naval Research Laboratory scientists differed in some respects from those used by Tracerlab and by the Radiochemistry Group at Los Alamos.[37] Hence, when Joe-1 was detected, the Navy's data on this event did indeed provide an independent means of validating the results obtained by other laboratories.

Although Friedman, King, and Lockhart found their discussions with AFOAT-1 scientists to be a "one-way street" regarding the exchange of information, AFOAT-1 considered the Navy's scavenging work to be worthy of further support. In the AFOAT-1 research plan that Doyle Northrup submitted to the Research and Development Board in February 1949, a total of $150,000 ($50,000 in fiscal 1949) was included for the Naval Research Laboratory to "develop techniques for the collection and measurement of fission products scavenged from the atmosphere by rain, snow and ice."[38] In March, the fiscal 1949 portion of the plan was conditionally approved by the Board, and the Naval Research Laboratory received $50,000 to continue this development, dubbed Project Rainbarrel. King lost no time in getting started, and in the following month two of the Navy's monitoring stations were provided with rainwater collectors.

At that time, the Navy was operating ground-based monitoring stations at naval facilities located at Manila, in the Philippines; Honolulu, Hawaii; Kodiak, Alaska; and Washington, D.C. Each of these stations was equipped with an apparatus that provided a continuous record of the local level of gamma radiation and a high-speed blower that drew air through a filter to collect airborne radioactive material. In April, the stations at Kodiak and Washington were provided with a rainwater collector consisting of a rooflike aluminum structure 2,500 square feet in area, supported on 10-foot high posts. Runoffs positioned along the perimeter of the structure allowed water to be collected in storage tanks. Periodically, the collected water was treated to precipitate rain-scavenged material, and samples of the precipitate were sent to the Naval Research laboratory for analysis.[39]

In the months that followed, King, Lockhart, and Friedman honed their analytical and interpretive skills so that by midsummer the Navy's four-station monitoring system was operating in a routine surveillance mode.

THE BRITISH MONITORING NET

On April 4, 1949, the North Atlantic Treaty was signed in Washington, an event that seemed to symbolize official recognition that a condition popularly known as the Cold War prevailed between Russia and the West. On May 12, the Berlin Blockade ended, but Soviet-American relations remained fractious. On August 24, the North Atlantic Treaty Organization, or NATO, was formed, consisting of the United States, Canada, and ten European nations.

The Treaty eventually had a considerable effect on America's nuclear surveillance capability by facilitating the placement of AFOAT-1's monitoring stations in NATO countries located near the Soviet Union and by providing the basis for the United States to cooperate with these countries in developing monitoring systems of their own. In the summer of 1949, however, all this lay in the future. Indeed, at that time, the provisions of the MacMahon Bill of 1946 greatly inhibited international collaboration on atomic matters. This had a particularly adverse effect on the special relationship between the United States and Britain. According to one commentator, the bill placed "such restrictions on the sharing of [atomic] information as effectively to scuttle Anglo-American interchange."[40]

The breakdown of the Anglo-American partnership was perhaps most keenly felt by those members of the American intelligence community concerned with evaluating Soviet atomic progress who had long-standing ties with their British counterparts. These ties, in fact, originated in 1943 when Major Horace K. Calvert, who headed the London office of the Manhattan Engineer District's Foreign Intelligence Section, established a close working relationship with Lieutenant Commander Eric Welsh of the British Secret Intelligence Service, who was responsible for gathering intelligence on the German atomic program.[41] After the war, when American and British intelligence services turned their attention to the Soviet Union, the collaboration continued. When the Foreign Intelligence Section was absorbed into the Central Intelligence Agency's Nuclear Energy Group,

Commander Welsh continued to work closely with members of the Group. A representative of the British Secret Intelligence Service, who reported to Welsh, was provided with an office in the agency's Washington headquarters.[42] This intimate relationship ended, at least formally, when the MacMahon Bill was implemented beginning in 1947.

The "Control of Information" provision of the bill, in effect, interposed the newly formed Atomic Energy Commission (AEC) into Anglo-American collaboration on atomic matters, including those pertaining to atomic intelligence. Throughout 1947, attempts by the American and British intelligence services to work within the limitations imposed by the bill became increasingly unproductive. This situation was exacerbated by the attitude of Atomic Energy Commissioner Lewis Strauss.

William Golden, Strauss's assistant, later recalled that "when security matters came up at the AEC, Strauss was keenly interested in them." And if such matters involved the British, Strauss generally opposed any collaboration that would expose American secrets because, according to Golden, he felt "the British were very lax about security." Whether Strauss's belief was due to prescience or merely to the fact that he was, as Golden later averred, "an extremely security-minded person," these misgivings were to prove justified.[43]

In December 1947, the members of the Combined Policy Committee's Subgroup on Technical Cooperation attempted to overcome the roadblocks the MacMahon Bill had introduced into Anglo-American atomic collaboration by presenting to the Committee a memorandum listing nine areas in which cooperation would be mutually advantageous. (The Combined Policy Committee, which included American, British, and Canadian members, had been established during the war by Roosevelt and Churchill to coordinate atomic energy plans.) The Committee duly forwarded this memorandum to the Atomic Energy Commission for approval.

One of the American members of the Subgroup that prepared the memorandum was Vannevar Bush, who was aware that the Air Force was just beginning the process of forming a monitoring organization, global in scope, whose surveillance activities would be facilitated by the Canadian and British cooperation. Hence, it is not surprising to find that the fifth area listed in the memorandum was the possibility of creating a jointly operated monitoring system to detect a future Soviet atomic bomb test. When the commissioners met on January 7, 1948, to consider this memorandum, Strauss expressed "grave doubts as to the desirability of operating the long-range detection program jointly with the United Kingdom and Canada." In response to Strauss's objection, the Commission agreed to limit Canadian and British participation in this area to "meteorological and geophysical cooperation" and noted that this collaboration was "not now intended to include joint operation of detection stations." Moreover, to ensure "cooperation" did not stray beyond the prescribed areas, the Commission indicated that its prior approval of atomic information to be revealed in Anglo-American conferences would be required.[44]

Although certain kinds of meteorological and geophysical data were important to the monitoring program that had been projected by the Air Force, the limitation of the Anglo-American effort on monitoring to cooperation in these two areas fell short of the type of collaboration envisaged by Bush and needed by the Air Force. This need became especially apparent after the completion of Fitzwilliam. Although most of the WB-29 bases and the radiological ground stations used in this operation were in locations suitable for monitoring the Soviet Union, they covered only the Pacific and south Atlantic regions. More planes and ground stations would be needed to cover the north Atlantic region. The British had aircraft and bases that were well positioned geographically to carry out this task, but to do so effectively they would need to know what the United States Air Force had learned about radiological monitoring methods. In this regard, the Fitzwilliam results, describing the best methods of collecting and analyzing fission products, were especially important. The restrictions imposed by the Commission on Anglo-American cooperation in the area of monitoring, however, precluded any disclosure of such information.

In the spring of 1948, the British were "notified" of the impending Sandstone tests and Operation Fitzwilliam, apparently through informal channels. Royal Air Force bombers were hurriedly equipped with filters, and, during Sandstone, these aircraft succeeded in collecting airborne bomb debris for analysis. When Strauss learned of this activity, he was incensed: "I would like to know who notified them," he later wrote.[45]

Shortly after the preliminary results of Fitzwilliam became available in August, a meeting of military representatives of the United States, Canada, and the United Kingdom was planned, and, because one topic on the agenda was related to radiological monitoring, the information that would be revealed during the discussions had to be cleared by the Atomic Energy Commission. On the morning of September 2, the commissioners considered this matter. Strauss had been unable to attend this meeting and his assistant, William Golden, attended as Strauss's proxy. The nature of the proposed discussion on monitoring exceeded the topical limits imposed by the Commission in January. In a report to Strauss, Golden noted that "the Commission did not approve the disclosure of information in these talks."[46]

The meeting of the military representatives of the three countries also took place on September 2, beginning a few hours after the meeting of the commissioners ended. Although the American representatives had been informed that the Commission had not approved disclosure of the information they proposed to present, the discussions proceeded as planned. In his report to Strauss, Golden averred that, after several days had elapsed, he received documents indicating that, in regard to monitoring, "these international discussions were held and the disclosures were made without let or hindrance." He noted that somehow clearance for the talks had been obtained, but he had been unable to find out "how and from whom." He wrote: "The fact of our making these disclosures, with little or nothing in return, and the circumstances surrounding the talks are beyond my comprehension. . . . The method of disclosure of this unique information from Sandstone and elsewhere may be characterized as, at best, grossly improvident."[47]

Golden's assessment was wrong to the extent that, in return for the information given, a *quid pro quo* was expected and received by the United States—namely, that its allies would fill the north Atlantic gap in AFOAT-1's interim net. But he was right in that the involvement of the British (as will be later described) *did* turn out to be "grossly improvident" from the standpoint of AFOAT-1's security.

The British lost no time in making use of the information given them. Radiological ground stations were established at air bases in Scotland, Northern Ireland, and Gibraltar. Royal Air Force bombers at these bases were equipped with filters similar in efficiency to those used on the American WB-29s, but, unlike the arrangement used in WB-29s, the filters could not be changed in flight since they were contained in tubelike structures mounted under each wing of the aircraft. (This made the British data somewhat less detailed geographically than those obtained by the WB-29s.) The ground stations were equipped with an apparatus containing Geiger counters which was used to measure radioactivity on the filter after each flight. Filters exhibiting an anomalously high count rate were rushed to the British atomic energy facility at Harwell, England, where, through radiochemical techniques, they were analyzed.[48]

By the summer of 1949, the Royal Air Force was monitoring flights on a routine basis in an area bounded in the east by the longitude of Greenland, in the west by the European coast, in the south by the latitude of Gibraltar, and in the north by the Arctic Circle. Flights from Gibraltar were code named Nocturnal, while those from bases in Britain were code named Bismuth. The north Atlantic gap in interim net had been closed.

NOTES

1. "Project Firstrate, AWS Participation in Fitzwilliam," p. 7. In "Final Report on Operation Fitzwilliam," Vol. I, Tab D.

2. F. C. Henriques interview, August 18, 1981.

3. D. L. Northrup, "Technical Memo No. 31," to A. F. Hegenberger, February 2, 1949, p. 3.

4. Strauss, "History of the Long Range Detection Program," July 21, 1948, p. 5.

5. L. L. Strauss to Commissioners, August 9, 1945. Strauss Papers, AEC Series, Box 76, HHPL.

6. L. L. Strauss to J. F. Forrestal, January 3, 1949. Strauss Papers, AEC Series, Box 76, HHPL.

7. Northrup, "Technical Memo No. 31," February 2, 1949, pp. 11–15.

8. Ibid., pp. 21–24.

9. Ibid., pp. 18–20.

10. Ibid., pp. 12–14.

11. Northrup, "Technical Memo No. 34," June 3, 1949.

12. "Project Firstrate, AWS Participation in Fitzwilliam," p. 11. In "Final Report on Operation Fitzwilliam," Vol. I, Tab D.

13. For uranium, the percent fission yield for xenon-133 is twenty-six times greater than that for krypton-85, and a similar ratio holds for plutonium. See B. J. Wilson, ed., *Radiochemical Manual*, 2nd ed., 1966, p. 255.

14. A. C. Trakowski to Colonel M. Duffy, December 10, 1948, pp. 1,2. Folder on Long Range Detection FY1949, History Office, AFGL.

15. P. A. Humphries interview, January 3, 1990.

16. C. Peebles, *The Moby Dick Project*, 1991, p. 196.

17. W. B. Mann interview, April 11, 1990.

18. Trakowski to Colonel Duffy, December 10, 1948, p. 2.

19. C. P. Boner to Chairman, RDB Committee on Atomic Energy, Attachment A, p. 1. RG 330, Secretary of Defense, Entry 341, Research and Development Board, Box 10, Folder 100 Atomic Energy/Long Range Detection, NA.

20. J. Richelson, *American Espionage and the Soviet Target*, 1987, p. 120. In the 1950s American U-2 "spy planes" equipped with cryogenic air-samplers were used to collect krypton-85, as were high-altitude balloons. See J. D. Miller, *Lockheed U-2*, 1983, p. 103 and Peebles, *The Moby Dick Project*, p. 197.

21. For a more detailed description of ground-station equipment transferred to the Whitesmith net, see "Project Firstrate, AWS Participation in Fitzwilliam," pp. 12–16. In "Final Report on Operation Fitzwilliam," Vol. I, Tab D.

22. For a list of ground stations, see "Project Firstrate, AWS Participation in Fitzwilliam," Appendix V, p. 84.

23. R. W. Spence, letter to C. A. Ziegler, November 30, 1988. Copy in the authors' collection.

24. R. W. Spence interview, November 1, 1990.

25. Ibid.

26. Ibid.

27. G. M. Leies interview, September 11, 1990.

28. H. A. Schade to Chief of Naval Operations, January 18, 1949, pp. 1, 2. Copy in the authors' collection. Our thanks to Dr. Luther B. Lockhart for providing us with a declassified copy of this memorandum.

29. H. Friedman interview, July 27, 1989.

30. L. B. Lockhart interview, April 6, 1990.

31. P. King, L. Lockhart, H. Friedman, and I. H. Blifford, "Recovery and Identification of Atomic Bomb Fission Products Scavenged from the Atmosphere by Rainfall Following an Air Burst," January 18, 1949, NRL Report CN3378. Copy in the authors' collection. Our thanks to Dr. Luther B. Lockhart for providing us with a declassified copy of this report.

32. Schade to Chief of Naval Operations, January 18, 1949, pp. 1, 2.

33. Ibid., p. 2.

34. H. Friedman interview, July 27, 1989.

35. "Project Firstrate, AWS Participation in Fitzwilliam," pp. 10, 16; and "Design of Operation and Summary of Results," p. 15. In "Final Report on Operation Fitzwilliam," Vol. I, Tab D.

36. "Project Firstrate, AWS Participation in Fitzwilliam," p. 84. In "Final Report on Operation Fitzwilliam," Vol. I, Tab D.

37. R. W. Spence interview, November 1, 1990.

38. D. L. Northrup, "Technical Memo No. 31," February 2, 1949, p. 13.

39. P. King and H. Friedman, "Collection and Identification of Fission Products of Foreign Origin," NRL Report CN3536, September 22, 1949, pp. 1–8. President's Secretary's File, Box 200, NSC-Atomic, HTPL.

40. Herken, *The Winning Weapon*, p. 148.

41. During the war, Eric Welsh had been notably successful in obtaining information about Germany's atomic program through his agent in Berlin, Paul Rosebaud. See A. Kramish, *The Griffin*, 1986, p. 54. This information was so complete that, by 1943, Allied concern over the threat of a German atomic bomb had largely disappeared. See Hinsley, *British Intelligence in the Second World War*, Vol. 2, 1981, pp. 125–28.

42. W. B. Mann interview, April 11, 1990.

43. W. T. Golden interview, January 4, 1990.

44. "Atomic Energy Commission meeting No. 139," January 7, 1948, pp. 10, 11. Minutes of Atomic Energy Commission Meetings, Archives, DOE.

45. L. L. Strauss to W. Webster, December 1, 1948. Strauss Papers, AEC Series, Box 113, Folder on Tests and Testing 1947–1954, HHPL.

46. W. T. Golden to L. L. Strauss, September 10, 1948, p. 1. Strauss Papers, AEC Series, Box 113, Folder on Tests and Testing 1947–1954, HHPL.

47. Golden to Strauss, September 10, 1948, p. 1.

48. W. G. Penney, "An Interim Report of British Work on Joe," September 22, 1949, pp. 1–6. President's Secretary's File, Box 200, NSC-Atomic, HTPL.

CHAPTER 11

THE DETECTION OF JOE-1 AND BEYOND

By mid-1949, the organizational complex created by AFOAT-1 to carry out its monitoring task had involved that agency in relationships with eleven government agencies, three industrial firms, a foundation, a university, and one foreign government. The nature of these relationships varied. In a few cases, the relationship was informal—a cooperative arrangement based on mutual interests. Most, however, were formal client relationships in which the organizations carried out research on projects sponsored by AFOAT-1. Some organizations within the latter group were also engaged in the day-to-day operation of the interim net, while others were available on an "on-call" basis to provide special services related to the net's operational needs. As will be seen, these organizational resources were mobilized to the fullest by the leaders of AFOAT-1 when they faced the first challenge for which their agency was created: the detection of Joe-1.

BEQUEATH

In February 1949, in accord with normal security practice, the code name for monitoring the Soviet Union on a routine basis was changed from Whitesmith to Bequeath. By the spring of 1949, monitoring had indeed become routine, and the radiological component of the interim net was operating smoothly. Key elements of this component were the Air Weather Service squadrons whose WB-29s were engaged in the collection of airborne dust. Although ostensibly part of the regular weather reconnaissance flights, monitoring missions involving nonstop flights of up to eighteen hours stressed both aircraft and crews. At the start of a mission, the WB-29s were loaded to their maximum takeoff weight with 20 tons of fuel in the wing tanks, 6 tons in portable tanks shackled in the bomb bays, and 2 tons of survival gear. The planes also carried a variety of meteorological instruments that required the crew's attention during the flight, a task made more onerous by the apparatus that had been installed to collect atmospheric dust.

For security reasons, the flight crews had not been told the true radiological purpose of the dust collectors, and they considered them to be merely part of the array of weather instruments aboard the aircraft. In fact, the assumption that the purpose of the dust collectors was meteorological seemed reasonable because it had long been known that atmospheric dust from natural and man-made sources can strongly affect insolation and the weather. The flight crews were thus excluded from the small group of Air Weather Service officers who knew that the weather reconnaissance flights were part of an atomic surveillance system that targeted the Soviet Union.[1]

Promptly dubbed the "bug-catcher" by the flight crews, the dust collection apparatus consisted of a metal housing that projected upward through the midsection of the fuselage. The housing was slotted to receive a 9 x 22 inch filter sandwiched between two layers of open-mesh wire screening. From inside the fuselage, the sandwich could be pushed up into the slots in the housing where the surface of the filter was exposed to the slipstream. The filters had to be changed every two to three hours, a task that made the already arduous flights more stressful because of the location of the bug-catcher. It was located in an unpressurized section of the aircraft where, since most of the missions were flown at an altitude of about 18,500 feet, the temperature of the rarified air was minus 60 degrees Fahrenheit.

Every few hours the crewman designated to change the filter would shout "going out!" to alert the man assigned to watch him. Should his oxygen mask slip or the hose to the walkaround oxygen bottle strapped to his leg catch on some projection and be pulled from its socket, his safety could depend on getting prompt help from the watcher. Heavy gloves and a flight suit, donned over several suits of underwear and a pile-lined parka, greatly hampered movement and the performance of even the simplest tasks needed to operate the various weather instruments.

The repeated depressurization and repressurization of the cabin that occurred when a crewman serviced the bug-catcher was physically stressful for all the crew members. As the flight continued, the temperature in the cabin became progressively lower, a condition that layers of clothing and feebly heated flight suits did little to ameliorate. Crews emerged from their planes exhausted after thirteen- to eighteen-hour missions performed under these harsh conditions. Nor were the flights without other hazards. On missions near the Soviet Union the planes occasionally drifted off course into Soviet airspace where they ran the risk of being shot down.[2] Virtually all monitoring missions required long overwater flights or flights over some of the most inhospitable territory on earth, such as the Arctic regions, where the chances of surviving a forced landing were slight. All of these hazards existed for flight crews of the 375th Weather Reconnaissance Squadron which, of the four squadrons involved in the interim net, was based closest to the Soviet border.

Filter-equipped WB-29s of Colonel Karl Rauk's 375th Weather Reconnaissance Squadron flew routinely, usually every other day, along two synoptic weather tracks. The first, code named Ptarmigan, required a 3,500-mile journey from Eielson Air Force Base at Fairbanks, Alaska, to the Pole and back; the second, code named Loon Charlie, stretched 3,600 miles from Eielson to the Air Force base at Yokota, Japan. A turnaround detachment of the 375th was stationed at Yokota to provide flight crews to fly the reverse track of Loon Charlie so that, within a day or two of its arrival at Yokota, a WB-29 would be on its way back to Eielson.[3]

When traced on a map, Ptarmigan and Loon Charlie extended outward from Eielson, forming two arms of a giant V-shape whose open end faced the Soviet Union and which stretched from 35 degrees north latitude to the Pole. Since virtually all Soviet territory was north of the 35th parallel, these two tracks were positioned to intercept airborne dust carried eastward from any point within the borders of the Soviet Union. Indeed, as will be described later, it was a plane of the 375th that picked up the first faint traces of the test of a Soviet bomb, a bomb that was the product of years of intensive research and development.

BORODINO AND FIRST LIGHTNING

Both Western and Soviet sources appear to agree on the highlights of the Soviet effort to make an atomic bomb.[4] Following the discovery of nuclear fission in Germany in 1938, Soviet scientists alerted their government to the possibility of utilizing this phenomenon in a bomb of unprecedented power. Their subsequent research in this area roughly paralleled that of Western scientists until the German invasion in the spring of 1941 ended the program. Through espionage, Soviet leaders remained aware of atomic developments in Britain and the United States, and early in 1943, after the immediate German threat had been contained, atomic research was started anew under the leadership of the physicist Igor V. Kurchatov. Because of wartime exigencies, however, the program remained quite small compared to that of the United States. Although their own research was necessarily on a modest scale, Kurchatov and his team were kept abreast of the more important findings of the American atomic bomb program by Lavrentiy Beria, head of the secret police, who was privy to the reports of spies operating in the United States.

In August 1945, Soviet Premier Josef Stalin called for an all-out effort to break the American atomic monopoly. Code named Operation Borodino, a massive effort to make a bomb in the shortest possible time was launched, with Beria as the titular head. He controlled the prisons and, according to one commentator, "Half the research on nuclear weapons was done in prison institutes, while most of the construction and [uranium] mining was done by prison labor."[5] Kurchatov continued to lead the revitalized scientific program. As a brilliant scientist-administrator, he occupied a position in Operation Borodino similar in some ways to that of J. Robert Oppenheimer in America's Manhattan Project.

In defeated Germany, scientists, technicians, and equipment were swept up by Soviet intelligence teams and sent to the Soviet Union where they were employed on Operation Borodino. German technology for processing uranium ore was especially valuable to the Soviets. From its beginnings, Soviet atomic research had been plagued by a lack of high-grade uranium ore. After 1945, a tremendous effort to find sources of such ore and to speed up the processing of the low-grade ores, available in Eastern Europe and the Soviet Union, was undertaken. In 1946, the uranium problem had been solved to the extent that Kurchatov's team was able to construct a reactor that became operational in December of that year. (Apparently, this was an exact copy of America's Hanford 305 reactor.)[6] In the spring of 1947, the Soviets began construction of facilities for producing plutonium at a site near Sverdlovsk. In April 1949, shortly after the United States deployed the Interim Surveillance Research Net, a second plutonium reactor became operational. By August, the first plutonium bomb had been constructed, and a site in the desert about 100 miles south of Semipalatinsk, in Kazakhstan, was selected to test the bomb in secret.

From the Soviet perspective, it was highly desirable to keep the attempt to detonate their first atomic bomb hidden from the world. In the previous four years, Stalin had responded to the American atomic monopoly with the classic strategy of bluff in the face of a stronger opponent. Beginning in 1946, Soviet spokesmen in the United Nations and other forums had voiced vaguely worded statements implying that their government's atomic program, undertaken only to explore peaceful uses of atomic energy, was on a par with that of the United States. Exposure of a Soviet bomb test would reveal the falsity of such pronouncements. More important, Western intelligence analysts would have a crucial "calibration point" that would allow them to calculate accurately the substantial lag between American and Soviet atomic weapons capability. This gap would redound negatively on Soviet policy. Even worse, the probable American response would be to accelerate the development of atomic weapons, making it that much harder for Operation Borodino to catch up.

Like all atomic facilities, the test site, called the *Poligon*, was patrolled by select guard units. The location of the site and technical details, like the height of the 165-foot tower on which the bomb was placed, were such as to minimize the possibility that detonation of the bomb would be betrayed by seismic effects or the radioactive dust that would fall beyond the borders of the Soviet Union. Because of the prevailing west winds over the *Poligon*, bomb debris produced by the explosion would travel 3,000 miles and more before reaching areas subject to monitoring by American aircraft.

The Soviets were familiar with air-sampling methods and had used them to track American atomic progress. But it was not necessary for them to develop methods of long-range detection for this purpose since the area close to American test sites in the Pacific could be monitored by their vessels operating in international waters. Hence, they apparently believed (as did American scientists, until Fitzwilliam) that at distances of several thousand miles from the detonation site the airborne radioactivity produced by the explosion was not likely to be identified unambiguously as the result of a bomb test.

For Kurchatov and his team, the test, code named *Pervaya Molniya* or First Lightning, represented the culmination of nearly seven years of tireless, painstaking scientific work. At dawn on August 29, 1949, Kurchatov, Beria, and other notables were in a bunker at the *Poligon* awaiting the detonation. This occurred at 6:00 A.M. and, according to one history, "The blast was immediately evident to everyone in the bunker. With a sweeping gesture Kurchatov began shouting, 'There it is, there it is, there it is!' "[7] As the characteristic mushroom cloud formed over the *Poligon*, dust particles kicked up by the explosion were pushed into the stratosphere where high-altitude wind currents carried them eastward.

LOON CHARLIE

The first indication of the Soviet bomb test was picked up by a WB-29 of the 375th, flying the reverse Loon Charlie track from the Air Force base at Misawa, Japan, to Eielson. Such flights normally originated at Yokota, and it is pertinent to describe the events that caused this flight to begin at Misawa since they illustrate the hazards faced by the men who flew the war-worn WB-29s.

On August 29, the scheduled Loon Charlie WB-29 landed at the United States Air Force base in Yokota, Japan, and on the afternoon of the following day Crew 5A, from the 375th's turn-around detachment, prepared to fly the aircraft on the reverse Loon Charlie track to Eielson. An engine failed, however, during the runup prior to takeoff, and the eleven-man crew had to transfer to a standby aircraft. A typhoon warning had been posted that morning, and the delay caused by the transfer pushed the mission departure time close to the expected onset of the storm.

Shortly after takeoff, the pilot and aircraft commander, First Lieutenant Robert C. Johnson, experienced his second engine failure of the day. The number three engine burst into flames which, fortunately, the automatic extinguishers were able to smother. He turned the plane backward to Yokota, but, by then, the airport was closing down because of the storm. After one failed attempt to make an emergency landing on instruments at

Yokota, during which the heavily burdened aircraft came perilously close to stalling, he was directed to fly to Misawa, 350 miles to the north. The WB-29 was grossly overloaded for a three-engine flight, and it was impossible to lighten it by jettisoning fuel because most of the assigned route was over heavily populated areas. Although the crew landed safely at Misawa, the engine temperature gauges had been red-lined for much of the flight and the WB-29 had to be taken out of service to replace all four engines.

On September 1, the Loon Charlie WB-29 from Eielson, flown by Captain Errol D. Barclay, was diverted from Yokota to Misawa, and on September 3, Johnson and Crew 5A used this aircraft to fly the reverse Loon Charlie track. After their harrowing experiences on August 30, Johnson and his crew were thankful that the 13 1/2 hour flight seemingly proved uneventful.[8]

But unknown by the flight crew, a significant event *had* occurred during the mission. Four and a half hours after leaving Misawa, their WB-29, flying at 18,000 feet, had intercepted dust particles that had been produced by the Soviet bomb test. At the time, their aircraft was skirting the coast of the Kamchatka Peninsula. Upon landing at Eielson, the weather observation officer on the flight, First Lieutenant Robert L. Lulofs, gave the envelope containing the filters that had been exposed for three-hour intervals along the flight-path to the debriefing officer who delivered it to Base Security. From there, the envelope was taken to a trailer-type shack located in a secure area of the base that was off-limits to Squadron personnel. The shack contained various ground-station monitoring instruments, including a 600-pound lead cylinder 1 foot in diameter and 2 feet long that housed the wraparound counter.

Sergeant Eugene W. Tews was on duty in the shack when the envelope from Johnson's Loon Charlie flight arrived and he promptly began using the wraparound counter to measure the radioactivity on each of the filters. A month earlier, an improved procedure for operating the wraparound counter had been put into effect at all ground stations, which allowed AFOAT-1 to reduce the level of significance for aerial filters from 100 to 50 counts per minute. Tews found that the second filter exposed on the just-completed Loon Charlie mission registered 85 counts per minute. Under the new criterion, this barely qualified as an official "alert." Tews immediately notified his superior, Captain Carroll L. Hasseltine, of the situation.[9]

JOE-1/VERMONT

Meanwhile, 3,400 miles away in Washington, D.C., the Labor Day weekend had begun. Most Washington officials were at home or at vacation retreats, and most government offices were empty. But in the yellow-brick building on G Street that housed AFOAT-1's headquarters, the Data Analysis Center was manned around the clock. The Center, located on the heavily guarded third floor of the building, was an oversize conference room entered through a steel door, containing a bank of teletypes at one end, a large, 7-foot diameter globe at the other, and a long table surrounded by chairs in the middle. Brilliantly lit by overhead fluorescent lights, the walls of the windowless room were covered with maps and meteorological charts studded with colored pins. In the months since the radiological component of the interim net had been operational, the Center had logged 111 alerts. Each had been treated as evidence of a possible Soviet bomb test, but each had turned out to be due to natural causes such as volcanic activity, earthquakes, or normal statistical variations in background radioactivity. Alert No. 112, which appeared on the teletype early Saturday evening, September 3, was to prove different from these previous alerts.[10]

Summoned to the Center by the alert, Doyle Northrup and key members of his staff made a preliminary evaluation of the available information. Although the report from Alaska qualified only as a "minimal alert," Northrup requested Sergeant Tews, at Eielson, to periodically remeasure the radioactivity on the filters. As the reports of these successive measurements were received at the Center, they were plotted by Dr. William Urry and Dr. Donald Rock. From the rate of radioactive decay it was soon apparent that the dust picked up by the filters might contain fresh fission products. But was it bomb debris? And, if so, was the debris from the leading edge of the radioactive dust cloud or at the tail end? To answer the first question, it was necessary to collect more filter samples and send them to Tracerlab's Berkeley, California, laboratory for detailed analysis. To answer the second question, more monitoring flights along the estimated path of the cloud were needed.

Special WB-29 missions were vectored from Alaska to Hawaii and from California to Alaska, and flights from Alaska searched the area over the Beaufort Sea. On Monday, September 5, the Center received an unexpected report from Japan. The pilot of a WB-29 of the 514th Weather Reconnaissance Squadron, flying the weather track from North Air Force Base, Guam, to Yokota, had turned in a filter that registered 1,000 counts per minute. Northrup immediately directed the Yokota ground station to send this filter to Tracerlab by special aircraft. That same day the Center received reports from the Air Weather Service Alaskan ground stations that their instruments were registering anomalously high levels of radioactivity.[11]

Air Force couriers bearing filters began arriving at Tracerlab's Berkeley laboratory on September 6. Dr. Lloyd Zumwalt, who headed the analytical effort at Berkeley, later recalled that "it was very exciting to analyze these things. . . . We worked on them through the night." The analysis quickly revealed that fresh fission products of barium, cerium, and molybdenum were present on the filters and that they appeared to have the same "birthday," indicating that they were probably from a bomb rather than, say, a nuclear reactor accident. Also, according to Zumwalt, "the fission product yield curve was more like what you should expect from plutonium than from uranium. . . . We communicated that to Urry and he was very excited."[12]

In fact, Urry received Zumwalt's telephone call at 3:30 on the morning of September 7. Major General Morris R. Nelson, who had succeeded Hegenberger as head of AFOAT-1, upon receipt of this electrifying news, decided that an all-out effort should be made to collect as many samples as possible. By the following day, enough data were available from monitoring flights over the northern United States and Canada to indicate that the air mass containing debris from the Soviet bomb would soon depart North America en route to the northern regions of the Atlantic. General Nelson, in conference with Northrup, decided to alert the British. On September 9, they called upon Carroll Wilson, general manager of the Atomic Energy Commission, to arrange Commission approval of this action.[13]

"This was a sticky matter for Wilson," according to the Commission's official history, because "To alert the British might constitute a technical violation of the Atomic Energy Act." But, as the history goes on to point out, Wilson realized that "to withhold the information even for twenty-four hours might preclude the possibility of obtaining British samples. Perhaps he could justify the action under the technical cooperation program, but there was no time to find out. Wilson picked up the telephone at six o'clock and called Alexander K. Longair, the British representative on technical cooperation in Washington."[14] Longair hurried over to AFOAT-1's Data Analysis Center where he was briefed by Nelson and Northrup. He was then taken to the Pentagon where he spent the remainder of the evening explaining the situation to his London contacts via teletype.

On the morning of September 10, British atomic energy authorities requested the Royal Air Force to make special monitoring flights north of Britain. Until September 10, the regular monitoring flights by filter-equipped aircraft of the Royal Air Force over the weather tracks Nocturnal and Bismuth had not revealed any anomalous increase in airborne radio-activity. But on that day, the special flight from Scotland, along a track due north to the Arctic Circle and back, gave immediate results. During the next six days, nineteen special flights were made in addition to Nocturnal and Bismuth missions, and the exposed filters were sent to the British atomic energy facility at Harwell for analysis. Within a few days, fission products of the elements barium, cerium, and iodine had been identified on the filter samples and the time of fissioning (between August 26 and 29) was estimated.[15]

Meanwhile, much had been accomplished in the United States during the period when the radioactive cloud traversed North America. To follow the cloud's trajectory, the Air Weather Service had vectored ninety-two special monitoring flights, collecting over 500 filter samples. On September 8, AFOAT-1 staff members had conferred with Naval Research Laboratory scientists and requested that the Laboratory perform a completely independent analysis of the radioactive material on some of the aerial filters exposed on WB-29 flights.

Up until that time, monitoring instruments in Navy ground stations had not registered an anomalous increase in radioactivity. Such an increase, however, was recorded on September 9 by the station at Kodiak, Alaska. Subsequently, the rainwater collectors at Kodiak and at Washington, D.C., yielded large samples of bomb debris. The Naval Research Laboratory had begun to analyze the aerial filter samples provided by the Air Force, and on September 14 Laboratory scientists provided General Nelson and his staff with an oral report on their preliminary findings. These scientists had identified the fission products of five elements, but they pointed out that it would be advisable for them to discontinue the analysis of the Air Force filter samples, so that they could begin analysis of the larger samples that had become available from the Navy's rainwater collectors. Subsequently, fission products of the elements ruthenium, cerium, yttrium, and silver were identified in the rainwater samples. Their date of origin was estimated as "not earlier than 24 August."[16]

Northrup and his scientific staff were well aware of the crucial nature of the radiochemical findings. He later recalled that "AFOAT-1 believed it imperative to obtain confirmatory radiochemical analyses."[17] On September 10, Urry sent a filter sample to Dr. Roderick Spence's Radiochemistry Group at the Los Alamos Laboratory for independent analysis. Subsequently, Spence reported that fission products of barium, cerium, molybdenum, silver, and zirconium had been found, "their age probably being one month or less."[18]

Spence was also of help to Tracerlab. Zumwalt later recalled that Spence "got over the point that we should analyze for neptunium because that would give information as to whether or not there was a lot of uranium around the bomb, what they call 'tampering' [surrounding the plutonium core with a layer of uranium that facilitates the chain reaction by reflecting neutrons back into the active mass]." Unfortunately, according to Zumwalt, "that was one thing we hadn't set up for, to analyze neptunium."[19] To provide Tracerlab with a calibration sample of the needed radioisotope of neptunium, Spence arranged for an irradiation in the Los Alamos reactor over the weekend of September 10. On Monday, the sample was delivered to Zumwalt, and within a few days he was able to report that the Soviet bomb did have a uranium tamper.

By September 14, there was no doubt in the minds of those who could understand the data that a Soviet bomb test had been detected. Indeed, insiders had dubbed it "Joe-1," although the official code name later assigned was Vermont. Enough evidence had been collected, chiefly by Tracerlab's analyses, to suggest that the Soviet bomb was probably similar in construction to the American Trinity bomb (a plutonium implosion type with a uranium tamper) and that, like the latter, it had been detonated atop a tower.

This view, however, was not universally accepted by government leaders. According to the official history of the Atomic Energy Commission, "By Wednesday, September 14, most of those in the know were convinced that a Soviet test had occurred. A notable exception was [Defense] Secretary Johnson, who, despite [his deputy for atomic energy William] Webster's argument that 95 percent of the experts accepted the fact, preferred to side with the 5 percent who doubted the evidence."[20]

In the face of continued skepticism on the part of a few officials, it was apparent that Northrup was on the right track in his efforts to obtain independent analyses from several sources. Portions of Tracerlab's data had been confirmed by Navy and Atomic Energy Commission scientists, but the British had not reported their findings, and Longair was asked to expedite this matter. Longair elected to channel the request for information through his colleague at the British Embassy, Dr. Wilfred B. Mann, a physicist who was the representative of an atomic intelligence unit at the Ministry of Supply's Directorate of Atomic Energy. Late on the evening of September 14, Longair called Mann at his Washington apartment. Mann later recalled that the "time was 11:30 P.M. I had one foot in the bathtub when the phone rang. It was Alex Longair . . . to tell me that I had been invited to attend a session in a "war room" which was then situated not far from the White House."[21]

Mann arrived at the Data Analysis Center at 1:00 A.M. for a briefing and then went to the Pentagon where he conducted a teletype discussion with colleagues in London including his boss, Lieutenant Commander Eric Welsh. Welsh had been "seconded" from the Secret Intelligence Service, or MI6, to head the intelligence unit of the Directorate of Atomic Energy. Arrangements were made for Welsh to come to Washington accompanied by a Directorate scientist, Dr. William G. Penney, to brief General Nelson and his technical staff on the results of the British monitoring effort.[22]

In addition to obtaining independent analyses by Navy, Atomic Energy Commission, and British scientists, General Nelson decided it would be wise to have all the evidence examined by a panel of eminent scientists unconnected with the Air Force. Defense Department officials concurred, and, during the week of September 11, Nelson and Northrup visited Dr. Vannevar Bush, who had returned to the Carnegie Institution after nearly ten years of government service. They asked him to convene and chair an advisory group consisting of Dr. Robert Bacher, former Atomic Energy commissioner, Dr. J. Robert Oppenheimer, former head of Los Alamos Laboratory, Dr. Karl T. Compton, Research and Development Board chairman, and Admiral William S. Parsons of the Military Liaison Committee. Bush and the other designated scientists, except for Compton, agreed, and they were officially empaneled by General Hoyt S. Vandenberg, Air Force chief of staff, to review AFOAT-1's data and conclusions at a meeting scheduled for September 19.

The data Nelson and his scientific staff arranged to have presented to the panel were almost entirely radiological. An exception was the meteorological information used to backtrack the radioactive cloud to determine the location of the Soviet explosion. A preliminary examination of sonic and seismic records had not been helpful in this regard; hence, AFOAT-1 depended on the meteorologists of the Weather Bureau Special Projects Section, headed by Dr. Lester Machta, to accomplish this task. A member of the Section, Dr. Kenneth M. Nagler, later recalled their activities in the period prior to the panel meeting: "Much of our work that week was done in a very secure but small room in the Air Weather Service building at Andrews Air Force Base. There we had prompt access to the radioactivity intercept data as well as to weather data and analyses. In that crowded room I recall using a drawer in a map-filing cabinet as a table for the analysis and trajectory work."[23]

The Air Weather Service had done some backtracking of air masses during Fitzwilliam, and it was expected that, in the absence of sonic and seismic data, this method would be used to locate the site of a Soviet explosion. The approach adopted by Machta and his group made use of the fact that the heights of "standard" pressure levels, such as the 500-millibar level, were charted routinely several times a day by meteorologists. Such charts, based on "weather balloon" data from American and foreign stations scattered across a huge geographic area, were readily available. Graphic representation of the heights of a given pressure level on the chart were indicative of the horizontal airflow, especially at and above the 500-millibar level (about 18,500 feet) where the airflow is relatively unaffected by ground friction. Indeed, one reason why monitoring flights were usually carried out at altitudes of about 18,500 feet was to facilitate the use of the standard 500 millibar charts in backtracking any radioactive bomb debris that might be picked up at this level. In the case of Joe-1, according to Machta, "a reasonable assumption was made that there was radioactivity in the initial nuclear cloud at 500 mb [millibars] and that its transport on the 500 mb surface brought the radioactivity to the point of detection near Kamchatka."[24]

To determine the site of Joe-1, it was, of course, necessary to know where to stop the backward projection of the trajectories of the air masses by knowing the time of the detonation. The best estimate of the time was provided by Tracerlab analyses. Based on eleven determinations, derived from the ratios of fission products, the date of origin was established as within the period August 26 to 29. A separate determination based on the empirical decay rate of radioactivity on the filters, as calibrated from Fitzwilliam data, yielded a time of 0000 GMT 29 August.[25] Northrup, in his report to the panel on September 19, indicated the time of the explosion as "between 26 and 29 August."[26] But earlier, apparently to provide a more precise figure for backtracking, he had given Machta a time that was the average of the eleven determinations derived from fission product ratios. This time was 1500 GMT 27 August.[27]

The use of this time estimate, which was thirty-four hours earlier than the actual time of Joe-1, caused Machta and his group to carry the backtracking process too far to the east. Their projection thus suggested that the most probable site of the explosion lay in the region at the north end of the Caspian Sea, although areas of lesser probability in their projection did include the actual site near Semipalatinsk. Despite uncertainties in their methodology, the result provided by Weather Bureau scientists more than satisfied the criterion given to AFOAT-1 by the Joint Chiefs of Staff—namely, that "the explosion can be determined to be within the USSR or the Soviet sphere."[28]

Although the information was not available until some time after the president's announcement of Joe-1, AFOAT-1 scientists were able to determine the site of the explosion and other parameters, such as the yield, more precisely by sonic means. Northrup later wrote:

Although acoustic records were carefully examined at the time of Joe-1, no signals relating to the event were discovered. However, subsequent review of these records revealed weak signals at two stations. These signals were very useful because they helped after the fact to establish the location, time and size [yield] of Joe-1 with greater precision than was otherwise possible.[29]

If the sonic results obtained during Fitzwilliam are a useful guide to the performance of the then existing sonic methods, this statement by Northrup indicates that AFOAT-1 was able to determine the site of Joe-1 to within 100 miles and the time to within ten minutes.

The review panel consisting of Bush, Bacher, Oppenheimer, and Parsons met on September 19. Much of the evidence was presented to them in the form of oral presentations because the involved scientists had no time to prepare lengthy written reports. Penney described the results obtained by the British, Spence reported on the work of his Radiochemistry Group at Los Alamos, and James R. Smith of the Naval Research Laboratory presented the findings of Navy scientists King and Friedman. Other presentations were made by AFOAT-1 staff members. According to Northrup,

The key point which the panel had to consider was whether the debris originated from an atomic bomb explosion or from some other source such as a reactor accident. It was the large number of comprehensive analyses providing internally consistent data presented by Tracerlab personnel Dr. Frederick Henriques, Dr. A. J. Stevens and Dr. Lloyd Zumwalt which provided the deciding factor.[30]

Northrup had given the panel a three-page report, Technical Memo No. 37, summarizing all the Joe-1 evidence. On September 20, the four panel members sent Northrup's report together with a covering memorandum to

General Vandenberg in which they indicated that, regarding the findings of Northrup and his staff, "We unanimously agree with their conclusions as presented in Technical Memo No. 37."[31] That same day, Vandenberg submitted these documents to Defense Secretary Johnson with a memorandum in which he stated:

I believe an atomic bomb has been detonated over the Asiatic land mass during the period 26 August 1949 to 29 August 1949. I base this on positive information that has been obtained from the system established by the Air Force for the long range detection of foreign atomic activities. . . . Conclusions by our scientists based on physical and radiochemical analyses of collected data have been confirmed by scientists of the AEC, United Kingdom and Office of Naval Research.[32]

On September 21, President Truman read Vandenberg's memorandum. He had been receiving reports, mostly verbal, on the detection of Joe-1 for nearly two weeks. Truman was disquieted by the reports, not least because the American atomic monopoly was the keystone of his policy of imposing a strict budgetary ceiling on defense spending. This, and many other aspects of American policy, would now have to change.

During the previous week, some of Truman's advisors had pressed him to announce the Soviet test on the somewhat specious grounds that, since an estimated 300 people in the Air Force, the administration, and the British Directorate of Atomic Energy knew of the Soviet test, it would be impossible to avoid a leak of information. However, he had delayed public acknowledgment of the Soviet atomic explosion, partly because he wanted further validation of the reports he had received and partly for a political reason: he wanted to be sure that the devaluation of the British pound, declared on September 18, was not viewed by the public as a panic reaction to the emergence of a second atomic power.

Another important reason for delay was to allow the Soviets time to make their own announcement. The American intelligence community hoped for this outcome because their successful penetration of Soviet atomic security could then remain unknown. To a degree, any announcement by the president would compromise American technical intelligence methods.

It now seemed clear that the Soviets, far from crowing about the achievement, apparently believed that they had been successful in keeping it hidden.[33] Truman now knew he would have to go public with the information, if only to marshall national support for the considerable increase in defense spending that he would be bound to propose and which the new situation seemed to demand.

On September 23, shortly before 11:00 A.M., White House Press Secretary Charles Ross distributed copies of a presidential statement to waiting reporters. By 11:05 A.M., the gist of the president's announcement was on the wires: ". . . within recent weeks an atomic explosion occurred within the U.S.S.R."[34]

AFTERWORD

For the most part, the public in the United States greeted the advent of a second atomic power with concern, but not undue surprise since many Americans had long anticipated this eventuality. Top Washington officials, however, *were* shocked and surprised. As insiders, they were privy to Murray Hill Area, and they had believed the forecasts, stemming from this program, of a longer duration for the American atomic monopoly.

Reporters and assorted pundits of the day pondered over every word and nuance of the presidential announcement. Their chief concern was identifying the source of the president's knowledge. But in stark contrast to the elaborations and explanations proffered by administration officials on other aspects of the announcement, reporters' queries on how the president learned of the Soviet explosion were met with stony-faced silence or facile evasions.

Only the president, a few top officials, and those members of the intelligence services charged with the task of interpreting the significance of the information produced by Bequeath and the analytic methods of long-range detection knew with certainty that the timely identification of the Soviet atomic bomb test was one of the first American intelligence coups of the Cold War (and, in the light of later events, arguably one of the most influential as well).

The importance of the breach in Soviet atomic security created by Bequeath was greatly enhanced in the eyes of administration officials and military leaders when viewed against the backdrop of other intelligence operations directed against the Soviet Union. These efforts had produced few hard facts on atomic matters. In June 1949, David E. Lilienthal, chairman of the Atomic Energy Commission, confided to his diary that reports by American intelligence agencies on Soviet atomic progress were at most guesswork, based on "the meager stuff" they had been able to collect.[35] General Omar N. Bradley, who was chairman of the Joint Chiefs of Staff at the time of the Soviet explosion, recalled that prior to the detection of Joe-1, "Sound and reliable intelligence on Soviet nuclear activities was almost impossible to come by."[36]

Thus, the quiet rain of Asian dust that fell from the stratosphere was like manna from heaven to the intelligence community, not only because of the information it yielded when subjected to the analytical techniques that had been developed, but also because of what could be inferred about Soviet atomic progress from this information.

If they remained uncompromised, these techniques held great promise for the future. Hence, everything about them continued to be veiled in secrecy. A small rip in this veil appeared almost immediately: Colorado Senator Edwin C. Johnson, a member of the Congressional Committee on Atomic Energy, revealed that the Soviet bomb contained plutonium. According to *Time* magazine, he "unwarily blurted it out" during a televised

A major change in our [Embassy's] organization occurred in that Autumn of 1949 when Kim Philby arrived in early October to succeed Peter Dwyer as the MI6 representative in Washington—a replacement more momentous in the light of history than it seemed at the time. They overlapped for a brief period in the aftermath of the Russian atom bomb so that Philby was able to put himself fully in the picture of what had happened in the previous three weeks.[43]

Philby was not only briefed on the details of how Joe-1 was detected but, in addition, could read in the files turned over to him the messages exchanged between Washington and London on this topic. The fact that Philby was a Soviet spy was not known until much later.[44] Thus, in the years that followed Joe-1, the security surrounding all aspects of the long-range detection system was carefully maintained.

As the years passed, and no further information from official sources was forthcoming, the detection of Joe-1 was described in books and articles largely on the basis of the suppositions just described. Thus, some beliefs about this event evolved whose correspondence with reality, in the light of present-day knowledge, is sufficiently tenuous to warrant calling them myths. Some of these beliefs seem to have acquired a life of their own, for they continued to appear in works published long after the documents that prove them false had been declassified.[45]

Not surprisingly, the successful detection of Joe-1 influenced the subsequent development of AFOAT-1 and the monitoring system. Seven months before Joe-1, Doyle Northrup had correctly predicted that "The electrifying news that Russians had unquestionably exploded a successful atomic bomb would knife through the less essential activities and prod this country into a rate of preparedness for atomic war greater than the present rate by at least an order of magnitude."[46] As part of this accelerated preparedness, the research and development programs proposed by AFOAT-1 encountered less resistance in the two years that followed Joe-1 than they had in the year preceding this event. Notably, much sonic and seismic research was carried out, and the range of sonic methods was increased. The "complete" net, including seismic and sonic as well as radiological components, was tested during the Greenhouse atomic tests in the Pacific in the spring of 1951. All the components of the net were operational by the scheduled date of mid-1951.

Indeed, the report on the detection of Joe-2, which was detonated in the fall of 1951, reads like a textbook example of how the complete net was supposed to work:

An acoustic signal of unusual intensity, apparently originating within the USSR was picked up by stations of the Air Force's Atomic Energy Detection System on 24 September 1951. The source of the signal was tentatively located ... centered at 49°N 81°E, which is about 100 miles south southeast of Semipalatinsk, USSR. . . . The time of origin of the signal was 1015Z [Z=zulu time=GMT] 24 September (about 1515 local time)... Considering thissignalasapossibleatomicexplosion,weatherprognosticationsweremadetodetermine when and where the air masses would come out beyond the borders of the USSR.[47]

The report goes on to say that filter-equipped aircraft were vectored to appropriate positions along the Soviet border to intercept these air masses. Samples of the airborne radioactivity obtained on these flights were analyzed to reveal "fresh fission products of an age corresponding to 24 September, which could only have come from a high order atomic explosion."

By the mid-1950s, what had begun as a network of organizations coordinated by AFOAT-1 had become institutionalized to the point where many of the operational tasks that had been initially carried out by other agencies were performed by divisions within AFOAT-1. As an organizational entity, AFOAT-1 had, in turn, increasingly assumed the more familiar form of a government bureaucracy. To signal these structural changes, its designation was changed from AFOAT-1 to AFTAC (Air Force Technical Applications Center) in 1958. The intelligence "product" of AFTAC, always of great military importance, assumed a political dimension as well after 1963, when AFTAC became responsible for monitoring compliance with a series of treaties on atomic bomb testing that were negotiated in the period 1963–1974.

In the four decades that followed Joe-1, large amounts of money were budgeted for nuclear surveillance. These funds were allotted to or indirectly controlled by AFTAC to develop new and improved methods and systems in an effort to detect and analyze foreign atomic activities conducted underground, underwater, in the atmosphere, and in space. Additional funds were expended to operate these complex systems, including the atomic operations of the so-called "black" Air Force, involving satellites, high-altitude spy planes, and balloons.[48]

It remains to record some of the consequences that flowed from the development of the monitoring system, the detection of Joe-1, and the institutionalizing of a nuclear surveillance capability. For example, the timely detection of Joe-1 by the interim net influenced the decision to develop the hydrogen bomb, a decision that affected not only the government's military position, but also its stance on scientific, diplomatic, and economic matters. Later, the surveillance system allowed government officials to track the buildup of the Soviet Union's atomic stockpile and to assess the ability of that country to wage atomic war. It also provided essential counterstrike information on the location of Soviet atomic facilities that was crucial to the maintenance of the delicate balance of power between the two nations. Information derived from the nuclear surveillance system was thus a vital ingredient in the intelligence summaries that guided American policy-makers throughout the Cold War, not only in day-to-day military and budgetary decisions, but also in the arena of international politics, especially during confrontations like the 1962 Cuban missile crisis.

The possession of a nuclear surveillance capability also affected decisions that have had an ecological impact. By the mid-1950s, the world had become acutely aware that repeated above-ground bomb testing by an ever-lengthening list of atomic powers was leading to ecological disaster on a global scale. By the late 1950s, serious international negotiations began which led to the 1963 Limited Test Ban Treaty, now ratified by 125 nations. This treaty, which prohibits atomic explosions in the atmosphere, underwater, and in space, has forestalled the serious environmental degradation that would otherwise have occurred. The fact that the United States and the Soviet Union already possessed adequate means for monitoring compliance with such an agreement was perhaps the single most important factor in the timely success of the negotiations.[49]

The ecological effects that may be traced to the surveillance system were not uniformly benign, however. In December 1949, apparently as part of a project to develop methods for locating atomic facilities within the Soviet Union, an experiment called Green Run, involving a purposeful release of radioactivity into the environs of the Atomic Energy Commission's Hanford facility, was carried out.[50] From an ecological perspective, the negative effect of Green Run appears to have been vastly outweighed by the positive effect of the Test Ban Treaty. On balance, therefore, it seems that the surveillance system has had a beneficial effect on the environment.

The secrecy that shrouded the surveillance system prevented its relevance in some areas from becoming known. For example, the widely publicized 1954 Oppenheimer Hearings have been extensively analyzed in books and articles, but the degree to which Oppenheimer's role in the development of the surveillance system was partly responsible for the distrust with which he was regarded in some government circles has remained unrecognized.[51] Other examples of unrecognized effects fall in the area of "technology transfer." One analyst has defined this term as "the process whereby technical information originating in one institutional setting is adapted for use in another . . . it implies the adaptation of new technology through creative transformation and application to a different end use."[52]

Occurrences of technology transfer stemming from the development of the nuclear surveillance system were, in some cases, facilitated by the fact that some AFOAT-1 research projects were carried out by industrial firms that were able to put a few of the ideas resulting from this research to commercial use. For example, Tracerlab was able to incorporate modified versions of some of the instruments, such as large-area radiation detectors, developed for AFOAT-1, into their product line for industrial applications. Two of the more important transfers of technology, however, did not involve industrial firms.

The first of these transfers occurred when scientists at the Air Force Cambridge Research Center initiated a study of high-altitude winds using constant-altitude balloons. This study was based on an adaptation of the balloon technology originally developed as part of the Mogul project on sonic methods of detecting atomic explosions. In 1951, when the study was launched, little was known about such winds. This research provided the first systematic data on air currents at various levels between 50,000 and 100,000 feet.[53] Such information was an essential ingredient in developing the kind of meteorological modeling that has greatly enhanced the reliability of modern weather prediction.

The second transfer resulted from Naval Research Laboratory scientists adapting the technology originally developed for the Navy's monitoring system to study natural and man-made atmospheric radioactivity. This research project, which extended over a decade, provided new information about natural processes that modified the gross radioactivity of the atmosphere and the mixing rates of atmospheric gases from the Northern and Southern Hemisphere.[54] This Navy study, as well as that of the Air Force described above, resulted in important contributions to scientific knowledge. They provide examples of how technology developed as part of highly classified projects can be transferred to unclassified basic research via scientists who work in both areas.

In summary, it can be said that the nuclear surveillance system has had a significant but often unrecognized influence in the life of the nation in military, diplomatic, economic, ecological, and scientific areas. This influence continues today, for the organization responsible for operating this system, AFTAC, has remained an important member of the intelligence community.

Currently, AFTAC, headquartered at Patrick Air Force Base, Florida, operates a worldwide Atomic Energy Detection System (AEDS), through three intermediate headquarters located in Hawaii, California, and Germany. These, in turn, supervise twenty manned detachments, five operating locations, and seventy equipment stations scattered through thirty-five countries. In addition to these seismic, sonic, hydroacoustic, and electromagnetic monitoring stations, aircraft and satellites are employed as part of the surveillance network. The aircraft used for aerial sampling include low-altitude types like B-52s and high-altitude "spy planes" such as the U-2, so that various atmospheric levels up to 70,000 feet can be monitored. At least two satellite systems have, as one of their functions, the detection of atomic explosions in space: the DSP (Defense Support Program) satellites in geosynchronous orbit and the navigational GPS (Global Positioning System) satellites. Together, they provide AFTAC with complete coverage of the earth and near-earth space. AFTAC's Technical Operations Division at McClellan Air Force Base, California, acts as a logistic depot for the engineering, maintenance, and supply of the AEDS. It also houses the Central Laboratory consisting of three divisions: the Applied Physics Laboratory for instrumental analysis, the Radiation Analysis Laboratory for radiochemical processing of samples, and the Gas Analysis Laboratory, which measures various radioactive gases of interest to AFTAC.[55]

In context, AFTAC is one of five Air Force organizations that perform intelligence missions. These five are part of more than twenty-five intelligence organizations maintained by the United States. Collectively, they now spend over $20 billion annually. Intelligence agencies such as AFTAC, which depend on high-technology data-gathering systems, tend to control a disproportionate share of these funds. Until recently, more than half the funds budgeted for intelligence were spent gathering information on the Soviet Union. With the demise of the Soviet empire, an extensive redirection of American intelligence activities was initiated, including cutbacks in many areas. However, because the list of countries that possess or aspire to possess atomic bombs continues to grow longer, it seems likely that America's nuclear surveillance system will continue to remain an important component of our national intelligence apparatus.

NOTES

1. According to Doyle Northrup, "The flight crews had not been briefed on the mission of detection . . . for the mission was highly classified" (Northrup, "Detection of the First Soviet Nuclear Test on August 29, 1949," February 6, 1962, p. 10).

2. On October 22, 1949, a few weeks after the flight that detected Joe-1, a B-29 on reconnaissance over the Sea of Japan was attacked by Soviet fighters but managed to avoid being shot down. This was the first of a number of Soviet attacks on American planes that took place over the following three decades. See Richelson, *American Espionage and the Soviet Target*, p. 121.

3. J. F. Fuller, *Thor's Legions*, 1990, p. 246.

4. See, for example: Kramish, *Atomic Energy in the Soviet Union*; I. Golovin, *I. V. Kurchatov*, 1968; D. Holloway, "Entering the Nuclear Arms Race," *Social Studies of Science*, Vol. 11, 1981, pp. 159–197; D. Holloway, *The Soviet Union and the Arms Race*, 1983; T. Szulc, "The Untold Story of How Russia Got the Bomb," *Los Angeles Times*, August 26, 1984; K. Smirnov, interview with A. P. Alexander on the early Soviet bomb program, *Izvestiya*, July 23, 1988; and Zaloga, *Target America*, pp. 29–63.

5. Holloway, *The Soviet Union and the Arms Race*, p. 22.

6. Kramish, *Atomic Energy in the Soviet Union*, p. 112.

7. Zaloga, *Target America*, p. 61.

8. This account of the tribulations endured by Crew 5A is based partly on Fuller, *Thor's Legions*, pp. 246–247, and partly on L. Root, "A Routine Patrol" (n.d.), unpublished manuscript in the authors' collection. Our thanks to Mrs. Root for providing us with a copy of her manuscript, which was based on an interview with (former First Lieutenant) Robert C. Johnson on September 3, 1974. See also the history of the 375th Weather Reconnaissance Squadron, Reel:SQ-WEA-375–H1 (406)/48 through (3–4)/50, Index 1945, AFHRC.

9. Northrup, "Detection of the First Soviet Nuclear Test on August 29, 1949," February 6, 1962, pp. 9, 10.

10. Root, "A Routine Patrol," p. 18; W. B. Mann, *Was There a Fifth Man?*, 1982, p. 67; Northrup, "Detection of the First Soviet Nuclear Test on August 29, 1949," p. 7.

11. Fuller, *Thor's Legions*, p. 247; Northrup, "Detection of the First Soviet Test on August 29, 1949," p. 9.

12. L. R. Zumwalt interview, July 29, 1985.

13. Northrup, "Detection of the First Soviet Nuclear Test on August 29, 1949," pp. 9, 10.

14. Hewlett and Duncan, *Atomic Shield*, p. 364.

15. Penney, "An Interim Report of British Work on Joe," September 22, 1949, pp. 1, 4.

16. P. King and H. Friedman, "Collection and Identification of Fission Products of Foreign Origin," NRL Report CN3536, September 22, 1949, p. 1; and T. A. Solberg to Assistant for Atomic Energy, DCS/O, September 26, 1949. President's Secretary's Files, Box 200, NSC-Atomic, HTPL.

17. Northrup, "Detection of the First Soviet Nuclear Test on August 29, 1949," p. 12.

18. R. W. Spence, "Identification of Radioactivity in Special Samples," October 4, 1949, p. 5. President's Secretary's Files, Box 200, NSC-Atomic, HTPL.

19. L. R. Zumwalt interview, July 29, 1985.

20. Hewlett and Duncan, *Atomic Shield*, p. 364.

21. Mann, *Was There a Fifth Man?*, p. 67.

22. W. B. Mann interview, April 11, 1990.

23. K. M. Nagler, "Editor's Note." In L. Machta, "Finding the Site of the First Soviet Nuclear Test in 1949," *Bulletin of the American Meteorological Society* 73, 1992, p. 1806.

24. Machta, "Finding the Site of the First Soviet Nuclear Test in 1949," p. 1800.

25. L. R. Zumwalt, A. J. Stevens, and F. C. Henriques, "Special Filter Paper Analyses and Preliminary Information Derived Therefrom," September 18, 1949, p. 16. President's Secretary's Files, Box 200, NSC-Atomic, HTPL. This report provides various times for Joe-1 within the period August 26 to 29, but no text to indicate which day and hour was favored by Tracerlab scientists. Nor did Northrup indicate a preference in his report to the panel on September 19, 1949. A paper by one of us (Ziegler, "Waiting for Joe-1," p. 219) quotes only one of these times, 0000 GMT 29 August, as *the* time determined by Tracerlab since it was assumed, based on another source (Hewlett and Duncan, *Atomic Shield*, p. 366), that the panel had selected this time as most probable. A subsequent paper (Machta, "Finding the Site of the First Nuclear Test in 1949," p. 1800) indicates that AFOAT-1 scientists then believed that 1500 GMT 27 August was the most probable time. Apparently, only after the sonic data had been reanalyzed did they realize that Joe-1 was detonated on August 29.

26. D. L. Northrup to Major General Nelson, "Atomic Detection System Alert No. 112," Technical Memo No. 37, September 19, 1949. President's Secretary's Files, Box 200, NSC-Atomic, HTPL.

27. Machta, "Finding the Site of the First Soviet Nuclear Test in 1949," p. 1800.

28. W. G. Lalor to Chairman, Research and Development Board, March 28, 1949, p. 2. RG 330, Secretary of Defense, RDB 47–53, Box 7, Folder 100 Atomic Energy/Long Range Detection, NA.

29. Northrup, "Detection of the First Soviet Nuclear Test on August 29, 1949," pp. 16, 17.

30. Ibid., p. 15. For a detailed technical description of the analyses carried out by Tracerlab, see L. R. Zumwalt, "Analysis of Fission Products from Russia's First Atomic Bomb Test," in J. W. Behrens and A. D. Carlson, eds., *50 Years with Nuclear Fission*, Vol. I, 1989, pp. 343–350.

31. V. Bush to General H. S. Vandenberg, September 20, 1949. President's Secretary's Files, Box 200, NSC-Atomic, HTPL.

32. General H. S. Vandenberg to Secretary of Defense, September 20, 1949. President's Secretary's Files, Box 200, NSC-Atomic, HTPL.

33. Regarding the Soviet expectation that their atomic test would remain unknown, Arnold Kramish, who at the time was a member of the Atomic Energy Commission's intelligence unit, later recalled that following Truman's announcement, "The initial Soviet reaction . . . was quite flustered, and for a number of days they were completely unprepared to exploit their achievement sensibly." See Kramish, *Atomic Energy in the Soviet Union*, p. 123.

34. For the complete text of the president's announcement, see *New York Times*, September 24, 1949.

35. Lilienthal, *The Journals of David E. Lilienthal*, Vol. II, p. 376.

36. O. N. Bradley and C. Blair, *A General's Life*, 1983, p. 376.

37. *Time*, December 5, 1949, p. 50.

38. *Newsweek*, May 3, 1948, p. 11.

39. *Time*, October 3, 1949, p. 7.

40. Gutenberg and Richter, "Seismic Waves from Atomic Bomb Tests," p. 776.

41. *Scientific American* 181, November 1949, pp. 26–28. The discovery of fission products in strawboard referred to in this editorial was mentioned in Chapter 3 (Webb, "Fogging of Photographic Film by Radioactive Contaminants," p. 375). Only one fission product, cerium-141, was detected at a distance of 1,000 miles from the site of the Trinity bomb test.

42. Weeks and Weeks, "Effort to Observe Anomalous Gamma Rays Connected With the Atomic Bomb Test of July 1, 1946," p. 565; Fearson et al., "Results of Atmospheric Analysis Done at Tulsa Oklahoma During the Period Neighboring the Time of the Second Bikini Atomic Bomb Test," p. 564; Hess and Lugar, "Ionization of the Atmosphere in the New York Area Before and After the Bikini Atom Bomb Test," p. 565; Herzog, "Gamma Ray Anomaly Following the Atomic Bomb Test of July 15, 1946," p. 279; Gutenberg, "Interpretation of Records Obtained from the New Mexico Atomic Bomb Test, July 16, 1945," p. 327.

43. Mann, *Was There a Fifth Man?*, p. 70.

44. Philby was recalled to London in June 1951 in the wake of the defection of the British diplomats Guy Burgess and Donald Maclean to the Soviet Union. He was suspected at that time of being a Soviet spy by American intelligence, but this was not confirmed until he fled to Moscow in 1962. See V. W. Newton, *The Cambridge Spies*, 1991, pp. 336, 341.

45. Regarding myths and misconceptions, some books published in the 1950s indicated that the detection of Joe-1 was completely accidental and provided

details of this fortuitous happening (see, for example, Jungk, *Brighter Than a Thousand Suns*, p. 254; J. R. Shepley and C. Blair, *The Hydrogen Bomb*, 1954, pp. 1, 2). A great deal of ink has been expended in describing the putative reason for the use of "atomic explosion" rather than "atomic bomb explosion" in President Truman's September 23, 1949, announcement. A typical view is that Truman had "reservations about the panel report [on Joe-1]" (Hewlett and Duncan, *Atomic Shield*, p. 368). This particular myth appears to have been laid to rest by the memoirs of Major General Nichols (see K. D. Nichols, *The Road to Trinity*, 1987, p. 272). The fact that Truman's announcement indicated that Joe-1 had occurred "within recent weeks" led some authors to conclude that the monitoring system was incapable of providing a more precise time, while others (see, for example, A. Sinclair, *The Red and the Blue*, 1986, p. 119) asserted that the Soviets exploded an atomic device prior to Joe-1. The latter is most unlikely (see Ziegler, "Waiting for Joe-1," p. 221).

46. Northrup to Major General Hegenberger, "Technical Memo No. 31," February 2, 1949, p. 3.

47. H. M. Caldwell to A. Dulles, Acting Director of Central Intelligence, October 1, 1951. President's Secretary's Files, Box 200, NNC-Atomic, HTPL.

48. For more on the "black" Air Force, see W. E. Burrows, *Deep Black*, 1986.

49. Divine, *Blowing on the Wind*, pp. 241–261. Apparently, the Soviets had long possessed a monitoring system of their own. For example, Dr. Georgiy N. Flyorov (spelled Flerov in some English texts), who worked with Kurchatov to produce the first Soviet bomb, revealed in a televised interview that, prior to Joe-1, the Soviets had assessed American atomic progress via "air analysis." See Transcript, Nova Show No. 2004, aired February 2, 1993, p. 5, available from Journal Graphics, Denver, Colorado. Although it remains unconfirmed, some analysts have claimed that Soviet air monitoring during America's hydrogen bomb test (Mike) probably provided them with the secret of the Teller-Ulam invention that made this bomb possible. See D. Hirsch and W. G. Mathews, "The H-Bomb: Who Really Gave Away the Secret?," *Bulletin of the Atomic Scientists* 46, January 1990, pp. 23–30.

50. See K. D. Steele, "Hanford's Bitter Legacy," *Bulletin of the Atomic Scientists* 44, January 1988, pp. 17–23.

51. As pointed out elsewhere in this book, references to the monitoring system appear in the transcript of the 1954 Oppenheimer Hearing (especially in General Wilson's testimony), but they are difficult to identify as such because much has been deleted for security reasons. Oppenheimer's negative opinions regarding monitoring, when viewed in the context of the then existing knowledge on that topic, seem similar to those held by other scientists. Some observers, however, perceived them as sinister. Wilson's feelings about Oppenheimer's behavior regarding monitoring and other matters impelled him to discuss the matter with General Cabell, head of Air Force intelligence (Roland to Strauss, February 4, 1954). General Canterbury, then head of AFOAT-1, and Doyle Northrup were also apparently disturbed by Oppenheimer's attitude on monitoring. In the fall of 1953, Strauss recorded that both men had told him that "Dr. Oppenheimer had sought during the past two years to have the long range detection system curtailed or discontinued" (L. L. Strauss, memorandum to the files, October 6, 1953. Strauss Papers, Box 75, Oppenheimer Folder, HHPL). Strauss apparently saw his own suspicions confirmed by the remarks of Wilson, Canterbury, and Northrup, an

outcome that did not redound in Oppenheimer's favor since Strauss was influential in orchestrating the hearings.

52. S. I. Doctors, *The Role of Federal Agencies in Technology Transfer*, 1969, p. 3.

53. Peebles, *The Moby Dick Project*, pp. 104–106.

54. L. B. Lockhart, R. A. Baus, P. King, and I. H. Blifford, "Atmospheric Radioactivity Studies at the U.S. Naval Research Laboratory," *Journal of Chemical Education* 36, 1959, pp. 291–295.

55. "Air Force Technical Application Center," *Air Force Magazine*, May 1987, pp. 165, 166; Richelson, *The U.S. Intelligence Community*, pp. 76, 85–87, 214–224.

BIBLIOGRAPHY

Alvarez, Luis W. *Alvarez: Adventures of a Physicist*. New York: Basic Books, 1987.

Bates, Charles C., and Fuller, John F. *America's Weather Warriors: 1814–1985*. College Station, TX: Texas A&M University, 1986.

Behrens, James W., and Carlson, Allan D., eds. *50 Years with Nuclear Fission*. Vol. I. La Grange Park, IL: American Nuclear Society, 1989.

Bradley, Omar N., and Blair, C. *A General's Life*. New York: Simon & Schuster, 1983.

Brown, Anthony C., and MacDonald, Charles B. *The Secret History of the Atomic Bomb*. New York: Dial Press, 1977.

Bundy, McGeorge. *Danger and Survival: Choices about the Bomb in the First Fifty Years*. New York: Random House, 1988.

Burrows, William E. *Deep Black: Space Espionage and National Security*. New York: Random House, 1986.

Clark, Ronald W. *The Greatest Power on Earth*. New York: Harper & Row, 1980.

Condit, Kenneth W. *History of the Joint Chiefs of Staff: The Joint Chiefs of Staff and National Policy*. Wilmington, MA: M. Glazier, 1979.

Darling, Arthur B. *The Central Intelligence Agency: An Instrument of Government to 1950*. University Park, PA: Pennsylvania State University Press, 1990.

Divine, Robert A. *Blowing on the Wind: The Nuclear Test Ban Debate, 1954–1960*. New York: Oxford University Press, 1978.

Doctors, S. I. *The Role of Federal Agencies in Technology Transfer*. Cambridge, MA: Massachusetts Institute of Technology Press, 1969.

Effects of Atomic Weapons. Washington, DC: Combat Forces Press, 1950.

Freedman, Lawrence. *U.S. Intelligence and the Soviet Strategic Threat*. 2nd ed. Princeton, NJ: Princeton University Press, 1986 (original edition 1977).

Fuller, John F. *Thor's Legions*. Boston: American Meteorological Society, 1990.

Gilpin, Robert. *American Scientists and Nuclear Weapons Policy*. Princeton, NJ: Princeton University Press, 1962.

Glynn, Patrick. *Closing Pandora's Box: Arms Races, Arms Control, and the History of the Cold War*. New York: Basic Books, 1992.

Golovin, Igor N. *I. V. Kurchatov: A Socialist-Realist Biography of the Soviet Nuclear Scientist*. Bloomington, IN: Selbstverlag Press, 1968.

Goudsmit, Samuel A. *Alsos*. New York: Harry Schuman, 1983 (original edition 1947).

Groves, Leslie R. *Now It Can Be Told*. New York: Harper & Row, 1962.

Hacker, Barton C. *The Dragon's Tail*. Berkeley, CA: University of California Press, 1987.

Hanson, Chuck. *U.S. Nuclear Weapons*. Arlington, TX: Aerofax, 1988.

Helmreich, Jonathan E. *Gathering Rare Ores: The Diplomacy of Uranium Acquisitions, 1943–1954*. Princeton, NJ: Princeton University Press, 1986.

Herken, Gregg. *The Winning Weapon: The Atomic Bomb in the Cold War, 1945–1950*. New York: Vintage Books, 1982.

Hewlett, Richard G., and Anderson, Oscar E. *The New World, 1939/1946*. University Park, PA: Pennsylvania State University Press, 1962.

Hewlett, Richard G., and Duncan, Francis. *Atomic Shield, 1947/1952*. University Park, PA: Pennsylvania State University Press, 1969.

Hinsley, F. H. *British Intelligence in the Second World War*. Vol. 2. London: Her Majesty's Stationery Office, 1981.

Holloway, David. *The Soviet Union and the Arms Race*. New Haven, CT: Yale University Press, 1983.

Jungk, Robert. *Brighter Than a Thousand Suns*. London: Victor Gollancz Ltd., 1958.

Kramish, Arnold. *Atomic Energy in the Soviet Union*. Stanford, CA: Stanford University Press, 1959.

———. *The Griffin*. Boston: Houghton Mifflin Co., 1986.

Kunetka, James W. *Oppenheimer, the Years of Risk*. Englewood Cliffs, NJ: Prentice-Hall, 1982.

Lawren, William. *The General and the Bomb*. New York: Dodd, Mead & Co., 1988.

Leary, William M., ed. *The Central Intelligence Agency: History and Documents*. University, AL: University of Alabama Press, 1984.

Libby, Leona M. *The Uranium People*. New York: Scribners, 1979.

Lilienthal, David E. *The Journals of David E. Lilienthal: The Atomic Energy Years, 1945–1950*. Vol. II. New York: Harper & Row, 1964.

Mann, Wilfred B. *Was There a Fifth Man?* New York: Pergamon Press, 1982.

Marshak, Robert E., Nelson, Eldred C., and Schiff, Leonard I. *Our Atomic World*. Albuquerque: University of New Mexico Press, 1946.

Miller, Jay D. *Lockheed U-2*. Austin, TX: Aerofax, 1983.

Newton, Verne W. *The Cambridge Spies: The Untold Story of Maclean, Philby, and Burgess in America*. New York: Madison Books, 1991.

Nichols, Kenneth D. *The Road to Trinity: A Personal Account of How America's Nuclear Policies Were Made*. New York: William Morrow, 1987.

O'Toole, G.J.A. *Encyclopedia of American Intelligence and Espionage*. New York: Facts on File, 1988.

Pash, Boris T. *The Alsos Mission*. New York: Award House, 1969.

Peebles, Curtis. *The Moby Dick Project: Reconnaissance Balloons over Russia*. Washington, DC: Smithsonian Institution Press, 1991.

Pfau, Richard. *No Sacrifice Too Great*. Charlottesville, VA: University of Virginia Press, 1984.

Powers, Thomas. *Heisenberg's War: The Secret History of the German Bomb*. New York: Knopf, 1993.

Prados, John. *The Soviet Estimate: U.S. Intelligence Analysis & Soviet Strategic Forces*. Princeton, NJ: Princeton University Press, 1986 (original edition 1982).

Pringle, Peter, and Spigelman, James. *The Nuclear Barons*. New York: Holt, Rinehart & Winston, 1981.

Rhodes, Richard. *The Making of the Atomic Bomb*. New York: Simon & Schuster, Touchstone edition, 1988 (original edition 1986).

Richelson, Jeffrey. *American Espionage and the Soviet Target*. New York: William Morrow, 1987.

_____ . *The U.S. Intelligence Community*. 2nd ed. Cambridge, MA: Ballinger, 1989.

Shepley, James R., and Blair, Clay. *The Hydrogen Bomb*. Westport, CT: Greenwood Press, 1954.

Sherwin, Martin J. *A World Destroyed*. New York: Knopf, 1975.

Sinclair, Andrew. *The Red and the Blue: Cambridge, Treason and Intelligence*. Boston: Little, Brown & Co., 1986.

Sternglass, Ernest. *Secret Fallout*. New York: McGraw-Hill, 1981 (original edition 1972).

Strauss, Lewis L. *Men and Decisions*. Garden City, NY: Doubleday, 1962.

Syles, Robert T. *The Nuclear Oracles*. Ames: Iowa State University Press, 1947.

Transcript of Hearing in the Matter of J. Robert Oppenheimer. Washington, DC: Government Printing Office, 1954.

Truman, Harry S. *Years of Trial and Hope*. Garden City, NY: Doubleday, 1956.

Truslow, Edith C., and Smith, Ralph C. *Project Y: The Los Alamos Story*. Los Angeles, CA: Tomash Publishers, 1983.

Ulam, Stanislaw M. *Adventures of a Mathematician*. New York: Scribners, 1976.

Williams, Robert C., and Cantelon, Philip L., eds. *The American Atom: A Documentary History of Nuclear Policies from the Discovery of Fission to the Present, 1939–1984*. Philadelphia: University of Pennsylvania Press, 1984.

Wilson, Brian J., ed. *Radiochemical Manual*. 2nd ed. London: Her Majesty's Stationery Office, 1966.

Yergin, Daniel. *Shattered Peace*. Boston: Houghton Mifflin Co., 1977.

Zaloga, Steven J. *Target America: The Soviet Union and the Strategic Arms Race, 1945–64*. Novato, CA: Presidio Press, 1993.

INDEX

About the Authors

CHARLES A. ZIEGLER is Lecturer in Social Anthropology at Brandeis University. He holds advanced degrees in anthropology and physics and has published in both fields.

DAVID JACOBSON is Associate Professor of Anthropology at Brandeis University. He has written books and articles on ethnography and social networks.